STATISTICAL ANALYSIS IN PSYCHOLOGY AND EDUCATION

McGRAW-HILL SERIES IN PSYCHOLOGY
Consulting Editor
Norman Garmezy

Adams Human Memory
Berlyne Conflict, Arousal, and Curiosity
Bernstein and Nietzel Introduction to Clinical Psychology
Blum Psychoanalytic Theories of Personality
Bock Multivariate Statistical Methods in Behavioral Research
Brown The Motivation of Behavior
Campbell, Dunnette, Lawler, and Weick Managerial Behavior, Performance, and Effectiveness
Crites Vocational Psychology
D'Amato Experimental Psychology: Methodology, Psychophysics, and Learning
Dollard and Miller Personality and Psychotherapy
Ferguson Statistical Analysis in Psychology and Education
Fodor, Bever, and Garrett The Psychology of Language: An Introduction to Psycholinguistics and Generative Grammar
Forgus and Melamed Perception: A Cognitive-Stage Approach
Franks Behavior Therapy: Appraisal and Status
Gilmer and Deci Industrial and Organizational Psychology
Guilford Psychometric Methods
Guilford The Nature of Human Intelligence
Guilford and Fruchter Fundamental Statistics in Psychology and Education
Guion Personnel Testing
Hetherington and Parke Child Psychology: A Contemporary Viewpoint
Hirsh The Measurement of Hearing
Hjelle and Ziegler Personality Theories: Basic Assumptions, Research, and Applications
Horowitz Elements of Statistics for Psychology and Education
Hulse, Egeth, and Deese The Psychology of Learning
Hurlock Adolescent Development
Hurlock Child Development
Hurlock Developmental Psychology: A Life-Span Approach
Krech, Crutchfield, and Ballachey Individual in Society
Lakin Interpersonal Encounter: Theory and Practice in Sensitivity Training
Lawler Pay and Organizational Effectiveness: A Psychological View
Lazarus, A. Behavior Therapy and Beyond
Lazarus, R. Patterns of Adjustment
Lewin A Dynamic Theory of Personality
Maher Principles of Psychopathology
Marascuilo Statistical Methods for Behavioral Science Research
Marx and Hillix Systems and Theories in Psychology
Morgan Physiological Psychology
Novick and Jackson Statistical Methods for Educational and Psychological Research
Nunnally Introduction to Statistics for Psychology and Education
Nunnally Psychometric Theory
Overall and Klett Applied Multivariate Analysis
Porter, Lawler, and Hackman Behavior in Organizations

Robinson and Robinson The Mentally Retarded Child
Ross Psychological Disorders of Children: A Behavioral Approach to Theory, Research, and Therapy
Shaw Group Dynamics: The Psychology of Small Group Behavior
Shaw and Costanzo Theories of Social Psychology
Shaw and Wright Scales for the Measurement of Attitudes
Sidowski Experimental Methods and Instrumentation in Psychology
Siegel Nonparametric Statistics for the Behavioral Sciences
Steers and Porter Motivation and Work Behavior
Vinacke The Psychology of Thinking
Winer Statistical Principles in Experimental Design

STATISTICAL ANALYSIS IN PSYCHOLOGY AND EDUCATION

FIFTH EDITION

George A. Ferguson
Professor of Psychology McGill University

McGraw-Hill Book Company
New York St. Louis San Francisco Auckland Bogotá Hamburg
Johannesburg London Madrid Mexico Montreal New Delhi
Panama Paris São Paulo Singapore Sydney Tokyo Toronto

STATISTICAL ANALYSIS IN PSYCHOLOGY AND EDUCATION

1 2 3 4 5 6 7 8 9 0 DODO 8 9 8 7 6 5 4 3 2 1 0

This book was set in Times Roman by Ruttle, Shaw & Wetherill, Inc.
The editors were Rhona Robbin and Susan Gamer;
the production supervisor was Donna Piligra.
The cover was designed by Antonia Goldmark.
R. R. Donnelley & Sons Company was printer and binder.

Library of Congress Cataloging in Publication Data

Ferguson, George Andrew.
Statistical analysis in psychology and education.

(McGraw-Hill series in psychology)
Bibliography: p.
Includes index.
1. Statistics. 2. Psychometrics. 3. Educational statistics. I. Title.
QA276.12.F45 1981 519.5 80-19584
ISBN 0-07-020482-9

CONTENTS

PREFACE

The object of this book, like that of the previous editions, is to introduce students and research workers in psychology and education to the concepts and applications of statistics. Emphasis is placed on the analysis and interpretation of data obtained from the conduct of experiments. Students and workers in other branches of behavioral and biological science, in medicine, and elsewhere, may also find the book useful.

This book has been designed as a text for both one-semester and full-year courses in statistics. In either case the instructor has some freedom of choice in the selection of material. For a one-semester course the selection will include most of Chapters 1 to 13.

I have attempted not only to introduce the student to the practical technology of statistics but also to explain in a nonmathematical and frequently intuitive way the nature of statistical ideas. This is not always easy. Obviously, the extent to which an understanding of statistics can be communicated without some mathematical knowledge is limited. Skill in high school or first-year college algebra will prove most helpful.

The writing of a book of this type demands numerous compromises between a tidy, logical arrangement of material, sound pedagogy, and common usage, which are not always compatible. The desire for completeness has led to the inclusion of occasional sections which are not essential in an introductory text. Instructors can readily identify these sections and omit them if they choose.

Many changes both consequential and trivial have been made in this edition. Sections of Part I, Basic Statistics, have been revised and rewritten. Statistical notation is treated separately in a short chapter (Chapter 3). Material on the interpretation of the correlation coefficient has been greatly expanded. In Part II the chapter on the analysis of covariance (Chapter 20) has been revised. In Part IV the material on multivariate statistics has been revised and expanded. Obsolete computational procedures have been eliminated. The chapter on factor analysis found in the Third Edition has been included with revisions in the present edition as Chapter 28. Some of the changes in this edition have been made to adapt to the changing usage of statistical procedures and to the availability of computers.

The usefulness of this book is enhanced by the kindness of authors and publishers who have permitted the adaptation and reproduction of tables and other materials published originally by them. I should like to express my gratitude to Francis G. Cornell, Allen L. Edwards, J. P. Guilford, Harry H. Harman, H. Leon Harter, S. K. Katti, M. G. Kendall, John F. Kenny, Don Lewis, Quinn McNemar, Edwin G. Olds, George W. Snedecor, Herbert Sorenson, L. R. Verdooren, James E. Wert, and Frank Wilcoxon; and to the Scottish Council for Research in Education, the University of London Press, Ltd., Charles Griffin and Company, Ltd., Prentice-Hall, Inc., John Wiley and Sons, Inc., Van Nostrand Reinhold Company, Rinehart and Company, Inc., Iowa State College Press, The University of Chicago Press, *Biometrika,* and the *Annals of Mathematical Statistics*. I owe special thanks to the Literary Executor of the late Sir Ronald A. Fisher, F.R.S., to Dr. Frank Yates, F.R.S., and to Oliver and Boyd Ltd., Edinburgh, for permission to reprint Tables III, IV, and VII of their book *Statistical Tables for Biological, Agricultural and Medical Research.*

I should like to express my indebtedness to the late Sir Godfrey H. Thomson, to W. G. Emmett, and to D. N. Lawley, all of the University of Edinburgh, and to the late R. W. B. Jackson of the Ontario Institute for the Study of Education. These men were responsible for my persisting interest in the application of statistical methods to educational and psychological problems.

This book in its various editions has benefited greatly from many helpful criticisms offered by many individuals. In particular I should like to thank Julian C. Stanley of Johns Hopkins University, Lyle Jones of the University of North Carolina, Phillip H. Dubois of Washington University, and Michael Corballis, now at the University of Auckland. Chester Olson, formerly at McGill and now at Camrose Lutheran College, has made many helpful suggestions. In the preparation of the present edition I have received much assistance and criticism from Lawrence R. Gordon of the University of Vermont. I regret that limitations of time, ingenuity, and wisdom have prevented me from incorporating some insightful recommendations and reconciling conflicting suggestions.

George A. Ferguson

I

BASIC STATISTICS

1

BASIC IDEAS IN STATISTICS

1.1 INTRODUCTION

This book is concerned with the elementary statistical treatment of experimental data in psychology, education, and related disciplines. The data resulting from any experiment are usually a collection of observations or measurements. The conclusions to be drawn from the experiment cannot be reliably ascertained by simple direct inspection of the data. Classification, summary description, and rules of evidence for the drawing of valid inference are required. Statistics provides the methodology whereby this can be done.

Implicit in any experiment is the presumption that it is possible to argue validly from the particular to the general and that new knowledge can be obtained by the process of inductive inference. Statisticians do not assume that such arguments can be made with certainty. On the contrary, they assume that some degree of uncertainty must attach to all such arguments; that some of the inferences drawn from the data of experiments are wrong. They further assume that the uncertainty itself is amenable to precise and rigorous treatment, that it is possible to make rigorous statements about the uncertainty which attaches to any particular inference. Thus in the uncertain milieu of experimentation they apply a rigorous method.

A knowledge of statistics is an essential part of the training of all students in psychology and education. There are many reasons for this. *First,* an understanding of the modern literature of psychology and education requires a knowledge of statistical method and modes of thought. Many current books and journal articles either report experimental findings in statistical form or present theories or arguments using statistical concepts. These concepts play an increasing role in our thinking about psychological problems, quite apart from the treatment of data. *Second,* training in psychology and education at an advanced level requires that the students themselves design and conduct experiments. The design of an experiment is inseparable from the statistical treatment of the results. Experiments must be designed to facilitate the treatment of results in such a way as to permit clear interpretation, and to fulfill the purposes which motivated the experiment in the first place. If the design of an experiment is faulty, no amount of statistical manipulation can lead to the drawing of valid inferences. Experimental design and statistical procedures are two sides of the same coin. Thus not only must advanced students conduct experiments and interpret results, they must plan their experiments in such a way that the interpretation of results can conform to known rules of scientific evidence. *Third,* training in statistics is training in scientific method. Statistical inference is scientific inference, which in turn is inductive inference, the making of general statements from the study of particular cases. These terms are, for all practical purposes, and at a certain level of generality, synonymous. Statistics attempts to make induction rigorous. Induction is regarded by some scholars as the only way in which new knowledge comes into the world. While this statement is debatable, the role in modern society of scientific discovery through induction is obviously of the greatest importance. For this reason no serious student of psychology, or any other discipline, can afford not to know something of the rudiments of the scientific approach to problems. Statistical procedures and ideas play an important role in this approach.

1.2 QUANTIFICATION IN PSYCHOLOGY AND EDUCATION

Although this book is largely concerned with elementary statistical procedures and ideas, some mention may be made of the general role of quantitative method in psychology and education.

The attempt to quantify has a long and distinguished history in experimental psychology. Since the experimental work in psychophysics of E. H. Weber and Gustav Fechner in the nineteenth century, determined attempts have been made to develop psychology as an experimental science. The early psychophysicists were concerned with the relationship between the "mind" and the "body" and developed certain mathematical functions which they held to be descriptive of that relationship. While much of their thinking on the mind-body problem has been discarded, their methods and techniques, with development and elaboration, are still used. Shorn of its philosophical and theoretical encumbrances, the work of the early psychophysicists was reduced in effect to the study of the relationship between measurements, obtained in two different ways, of what was presumed to be the same property. Thus, for example, they studied the relationship between weight, length, and temperature defined by the responses of human subjects as instruments, and weight, length, and temperature defined by other measuring instruments (scales, foot rules, and thermometers). A psychophysical law, so-called, is a statement of the relationship between measurements obtained by these two methods. Modern psychophysics is concerned to a considerable extent with the scaling of the responses of the human subject as instrument and with the use of the human subject as instrument in dealing with a wide variety of practical problems. It is concerned with the capacities of human subjects as components in complex man-machine systems.

The early psychophysicists invented certain experimental methods and developed statistical procedures for handling the data obtained by these methods. It is of interest to note that one method, the *constant process*, developed by G. E. Müller and F. M. Urban, has, with modification, found application in biological assay work in assessing the potency of hormones, toxicants, and drugs of all types. It is known in biology as the *method of probits*.

Statistical methods have found extensive application in the psychological and educational testing field and in the study of human ability. Since the time of Binet, who developed the first extensively used and successful test of intelligence, a comprehensive body of theory and technique has been developed which is primarily of a statistical nature. This body of theory and technique is concerned with the construction of instruments for measuring human ability, personality characteristics, attitudes, interests, and many other aspects of behavior; with the nature and magnitude of the errors involved in such measurement; with the logical conditions which such measuring instruments must satisfy; with the quantitative prediction of human behavior; and with other related topics.

Not only have research workers in psychology and education made substantial contributions to statistical methodology, but problems encountered by them have stimulated mathematicians to contribute to such methodology. An example of this process is factor analysis, a method concerned with the discovery of structure in a set of variables and the reduction of the set to a smaller number of more basic variables, called factors. This method had its origins in the work of C. Spearman, a British psychologist, who in 1904 published the first paper on the topic in the *American Journal of Psychology*. Since that time many psychologists and educationists, as well as mathematicians working in collaboration with them, have contributed to the development of factor analysis. An important factor analytic method, the method of principal components, was developed by the statistical mathematician Harold Hotelling. His interest in this subject was stimulated by his association with workers in the field of education. His original paper describing the method was published in 1933 in the *Journal of Educational Psychology*.

The student may well ask why psychologists and educationists have made useful contributions to statistical methodology, a field which may be perceived by some to be the province of the mathematician. This question has a general and obvious answer. In many fields, including psychology and education, problems arising from the study of real sets of data have stimulated the invention of methods of statistical analysis and interpretation. A high proportion of statistical procedures in common use evolved from the study of practical problems. Many of the leading contributors to statistics, including those who have contributed to its mathematical foundations, have worked in close touch with data analysis. Historically the concern of statisticians with applied problems has been highly productive, and will probably remain so.

1.3 STATISTICS AS THE STUDY OF POPULATIONS

Statistics is a branch of scientific methodology. It deals with the collection, classification, description, and interpretation of data obtained by the conduct of surveys and experiments. Its essential purpose is to describe and draw inferences about the numerical properties of populations. The terms population and numerical property require clarification.

In everyday language the term *population* is used to refer to groups or aggregates of people. We speak, for example, of the population of the United States, or of the state of Texas, or of the city of New York, meaning by this all the people who occupy defined geographical regions at specified times. This, however, is a particular usage of the term population. Statisticians employ the term in a more general sense to refer not only to defined groups or aggregates of people but also to defined groups or aggregates of animals, objects, materials, measurements, or "things" or "happenings" of any kind. Thus statisticians may define, for their particular purposes, populations of laboratory animals, trees, nerve fibers, liquids, soil, manufactured articles, automobile accidents, microorganisms, birds' eggs, insects, or fishes in the sea. On occasion they may deal with a population of measurements. By this is meant an indefinitely large aggregate of measurements which, hypothetically, might be obtained under specified experimental conditions. To illustrate, a series of measurements might be made of the length of a desk. Some or all of these measurements may differ one from another because of the presence of errors of measurement. This series of measurements may be regarded as part of an indefinitely large aggregate or population of measurements which might, hypothetically, be obtained by measuring the length of the desk over and over again an indefinitely large number of times.

The general concept implicit in all these particular uses of the word population is that of group or aggregation. The statistician's concern is with properties which are descriptive of the group or aggregation itself rather than with properties of particular members. Thus measurements may be made of the height and weight of a group of individuals. These measurements may be added together and divided by the number of cases to obtain the mean height and weight. These means describe a property of the group as a whole and are not descriptive of particular individuals. To illustrate further, a child may have an IQ of 90 and belong to a high socioeconomic group. Another child may have an IQ of 120 and belong to a low socioeconomic group. These facts as such about individual children do not directly concern the statistician. If, however, questions are raised about the proportion of children in a particular population or subpopulation with IQs above or below a specified value, or if more general questions are raised about the relationship between intelligence and socioeconomic level, then these are questions of a statistical nature, and the statistician has techniques which assist their exploration.

The distinction is sometimes made between *finite* and *infinite* populations. The children attending school in the city of Chicago, the inmates of penitentiaries in Ontario, the cards in a deck are examples of finite populations. The members of such a population can presumably be counted, and a finite number obtained. The possible rolls of a die and the possible observations in many scientific experiments are examples of infinite or indefinitely large populations. The number of rolls of a die or the number of scientific observations may, at least theoretically, be increased without any finite limit. In many situations the populations which the statistician proposes to describe are finite, but so large that for all practical purposes they may be regarded as infinite. The 250 million or so people living in the United States constitute a large but finite population. This population is so large that for many types of statistical inference it may be assumed to be infinite. This would not apply to the cards in a deck, which may be thought of as a small finite population of 52 members.

1.4 STATISTICS AS THE STUDY OF VARIATION

Statistics is sometimes viewed as the study of variation, because it provides a technology for the exploration of variation in the events of nature and for the making of inferences about the causal circumstances which underlie that variation. Emphasis on the study of variation originated with Darwin in "The Origin of Species" (1859). Variation was a central concept in the theory of natural selection because evolution could not occur without it. In Darwin's words,

> The many slight differences which appear in the offspring from the same parent . . . may be called individual differences. . . . These individual differences are of the highest importance for us, for they are often inherited, as must be familiar to everyone; and they thus afford materials for natural selection to act on and accumulate.

The matter is more clearly stated in an editorial in the first issue of the journal *Biometrika* (1901), probably written by Karl Pearson:

> The starting point of Darwin's theory of evolution is precisely the existence of those differences between individual members of a race or species which morphologists for the most part rightly neglect. The first condition necessary, in order that any process of Natural Selection may begin among a race, or species, is the existence of differences among its members; and the first step in an enquiry into the possible effect of a selective process upon any character of a race must be an estimate of the frequency with which individuals, exhibiting any given degree of abnormality with respect to that character, occur. The unit, with which such an enquiry must deal, is not an individual but a race, or a statistically representative sample of a race; and the result must take the form of a numerical statement, showing the relative frequency with which the various kinds of individuals composing the race occur.

Darwin made no direct contribution to statistical method. He did, however, create a theoretical context, based on observation and report, which made the study of variation meaningful and required, as it were, the development of statistical methods for its rigorous study. Darwin's disciple Galton understood fully the concept of variation. He was responsible for the initial applications of the so-called "normal" curve, or distribution, in psychological enquiry and made important contributions to the development of methods of correlation. He greatly influenced Karl Pearson, his disciple and biographer. Pearson perceived his role as that of helping to build a mathematical basis for evolutionary theory; he did, in fact, build the foundations of modern statistics. Between 1894 and 1916 he published 19 papers and monographs on statistical subjects, some of great importance and comprehensiveness. All these papers were titled "Contributions to the Mathematical Theory of Evolution."

Despite influences from many sources, modern statistics is largely a direct emergent of the biological revolution of the nineteenth century which Darwin helped to create. The central concept in evolutionary theory which gave rise to this line of development was the concept of variation. Since Pearson the development of statistical method has been closely associated with the attempt to find solutions to biological problems. R. A. Fisher, who developed the analysis of variance and made numerous other contributions between 1920 and 1960, devoted his life primarily to statistical problems of experimentation in the biological sciences and to the mathematical foundations of genetics.

1.5 SAMPLES AND SAMPLING

Because of the large size of many populations, it may be either impracticable or impossible for the investigator to produce statistics based on all members. If, for example, interest is in investigating the attitudes of adult Canadians toward immigrants, it would obviously be a prohibitively expensive and time-consuming task to measure the attitudes of all adult Canadians and produce statistics based on a study of the complete population. If a population is indefinitely large, it is of course impossible, ipso facto, to produce complete population statistics. Under circumstances such as these the investigator selects what is spoken of as a sample. A sample is any subgroup or subaggregate drawn by some appropriate method from a population, the method used in drawing the sample being important. Methods used in drawing samples will be discussed in later chapters of this book. Having drawn his sample, the investigator utilizes appropriate statistical methods to describe its properties. He then proceeds to make statements about the properties of the population from his knowledge of the properties of the sample; that is, he proceeds to generalize from the sample to the population. To return to the example above, an investigator might draw a sample of 1,000 adult Canadians, the term *adult* being assigned a precise meaning, measure their attitudes toward immigrants by using an acceptable technique of measurement, and calculate the required statistics. Questions may then be raised about the attitudes of all adult Canadians from the information obtained from a study of a sample of 1,000.

The fact that inferences can be made about the properties of populations from a knowledge of the properties of samples is basic in research thinking. Such statements are of course subject to error. The magnitude of the error involved in drawing such inferences can, however, in most cases be estimated by appropriate procedures. Where no estimate of error of any kind can be made, generalizations about populations from sample data are worthless.

Information about properties of particular samples, quite apart from any generalizations about the population, is of little intrinsic interest in itself. Consider a case where the investigator's interest is in the relative effects of two types of psychotherapy when applied to patients suffering from a particular mental disorder. She may select two samples of patients, apply one type of treatment to one sample and the other type of treatment to the other sample, and collect data on the relative rates of recovery of patients in the two samples. Clearly, in this case her interest is in finding out whether the one treatment is better or worse than the other when applied to the whole class of patients suffering from the mental disorder in question. She is interested in the sample data only insofar as these data enable her to draw inferences with some acceptable degree of assurance about this general question. Her experimental procedures must be designed to enable the drawing of such inferences, otherwise the experiment serves no purpose. On occasion research reports are found where the investigator states that the experimental results obtained should not be generalized beyond the particular sample of individuals who participated in the study. The adoption of this view means that the investigator has missed the essential nature of experimentation. Unless the intention is to generalize from a sample to a population, unless the procedures used are such as to enable such generalizations justifiably to be made, and unless some estimate of error can be obtained, the conduct of experiments is without point.

Statistical procedures used in describing the properties of samples, or of populations where complete population data are available, are referred to by some writers as *descriptive statistics*. If we measure the IQ of the complete population of students in a particular university and compute the mean IQ, that mean is a descriptive statistic because it describes a characteristic of the complete population. If, on the other hand, we measure the IQ of a sample of 100 students and compute the mean IQ for the sample, that mean is also a descriptive statistic because it describes a characteristic of that sample.

Statistical procedures used in the drawing of inferences about the properties of populations from sample data are frequently referred to as *inferential statistics*. If, for example, we wish to make a statement about the mean IQ in the complete population of students in a particular university from a knowledge of the mean computed on the sample of 100 and to estimate the error involved in this statement, we use procedures from inferential statistics. The application of these procedures provides information about the accuracy of the sample mean as an estimate of the population mean; that is, it indicates the degree of assurance we may place in the inferences we draw from the sample to the population.

In this section no discussion is advanced on methods of drawing samples or the conditions which these methods must satisfy to allow the drawing of valid inferences from the sample to the population. Further, no precise meaning has been assigned to the term "error." These topics will be elaborated at a later stage.

1.6 PARAMETERS AND ESTIMATES

A clear distinction is usually drawn between parameters and estimates. A *parameter* is a property descriptive of the population. The term *estimate* refers to a property of a sample drawn at random from a population. The sample value is presumed to be an estimate of a corresponding population parameter. Suppose, for example, that a sample of 1,000 adult male Canadians of a given age range is drawn from the total population, the height of the members of the sample measured, and a mean value, 68.972 inches, obtained. This value is an estimate of the population parameter which would have been obtained had it been possible to measure all the members in the population. Usually parameters or population values are unknown. We estimate them from our sample values. The distinction between parameter and estimate is reflected in statistical notation. A widely used convention in notation is to employ Greek letters to represent parameters and Roman letters to represent estimates. Thus the symbol σ, the Greek letter sigma, may be used to represent the standard deviation in the population, the standard deviation being a commonly used measure of variation. The symbol s may be used as an estimate of the parameter σ. This convention in notation is applicable only within broad limits. By and large we shall adhere to this convention in this book, although in certain instances it will be necessary to depart from it. By common practice and tradition a Greek letter may be used on occasion to denote a sample statistic.

1.7 VARIABLES AND THEIR CLASSIFICATION

The term *variable* refers to a property whereby the members of a group or set differ one from another. The members of a group may be individuals and may be found to differ in sex, age, eye color, intelligence, auditory acuity, reaction time to a stimulus, attitudes toward a political issue, and many other ways. Such properties are variables. The term *constant* refers to a property whereby the members of a group do not differ one from another. In a sense a constant is a particular type of variable; it is a variable which does not vary from one member of a group to another or within a particular set of defined conditions.

Labels or numerals may be used to describe the way in which one member of a group is the same as or different from another. With variables like sex, racial origin, religious affiliation, and occupation, labels are employed to identify the members which fall within particular classes. An individual may be classified as male or female; of English, French, or Dutch origin; Protestant or Catholic; a shoemaker or a farmer; and so on. The label identifies the class to which the individual belongs. Sex for most practical purposes is a two-valued variable, individuals being either male or female. Occupation, on the other hand, is a multivalued variable. Any particular individual may be assigned to any one of a large number of classes. With variables like height, weight, intelligence, and so on, measuring operations may be employed which enable the assignment of descriptive numerical values. An individual may be 72 inches tall, weigh 190 pounds, and have an IQ of 90.

The particular values of a variable are referred to as *variates,* or *variate values.* To illustrate, in considering the height of adult males, height is the variable, whereas the height of any particular individual is a variate, or variate value.

In dealing with variables which bear a functional relationship one to another, the distinction may be drawn between *dependent* and *independent* variables. Consider the expression $Y = f(X)$. This expression says that a given variable Y is some unspecified function of another variable X. The symbol f is used generally to express the fact that a functional relationship exists, although the precise nature of the relationship is not stated. In any particular case the nature of the relationship may be known; that is, we may know precisely what f means. Under these circumstances, for any given value of X a corresponding value of Y can be calculated; that is, given X and a knowledge of the functional relationship, Y can be predicted. It is customary to speak of Y, the predicted variable, as the dependent variable because the prediction of it depends on the value of X and the known functional relationship, whereas X is spoken of as the independent variable.

Given an expression of a kind $Y = X^3$ for any given value of X, an exact value of Y can readily be determined. Thus if X is known, Y is also known exactly. Many of the functional relationships found in statistics permit probabilistic and not exact prediction to occur. Such relationships may provide the most probable value of Y for any given value of X, but do not permit the making of perfect predictions.

A distinction may be drawn between *continuous* and *discrete* (or *discontinuous*) variables. A continuous variable may take any value within a defined range of values. The possible values of the variable belong to a continuous series. Between any two values of the variable an indefinitely large number of in-between values may occur. Height, weight, and chronological time are examples of continuous variables. A discontinuous or discrete variable can take specific values only. Size of family is a discontinuous variable. A family may comprise 1, 2, 3, or more children, but values between these numbers are not possible. The values obtained in rolling a die are 1, 2, 3, 4, 5, and 6. Values between these numbers are not possible. Although the underlying variable may be continuous, all sets of real data in practice are discontinuous or discrete. Convenience and errors of measurement impose restrictions on the refinement of the measurement employed.

Another classification of variables is possible which is of some importance and is of particular interest to statisticians. This classification is based on differences in the type of information which different operations of classification or measurement yield. To illustrate, consider the following situations. An observer using direct inspection may rank-order a group of individuals from the tallest to the shortest according to height. On the other hand, he may use a foot rule and record the height of each individual in the group in feet and inches. These two operations are clearly different, and the nature of the information obtained by applying the two operations is different. The former operation permits statements of the kind: individual A is taller or shorter than individual B. The latter operation permits statements of how much taller or shorter one individual is than another. Differences along these lines serve as a basis for a classification of variables, the class to which a variable belongs being determined by the nature of the information made available by the measuring operation used to define the variable. Four broad classes of variables may be identified. These are referred to as (1) nominal, (2) ordinal, (3) interval, and (4) ratio variables.

A *nominal variable* is a property of the members of a group defined by an operation which permits the making of statements only of equality or difference. Thus we may state that one member is the *same as* or *different from* another member with respect to the property in question. Statements about the ordering of members, or the equality of differences between members, or the number of times a particular member is greater than or less than another are not possible. To illustrate, individuals may be classified by the color of their eyes. Color is a nominal variable. The statement that an individual with blue eyes is in some sense "greater than" or "less than" an individual with brown eyes is meaningless. Likewise the statement that the difference between blue eyes and brown eyes is equal to the difference between brown eyes and green eyes is meaningless. The only kind of meaningful statement possible with the information available is that the eye color of one individual is the same as or different from the eye color of another. A nominal variable may perhaps be viewed as a primitive type of variable, and the operations whereby the members of a group are classified according to such a variable constitute a primitive form of measurement. In dealing with nominal variables, numerals may be assigned to represent classes, but such numerals are labels, and the only purpose they serve is to identify the members within a given class.

An *ordinal variable* is a property defined by an operation which permits the rank ordering of the members of a group; that is, not only are statements of equality and difference possible, but also statements of the kind *greater than* or *less than*. Statements about the equality of differences between members or the number of times one member is greater than or less than another are not possible. If a judge is required to order a group of individuals according to aggressiveness, or cooperativeness, or some other quality, the resulting variable is ordinal in type. Many of the variables used in psychology are ordinal.

An *interval variable* is a property defined by an operation which permits the making of statements of equality of intervals, in addition to statements of sameness or difference or greater than or less than. An interval variable does not have a "true" zero point, although a zero point may for convenience be arbitrarily defined. Fahrenheit and centigrade temperature measurements constitute interval variables. Consider three objects, A, B, and C, with temperatures 12°, 24°, and 36°, respectively. It is appropriate to say that the difference between the temperature of A and B is equal to the difference in the temperature of B and C. It is appropriate also to say that the difference between the temperature of A and C is twice the difference between the temperature of A and B or B and C. It is not appropriate to say that B has twice the temperature of A, or that C has three times the temperature of A. In common usage, if the temperature yesterday was 64° and today it was 32°, we would not say that it was twice as hot yesterday, or that the temperature was twice as great, as it was today. Calendar time is also an interval variable with an arbitrarily defined zero point.

A *ratio variable* is a property defined by an operation which permits the making of statements of equality of ratios in addition to all other kinds of statements discussed above. This means that one variate value, or measurement, may be spoken of as double or triple another, and so on. An absolute 0 is always implied. The numbers used represent distances from a natural origin. Length, weight, and the numerosity of aggregates are examples of ratio variables. One object may be twice as long as another, or three times as heavy, or four times as numerous. Many of the variables used in the physical sciences are of the ratio type. In psychological work, variables which conform to the requirements of ratio variables are uncommon.

The essential difference between a ratio and an interval variable is that for the former the measurements are made from a true zero point, whereas for the latter the measurements are made from an arbitrarily defined zero point or origin. Because of this, for a ratio variable, ratios may be formed directly from the variate values themselves, and meaningfully interpreted. For an interval variable, ratios may be formed from differences between the variate values. The differences constitute a ratio variable, because the process of subtraction eliminates, or cancels out, the arbitrary origin. Differences are the same regardless of the location of the zero point or origin.

Statistical methods exist for the analysis of data composed of nominal variables, ordinal variables, and interval and ratio variables. From the viewpoint of practical statistical work in psychology and education the distinction between interval and ratio variables is perhaps unimportant, and it is convenient to think of three, and not four, classes of variables, with three corresponding classes of statistical method. Procedures for the analysis of interval and ratio variables constitute by far the largest, and most important, class of statistical method.

The importance of the concept of a variable cannot be overemphasized. All statistics is concerned with variables and the relations between them. Variables are the stuff of which statistics is made. Variables may be added together, partitioned into additive bits, related one to another, and interrelated in complex networks.

An investigator may study variables, and their interrelations, as they exist in nature, for example, age, IQ, examination marks, blood pressure, and anxiety level. Such investigations are sometimes called correlational studies. Some of the more important variables in nature are nominal, such as being male or female and being alive or dead. On the other hand, in many investigations one or more variables may be created by the investigator who decides the values the variable will take and the frequency of occurrence of these values. Such investigations are called experiments. For example, a simple experiment may involve two groups of subjects; a particular dosage of a drug is administered to one group, but not to the other. Then some measure of motor performance is obtained from the members of the two groups. Two variables are involved here: the independent variable—receiving or not receiving the drug—and the dependent variable—motor performance. The purpose of the experiment is to explore the relation between these two variables, that is, to study how motor performance *depends* on the presence or the absence of the drug. Note that the independent variable is a simple nominal variable with two categories denoting group membership. A subject is or is not a member of the group receiving the drug. The basic point here is that in both correlational and experimental studies, variables and the relations between them are the object of inquiry.

1.8 DATA ANALYSIS

Since the first edition of this book in 1959, remarkable changes have occurred in computational methods. Also, enormous changes are anticipated in the future, as increasing computational power is incorporated in smaller computers at decreasing cost. These technical developments have eliminated much of the drudgery associated with statistical work. Many earlier statisticians devoted substantial parts of their working lives to arithmetical labor. Much of this was elementary. Many methods that were devised to reduce computational time are now obsolete. These have been removed from the various editions of this book, except in a few instances where they may assist the student in understanding some aspect of the statistic under discussion.

The speed and ease with which numerical computation can now be done has led to the increased frequency of use of complex statistical methods, many of which were known earlier but were not often used because of the arithmetical labor required. In this context, multivariate statistical methods deserve mention. These are methods that require the analysis of data comprising several, perhaps many, variables. In psychology, education, and elsewhere, the attempt to explain or predict a particular phenomenon may entail studying processes that are highly complex and that involve the functioning of collections of interacting variables. Inspection of current research literature suggests that the frequency of use of multivariate methods is increasing rapidly.

The ready availability of small and powerful computers has enhanced, not diminished, the importance of understanding statistical concepts. Also, this development has added to the importance of understanding the task a particular method is designed to accomplish, the problems it purportedly solves, and the assumptions it requires. This understanding can in many instances be readily acquired by the student without any extended exploration of either the mathematical apparatus underlying a method or the details of the computational procedures employed.

Earlier statisticians with primitive computing devices spent much time and labor exploring by various methods, graphic and otherwise, the information contained in collections of data. This investment of effort not uncommonly led to a useful intuitive understanding of what the data had to say. Now that powerful computing devices are available, this intuitive level of understanding frequently is not attained. Data are fed to computers, and solutions are regurgitated. Not uncommonly the investigator only partially understands the data. A strong case can be made for an initially rather simpleminded, exploratory approach to data analysis. In the study of many sets of data, simple preliminary insights into the story the data have to tell are useful in guiding subsequent forms of analysis. The computer may, of course, be used to assist such exploratory data analyses.

Tukey (1977) stresses exploratory data analysis. Van Dantzig (1978) provides a simple and readable overview of the topic. Exploratory data analysis is descriptive, involving simple data manipulations, summary methods, and graphic descriptions. Its purpose is to make the nature and structure of the data understandable to the investigator, and, thereby, to lead to the subsequent use of appropriate statistical models. Tukey speaks of exploratory data analysis as analogous to detective work; it may compel the investigator to notice aspects of the data that were not expected. All scientists know that careful exploration of sets of data will on occasion lead to findings they did not anticipate. The importance of an experiment may go beyond the original intent.

Tukey distinguishes exploratory from confirmatory data analysis. By confirmatory data analysis he means the confirmation, or disconfirmation, of the original hypothesis that led to the experiment in the first place. Tukey argues that exploratory and confirmatory data analyses complement each other.

The present writer has long held the view that in the analysis of any set of data, some understanding of the data, obtained perhaps by calculating a few simple statistics, is a necessary preliminary to the planning of more elaborate analyses. Such analyses frequently involve complex models and assumptions. The availability of powerful computers, which automatically apply so-called canned programs, enhances the importance of simple methods that assist the investigator in getting to know the data.

BASIC TERMS AND CONCEPTS

Population

Population: finite; infinite

Sample

Descriptive statistics

Sampling statistics

Parameter

Estimate

Variable

Variate value

Variable: dependent; independent

Variable: continuous; discrete

Variable: nominal; ordinal; interval; ratio

Correlational study

Experiment

2

FREQUENCY DISTRIBUTIONS AND THEIR GRAPHIC REPRESENTATION

2.1 INTRODUCTION

The data obtained from the conduct of experiments or surveys are frequently collections of numbers. Simple inspection of a collection of numbers will ordinarily communicate very little to the understanding of the investigator. Some form of classification and description of these numbers is required to assist interpretation and to enable the information which the numbers contain to emerge. Under certain circumstances advantages attach to the classification of data in the form of frequency distributions. Such classification may help the investigator to understand important features of the data. This chapter discusses the arrangement of data in the form of frequency distributions, the graphic representation of frequency distributions, and the ways in which one frequency distribution may differ from another. Chapters 3, 4, and 5 to follow, discuss the statistics used to describe the properties of frequency distributions, or the properties of the collections of numbers which these distributions comprise.

2.2 FREQUENCY DISTRIBUTIONS

A coin is tossed 10 times, and the following results are obtained: H H T H T H H H T H. Here the number of times heads occurs, the frequency of heads, is 7, and the number of times tails occurs, the frequency of tails, is 3. These data could be arranged in the form

	f
H	7
T	3
Total	10

The symbol f denotes the frequency. This arrangement of the data is a frequency distribution. It is an arrangement of the data that shows how often heads and tails occur.

A die is rolled 24 times and the following results recorded:

6	3	1	4	1	6
5	2	4	3	5	5
4	1	5	2	2	6
5	3	3	4	5	5

The numbers appearing when a die is rolled constitute a variable X, which is limited to the values 1, 2, 3, 4, 5, and 6. In the above data, 6 occurs 3 times, 5 occurs 7 times, and so on. These data may be arranged as follows:

X	f
6	3
5	7
4	4
3	4
2	3
1	3
Total	24

This arrangement is a frequency distribution. It shows the frequency of occurrence of the values 1, 2, 3, 4, 5, and 6.

Consider the data of Table 2.1. These are the IQs of 100 children obtained from the administration of a psychological test. These scores range from 67 to 134. By counting the number of times each score occurs, an arrangement of the data as shown in Table 2.2 is obtained. This arrangement shows how many times each score occurs, and is a frequency distribution. Note, however, that the number of groupings of scores is large. Usually it is advisable to reduce the number of classes by arranging the data into arbitrarily defined groupings of the variable; thus all scores within the range 65 to 69, that is, all scores with the values 65, 66, 67, 68, and 69, may be grouped together. All scores within the ranges 70 to 74, 75 to 79, and so on, may be similarly grouped. Such groupings of data are usually done by entering a tally mark for each score opposite the range of the variable within which it falls and counting those tally marks to obtain the number of cases within the range. This procedure is shown in Table 2.3.

Table 2.1

Intelligence quotients made by 100 pupils on a mental test

109	111	82	105	134
113	90	79	100	117
80	90	121	75	93
99	90	92	96	82
101	104	80	81	83
104	93	109	72	110
111	91	109	111	81
122	83	92	101	77
99	103	93	91	67
108	93	84	84	100
102	84	96	89	81
107	95	91	107	102
109	93	82	103	116
86	78	73	104	104
103	108	76	94	108
72	87	121	80	127
105	103	106	119	90
93	89	110	103	100
99	79	117	114	117
93	82	98	89	119

SOURCE: From R. W. B. Jackson and George A. Ferguson. "Manual of Educational Statistics." University of Toronto, Department of Educational Research, Toronto, 1942.

Table 2.2

Frequency distribution of intelligence quotients of Table 2.1 with as many classes as score values

Score	*f*	Score	*f*	Score	*f*	Score	*f*
134	1	117	3	100	3	83	2
133	. . .	116	1	99	3	82	4
132	. . .	115	. . .	98	1	81	3
131	. . .	114	1	97	. . .	80	3
130	. . .	113	1	96	2	79	2
129	. . .	112	. . .	95	1	78	1
128	. . .	111	3	94	1	77	1
127	1	110	2	93	7	76	1
126	. . .	109	4	92	2	75	1
125	. . .	108	3	91	3	74	. . .
124	. . .	107	2	90	4	73	1
123	. . .	106	1	89	3	72	2
122	1	105	2	88	. . .	71	. . .
121	2	104	4	87	1	70	. . .
120	. . .	103	5	86	1	69	. . .
119	2	102	2	85	. . .	68	. . .
118	. . .	101	2	84	3	67	1

SOURCE: From R. W. B. Jackson and George A. Ferguson, "Manual of Educational Statistics." University of Toronto, Department of Educational Research, Toronto, 1942.

Table 2.3
Frequency distribution of the intelligence quotients of Table 2.1

Class interval	Tally	Frequency
130–134	/	1
125–129	/	1
120–124	///	3
115–119	~~////~~ /	6
110–114	~~////~~ //	7
105–109	~~////~~ ~~////~~ //	12
100–104	~~////~~ ~~////~~ ~~////~~ /	16
95–99	~~////~~ //	7
90–94	~~////~~ ~~////~~ ~~////~~ //	17
85–89	~~////~~	5
80–84	~~////~~ ~~////~~ ~~////~~	15
75–79	~~////~~ /	6
70–74	///	3
65–69	/	1
Total		100

SOURCE: From R. W. B. Jackson and George A. Ferguson, "Manual of Educational Statistics." University of Toronto, Department of Educational Research, Toronto, 1942.

The arbitrarily defined groupings of the variable are called *class intervals*. In Table 2.3 the class interval is 5. This arrangement of data is also a frequency distribution, and the number of cases falling within each class interval is a frequency. The only difference between Tables 2.2 and 2.3 is in the class interval, which is 1 in the former table and 5 in the latter.

In general, a frequency distribution is any arrangement of the data that shows the frequency of occurrence of different values of the variable or the frequency of occurrence of values falling within arbitrarily defined ranges of the variable known as class intervals.

2.3 CONVENTIONS REGARDING CLASS INTERVALS

In the arrangement of data with a class interval of 1, as shown in Table 2.2, the original observations are retained and may be reconstructed directly from the frequency distribution without loss of information. If the class interval is greater than 1, say, 3, 5, or 10, some loss of information regarding individual observations is incurred; that is, the original observations cannot be reproduced exactly from the frequency distribution. If the class interval is large in relation to the total range of the set of observations, this loss of information may be appreciable. If the class interval is small, the classification of data in the form of a frequency distribution may lead to very little gain in convenience over the utilization of the original observations.

The rules listed below are widely used in the selection of class intervals and lead in most cases to a convenient handling of the data.

1 Select a class interval of such a size that between 10 and 20 such intervals will cover the total range of the observations. For example, if the smallest observation in a set were 7 and the largest 156, a class interval of 10 would be appropriate and would result in an arrangement of the data into 16 intervals. If the smallest observation were 2 and the largest 38, a class interval of 3 would result in an arrangement of 14 intervals. If the observations ranged from 9 to 20, a class interval of 1 would be convenient.

2 Select class intervals with a range of 1, 3, 5, 10, or 20 points. These will meet the requirements of most sets of data.

3 Start the class interval at a value which is a multiple of the size of that interval. For example, with a class interval of 5, the intervals should start with the values 5, 10, 15, 20, etc.

4 Arrange the class intervals according to the order of magnitude of the observations they include, the class interval containing the largest observations being placed at the top.

2.4 EXACT LIMITS OF THE CLASS INTERVAL

Where the variable under consideration is continuous, and not discrete, we select a unit of measurement and record our observations as discrete values. When we record an observation in discrete form and the variable is a continuous one, we imply that the value recorded represents a value falling within certain limits. These limits are usually taken as one-half unit above and below the value reported. Thus when we report a measurement to the nearest inch, say, 16 inches, we mean that if a more accurate form of measurement had been used, the value obtained would fall within the limits 15.5 and 16.5 inches.

Strictly speaking the limits are 15.5 to 16.499, where the latter figure is a recurring decimal, but for convenience we write the limits as 15.5 to 16.5. Similarly, a measurement made to the nearest tenth part of an inch, say, 31.7 inches, is understood to fall within the limits 31.65 and 31.75 inches. In a reaction-time experiment a particular observation measured to the nearest thousandth of a second might be, say, .196 second. This assumes that had a more accurate timing device been used the measurement would have been found to fall somewhere within the limits .1955 and .1965 second.

Class intervals are usually recorded to the nearest unit and thereby reflect the accuracy of measurement. For various reasons it is frequently necessary to think in terms of so-called *exact limits* of the class interval.

Table 2.4

Class intervals, exact limits, and midpoints for frequency distribution of intelligence quotients

1 Class interval	2 Exact limits	3 Midpoint of interval	4 Frequency
130–134	129.5–134.5	132.0	1
125–129	124.5–129.5	127.0	1
120–124	119.5–124.5	122.0	3
115–119	114.5–119.5	117.0	6
110–114	109.5–114.5	112.0	7
105–109	104.5–109.5	107.0	12
100–104	99.5–104.5	102.0	16
95–99	94.5–99.5	97.0	7
90–94	89.5–94.5	92.0	17
85–89	84.5–89.5	87.0	5
80–84	79.5–84.5	82.0	15
75–79	74.5–79.5	77.0	6
70–74	69.5–74.5	72.0	3
65–69	64.5–69.5	67.0	1
Total			100

SOURCE: From R. W. B. Jackson and George A. Ferguson, "Manual of Educational Statistics." University of Toronto, Department of Educational Research, Toronto, 1942.

These are sometimes spoken of as *class boundaries,* or *end values,* and sometimes as *real limits.* Consider the class interval 95 to 99 in Table 2.3. We grouped within this interval all measurements taking the values 95, 96, 97, 98, and 99. The limits of the lower value are 94.5 and 95.5, while those of the upper value are 98.5 and 99.5. The total range, or exact limits, which the interval is presumed to cover is then clearly 94.5 and 99.5, which means all values greater than or equal to 94.5 and less than 99.5.

The above discussion is applicable to continuous variables only. With discrete variables no distinction need be made between the class interval and the exact limits of the interval, the two being identical.

Table 2.4 shows the frequency distribution of the IQs of Table 2.1. Column 1 shows the class interval as usually written, while column 2 records the exact limits. In practice, of course, the exact limits are rarely recorded as in Table 2.4.

2.5 DISTRIBUTION OF OBSERVATIONS WITHIN THE CLASS INTERVAL

The grouping of data in class intervals results in a loss of information regarding the individual observations themselves. Scores may differ one from another within a limited range, and yet all be grouped within the same interval. In the calculation of certain statistics and in the preparation of graphs it becomes necessary to make certain assumptions regarding the values within the intervals. Two separate assumptions may be made, depending on the purposes we have in mind.

The first assumption states that the observations are uniformly distributed over the exact limits of the interval. This assumption is made in the calculation of such statistics as the median, quartiles, and percentiles and in the drawing of histograms. In Table 2.4 it will be observed that 16 cases fall within the interval 100 to 104, which has the exact limits 99.5 to 104.5. The assumption states that these 16 cases are distributed over the interval as follows:

Interval	Frequency
103.5–104.5	3.2
102.5–103.5	3.2
101.5–102.5	3.2
100.5–101.5	3.2
99.5–100.5	3.2
Total	16.0

The second widely used assumption states that all the observations are concentrated at the midpoint of the interval, that is, that all the observations for that interval are the same and equal to the value corresponding to the midpoint of the interval. The midpoint of any class interval is halfway between the exact limits of the interval. In the above example the midpoint of the interval 99.5 to 104.5 is 102. This second assumption is ordinarily made in the calculation of such statistics as means and standard deviations, and in the drawing of frequency polygons.

The determination of the midpoint of a class interval should present no difficulty. The midpoint may be conveniently obtained by adding one-half of the range of the class interval to the lower exact limit of that interval. Thus with the interval 100 to 104 the lower limit is 99.5, and one-half the class interval is 2.5. The midpoint is therefore 99.5 + 2.5, or 102. Consider a 10-point class interval written in the form 100 to 109. Here the lower limit is 99.5, and one-half the class interval is 5. The midpoint is then 99.5 + 5, or 104.5. Table 2.4, column 3, shows the midpoints of the corresponding class intervals.

2.6 CUMULATIVE FREQUENCY DISTRIBUTIONS

Situations occasionally arise where our concern is not with the frequencies within the class intervals themselves but rather with the number or percentage of values "greater than" or "less than" a specified value. Such information may be made readily available by the preparation of a cumulative frequency distribution. The cumulative frequencies are obtained by

Table 2.5

Cumulative frequencies and cumulative percentage frequencies for distribution of intelligence quotients

1 Class interval (IQs)	2 Frequency	3 Cumulative frequency	4 Cumulative percentage frequency
130–134	1	106	100.0
125–129	3	105	99.1
120–124	4	102	96.2
115–119	10	98	92.5
110–114	8	88	83.0
105–109	15	80	75.5
100–104	20	65	61.3
95–99	14	45	42.5
90–94	11	31	29.2
85–89	8	20	18.9
80–84	6	12	11.3
75–79	5	6	5.7
70–74	0	1	.9
65–69	1	1	.9
Total	106		

SOURCE: From R. W. B. Jackson and George A. Ferguson, "Manual of Educational Statistics." University of Toronto, Department of Educational Research, Toronto, 1942.

adding successively, starting from the bottom, the individual frequencies. Table 2.5 shows the cumulative frequencies and cumulative percentages for a distribution of IQs.

From a cumulative frequency distribution we can obtain the number of cases falling below particular score values. For example, in Table 2.5 we observe that 98 cases obtain a score of 119 or below, 88 cases obtain a score of 114 or below, and so on. Similarly the cumulative percentage frequencies, which are obtained by dividing the cumulative frequency by the total number of cases, show the percent of individuals falling below certain score values. Thus, in Table 2.5, 42.5 percent of individuals have scores of 99 or below, 22.4 percent have scores of 94 or below, and so on.

2.7 TABULAR REPRESENTATION

Statistical data are frequently arranged and presented in the form of tables. Many such tables are found in this book. Tables should be designed to enable the reader to grasp with minimal effort the information that they intend to convey. In constructing statistical tables for insertion in term papers, theses, or manuscripts, a number of general points should be kept in mind. First, every table should be self-explanatory. The title should be precise and should state clearly what the table is about. Second, columns of numbers should be appropriately labeled, and arranged in a logical sequence, where such a sequence exists. Third, any necessary explanatory footnotes should be incorporated at the bottom of the table. Fourth, the information contained in a table may be partitioned by the insertion of horizontal and, possibly, vertical lines. Such lines should partition the table in a meaningful way, facilitate the comprehension of the information the table contains, and contribute to its aesthetic quality.

Tables presented as part of a term paper, thesis, or manuscript should be appropriately numbered. They should be inserted in the text, close to where they are first mentioned, when possible.

The appropriate design of a table can become a matter of some complexity. This is particularly the case where it is necessary to present data which are cross-classified in a variety of ways.

2.8 GRAPHIC REPRESENTATION OF FREQUENCY DISTRIBUTIONS

Graphic representation is often of great help in enabling us to comprehend the essential features of frequency distributions and in comparing one frequency distribution with another. A graph is the geometrical image of a set of data. It is a mathematical picture. It enables us to think about a problem in visual terms. Graphs are used not only in the practical handling of real sets of data but also as visual models in thinking about statistical problems. Many problems can be reduced to visual form, and such reduction often facilitates their understanding and solution. Graphs have become a part of our everyday activity. Newspapers, popular magazines, trade publications, business reports, and scientific periodicals use graphic representation extensively. Graphic representation has been carefully studied, and much has been written on the subject. While graphic representation has many ramifications, we shall consider here only those aspects of the subject which are useful in visualizing the important properties of frequency distributions and the ways in which one frequency distribution may differ from another.

2.9 HISTOGRAMS

A histogram is a graph in which the frequencies are represented by areas in the form of bars. Table 2.6 presents measures of auditory reaction time for a sample of 188 subjects. Figure 2.1 shows the frequencies plotted in the form of a histogram. To prepare such a histogram proceed as follows. Obtain a piece of suitably cross-sectioned graph paper. Paper subdivided into tenths of an inch with heavy lines 1 inch apart is convenient. Draw a horizontal line to represent reaction time in seconds and a vertical line to represent frequencies. Select an appropriate scale, for both reaction time

Table 2.6

Frequency distribution of auditory reaction times for a sample of 188 University of Chicago undergraduates

Class interval, seconds	Midpoint of interval	Frequency	Cumulative frequency
.34–.35	.345	2	188
.32–.33	.325	2	186
.30–.31	.305	4	184
.28–.29	.285	5	180
.26–.27	.265	11	175
.24–.25	.245	17	164
.22–.23	.225	28	147
.20–.21	.205	69	119
.18–.19	.185	37	50
.16–.17	.165	12	13
.14–.15	.145	1	1
Total		188	

SOURCE: From L. L. Thurstone, "A Factorial Study of Perception," University of Chicago Press, Chicago, 1944.

and frequencies. In the present case if we allow $\frac{5}{10}$ inch for each class interval and $\frac{1}{20}$ inch for each unit of frequency, we obtain a graph roughly 6 inches long and 4 inches tall. The scale is arbitrary. The scale suggested in this case, however, results in a graph of convenient size. The midpoints of the interval are written along the horizontal base line and the frequency scale along the vertical. For each class interval the corresponding frequency is plotted and a horizontal line drawn the full length of the interval. To complete the graph we may join the ends of these lines to the corresponding ends of the intervals on the horizontal axis, although practice in this regard varies. Both the horizontal and vertical axes must be appropriately labeled. A concise statement of what the graph is about should accompany it. Observe that the width of each bar corresponds to the exact limits of the interval. Observe also that in this type of graph the frequencies are represented as equally distributed over the whole range of the interval.

2.10 FREQUENCY POLYGONS

In a histogram we assume that all the cases within a class interval are uniformly distributed over the range of the interval. In a frequency polygon we assume that all cases in each interval are concentrated at the midpoint of the interval. In this fact resides the essential difference between a histogram and a frequency polygon. Instead of drawing a horizontal line the full length of the interval, as in this histogram, we make a dot above the midpoint of each interval at a height proportional to the frequency. It is

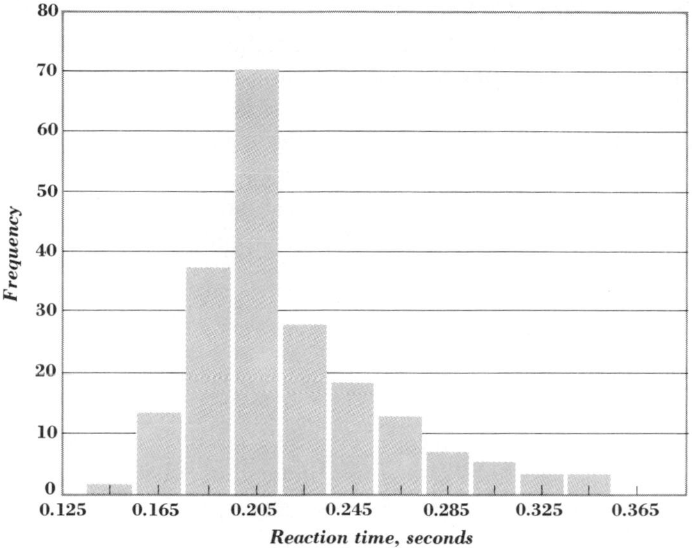

Figure 2.1 Histogram for data of Table 2.6. Auditory reaction times for 188 students.

customary to show an additional interval at each end of the horizontal scale and to join these dots to the dots of the adjacent interval. A frequency distribution based on the same data as the histogram in Figure 2.1 is shown in Figure 2.2.

Observe that the frequency distribution in Figure 2.2 is not a smooth

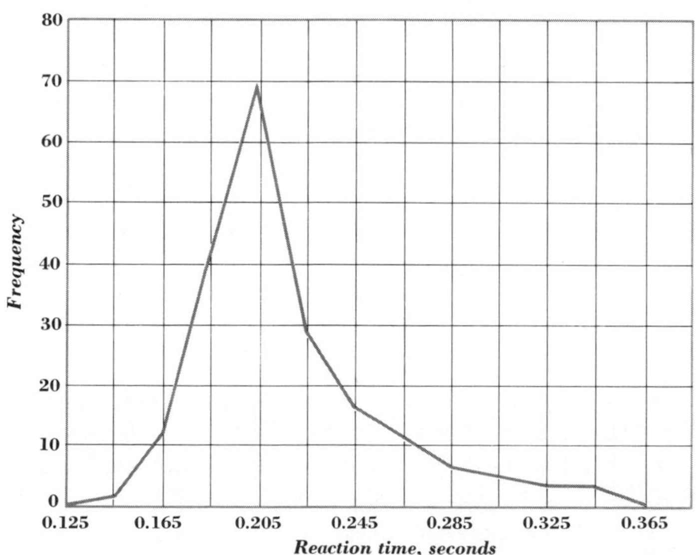

Figure 2.2 Frequency polygon for data of Table 2.6. Auditory reaction times for 188 students.

continuous curve, since the lines joining the various points are straight lines. If we subdivide our intervals into smaller intervals, we shall of course obtain irregular frequencies, there being too few members in each interval. Consider, however, a circumstance where our intervals become smaller and smaller and at the same time the total number of cases becomes larger and larger. If we carry this process to the extreme situation where we have an indefinitely small interval and an indefinitely large number of cases, we arrive at the concept of a continuous frequency distribution.

2.11 CUMULATIVE FREQUENCY POLYGONS

The drawing of a cumulative frequency polygon differs from that of a frequency polygon in two respects. *First,* instead of plotting points corresponding to frequencies, we plot points corresponding to cumulative frequencies. *Second,* instead of plotting points above the midpoint of each interval, we plot our points above the top of the exact limits of the interval. This is done because we wish our graph to visually represent the number of cases falling above or below particular values. In plotting the cumulative frequency distribution shown in Table 2.6, we would plot the cumulative frequency 188 against the top of the exact upper limit of the interval, that is, .355, the frequency 186 against .335, and so on. Figure 2.3 shows the cumulative frequency distribution for the data appearing in the last column of Table 2.6.

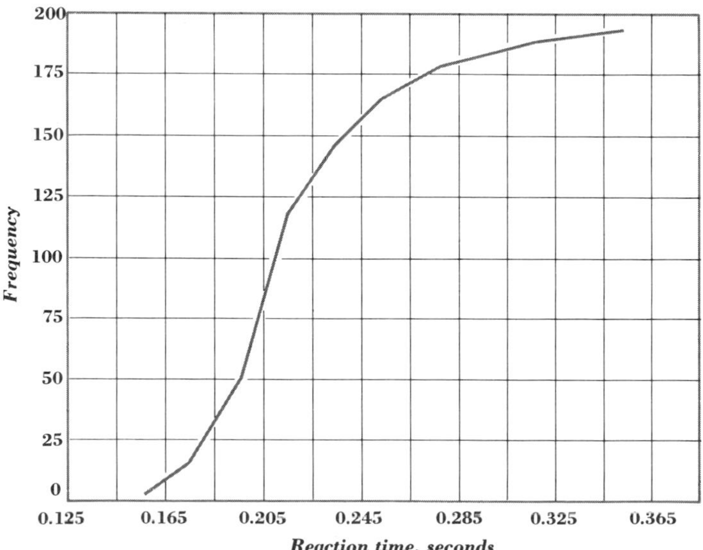

Figure 2.3 Cumulative frequency polygon for data of Table 2.6. Auditory reaction times for 188 students.

We may convert our raw frequencies to percentages such that all the frequencies added together add up to 100 instead of to the number of cases. We may then determine the cumulative percentage frequencies. Next we may graph these frequencies and obtain thereby a cumulative percentage polygon, or ogive. The advantage of this type of diagram is that from it we can read off directly the percentage of observations less than any specified value.

2.12 SOME CONVENTIONS FOR THE CONSTRUCTION OF GRAPHS

1 In the graphing of frequency distributions it is customary to let the horizontal axis represent scores and the vertical axis frequencies.

2 The arrangement of the graph should proceed from left to right. The low numbers on the horizontal scale should be on the left, and the low numbers on the vertical scale should be toward the bottom.

3 The distance along either axis selected to serve as a unit is arbitrary and affects the appearance of the graph. Some writers suggest that the units should be selected such that the ratio of height to length is roughly 3:5. This procedure seems to have some aesthetic advantages.

4 Whenever possible the vertical scale should be so selected that a zero point falls at the point of intersection of the axes. With some data this procedure may give rise to a most unusual-looking graph. In such cases it is customary to designate the point of intersection as the zero point and make a small break in the vertical axis.

5 Both the horizontal and vertical axes should be appropriately labeled.

6 Every graph should be assigned a descriptive title which states precisely what it is about.

2.13 HOW FREQUENCY DISTRIBUTIONS DIFFER

Comparison of a number of frequency distributions represented in either tabular or graphic form indicates that they differ one from another. An important problem in statistics is the identification and definition of properties or attributes of frequency distributions which describe how they differ. It is customary to designate four important properties of frequency distributions. These are central location, variation, skewness, and kurtosis. These properties may be viewed either as descriptive of the frequency distribution itself or as descriptive of the set of observations which the distribution comprises. These alternatives are in effect synonymous. A frequency distribution is a particular kind of arrangement of a set of obser-

Table 2.7

Hypothetical data illustrating frequency distributions of different shapes

1 Class interval	2 Symmetrical binomial	3 Leptokurtic	4 Platykurtic	5 Rectangular	6 Bimodal	7 U-shaped	8 Positively skewed	9 Negatively skewed	10 J-shaped
70–79	1	3	5	16	5	30	2	10	50
60–69	7	8	14	16	10	20	6	25	30
50–59	21	13	20	16	35	10	10	40	20
40–49	35	40	25	16	14	4	15	20	10
30–39	35	40	25	16	14	4	20	15	7
20–29	21	13	20	16	35	10	40	10	5
10–19	7	8	14	16	10	20	25	6	4
0–9	1	3	5	16	5	30	10	2	2
N	128	128	128	128	128	128	128	128	128

vations. Central location, variation, skewness, and kurtosis may be discussed either with direct reference to sets of observations or with reference to the observations arranged in frequency distribution form.

Central location refers to a value of the variable near the center of the frequency distribution. It is a middle point. Measures of central locations are called *averages*. These are discussed in detail in Chapter 4.

Variation refers to the extent of the clustering about a central value. If all the observations are close to the central value, their variation will be less than if they tend to depart more markedly from the central value. Measures of variation are discussed in Chapter 5.

Skewness refers to the symmetry or asymmetry of the frequency distribution. If a distribution is asymmetrical and the larger frequencies tend to be concentrated toward the low end of the variable and the smaller frequencies toward the high end, it is said to be *positively skewed.* If the opposite holds, the larger frequencies being concentrated toward the high end of the variable and the smaller frequencies toward the low end, the distribution is said to be *negatively skewed.*

Kurtosis refers to the flatness or peakedness of one distribution in relation to another. If one distribution is more peaked than another, it may be spoken of as more *leptokurtic*. If it is less peaked, it is said to be more *platykurtic*. It is conventional to speak of a distribution as leptokurtic if it is more peaked than a particular type of distribution known as the normal distribution, and platykurtic if it is less peaked. The normal distribution is spoken of as *mesokurtic,* which means that it falls between leptokurtic and platykurtic distributions.

Table 2.7 presents hypothetical data illustrating frequency distributions with different properties. The distribution in column 2 is a *symmetrical binomial,* a type of distribution which is of much importance in statistical work and will be considered in detail in a later chapter. The distribution in column 3 has central frequencies which are greater than those for the binomial. It is more peaked than the binomial, and as far as kurtosis is concerned, it can be said to be leptokurtic. The distribution in column 4 has smaller central frequencies than the binomial and larger frequencies toward the extremities. It can be spoken of as platykurtic. The distribution in column 5 has uniform frequencies over all class intervals and is described as *rectangular*. The distribution in column 6 has two humps, or modes. It is said to be *bimodal.* In the distribution in column 7 the largest frequencies occur at the extremities whereas the central frequencies are the smallest. Such a distribution is said to be *U-shaped.* The distributions in columns 2 to 7 are all symmetrical and have the same measures of central location although they differ in variation. Column 8 illustrates a *positively skewed* and column 9 a *negatively skewed* distribution. Extreme skewness leads to the type of distribution shown in column 10, which is described as *J-shaped.* Figure 2.4 illustrates a rectangular distribution, a bimodal distribution, a U-shaped distribution, and a J-shaped distribution.

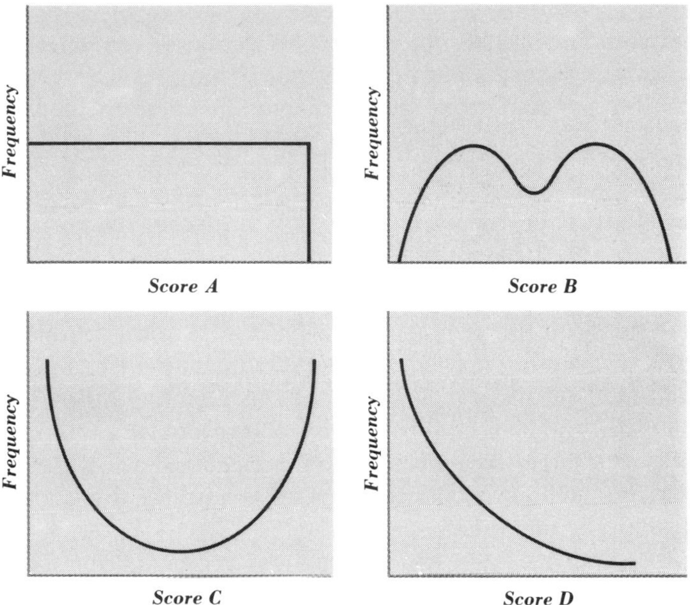

Figure 2.4 Four types of distributions: *A*, rectangular distribution; *B*, bimodal distribution; *C*, U-shaped distribution; *D*, J-shaped distribution.

2.14 THE PROPERTIES OF FREQUENCY DISTRIBUTIONS REPRESENTED GRAPHICALLY

The differing characteristics of frequency distributions can be readily represented in graphic form. Consider the three distributions in Figure 2.5. These distributions appear identical in shape. They are markedly different, however, in terms of the central values about which the observations in each distribution appear to concentrate; that is, they have different averages although they may be identical in all other respects. Distribution *A* has a lower average than *B* and *B* than *C*.

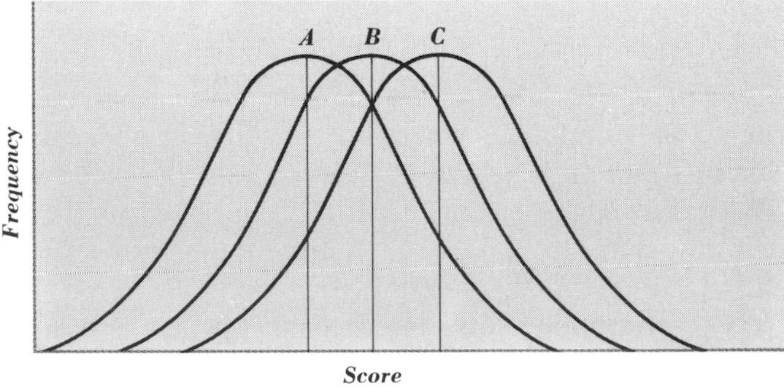

Figure 2.5 Three frequency distributions identical in shape but with different averages.

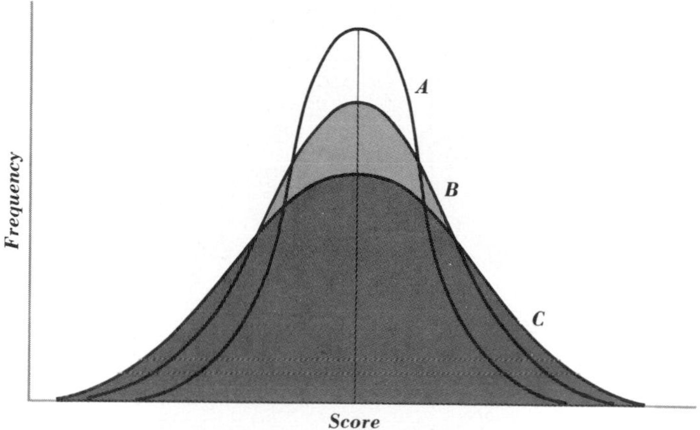

Figure 2.6 Three frequency distributions with the same average but with different variation.

Now consider the distributions in Figure 2.6. Inspection of these three distributions suggests that while the observations in each case appear to concentrate about the same average, they are nonetheless markedly different one from another. In the case of distribution *A* the observations appear to be more closely concentrated about the average than in the case of *B*, and the same applies to *B* in relation to *C*. Thus these distributions differ in variation. The observations in *A* are less variable than the observations in *B*, and those in *B* are less variable than those in *C*.

Examine now the distributions in Figure 2.7. These three distributions have different averages and possibly different measures of variation. They differ also in skewness. Distribution *B* is symmetrical about the average; that is, if we were to fold it over about the average, we should find that it had the same shape on both sides. Distributions *A* and *C* are asymmetrical, the shape to the left of the average being different from the shape

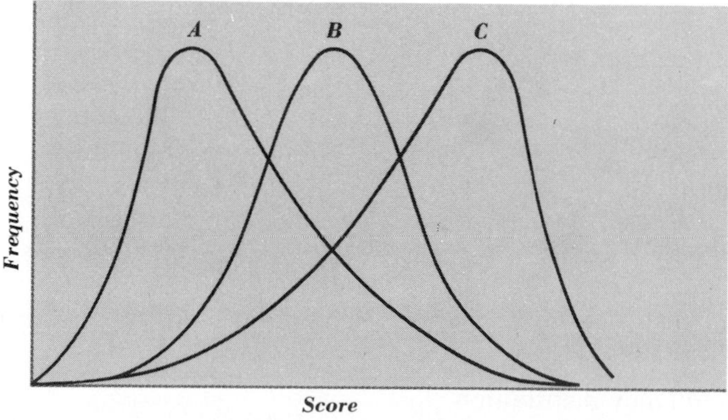

Figure 2.7 Three frequency distributions differing in skewness.

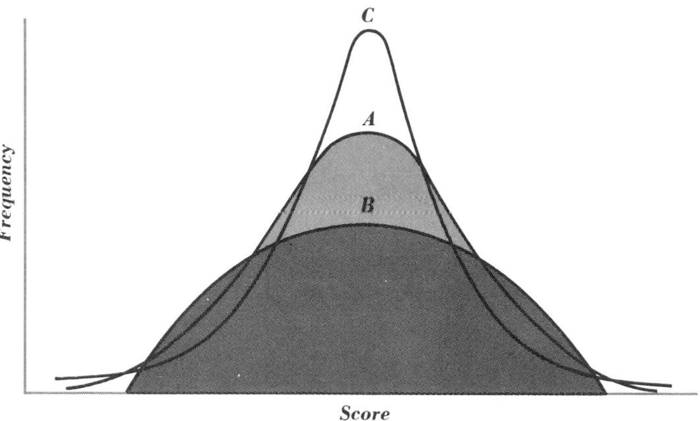

Figure 2.8 Three frequency distributions differing in kurtosis.

to the right. Distribution *A* is positively skewed, the longer tail extending toward the high end of the scale. Distribution *C* is negatively skewed, the longer tail extending toward the low end of the scale.

Consider now the graphic representation of kurtosis as shown in Figure 2.8. Distribution *A* is a symmetrical bell-shaped distribution known as the normal distribution. Distribution *B* is observed to be flatter on top than the normal distribution and is referred to as platykurtic, while distribution *C* is more peaked than the normal and is spoken of as leptokurtic.

In the above discussion the meaning which attaches to the descriptive properties of collections of measurements arranged in frequency distributions is largely intuitive and is derived from the inspection of distributions in tabular or graphic form. To proceed with the study of data interpretation we require precisely defined numerical measures of central location, variation, skewness, and kurtosis. Chapters 4 and 5 are concerned with the more precise and formal delineation of these properties, their numerical description, and calculation.

BASIC TERMS AND CONCEPTS

Frequency

Frequency distribution

Class interval

Exact limits of class interval

Midpoint of class interval

Cumulative frequency distribution

Histogram

Frequency polygon

Central location

Variation

Skewness: positive; negative

Symmetrical distribution

Rectangular distribution

EXERCISES

1 The following are marks obtained by a group of 40 university students on an English examination:

42	88	37	75	98	93	73	62
96	80	52	76	66	54	73	69
83	62	53	79	69	56	81	75
52	65	49	80	67	59	88	80
44	71	72	87	91	82	89	79

Prepare a frequency distribution and a cumulative frequency distribution for these data, using a class interval of 5.

2 Write down the exact limits and the midpoints of the class intervals for the frequency distribution obtained by answering Exercise 1 above.

3 Write down an acceptable set of class intervals, the exact limits of the intervals, and their midpoints, in preparing frequency distributions for the following data: (**a**) error scores ranging from 24 to 87 made by a sample of rats in running a maze, (**b**) IQs ranging from 96 to 137 for a group of schoolchildren, (**c**) scholastic aptitude test scores ranging from 227 to 896 obtained by a group of university students, (**d**) response latencies ranging from .276 to .802 second for a group of experimental subjects.

4 Prepare a frequency distribution for the following test scores:

2	11	6	4	18	1	9	2	2	15
8	16	12	11	17	3	3	5	3	7
11	9	5	16	16	16	4	9	5	7
4	10	4	4	15	15	5	5	11	18
5	10	9	8	7	7	2	6	13	1

Obtain a cumulative percentage frequency distribution for these data.

5 Prepare histograms for the data in Exercises 1 and 4 above.

6 Prepare frequency polygons for the data in Exercises 1 and 4 above.

7 Prepare cumulative frequency polygons for the data in Exercises 1 and 4 above.

8 The following are marks obtained by 40 students on a mathematics and English examination:

Mathematics				English			
22	92	86	52	49	38	74	78
31	62	75	40	84	88	76	72
55	94	37	42	86	69	55	42
76	88	41	76	31	91	88	72
48	88	76	29	65	66	72	78
49	72	64	72	56	99	92	84
50	65	66	59	63	86	72	67
85	62	58	42	81	59	24	77
7	25	66	54	91	86	88	72
38	88	76	62	51	84	62	89

Prepare frequency distributions for these marks, using a class interval of 10.

9 Represent the frequency distributions in Exercise 8 above as two frequency polygons. Plot both frequency polygons on the same graph.

10 Write down the midpoints for the following class intervals:

a 200–299 **b** .500–.574 **c** 25–49

d 4–7 **e** .05–.09

ANSWERS TO EXERCISES

1

Class interval	Frequency	Cumulative frequency
95–99	2	40
90–94	2	38
85–89	4	36
80–84	6	32
75–79	5	26
70–74	4	21
65–69	5	17
60–64	2	12
55–59	2	10
50–54	4	8
45–49	1	4
40–44	2	3
35–39	1	1
Total	40	

2

Interval	Exact limits	Midpoints
95–99	94.5–99.5	97
90–94	89.5–94.5	92
85–89	84.5–89.5	87
80–84	79.5–84.5	82
75–79	74.5–79.5	77
70–74	69.5–74.5	72
65–69	64.5–69.5	67
60–64	59.5–64.5	62
55–59	54.5–59.5	57
50–54	49.5–54.5	52
45–49	44.5–49.5	47
40–44	39.5–44.5	42
35–39	34.5–39.5	37

3

	Class interval	Exact limits	Midpoints
a	85–89	84.5–89.5	87
	80–84	79.5–84.5	82
	75–79	74.5–79.5	77
	. . .	. . .	. . .
b	135–137	134.5–137.5	136.0
	132–134	131.5–134.5	133.0
	129–131	128.5–131.5	130.0
	. . .	. . .	. . .
c	850–899	849.5–899.5	874.50
	800–849	799.5–849.5	824.50
	750–799	749.5–799.5	774.50
	. . .	. . .	. . .
d	.800–.849	.7995–.8495	.8245
	.750–.799	.7495–.7995	.7745
	.700–.749	.6995–.7495	.7245
	. . .	. . .	. . .

4

Class interval	Frequency	Cumulative frequency	Cumulative % frequency
17–18	3	50	100
15–16	7	47	94
13–14	1	40	80
11–12	5	39	78
9–10	6	34	68
7–8	6	28	56
5–6	8	22	44
3–4	8	14	28
1–2	6	6	12

8

Class interval	Frequency, mathematics	Frequency, English
90–99	2	4
80–89	5	11
70–79	7	10
60–69	7	6
50–59	6	4
40–49	6	2
30–39	3	2
20–29	3	1
10–19	. . .	. . .
0–9	1	. . .
Total	40	40

10 **a** 249.5 **b** .537 **c** 37
d 5.5 **e** .070

3

STATISTICAL NOTATION

3.1 INTRODUCTION

The representation of data in symbolic form and the specification of rules governing the use of symbols and their relations provides the student with the tools necessary for algebraic manipulation. The ability to use statistical symbols and to conduct elementary algebraic operations not only enriches the understanding of statistics, but also enhances the power to represent complex sets of data and their properties in simple ways.

The most commonly used form of notation in statistics is *summation notation*. The ideas present in the use of this notation are few in number and very simple. A fluent grasp of these ideas, acquired through practice, will prove of great value to the student.

3.2 SYMBOLIC REPRESENTATION OF VARIABLES

A set of variate values, measurements, or observations may be denoted by $X_1, X_2, X_3, \ldots, X_N$ or $Y_1, Y_2, Y_3, \ldots, Y_N$, where N denotes the number of variate values. The symbols X and Y are commonly used to denote variables. Other symbols may also be used. A variable, for example, may be the activity of a group of N rats in a maze, expressed as scores. The symbol X_1 represents the score of the first rat, X_2 the score of the second rat, and so on until X_N represents the score of the Nth rat. The numbers $1,2,3, \ldots, N$ are known as subscripts. These identify the particular member. If for $N = 5$ the scores are 10, 12, 19, 21, and 32, then $X_1 = 10$, $X_2 = 12$, $X_3 = 19$, $X_4 = 21$, and $X_5 = 32$. It is customary to denote any value of the variable by X_i

where the subscript i may take any value from 1 to N. The symbols j and k are also used as subscripts.

3.3 SUMMATION OF A VARIABLE

Consider now the sum $X_1 + X_2 + \ldots + X_N$, that is, all the values of the variable added together. This sum is represented by

[3.1] $$\sum_{i=1}^{N} X_i$$

Thus $$\sum_{i=1}^{N} X_i = X_1 + X_2 + \cdots + X_N$$

The symbol Σ is the Greek capital letter sigma and refers to the simple operation of adding things up. It is spoken of as a summation sign, signifying the summing of things. The symbols above and below the summation sign define the limits of the summation. Thus $\sum_{i=1}^{N} X_i$ means the addition of all those values of the variable X_i, where i takes the value extending from $i = 1$ to $i = N$. An expression of the kind $\sum_{i=1}^{5} X_i$ means the sum of the first five values of X_i, that is, the values extending from $i = 1$ to $i = 5$. The expression $\sum_{i=6}^{10} X_i$ means the sum of the next five values of X_i, that is, the values extending from $i = 6$ to $i = 10$. To illustrate, let the numbers 10, 12, 19, 21, and 32 be measures of activity for a group of five rats. The sum of these five measures, $10 + 12 + 19 + 21 + 32$, is represented symbolically by $\sum_{i=1}^{5} X_i$ and in this case is 94. When the limits of the summation are clearly understood from the context, which is frequently the case, it is customary to omit the notation above and below the summation sign and write ΣX_i or ΣX.

3.4 RULES FOR SUMMATION NOTATION

Three simple theorems may be stated which are useful in handling problems involving summation notation.

THEOREM 1

If every variate value in a group is multiplied by a constant number or factor, that factor may be removed from under the summation sign and written outside as a factor. Thus

$$\sum_{i=1}^{N} cX_i = cX_1 + cX_2 + \cdots + cX_N$$

[3.2]
$$= c(X_1 + X_2 + \cdots + X_N)$$

$$= c \sum_{i=1}^{N} X_i \qquad \square$$

This means that if we multiply each one of the measures 10, 12, 19, 21, 32 by any constant, say, 5, the sum of the resulting measures will be given directly by 5×94.

THEOREM 2

The summation of a constant over N terms is equal to Nc. Thus

[3.3]
$$\sum_{i=1}^{N} c = c + c + \cdots + c$$

$$= Nc \qquad \square$$

If $c = 5$ and N equals 4, it is obvious that $5 + 5 + 5 + 5 = 4 \times 5 = 20$.

THEOREM 3

The summation of the sum of any number of terms is the sum of the summations of these terms taken separately. Thus

$$\sum_{i=1}^{N} (X_i + Y_i + Z_i) = X_1 + Y_1 + Z_1 + X_2 + Y_2 + Z_2 + \cdots + X_N + Y_N + Z_N$$

[3.4]
$$= \sum_{i=1}^{N} X_i + \sum_{i=1}^{N} Y_i + \sum_{i=1}^{N} Z_i \qquad \square$$

An expression frequently encountered in statistics is $\sum_{i=1}^{N} X_iY_i$. This refers to the sum of the products of two sets of paired numbers. If, for example, 5, 6, 12, 15 are the scores X of four people on a test, and 2, 3, 7, 10 are the scores Y of the same four people on another test, then $\sum_{i=1}^{N} X_iY_i$ refers to the sum of products and is equal to $5 \times 2 + 6 \times 3 + 12 \times 7 + 15 \times 10$, or 262.

The following are examples which illustrate the application of the above theorems.

EXAMPLE 1

$$\sum_{i=1}^{N} (X + c) = \sum_{i=1}^{N} X + \sum_{i=1}^{N} c = \sum_{i=1}^{N} X + Nc$$

EXAMPLE 2

$$\sum_{i=1}^{N} (X + Y + c) = \sum_{i=1}^{N} X + \sum_{i=1}^{N} Y + \sum_{i=1}^{N} c = \sum_{i=1}^{N} X + \sum_{i=1}^{N} Y + Nc$$

EXAMPLE 3

$$\sum_{i=1}^{N} (X + c)^2 = \sum_{i=1}^{N} (X^2 + 2cX + c^2) = \sum_{i=1}^{N} X^2 + \sum_{i=1}^{N} 2cX + \sum_{i=1}^{N} c^2$$
$$= \sum_{i=1}^{N} X^2 + 2c \sum_{i=1}^{N} X + Nc^2$$

EXAMPLE 4

$$\sum_{i=1}^{N} (X + Y)^2 = \sum_{i=1}^{N} (X^2 + 2XY + Y^2) = \sum_{i=1}^{N} X^2 + \sum_{i=1}^{N} 2XY + \sum_{i=1}^{N} Y^2$$
$$= \sum_{i=1}^{N} X^2 + 2 \sum_{i=1}^{N} XY + \sum_{i=1}^{N} Y^2$$

EXAMPLE 5

$$\sum_{i=1}^{N} [(X + c)^2 - c^2] = \sum_{i=1}^{N} [X^2 + 2cX + c^2 - c^2]$$
$$= \sum_{i=1}^{N} X^2 + \sum_{i=1}^{N} 2cX = \sum_{i=1}^{N} X^2 + 2c \sum_{i=1}^{N} X$$

EXAMPLE 6

$$\sum_{i=1}^{N} [(X + Y)^2 - (X - Y)^2] = \sum_{i=1}^{N} [(X^2 + Y^2 + 2XY) - (X^2 + Y^2 - 2XY)]$$
$$= \sum_{i=1}^{N} [X^2 + Y^2 + 2XY - X^2 - Y^2 + 2XY]$$
$$= \sum_{i=1}^{N} 4XY = 4 \sum_{i=1}^{N} XY$$

EXAMPLE 7

$$\sum_{i=1}^{N} (X - 3Y)^2 = \sum_{i=1}^{N} (X^2 + 9Y^2 - 6XY) = \sum_{i=1}^{N} X^2 + \sum_{i=1}^{N} 9Y^2 - \sum_{i=1}^{N} 6XY$$
$$= \sum_{i=1}^{N} X^2 + 9 \sum_{i=1}^{N} Y^2 - 6 \sum_{i=1}^{N} XY$$

A further concrete illustrative example may prove helpful to the reader in grasping the nature of summation notation. Let the following be paired scores. The values of X_i below may be viewed as scores for four people on a test, and the values of Y_i may be viewed as scores for the same four people on another test.

$$\begin{array}{ll} X_1 = 5 & Y_1 = 2 \\ X_2 = 6 & Y_2 = 3 \\ X_3 = 12 & Y_3 = 7 \\ X_4 = 15 & Y_4 = 10 \end{array}$$

The following relate to these scores.

1 $\sum_{i=1}^{4} X_i = 5 + 6 + 12 + 15 = 38$

2 $\sum_{i=1}^{4} Y_i = 2 + 3 + 7 + 10 = 22$

3 $\sum_{i=1}^{4} 5X_i = 5 \times 5 + 5 \times 6 + 5 \times 12 + 5 \times 15 = 5 \times 38 = 190$

4 $\sum_{i=1}^{4} X_i^2 = 5 \times 5 + 6 \times 6 + 12 \times 12 + 15 \times 15 = 430$

5 $\sum_{i=1}^{4} (X_i - 5) = (5 - 5) + (6 - 5) + (12 - 5) + (15 - 5) = 18$

6 $\sum_{i=1}^{4} (X_i + Y_i) = (5 + 2) + (6 + 3) + (12 + 7) + (15 + 10)$

$$= 38 + 22 = 60$$

7 $\sum_{i=1}^{4} X_iY_i = 5 \times 2 + 6 \times 3 + 12 \times 7 + 15 \times 10 = 262$

The notation used in elementary statistics is simple, and skill in its manipulation can be acquired with a little practice. A good understanding of the nature of statistical method and its applications can be acquired with very little in the way of mathematical training at all. A little knowledge of arithmetic and a little elementary algebra go a long way in the study of statistics.

3.5 NOTATION FOR A FREQUENCY DISTRIBUTION

A frequency distribution is an arrangement of data that shows the frequency of occurrence of different values of a variable. With grouped data, the frequency distribution shows the frequency of occurrence of values within class intervals. A frequency distribution may be represented as

X_i	f_i
X_1	f_1
X_2	f_2
X_3	f_3
.	.
.	.
.	.
X_k	f_k

The symbol k denotes the number of different values of X_i, and f_i is a frequency where i extends from $i = 1$ to $i = k$. The summation of the variable may be written as

$$\sum_{i=1}^{N} X_i = \sum_{i=1}^{k} f_i X_i = f_1X_1 + f_2X_2 + f_3X_3 + \cdots + f_kX_k$$

Although the data comprised N values of the variable, only k different values occur with frequencies $f_1, f_2, f_3, \cdots, f_k$.

To illustrate, consider the values 1, 1, 1, 2, 2, 3, 3, 3, 3, 4, 4, 5, 5, 5, 5. Here $N = 15$ and $\sum_{i=1}^{15} X_i = 47$.

The 1s occur with a frequency of 3. Their contribution to the total sum is $1 \times 3 = 3$. The 2s occur with a frequency of 2. Their contribution is $2 \times 2 = 4$, and so on. Hence

$$\sum_{i=1}^{5} f_iX_i = (1 \times 3) + (2 \times 2) + (3 \times 4) + (4 \times 2) + (5 \times 4) = 47$$

Note than $\sum_{i=1}^{N} X_i$ and $\sum_{i=1}^{k} f_iX_i$ lead to the same numerical result where the X_i are a set of discrete values.

BASIC TERMS AND CONCEPTS

X_i

$\sum_{i=1}^{N} X_i$

Limits of summation

$\sum_{i=1}^{N} cX_i = c \sum_{i=1}^{N} X_i$

$\sum_{i=1}^{N} c = Nc$

$\sum_{i=1}^{N} (X_i + Y_i) = \sum_{i=1}^{N} X_i + \sum_{i=1}^{N} Y_i$

$$\sum_{i=1}^{N} X_i Y_i$$

$$\sum_{i=1}^{k} f_i X_i$$

EXERCISES

1 Write the following in summation notation:

a $X_1 + X_2 + \cdots + X_{15}$

b $Y_1 + Y_2 + \cdots + Y_N$

c $(X_1 + Y_1) + (X_2 + Y_2) + \cdots + (X_7 + Y_7)$

d $X_1Y_1 + X_2Y_2 + \cdots + X_NY_N$

e $X_1^3Y_1 + X_2^3Y_2 + \cdots + X_N^3Y_N$

f $(X_1 + c) + (X_2 + c) + \cdots + (X_8 + c)$

g $cX_1 + cX_2 + \cdots + cX_{25}$

h $X_1/c + X_2/c + \cdots + X_N/c$

i $cX_1^2Y_1 + cX_2^2Y_2 + \cdots + cX_N^2Y_N$

2 Write each of the following in full:

a $\sum_{i=1}^{2} X_i$ **b** $\sum_{i=1}^{3} X_iY_i$ **c** $\sum_{i=1}^{5} (X_i + Y_i)$

d $c\sum_{i=1}^{3} X_i^2Y_i$ **e** $\sum_{i=1}^{4} X_i + 4c$ **f** $\frac{1}{c}\sum_{i=1}^{5} X_i + c\sum_{i=1}^{5} Y_i$

3 Obtain

a $\sum^{4} 6$ **b** $\sum^{N} c$ **c** $\sum^{N} 2$

4 For $X_1 = 1$, $X_2 = 3$, and $X_3 = 5$, find the values of

a $\sum_{i=1}^{3} X_i$ **b** $\sum_{i=1}^{3} X_i^2$ **c** $\sum_{i=1}^{3} (X_i + 3)$

d $\left(\sum_{i=1}^{3} X_i\right)^2$ **e** $\sum_{i=1}^{3} X_i - 3$ **f** $\left[\left(\sum_{i=1}^{3} X_i\right) + 3\right]^2$

5 If $X_1 = 4$, $X_2 = 6$, $X_3 = 3$, and $X_4 = 7$, obtain the following:

a $\sum^{4} (X_i^2 - 4X_i)$ **b** $\sum^{4} (X_i^3 - X_i^2 + X_i)$

6 Show that

$$\sum_{i=1}^{N} (X_i + Y_i)(X_i - Y_i) = \sum_{i=1}^{N} X_i^2 - \sum_{i=1}^{N} Y_i^2$$

7 Show that

$$\sum_{i=1}^{N} (X_i + c)^2 = \sum_{i=1}^{N} X_i^2 + 2c \sum_{i=1}^{N} X_i + Nc^2$$

8 Consider the following paired observations:

$X_1 = 3 \quad Y_1 = 9$
$X_2 = 5 \quad Y_2 = 2$
$X_3 = 5 \quad Y_3 = 1$
$X_4 = 4 \quad Y_4 = 5$
$X_5 = 8 \quad Y_5 = 3$

Calculate

a ΣX_i^2 **b** ΣY_i^2 **c** $\Sigma (X_i - 5)^2$
d $\Sigma (Y_i - 4)^2$ **e** $\Sigma X_i Y_i$ **f** $\Sigma (X_i - 5)(Y_i - 4)$
g $\Sigma (5X_i - 4Y_i)^2$ **h** $\Sigma (X_i - 3^2)^2$ **i** $\Sigma (X_i / Y_i)$

In all instances the summation is understood to extend over the five paired observations.

9 Which of the following are true and which are false?

a $\sum_{i=1}^{N} X_i \sum_{i=1}^{N} Y_i = \sum_{i=1}^{N} X_i Y_i$

b $\left(\sum_{i=1}^{N} X_i\right)^2 = \sum_{i=1}^{N} X_i^2$

c $\sum_{i=1}^{N} (X_i + c)(X_i - c) = \sum_{i=1}^{N} X^2 - Nc^2$

d $\sum_{i=1}^{N} (X_i + Y_i)^2 = \sum_{i=1}^{N} X_i^2 + \sum_{i=1}^{N} Y_i^2 + 2 \sum_{i=1}^{N} X_i Y_i$

10 Show that

$$\Sigma \left(X - \frac{\Sigma X}{N}\right)^2 = \Sigma X^2 - \frac{(\Sigma X)^2}{N}$$

ANSWERS TO EXERCISES

1 a $\sum_{i=1}^{15} X_i$ **b** $\sum_{i=1}^{N} Y_i$ **c** $\sum_{i=1}^{7} X_i + \sum_{i=1}^{7} Y_i$

d $\sum_{i=1}^{N} X_i Y_i$ **e** $\sum_{i=1}^{N} X_i^3 Y_i$ **f** $\sum_{i=1}^{8} X_i + 8c$

g $c \sum_{i=1}^{25} X_i$ **h** $\frac{1}{c} \sum_{i=1}^{N} X_i$ **i** $c \sum_{i=1}^{N} X_i^2 Y_i$

2 a $X_1 + X_2$
b $X_1 Y_1 + X_2 Y_2 + X_3 Y_3$

c $(X_1 + Y_1) + (X_2 + Y_2) + (X_3 + Y_3) + (X_4 + Y_4) + (X_5 + Y_5)$

d $cX_1^2Y_1 + cX_2^2Y_2 + cX_3^2Y_3$

e $(X_1 + c) + (X_2 + c) + (X_3 + c) + (X_4 + c)$

f $\frac{X_1}{c} + \frac{X_2}{c} + \frac{X_3}{c} + \frac{X_4}{c} + \frac{X_5}{c} + cY_1 + cY_2 + cY_3 + cY_4 + cY_5$

3 **a** 24 **b** Nc **c** $2N$

4 **a** 9 **b** 35 **c** 18
d 81 **e** 6 **f** 144

5 **a** 30 **b** 560

6 $$\sum_{i=1}^{N} (X_i + Y_i)(X_i - Y_i) = \sum_{i=1}^{N} (X_i^2 - Y_i^2 - X_iY_i + X_iY_i)$$
$$= \sum_{i=1}^{N} (X_i^2 - Y_i^2) = \sum_{i=1}^{N} X_i^2 - \sum_{i=1}^{N} Y_i^2$$

7 $$\sum_{i=1}^{N} (X_i + c)^2 = \sum_{i=1}^{N} (X_i^2 + 2X_ic + c^2)$$
$$= \sum_{i=1}^{N} X_i^2 + 2c \sum_{i=1}^{N} X_i + Nc^2$$

8 **a** 139 **b** 120 **c** 14
d 40 **e** 86 **f** -14
g 1,955 **h** 94 **i** 11.30

9 **a** False **b** False **c** True **d** True

10 $$\sum \left(X - \frac{\Sigma X}{N}\right)^2 = \sum \left[X^2 + \frac{(\Sigma X)^2}{N^2} - \frac{2X\Sigma X}{N}\right]$$
$$= \Sigma X^2 + \frac{N(\Sigma X)^2}{N^2} - \frac{2\Sigma X \Sigma X}{N}$$
$$= \Sigma X^2 + \frac{(\Sigma X)^2}{N} - \frac{2(\Sigma X)^2}{N}$$
$$= \Sigma X^2 - \frac{(\Sigma X)^2}{N}$$

4

AVERAGES

4.1 INTRODUCTION

Chapter 2 discussed the organization of collections of numbers in the form of frequency distributions and how such frequency distributions could be represented graphically. We now proceed to a consideration of how a collection of numbers, whether organized in the form of a frequency distribution or not, may be described. What are the descriptive properties of collections of numbers or their corresponding frequency distributions? One such property is the *average*.

The average is a measure of central location. The word "average" is commonly used to refer to a value obtained by adding together a set of measurements and then dividing by the number of measurements in the set. This is one type of average only and is called the arithmetic mean. In general, an average is a central reference value which is usually close to the point of greatest concentration of the measurements and may in some sense be thought to typify the whole set. For certain purposes a particular measurement may be viewed as a certain distance above or below the average. Averages in common use in psychology and education are the arithmetic mean and the median. Other measures of central location include the mode, the geometric mean, and the harmonic mean. This chapter contains a discussion of the arithmetic mean, the median, and the mode. Discussion of the geometric mean and the harmonic mean has been omitted because of the infrequency of their use in practice. By far the most important and widely used measure of central location is the arithmetic mean. This statistic is an appropriate measure of central location for interval and ratio variables. The median and mode are sometimes viewed as appropriate measures for ordinal and nominal variables, respectively, although they can also be used with interval and ratio variables.

4.2 THE ARITHMETIC MEAN

By definition the arithmetic mean is the sum of a set of measurements divided by the number of measurements in the set. Consider the following measurements: 7, 13, 22, 9, 11, 4. The sum of these measurements is 66. The arithmetic mean is, therefore, 66 divided by 6, or 11.

In general, if N measurements are represented by the symbols X_1, X_2, $X_3, \ldots, X_N$, the arithmetic mean in algebraic language is

[4.1]
$$\bar{X} = \frac{X_1 + X_2 + X_3 + \cdots + X_N}{N} = \frac{\sum_{i=1}^{N} X_i}{N}$$

The symbol $\bar{X}$, spoken of as X bar, is used to denote the arithmetic mean of the values of X. The Greek letter sigma, $\sum_{i=1}^{N}$, describes the operation of summing the N measurements. The summation extends from $i = 1$ to $i = N$. Quite commonly the arithmetic mean is written simply as

[4.2]
$$\bar{X} = \frac{\Sigma X}{N}$$

The limits of the summation are omitted. The summation is understood to extend over all available values of X.

A bar above a symbol always denotes an arithmetic mean. Thus $\bar{Y}$ denotes the mean of a variable Y. The use of a bar to denote the mean is a widely used and preferred notational practice.

The reader should note also that $\Sigma X = N\bar{X}$. Thus the sum of a variable X is N times the mean of X. An awareness of this simple fact is useful in a variety of situations.

4.3 CALCULATING THE MEAN FROM FREQUENCY DISTRIBUTIONS

Consider a situation where different values of X occur more than once. The arithmetic mean is then obtained by multiplying each value of X by the frequency of its occurrence, adding together these products, and then dividing by the total number of measurements. Consider the following measurements: 11, 11, 12, 12, 12, 13, 13, 13, 13, 13, 14, 14, 15, 15, 15, 16, 16, 17, 17, 18. The value 11 occurs with a frequency of 2, 12 with a frequency of 3, 13 with a frequency of 5, and so on. These data may be written as follows:

X_i	f_i	f_iX_i
18	1	18
17	2	34
16	2	32
15	3	45
14	2	28
13	5	65
12	3	36
11	2	22
Total	20	280

This is a frequency distribution with a class interval of 1. The symbol f_i is used to denote the frequency of occurrence of the particular value X_i. Multiplying each value X_i by the frequency of its occurrence and adding together the products f_iX_i, we obtain the sum 280. The arithmetic mean is then 280 divided by 20, or 14.0.

In general, where $X_1, X_2, X_3, \ldots, X_k$ occur with frequencies $f_1, f_2, f_3, \ldots, f_k$, where k is the number of *different* values of X, the arithmetic mean

[4.3] $$\bar{X} = \frac{f_1X_1 + f_2X_2 + f_3X_3 + \cdots + f_kX_k}{N} = \frac{\sum_{i=1}^{k} f_iX_i}{N}$$

Observe that here the summation is over k terms, the number of *different* values of the variable X. Observe also that $\sum_{i=1}^{N} X_i = \sum_{i=1}^{k} f_iX_i$. The above discussion suggests a simple method for calculating the mean from data grouped in the form of a frequency distribution regardless of the size of the class interval. The midpoint of the interval may be used to represent all values falling within the interval. We assume that the variable X takes values corresponding to the midpoints of the intervals, and these are weighted by the frequencies. We multiply the midpoints of the intervals by the frequencies, sum these products, and divide this sum by N to obtain the mean. More explicitly, the steps involved are as follows. *First,* calculate the midpoints of all intervals. *Second,* multiply each midpoint by the corresponding frequency. *Third,* sum the products of midpoints by frequencies. *Fourth,* divide this sum by N to obtain the mean. To illustrate, consider Table 4.1.

The midpoints of the intervals X_i appear in column 2. The frequencies f_i appear in column 3. The products of the midpoints by the frequencies f_iX_i are shown in column 4. The sum of these products $\sum_{i=1}^{k} f_iX_i$ is 1,612, N is 76, and the mean $\bar{X}$ is obtained by dividing 1,612 by 76 and is 21.21.

Table 4.1
Calculating the mean for distribution of test scores

1 Class interval	2 Midpoint X_i	3 Frequency f_i	4 Frequency × midpoint f_iX_i
45–49	47	1	47
40–44	42	2	84
35–39	37	3	111
30–34	32	6	192
25–29	27	8	216
20–24	22	17	374
15–19	17	26	442
10–14	12	11	132
5–9	7	2	14
0–4	2	0	0
Total		76	1,612

$\sum_{i=1}^{k} f_iX_i = 1{,}612 \qquad \bar{X} = 1{,}612/76 = 21.21$

4.4 DEVIATION FROM THE MEAN

In statistical work frequent use is made of the difference between a particular score X_i and the mean. Such a difference is spoken of as a deviation from the mean. Thus

$$x_i = X_i - \bar{X}$$

Here x_i, lowercase x_i, is used to denote such a deviation. Similarly y denotes a deviation from the mean of Y. Measurements or scores in their original form, denoted by X or Y, are sometimes known as *raw scores*. Measurements or scores expressed as deviations about the mean, denoted by x or y, are called *deviation scores*. The use of deviation scores instead of raw scores involves a change of origin. Raw scores have a mean $\bar{X}$ or $\bar{Y}$. Deviation scores have a mean of 0.

4.5 SOME PROPERTIES OF THE ARITHMETIC MEAN

The arithmetic mean has a number of interesting and useful properties. The first of these is a very simple property. *The sum of deviations of all the measurements in a set from their arithmetic mean is* 0. The arithmetic mean of the measurements 7, 13, 22, 9, 11, and 4 is 11. The deviations of these measurements from this mean are −4, 2, 11, −2, 0, and −7. The sum of these deviations is 0. Proof of this result is as follows:

[4.4] $$\sum_{i=1}^{N} (X_i - \bar{X}) = \sum_{i=1}^{N} X_i - \sum_{i=1}^{N} \bar{X} = N\bar{X} - N\bar{X} = 0$$

Since $\bar{X} = \left(\sum_{i=1}^{N} X\right) / N$, it follows that $\sum_{i=1}^{N} X = N\bar{X}$. Also, adding $\bar{X}$, the mean, N times is the same as multiplying $\bar{X}$ by N; thus if $\bar{X}$ is 11 and N is 6, we observe that $11 + 11 + 11 + 11 + 11 + 11 = 6 \times 11 = 66$.

In many situations in statistics, use is made of the square of deviations from the mean; that is, quantities of the kind $(X_i - \bar{X})^2$ are considered. A second useful property of the mean involves the sum of squares of deviations from the mean, that is, the quantity $\sum_{i=1}^{N} (X_i - \bar{X})^2$. *The sum of squares of deviations from the arithmetic mean is less than the sum of squares of deviations from any other value.* The deviations of the measurements 7, 13, 22, 9, 11, 4 from the mean 11 are −4, 2, 11, −2, 0, −7. The squares of these deviations are 16, 4, 121, 4, 0, 49. The sum of squares is 194. Had any other origin been selected, the sum of squares of deviations would be greater than the sum of squares about the mean. Select a different origin, say, 13. The deviations are −6, 0, 9, 4, 2, −9. Squaring these, we have 36, 0, 81, 16, 4, 81. The sum of these squares is 218, which is greater than the sum of squares about the mean. Selection of any other origin will demonstrate the same result.

This property of the mean indicates that it is the centroid, or center of gravity, of the set of measurements. Indeed, the mean is the central value about which the sum of squares of deviations is a minimum. This result may be readily demonstrated. Consider deviations from an origin $\bar{X} + c$, where $c \neq 0$. A deviation of an observation from this origin is

[4.5] $$X_i - (\bar{X} + c) = (X_i - \bar{X}) - c$$

Squaring and summing over N observations, we obtain

[4.6] $$\sum_{i=1}^{N} [X_i - (\bar{X} + c)]^2 = \sum_{i=1}^{N} (X_i - \bar{X})^2 + \sum_{i=1}^{N} c^2 - 2c \sum_{i=1}^{N} (X_i - \bar{X})$$

Because the sum of deviations about the mean is 0, the third term to the right is 0. Also c^2 summed N times is Nc^2, and we write

[4.7] $$\sum_{i=1}^{N} [X_i - (\bar{X} + c)]^2 = \sum_{i=1}^{N} (X_i - \bar{X})^2 + Nc^2$$

This expression states that the sum of squares of deviations about an origin $\bar{X} + c$ may be viewed as comprising two parts, the sum of squares of deviations about the mean $\bar{X}$ and Nc^2. The quantity Nc^2 is always positive. Hence the sum of squares of deviations about an origin $\bar{X} + c$ will always be greater than the sum of squares about $\bar{X}$. Thus the sum of squares of deviations about the arithmetic mean is less than the sum of squares of deviations about any other value.

Any mean calculated on a sample of size N is an estimate of a population mean, which is the value that would have been obtained were it possible to measure all members of the population. The distinction between sample and population values was discussed in Chapter 1. The mean has the property that for most distributions it is a more accurate, or more efficient, estimate of the population mean than other measures of central location, such as the median and mode, are of the population values they purport to estimate. It is subject to less error. This is one reason why it is the most frequently used measure of central location. Proof of this result is beyond the scope of this book.

Reference has been made to a number of properties of the arithmetic mean. What importance attaches to these properties, or why should they be discussed? The fact that the sum of deviations about the mean is 0 greatly simplifies many forms of algebraic manipulation. Any term involving the sum of deviations about the mean will vanish. The fact that the sum of squares of deviations about the mean is a minimum in effect implies an alternative definition of the mean; namely, *the mean is that measure of central location about which the sum of the squares is a minimum.* In effect, the mean is a measure of central location in the *least-squares* sense. The method of least squares is of considerable importance in statistics and is used, for example, in the fitting of lines and curves. The mean may be regarded as a point located by the method of least squares. The fact that the sample mean provides a better estimate of a population parameter than other measures of central location is of primary importance. Throughout statistics we are concerned with the problem of making statements about population values from our knowledge of sample values. Obviously, the more accurate these statements are, the better.

4.6 THE MEDIAN

Another commonly used measure of central location is the *median*. The median is a value such that half the observations fall above it and half below it. Consider the following values of X arranged in rank order, where R corresponds to the rank and N is an odd number.

X	2	7	16	19	20	25	27
R	1	2	3	4	5	6	7

In this example the median is 19. It corresponds to the middle rank. Three observations fall above it and three below it. If another observation, say 31, is added, then N is an even number, and the median by common convention is arbitrarily taken as the arithmetic mean of the two middle values, 19 and 20, that is, $(19+20)/2$ or 19.5.

With some data problems arise in calculating the median. These have been discussed by Stavig (1978) and Stavig and Gibbons (1977). Consider the following values of X.

X	7	7	7	8	8	8	9	9	10	10
R	1	2	3	4	5	6	7	8	9	10

For these 10 observations we are required to locate a point such that half the observations fall above that point and half below. Two different procedures may be used. The choice of procedure depends on whether the variable is viewed as *continuous* or *discrete*. For the above data, if the variable is continuous, the three 8s may be assumed to occupy the interval 7.5 to 8.5. The median is then obtained by linear interpolation. In this instance we interpolate two-thirds of the way into the interval to obtain a point above and below which half the observations fall. The median is then $7.5 + .67 = 8.17$. If the variable is not continuous but is discrete and assumes only integral values, then the score corresponding to the middle rank, which is $(N + 1)/2$, is taken as the median. For the illustrative data above, if the data are discrete, the median is 8.

When certain values of the variable occur more than once, and the variable is *continuous* and not discrete, the median is calculated by the method used in calculating the median from data grouped in the form of a frequency distribution, as described in Section 4.7.

4.7 CALCULATING THE MEDIAN FROM FREQUENCY DISTRIBUTIONS

In calculating the median from data grouped in the form of a frequency distribution, the problem is to determine a value of the variable such that one-half the observations fall above this value and the other half below. The method will be illustrated with reference to the data in Table 4.2.

First, record the cumulative frequencies as shown in column 3. *Second,* determine $N/2$, one-half the number of cases, in this example 38. *Third,* find the class interval in which the 38th case, the middle case, falls. The 38th case falls within the interval 15 to 19, and the exact limits of this interval are 14.5 and 19.5. Clearly, the 38th case falls very close to the top of this interval because we know from an examination of our cumulative frequencies that 39 cases fall below the top of this interval, that is, below 19.5. *Fourth,* interpolate between the exact limits of the interval to find a value above and below which 38 cases fall. To interpolate, observe that 26 cases fall within the limits 14.5 and 19.5, and we assume that these 26 cases are uniformly distributed in rectangular fashion between these exact limits. Now to arrive at the 38th, or middle, case we require 25 of the 26 cases within this interval, because $2 + 11 + 25 = 38$. This means that we must find a point between 14.5 and 19.5 such that 25 cases fall below and 1 case above this point. The proportion of the interval we require is 25/26, which is $25/26 \times 5$ units of score, or 4.81. We add this to the lower limit of the interval to obtain the median, which is $14.50 + 4.81$, or 19.31.

Table 4.2
Frequency distribution of psychological test scores

1 Class interval	2 Frequency	3 Cumulative frequency
45–49	1	76
40–44	2	75
35–39	3	73
30–34	6	70
25–29	8	64
20–24	17	56
15–19	26	39
10–14	11	13
5–9	2	2
0–4	0	0
Total	76	

Let us summarize the steps involved:

1 Compute the cumulative frequencies.

2 Determine $N/2$, one-half the number of cases.

3 Find the class interval in which the middle case falls, and determine the exact limits of this interval.

4 Interpolate to find a value on the scale above and below which one-half the total number of cases falls. This is the median.

For the student who has difficulty in following the above, a simple formula may be employed.

[4.8] $$\text{Median} = L + \frac{N/2 - F}{f_m} h$$

where $L =$ exact lower limit of interval containing the median
$F =$ sum of all frequencies below L
$f_m =$ frequency of interval containing median
$N =$ number of cases
$h =$ class interval

In the present example $L = 14.5$, $F = 13$, $f_m = 26$, $N = 76$, and $h = 5$. We then have

$$\text{Median} = 14.5 + \frac{\frac{76}{2} - 13}{26} \times 5 = 19.31$$

This method assumes that the observations within the interval containing the median are uniformly distributed over the range of that interval, and simple linear interpolation is appropriate.

Data for a discrete variable may be encountered which have been arranged in the form of a frequency distribution. With such data the median is the midpoint of the interval containing the median.

4.8 PROPERTIES OF THE MEDIAN

The reader will recall that the arithmetic mean has the property that the sum of squares of deviations from it is less than the sum of squares of deviations about any other value. In effect the mean $\bar{X}$ is a value such that $\Sigma(X - \bar{X})^2$ is a minimum. The median has an analogous property. The sum of absolute deviations, (deviations without sign) about the median is less than the sum of absolute deviations about any other value. If we denote an absolute deviation from the median as $|X - \text{mdn}|$, then the median is a value such that $\Sigma|X - \text{mdn}|$ is a minimum.

Stavig (1978) points out that if a set of discrete values is treated as continuous, the median so calculated may not satisfy the requirement that $\Sigma|X - \text{mdn}|$ is a minimum. Consider the observations 7, 7, 7, 8, 8, 8, 9, 9, 10, 10. If the variable is viewed as discrete, the median is 8 and the sum of absolute deviations about it is 9. If the variable is treated as continuous, the median is 8.17 and the sum of absolute deviations is 10.83. Why does this discrepancy in the sum of absolute deviations occur? If the variable is viewed as continuous, any consideration of the sum of absolute deviations of the *original values* from the median is simply incorrect. An underlying continuous scale has been assumed. The median is a point on that scale corresponding to the middle rank. Every other rank has, however, a corresponding point on this scale. Consequently the appropriate sum of absolute deviations is the sum of absolute deviations of these points on the underlying scale from the median. Values corresponding to each rank on this scale may be readily calculated using a formula given by Stavig (1978), which is a modification of formula [4.8]: $\hat{X} = L + [(R - \cdot 5 - F)/f_m]h$. Here $\hat{X}$ is the score on the underlying scale, R is the rank from 1 to N without regard for ties, and the remaining terms are as in formula [4.8]. Note that this formula is the same as formula [4.8] except that $(R - .5)$ has been substituted for $N/2$. In the present example values of $\hat{X}$ corresponding to different values of R are 6.67, 7.00. 7.33, 7.67, 8.00, 8.33, 8.67, 9.00, 9.33, and 9.67. The median is 8.17. The sum of absolute deviations about this value is 8.33.

4.9 THE MODE

Another measure of central location is the *mode*. In situations where different values of X occur more than once, the mode is the most frequently occurring value. Consider the observations 11, 11, 12, 12, 12, 13, 13, 13,

13, 13, 14, 14, 14, 15, 15, 15, 16, 16, 17, 17, 18. Here the value 13 occurs five times, more frequently than any other value; hence the mode is 13.

In situations where all values of X occur with equal frequency, where that frequency may be equal to or greater than 1, no modal value can be calculated. Thus for the set of observations 2, 7, 16, 19, 20, 25, and 27 no mode can be obtained. Similarly, the observations 2, 2, 2, 7, 7, 7, 16, 16, 16, 19, 19, 19, 20, 20, 20, 25, 25, 25, 27, 27, 27 do not permit the calculation of a modal value. All values occur with a frequency of 3.

In the case where two adjacent values of X occur with the same frequency, which is larger than the frequency of occurrence of other values of X, the mode may be taken rather arbitrarily as the mean of the two adjacent values of X. Consider the observations 11, 11, 12, 12, 12, 13, 13, 13, 13, 14, 14, 14, 14, 15, 15, 16, 16, 17, 18. Here the values 13 and 14 both occur with a frequency of 4, which is greater than the frequency of occurrence of the remaining values. The mode may be taken as $(13 + 14)/2$, or 13.5.

Where two nonadjacent values of X occur such that the frequencies of both are greater than the frequencies in adjacent intervals, then each value of X may be taken as a mode and the set of observations may be spoken of as bimodal. Consider the observations 11, 11, 12, 12, 12, 13, 13, 13, 13, 13, 14, 14, 14, 15, 15, 15, 15, 16, 16, 16, 17, 17, 18. Here the value 13 occurs five times, and this is greater than the frequency of occurrence of the adjacent values. Also 15 occurs four times, and this is also greater than the frequency of occurrence of the adjacent values. This set of observations may be said to be bimodal.

With data grouped in the form of a frequency distribution, the mode is taken as the midpoint of the class interval with the largest frequency.

The mode is a statistic of limited practical value. It does not lend itself to algebraic manipulation. For distributions with two or more modes, such modes are obviously not measures of central location. The mode can be considered a measure of central location only for distributions that taper off systematically toward the extremities.

4.10 COMPARISON OF THE MEAN, MEDIAN, AND MODE

The arithmetic mean may be regarded as an appropriate measure of central location for interval and ratio variables. All the particular values of the variable are incorporated in its calculation. The median is an ordinal statistic. Its calculation is based on the ordinal properties of the data. If the observations are arranged in order, the median is the middle value. Its calculation does not incorporate all the particular values of the variable, but merely the fact of their occurrence above or below the middle value. Thus the sets of numbers 5, 7, 20, 24, 25 and 10, 15, 20, 52, 63 have the same median, namely, 20, although their means are quite different. The mode, the value or class with the greatest frequency, is a nominal statistic.

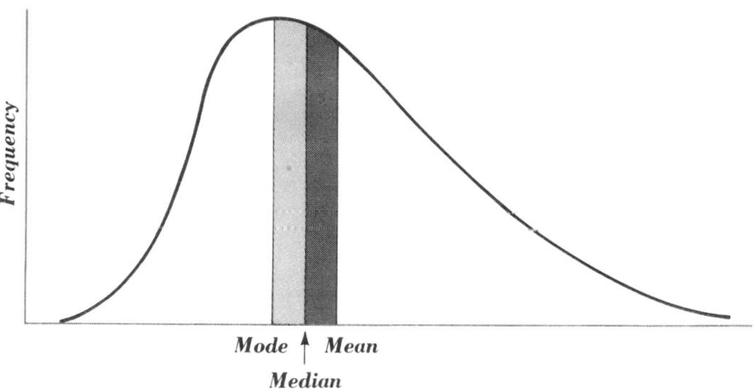

Figure 4.1 Relation between the mean, median, and mode in a positively skewed frequency distribution.

Its calculation does not depend on particular values of the variable or their order, but merely on their frequency of occurrence.

A comparison of the mean, median, and mode may be made when all three have been calculated for the same frequency distribution. If the frequency distribution is represented graphically, the mean is a point on the horizontal axis which corresponds to the centroid, or center of gravity, of the distribution. If a cutout of the distribution is made from heavy cardboard and balanced on a knife edge, the point of balance will be the mean. The median is a point on the horizontal axis where the ordinate divides the total area under the curve into two equal parts. Half the area falls to the left and half to the right of the ordinate at the median. The mode is a point on the horizontal axis which corresponds to the highest point of the curve.

If the frequency distribution is symmetrical, the mean, median, and mode coincide. If the frequency distribution is skewed, these three measures do not coincide. Figure 4.1 shows the mean, median, and mode for a positively skewed frequency distribution. We note that the mean is greater than the median, which in turn is greater than the mode. If the distribution is negatively skewed the reverse relation holds.

A question may be raised regarding the appropriate choice of a measure of central location. In practical situations this question is rarely in doubt. The arithmetic mean is usually to be preferred to either the median or the mode. It is rigorously defined, easily calculated, and readily amenable to algebraic treatment. It provides also a better estimate of the corresponding population parameter than either the median or the mode.

The median is, however, to be preferred in some situations. Observations may occur which appear to be atypical of the remaining observations in the set. Such observations may greatly affect the value of the mean. Consider the observations 2, 3, 3, 4, 7, 9, 10, 11, 86. Observation 86 is quite atypical of the remaining observations, and its presence greatly affects the value of the mean. The mean is 15, a value greater than eight of the nine observations. The median is 7. Under circumstances such as this it may prove advisable in treating the data to use statistical procedures that are based on the ordinal properties of the data in preference to procedures that incorporate all the particular values of the variable and may be grossly affected by atypical values. The median, an ordinal statistic, may under such circumstances be preferred to the mean. In the above example the set of observations is grossly asymmetrical. If the distribution of the variables shows gross asymmetry, the median may be the preferred statistic, because, regardless of the asymmetry of the distribution, it can always be interpreted as the middle value.

For a strictly nominal variable the mode, the most frequently occurring class or value, is the only "most typical" statistic that can be used. It is rarely used with interval, ratio, and ordinal variables where means and medians can be calculated.

BASIC TERMS AND CONCEPTS

Average

Arithmetic mean

Raw score

Deviation score

Sum of deviations about mean

Sum of squares of deviations about mean

Median

Mode

Bimodality

Method of least squares

Absolute deviation

EXERCISES

1 In 100 rolls of a die the frequencies of the six possible events are as follows:

X_i	f_i
6	17
5	14
4	20
3	15
2	15
1	19
	$N = 100$

Compute the arithmetic mean for this distribution.

2 The following is a frequency distribution of examination marks:

Class interval	f_i
90–94	1
85–89	4
80–84	2
75–79	8
70–74	9
65–69	14
60–64	6
55–59	6
50–54	4
45–49	3
40–44	3
	$N = 60$

Compute the arithmetic mean for this distribution.

3 A set of measurements has a mean of 80. A constant value 10 is added to each measurement. What is the mean of the resulting measurements?

4 In Exercise 3 above, if each measurement is multiplied by a constant 10, what is the mean of the resulting measurements?

5 For the measurements 1, 5, 10, 15, 19, find (**a**) the sum of squares of deviations from the mean and (**b**) the sum of squares of deviations about an arbitrary origin of 8.

6 The sum of squares of deviations of 10 observations from a mean of 50 is 225. What is the sum of squares of deviations from an arbitrary origin of 60?

7 On the assumption that the variables are continuous compute medians for the following data:

a 1, 2, 5, 7, 9
b 3, 7, 15, 26, 51
c 6, 25, 30, 35, 45, 64
d 1, 5, 6, 6, 6, 7, 10
e 1, 3, 6, 6, 6, 8, 10
f 12, 19, 24, 24, 29, 42
g 4, 4, 5, 5, 5, 6, 6
h 0, 0, 0, 7, 9, 25, 126

8 On the assumption that the variables are continuous compute medians for the data of Exercises 1 and 2 above.

9 On the assumption that the variables are discrete compute medians for the following data:

a 1, 2, 2, 2, 2, 3, 3
b 1, 1, 2, 2, 2, 3, 4, 4

c 1, 2, 2, 2, 3, 3, 3, 4
d 16, 17, 18, 19, 19, 19, 19

10 On the assumption that the variables are continuous compute medians for the data of Exercise 9 above.

11 Compute modes for the data of Exercises 1 and 2 above.

ANSWERS TO EXERCISES

1 3.46 **2** 66.58

3 90 **4** 800

5 a 212 **b** 232

6 1,225

7 a 5 **b** 15 **c** 32.50 **d** 6
e 6 **f** 24 **g** 5 **h** 7

8 Exercise 1, 4.00; Exercise 2, 67.36

9 a 2 **b** 2 **c** 2.5 **d** 19

10 a 2.125 **b** 2.167 **c** 2.500 **d** 18.625

11 a 6 **b** 5.5 **c** 6

12 Exercise 1, 4.00; Exercise 2, 67.00

5

MEASURES OF VARIATION, SKEWNESS, AND KURTOSIS

5.1 INTRODUCTION

Of great concern to the statistician is the variation in the events of nature. The variation of one measurement from another is a persisting characteristic of any sample of measurements. Measurements of intelligence, eye color, reaction time, and skin resistance, for example, exhibit variation in any sample of individuals. Anthropometric measurements such as height, weight, diameter of the skull, length of the forearm, and angular separation of the metatarsals show variation between individuals. Anatomical and physiological measurements vary; also the measurements made by the physicist, chemist, botanist, and agronomist. Statistics has been spoken of as the study of variation. Fisher (1970) has observed,

> The conception of statistics as the study of variation is the natural outcome of viewing the subject as the study of populations; for a population of individuals in all respects identical is completely described by a description of any one individual, together with the number in the group. The populations which are the object of statistical study always display variation in one or more respects.

The experimental scientist is frequently concerned with the different circumstances, conditions, or sources which contribute to the variation in the measurements he or she obtains. The analysis of variance (Chapter 15) developed by Fisher is an important statistical procedure whereby the variation in a set of experimental data can be partitioned into components which frequently may be attributed to different causal circumstances.

How may the variation in any set of measurements be described?

Consider the following measurements for two samples:

Sample *A*	10	12	15	18	20
Sample *B*	2	8	15	22	28

We note that the two samples have the same mean, namely, 15. Simple inspection indicates, however, that the measurements in sample *B* are more variable than those in sample *A*; they differ more one from another. Among the possible measures used to describe this variation are the range, the mean deviation, and the standard deviation. The most important of these is the standard deviation.

5.2 THE RANGE

The *range* is the simplest measure of variation. In any sample of measurements the range is taken as the difference between the largest and smallest measurements. The range for the measurements 10, 12, 15, 18, and 20 is 20 minus 10, or 10. The range for the measurements 2, 8, 15, 22, and 28 is 28 minus 2, or 26. The measurements in the second set quite clearly exhibit greater variation than those in the first set, and this reflects itself in a much greater range. The range has two disadvantages. First, for large samples it is an unstable descriptive measure. The sampling variance of the range for small samples is not much greater than that of the standard deviation but increases rapidly with increase in *N*. Second, the range is not independent of sample size, except under special circumstances. For distributions that taper to 0 at the extremities a better chance exists of obtaining extreme values for large than for small samples. Consequently, ranges calculated on samples composed of different numbers of cases are not directly comparable. Despite these disadvantages the range may be effectively used in the application of tests of significance with small samples.

5.3 THE MEAN DEVIATION

Consider the following measurements:

Sample *A*	8	8	8	8	8
Sample *B*	1	4	7	10	13
Sample *C*	1	5	20	25	29

Intuitively, the measurements in sample *A* are less variable than those in *B*, which in turn are less variable than those in *C*. Indeed, the measurements in *A* exhibit no variation at all. The means of the three samples are 8, 7, and 16. If we express the measurements as deviations from their sample means, we obtain

Sample A	0	0	0	0	0
Sample B	-6	-3	0	$+3$	$+6$
Sample C	-15	-11	$+4$	$+9$	$+13$

Inspection of these numbers suggests that as variation increases, the departure of the observations from their sample mean increases. We may use this characteristic to define a measure of variation. One such measure is the *mean deviation.* The mean deviation is the arithmetic mean of the absolute deviations from the arithmetic mean. An absolute deviation is a deviation without regard to algebraic sign. To obtain the mean deviation we simply calculate the deviations from the arithmetic mean, sum these, disregarding algebraic sign, and divide by N. For sample A above, the mean deviation is 0. For sample B the mean deviation is $(6+3+0+3+6)/5=\frac{18}{5}=3.6$. For sample C the mean deviation is $(15+11+4+9+13)/5=\frac{52}{5}=10.4$.

The mean deviation is given in algebraic language by the formula

[5.1] $$\text{MD}=\frac{\Sigma|X-\bar{X}|}{N}$$

Here $X-\bar{X}$ is a deviation from the mean and $|X-\bar{X}|$ is a deviation without regard to algebraic sign. The vertical bars mean that signs are ignored.

Hitherto, symbols above and below the summation sign Σ have been used to indicate the limits of the summation. In the above formula for the mean deviation these symbols have been omitted, the summation being clearly understood to extend over the N members in the sample. In this and subsequent chapters symbols indicating the limits of summation will, for convenience, be omitted where these are understood clearly from the context to extend over N sample members. Where any possibility of doubt could exist, the symbols above and below the summation sign will be inserted.

The mean deviation is infrequently used. It is not readily amenable to algebraic manipulation. This circumstance stems from the use of absolute values. In general, in statistical work the use of absolute values should be avoided, if at all possible.

The mean deviation is discussed here primarily for pedagogic reasons. It illustrates how one particular measure of variation may be defined.

5.4 THE SAMPLE VARIANCE AND STANDARD DEVIATION

Some of the deviations about the mean are positive; others are negative. The sum of deviations is 0. One method for dealing with the presence of negative signs is to use the absolute deviations, as in the calculation of the mean deviation as defined in formula [5.1]. This procedure has little to recommend it. An alternative and generally preferable procedure is to square the deviations about the mean, sum these squares, and use this sum of squares in the definition of a measure of variation. For example, the mean of the measurements 1, 4, 7, 10, and 13 is 7. The deviations from the mean are −6, −3, 0, +3, and +6. The squares of these deviations are 36, 9, 0, 9, and 36. The sum of squares is 90.

The sum of squares of deviations about the mean is used in the definition of a statistic known as the *variance*. Two methods exist for defining the variance. Both are in common use. One method defines the variance by dividing the sum of squares of deviations about the mean by N, the number of cases. Denote this statistic by s^2. Thus

[5.2] $$s^2 = \frac{\Sigma(X - \bar{X})^2}{N}$$

In the illustrative example in the paragraph above, the sum of squares of deviations about the mean was 90, and the variance, according to formula [5.2], is $s^2 = \frac{90}{5} = 18$.

An alternative method of defining the variance is to divide the sum of squares by $N - 1$ rather than N. Thus, according to this definition, the variances s^2 is given by

[5.3] $$s^2 = \frac{\Sigma(X - \bar{X})^2}{N - 1}$$

Both formulas, [5.2] and [5.3], provide alternative definitions of the variance. These formulas have no derivation but are obtained by a process of plausible reasoning. The reader will note that no notational distinction is made here between the variance defined by formula [5.2] and that defined by formula [5.3].

What is the essential difference between the variance as defined by formula [5.2] and the variance as defined by formula [5.3]? To understand the answer to this question the reader should recall that in Chapter 1 a distinction was made between sample values, or estimates, and population values, or parameters. Both formulas, [5.2] and [5.3], provide estimates of a population variance σ^2. For certain algebraic reasons when we divide $\Sigma(X - \bar{X})^2$ by N we obtain a biased estimate of σ^2. This estimate will show a systematic tendency to be less than σ^2. It is biased. When we divide $\Sigma(X - \bar{X})^2$ by $N - 1$, however, we obtain an unbiased estimate of σ^2. Such an estimate will show no systematic tendency to be either greater than or less than σ^2.

In all situations where an estimate of a population variance σ^2 is required, the statistic s^2 which divides the sum of squares by $N - 1$ and not N should be used. In the great majority of situations discussed in this book an unbiased estimate is required. In some situations involving descriptive statistics, convenience and simplicity dictate the use of an estimate which divides the sum of squares by N and not $N - 1$. *In general in this book the symbol s^2 will be used to refer to the unbiased estimate obtained by dividing the sum of squares by $N - 1$, as in formula* [5.3]. *In a few situations the definition of formula* [5.2] *will be used. All such instances will be clearly specified in the text.*

The reader should note that a population variance is defined as $\sigma^2 = \Sigma(X - \mu)^2/N_p$, where μ is the population mean and N_p is the number of members in the population. Thus in any situation where the variance of a complete population is required we divide the sum of squares by the number of members in the population.

While N is the number of measurements or observations, the quantity $N - 1$ is the number of deviations about the mean that are free to vary. To illustrate, consider the measurements 7, 8, and 15. The mean is 10, and the deviations about the mean are $-3, -2, +5$. The sum of deviations about the mean is 0; that is, $(-3) + (-2) + (5) = 0$. Because this is so, if any two of the deviations are known, the third deviation is fixed. It cannot vary. In this example, the sum of squares of deviations about the mean is $9 + 4 + 25 = 38$. Although this sum of squares is obtained by adding together three squared deviations, only two of these squared deviations can exhibit freedom of variation. The number of values that are free to vary is called the number of *degrees of freedom*. A quantity of the kind $\Sigma(X - \bar{X})^2$ is said to have associated with it $N - 1$ degrees of freedom, because $N - 1$ of the N squared deviations of which it is composed can vary. Some intuitive plausibility attaches to the idea that in the definition of a measure of variation we should divide the sum of squares by the number of values that can exhibit freedom of variation. The concept of degrees of freedom is a very useful and general concept in statistics and is elaborated in more detail later in this book.

In the above discussion the definition of the variance evolves initially from a consideration of deviations about the arithmetic mean. An alternative, and perhaps more elemental, approach is to begin by considering the differences between each value and every other value. With two measurements only, X_1 and X_2, we may consider the difference between them, $X_1 - X_2$. With three measurements, X_1, X_2, and X_3, we may consider the differences $X_1 - X_2$, $X_1 - X_3$, and $X_2 - X_3$. In general, for N measurements the number of such differences is $N(N - 1)/2$. To illustrate, for the measurements 1, 4, 7, 10, and 13, the differences between each measurement and every other measurement are $-3, -6, -9, -12, -3, -6, -9, -3, -6$, and -3. Note that the sign of the difference depends on the order of the measurements. If we obtain the sum of squares of the differences between each measurement and every other measurement and divide by the number of such differences, the result is closely related to s^2; in fact it is simply twice s^2. In our example the sum of squares of differences is 450. We divide this by 10 to obtain 45.0, which is seen to be twice the variance, 22.5, as calculated by formula [5.3]. In general, in algebraic notation it may be shown that

[5.4]
$$\frac{\Sigma(X_i - X_j)^2}{N(N-1)/2} = 2s^2$$

where the summation is understood to extend over $N(N-1)/2$ differences. This result means that s^2 is a descriptive index of how different each value is from every other value; in fact it is an average of the squared differences divided by 2.

The variance is a statistic in squared units. If $X - \bar{X}$ is a deviation in feet, then $(X - \bar{X})^2$ is a deviation in feet squared. For many purposes it is desirable to use a measure of variation which is not in squared units but in units of the original measurements themselves. We obtain this result by taking the square root of either formula [5.2] or formula [5.3]. This statistic is called the *standard deviation*. Thus

[5.5]
$$s = \sqrt{\frac{\Sigma(X - \bar{X})^2}{N}}$$

or

[5.6]
$$s = \sqrt{\frac{\Sigma(X - \bar{X})^2}{N-1}}$$

5.5 AN ILLUSTRATIVE APPLICATION

Our understanding of the nature of the variance and the standard deviation will be enriched by considering illustrative situations where these statistics are of interest. Consider a simple experiment designed to investigate the effect of a drug on a cognitive task such as coding. An experimental group of subjects, who receive the drug, and a control group, who do not receive the drug, are used. Each group contains 10 subjects. Let us assume the scores on the coding task for the two groups are as follows:

Experimental	5	7	17	31	45	47	68	85	96	99
Control	29	36	37	42	49	58	62	63	69	70

The mean score for the experimental group is 50.0, and that for the control, 51.5. The investigator might be led to conclude from inspecting these means that the drug had little or no effect on the performance of the subjects. The standard deviations for the two groups are, respectively, 35.63 and 14.86, the experimental group being much more variable in performance than the control group. Quite clearly the treatment appears to be exerting a substantial influence on the variation in performance, although its influence on level of performance is negligible. In the analysis of experimental data the investigator must attend to, and if possible interpret, differences in the standard deviation, or variance, as well as differences in the arithmetic mean.

5.6 CALCULATING THE SAMPLE VARIANCE AND THE STANDARD DEVIATION FROM UNGROUPED DATA

For purposes of calculation, it is convenient to write the variance and the standard deviation in a different form. The variance may be written as

$$\begin{aligned} s^2 &= \frac{\Sigma(X - \bar{X})^2}{N - 1} \\ &= \frac{\Sigma(X^2 + \bar{X}^2 - 2X\bar{X})}{N - 1} \\ &= \frac{\Sigma X^2 + N\bar{X}^2 - 2N\bar{X}^2}{N - 1} \\ &= \frac{\Sigma X^2 - N\bar{X}^2}{N - 1} \end{aligned}$$

In this derivation note that the summation of $\bar{X}^2$ over N is simply $N\bar{X}^2$; also the summation of $2X\bar{X}$ is $2\bar{X}\Sigma X = 2N\bar{X}^2$, since $\Sigma X = N\bar{X}$. The standard deviation is given by

$$s = \sqrt{\frac{\Sigma X^2 - N\bar{X}^2}{N - 1}}$$

Thus to calculate the standard deviation using this formula, we sum the squares of the original observations, subtract from this N times the square of the arithmetic mean, divide by $N - 1$, and then take the square root. For example, the five observations 1, 4, 7, 10, and 13 have a mean of 7. The squares of these observations are 1, 16, 49, 100, and 169. The sum of these squared observations is 335. The variance is then

$$s^2 = \frac{\Sigma X^2 - N\bar{X}^2}{N - 1} = \frac{335 - 5 \times 7^2}{5 - 1} = 22.50$$

and the standard deviation is $\sqrt{22.50} = 4.74$.

An alternative formula for the standard deviation which avoids the calculation of the arithmetic mean and may, therefore, be useful for certain computational purposes is

[5.7] $$s = \sqrt{\frac{N\Sigma X^2 - (\Sigma X)^2}{N(N - 1)}}$$

This formula requires one operation of division only.

5.7 THE EFFECT ON THE STANDARD DEVIATION OF ADDING OR MULTIPLYING BY A CONSTANT

If a constant is added to all the observations in a sample, the standard deviation remains unchanged. An examiner may conclude, for example, that an examination is too difficult. He may decide to add 10 points to all the marks assigned. The standard deviation of the original marks will be the same as the standard deviation of marks with the 10 points added. This result follows directly from the fact that if X is an observation, the corresponding observation with the constant c added is $X + c$. If $\bar{X}$ is the mean of the original observations, the mean with the constant added is $\bar{X} + c$. A deviation from the mean of the observations with the constant added is then $(X + c) - (\bar{X} + c)$, which is readily observed to be equal to $X - \bar{X}$. Since the deviations about the mean are unchanged by the addition of a constant, the standard deviation will remain unchanged. To illustrate, by adding a constant, say, 5, to the measurements 1, 4, 7, 10, and 13, we obtain 6, 9, 12, 15, and 18. The mean of the original measurements is 7, and the mean of the measurements with thc constant added is $7 + 5$, or 12. The deviations from the mean are in both instances the same, namely, $-6, -3, 0, +3$, and $+6$. The standard deviation in both instances is 4.74.

If all measurements in a sample are multiplied by a constant, the standard deviation is also multiplied by the absolute value of that constant. If the standard deviation of examination marks is 4 and all marks are multiplied by the constant 3, then the standard deviation of the resulting marks is $3 \times 4 = 12$. To demonstrate this result, we observe that if $\bar{X}$ is the mean of a sample of measurements, the mean of the measurements multiplied by c is $c\bar{X}$. A deviation from the mean is then $cX - c\bar{X} = c(X - \bar{X})$. By squaring, summing over N observations, and dividing by $N - 1$, we obtain

$$\frac{\Sigma(cX - c\bar{X})^2}{N-1} = \frac{c^2\Sigma(X - \bar{X})^2}{N-1} = c^2s^2$$

Thus if all measurements are multiplied by a constant c, the variance is multiplied by c^2 and the standard deviation by the absolute value of c. If c is a negative number, say, -3, s is multiplied by the absolute value 3. By way of illustration, the measurements 1, 4, 7, 10, 13 have a mean of 7, a variance of 22.50, and a standard deviation of 4.74. If the measurements are multiplied by the constant 5, we obtain 5, 20, 35, 50, 65. The mean is now 5×7, or 35. The deviations from the mean are $-30, -15, 0, +15, +30$. Squaring these we obtain 900, 225, 0, 225, 900. The sum of squares is 2,250, the variance is 562.50, and the standard deviation is 23.72, whereas 5 times the original standard deviation of 4.74 is 23.70. The slight discrepancy results from the rounding of decimals.

5.8 STANDARD SCORES

Hitherto we have considered scores or measurements in the form in which they were originally obtained. Such scores are represented by the symbol X with mean $\bar{X}$ and standard deviation s. Such scores in their original form are spoken of as *raw scores*. We have also considered deviations about the arithmetic mean, $x = X - \bar{X}$. These are known as *deviation scores* and have a mean of 0 and a standard deviation of s. If now we divide the deviation about the mean by the standard deviation, we obtain what is called a *standard score* represented by the symbol z. Thus

$$z = \frac{X - \bar{X}}{s} = \frac{x}{s}$$

Standard scores have a mean of 0 and a standard deviation of 1. As previously shown, if all measurements in a sample are multiplied by a constant, the standard deviation is also multiplied by the absolute value of that constant. Deviation scores, $x = X - \bar{X}$, have a standard deviation s. Each score has a constant $-\bar{X}$ added. This leaves s unchanged. If all the deviation scores are divided by s, which is the same thing as multiplying by the constant $1/s$, the standard deviation of the scores thus obtained is $s/s = 1$.

To illustrate, the following observations have been expressed in raw-score, deviation-score, and standard-score form.

Individual	X	x	z
A	3	-7	-1.11
B	6	-4	$-.63$
C	7	-3	$-.47$
D	9	-1	$-.16$
E	15	5	.79
F	20	10	1.58
Sum	60	.00	.00
Mean	10	.00	.00
s	6.32	6.32	1.00

Because standard scores have zero mean and unit standard deviation, they are readily amenable to certain forms of algebraic manipulation. Many formulations can be derived more conveniently using standard scores than using raw or deviation scores.

The use of standard scores means, in effect, that we are using the standard deviation as the unit of measurement. In the above example individual A is 1.11 standard deviations, or standard deviation units, below the mean, while individual F is 1.58 standard deviation units above the mean.

Standard scores are frequently used to obtain comparability of observations obtained by different procedures. Consider examinations in English and mathematics applied to the same group of individuals, and assume the means and standard deviations to be as follows:

Examination	$\bar{X}$	s
English	65	8
Mathematics	52	12

In effect, in relation to the performance of the individuals in the group, a score of 65 on the English examination is the equivalent of a score of 52 on the mathematics examination. To illustrate, a score one standard deviation above the mean, that is, 65 + 8, or 73, on the English examination can be considered to be the equivalent of a score one standard deviation above the mean, that is, 52 + 12, or 64, on the mathematics examination. If an individual makes a score of 57 on the English examination and a score of 58 on the mathematics examination, we may compare his relative performance on the two subjects by comparing his standard scores. On English his standard score is $(57 - 65)/8 = -1.0$, and on mathematics his standard score is $(58 - 52)/12 = .5$. Thus on English his performance is one standard deviation unit below the average, while on mathematics his performance is .5 standard deviation unit above the average. Quite clearly, this individual did much more poorly in English than in mathematics relative to the performance of the group of individuals taking the examinations, although this is not reflected in the original marks assigned. To attain rigorous comparability of scores, the distributions of scores on the two tests should be identical in shape. The meaning of this statement will become clear as we proceed.

The reader should note that the sum of squares of standard scores, Σz^2, is equal to $N - 1$. We observe that $z^2 = (X - \bar{X})^2/s^2$; hence

$$\Sigma z^2 = \frac{\Sigma(X - \bar{X})^2}{s^2} = \frac{\Sigma(X - \bar{X})^2}{\Sigma(X - \bar{X})^2/(N - 1)} = N - 1$$

The reader should note here that if s^2 is defined as $\Sigma(X - \bar{X})^2/N$, the sum of squares of standard scores is N and not $N - 1$.

5.9 ADVANTAGES OF THE VARIANCE AND STANDARD DEVIATION AS MEASURES OF VARIATION

The variance and standard deviation have many advantages over other measures of variation. Much statistical work involves their use. The variance has certain additive properties and may on occasion be partitioned into additive components, each of which may be related to some causal circumstance. The sample standard deviation is a more stable or accurate estimate of the population parameter than other measures of variation. Under certain assumptions it provides a more stable estimate of the standard deviation in the population than the sample mean deviation, for example, does of the mean deviation in the population. The variance and standard deviation are more amenable to mathematical manipulation than other measures. They enter into formulas for the computation of many types of statistics. They are widely used as measures of error. In later discussion on sampling statistics the reader will observe that the standard error is in effect the standard deviation of errors made in estimating population parameters from sample values. These errors result from the operation of chance factors in random sampling. A full appreciation of the importance and meaning of the variance and standard deviation in their many ramifications requires considerable familiarity with statistical ideas.

5.10 MOMENTS ABOUT THE MEAN

The mean and the standard deviation are closely related to a family of descriptive statistics known as *moments*. The first four moments about the arithmetic mean are as follows:

[5.8]
$$m_1 = \frac{\Sigma(X-\bar{X})}{N} = 0$$
$$m_2 = \frac{\Sigma(X-\bar{X})^2}{N} = \frac{N-1}{N}s^2$$
$$m_3 = \frac{\Sigma(X-\bar{X})^3}{N}$$
$$m_4 = \frac{\Sigma(X-\bar{X})^4}{N}$$

In general, the rth moment about the mean is given by

[5.9]
$$m_r = \frac{\Sigma(X-\bar{X})^r}{N}$$

The term "moment" originates in mechanics. Consider a lever supported by a fulcrum. If a force f_1 is applied to the lever at a distance x_1 from the origin, then f_1x_1 is called the moment of the force. Further, if a second force f_2 is applied at a distance x_2, the total moment is $f_1x_1 + f_2x_2$. If we square the distances x, we obtain the second moment; if we cube them, we obtain the third moment; and so on. When we come to consider frequency distributions, the origin is the analog of the fulcrum and the frequencies in the various class intervals are analogous to forces operating at various distances from the origin. Observe that the first moment about the mean is 0 and the second moment is $(N-1)/N$ times the unbiased sample variance. The third moment is used to obtain a measure of skewness, and the fourth moment, a measure of kurtosis.

5.11 MEASURES OF SKEWNESS AND KURTOSIS

The commonly used measure of skewness makes use of the third moment and is defined as

[5.10]
$$g_1 = \frac{m_3}{m_2\sqrt{m_2}}$$

The rationale for this statistic is based on the observation that when a distribution (or any set of numbers) is symmetrical, the sum of deviations above the mean, when raised to the third power, will balance the sum of deviations below the mean, when raised to the third power. Thus for a symmetrical distribution, $m_3 = 0$, and the $g_1 = 0$. If the distribution is asymmetrical, the sums of deviations above and below the mean, when raised to the third power, will not balance. Thus for an asymmetrical distribution $m_3 \neq 0$ and $g_1 \neq 0$. If the distribution, or set of numbers, is positively skewed, g_1 is positive; when negatively skewed g_1 is negative. The quantity $m_2\sqrt{m_2}$ is introduced in order to ensure that g_1 is comparable for distributions that differ in variability. Thus g_1 is independent of the scale of measurement. The skewness of a set of measurements in grams, meters, pounds, or units of score on a psychological test can be directly compared using g_1. The reader will recall that a standard score is defined as $z = (X - \bar{X})/s$. One reason for using standard scores is to achieve comparability of scores from one set of measurements to another. The use of $m_2\sqrt{m_2}$ in the definition of g_1 is directly analogous to the use of s in the definition of a standard score.

As an illustration of g_1, consider two sets of numbers, A and B.

A	6	8	10	12	14
B	6	8	10	11	15

These numbers expressed as deviations from the mean become

A	−4	−2	0	+2	+4
B	−4	−2	0	+1	+5

Set A is a symmetrical set of numbers, and set B is asymmetrical. These deviations raised to the third power are as follows:

A	−64	−8	0	+8	+64
B	−64	−8	0	+1	+125

For set A, $m_3 = 0$ and $g_1 = 0$. For set B, $m_3 = 10.80$, $m_2 = 9.20$, and $g_1 = .387$. Set B is a positively skewed set of numbers.

The commonly used measure of kurtosis involves the fourth moment, and is defined as

$$g_2 = \frac{m_4}{m_2^2} - 3 \qquad [5.11]$$

This definition is based on the observation that large deviations from the mean, when raised to the fourth power, will contribute substantially to the fourth moment. The concept of kurtosis is more closely linked to the relative thickness of the tails of distributions than to the idea that one distribution may be flatter or more peaked than another. Large deviations from the mean contribute much more to the fourth moment than smaller deviations. The term m_2^2 is used to achieve comparability. It serves the same purpose as $m_2\sqrt{m_2}$ does in the definition of g_1. The number 3 comes about because the ratio $m_4/m_2^2 = 3$ for a normal distribution. This means that $g_2 = 0$ for a normal distribution. For a leptokurtic distribution, g_2 is greater than zero, and for a platykurtic distribution, g_2 is less than zero.

As an illustration, consider the following sets of numbers.

A	6	8	10	12	14
B	5.64	9	10	11	14.36

Inspection suggests that both sets are platykurtic, but set *A* is more platykurtic than set *B*. The two sets of numbers have the same mean and the same standard deviations and both are symmetrical. Yet they differ in the property called kurtosis. Deviations about the mean, when raised to the *fourth power*, are as follows

A	256	16	0	16	256
B	361.36	1	0	1	361.36

For *A*, $m_4 = 108.80$; for *B*, $m_4 = 144.94$. For both set *A* and *B*, $m_2 = 8.00$. For set *A*, $g_2 = -1.30$; for set *B*, $g_2 = -.74$. Both sets are platykurtic, but set *A* is more platykurtic than set *B* as shown by the statistic g_2.

5.12 A SIMPLE DESCRIPTIVE SYSTEM

The mean, the standard deviation, and the measures of skewness and kurtosis constitute a simple system for describing collections of numbers and comparing one collection with another. When we have made these four statements, $\bar{X}$, s, g_1, and g_2, about any collection of numbers, we have said just about everything worth saying. The system is parsimonious and elegant. Few situations exist where, having made four precise and simple statements, just about everything has been said that is of any importance. It is true that on comparing two sets of numbers, $\bar{X}$, s, g_1, and g_2 may be equal, and yet the numbers may differ. Such differences may be explored using higher moments but for all practical purposes these differences will prove trivial.

BASIC TERMS AND CONCEPTS

Range

Mean deviation

Population variance, σ^2

Sample variance, s^2

Estimate: biased; unbiased

Degrees of freedom, *df*

Sample standard deviation, s

Standard score, z

Moments about the mean

Measure of skewness, g_1

Measure of kurtosis, g_2

EXERCISES

1 For the measurements 2, 5, 9, 10, 15, 19, compute (**a**) the range, (**b**) the mean deviation, (**c**) the sample variance, (**d**) the standard deviation.

2 The variance calculated on a sample of 100 cases is 15. What is the sum of squares of deviations about the arithmetic mean?

3 A biased variance estimate calculated on a sample of 5 cases is 10. What is the corresponding unbiased variance estimate?

4 The variance for a sample of N measurements is 20. What will the variance be if all measurements are (**a**) multiplied by a constant 5 and (**b**) divided by a constant 5?

5 Show that $\Sigma(X - \bar{X})^2 = \Sigma X^2 - N\bar{X}^2$.

6 Scholastic aptitude tests have a mean of 500 and a standard deviation of 100. A student's SAT-V score is 750 and his SAT-M score is 450. Express these scores in standard-score form.

7 Express the measurements 2, 4, 5, 6, 8 in standard-score form. What is the sum of squares of standard scores?

8 The mean and standard deviation of marks in a statistics examination for a class of 26 students are, respectively, 62 and 10. Three students make scores of 50, 75, and 90. What are their corresponding standard scores?

9 Calculate the second, third, and fourth moments for the observations 4, 6, 10, 14, 16. Compute also measures of skewness and kurtosis.

10 The following are measurements obtained for two groups of subjects:

Group I	2	3	5	10	20
Group II	2	4	8	12	14

Calculate measures of skewness for the two groups.

ANSWERS TO EXERCISES

1 **a** 17 **b** 4.67 **c** 39.20 **d** 6.26

2 1,485 **3** 12.50

4 **a** 500 **b** .80

5 $\Sigma(X - \bar{X})^2 = \Sigma(X^2 + \bar{X}^2 - 2X\bar{X}) = \Sigma X^2 + N\bar{X}^2 - 2N\bar{X}^2 =$
$\Sigma X^2 - N\bar{X}^2$

6 2.50; −.50

7 −1.342, −.477, .000, +.447, +1.342; $N - 1 = 4$

8 −1.2, 1.3, 2.8

9 $m_2 = 20.80$, $m_3 = 0$, $m_4 = 620.80$; $g_1 = 0$, $g_2 = -1.57$

10 .950; .000

6

PROBABILITY AND THE BINOMIAL DISTRIBUTION

6.1 INTRODUCTION

In Chapters 2 to 5 the reader was introduced to certain statistics which are used to describe the properties of frequency distributions, or the collections of numbers which frequency distributions comprise. In the present chapter the nature of probability is considered.

Probability is used to introduce the reader to one particular type of theoretical distribution known as the *binomial distribution*. The binomial distribution is only one of a number of theoretical distributions which are used in practical situations to estimate probabilities. Why are experimentalists concerned with questions of probability?

In experimental work a line of theoretical speculation may lead to the formulation of a particular hypothesis. An experiment is conducted, and data obtained. How are the data interpreted? Do the data support the acceptance or rejection of the hypothesis? What rules of evidence apply? Questions of this type involve considerations of probability. The answers are in probabilistic terms. The assertions of the investigator are not made with certainty but have associated with them some degree of doubt, however small.

Consider a hypothetical illustration. Two methods for the treatment of a disease are under consideration. Two groups of 20 patients suffering from the disease are selected. Method A is applied to one group, method B to the other. Following a period of treatment, 16 patients in group A and 10 patients in group B show marked improvement. How may this difference be evaluated? May it be argued from the data that treatment A is in general superior to treatment B? Here we proceed by adopting a trial hypothesis that no difference exists between the two treatments, that one treatment is no better than the other. We then estimate the probability of obtaining by random sampling under this trial hypothesis a difference equal to or greater than the one observed. If this probability is small, say, the chances are less than 5 in 100, we may consider this sufficient evidence for the rejection of the trial hypothesis and may be prepared to assert that one method of treatment is better than the other. If the probability is not small and the observed difference may be expected to occur quite frequently under the trial hypothesis, say, the chances are 20 in 100, then the evidence does not warrant the conclusion that one treatment is better than the other.

In general, the interpretation of the data of experiments is in probabilistic terms. The theory of probability is of the greatest importance in scientific work where questions about the correspondences between the deductive consequences of theory and observed data are raised. Probability theory had its origins in games of chance. It has become basic to the thinking of the scientist.

6.2 THE NATURE OF PROBABILITY

Diverse views of the nature of probability may be entertained. The topic is controversial. No inclusive summary of these different views will be attempted here. We shall discuss three approaches to probability: (1) the subjective, or personalistic, (2) the formal mathematical, and (3) the empirical relative-frequency approach. These different ways of regarding probability are not incompatible.

The term *probability* may be used subjectively to refer to an attitude of doubt with respect to some future event. For example, the assertions may be made that "It will probably rain tomorrow," or "The probability is small that I shall live to be 90 years old," or "There is a high probability that a particular horse will win the Kentucky Derby." Frequently, numerical terms are used in making assertions of this kind, such as, "The odds are even that it will rain tomorrow," or "I estimate that the chances are about 95 in 100 that I shall die before I am 90 years old," or "The chances are 3 to 1 that a particular horse will win the Kentucky Derby." All such assertions, whether numerical terms are used or not, refer to feelings of degrees of doubt or confidence with regard to future outcomes. This subjective usage is sometimes spoken of as psychological, or personalistic, probability.

A second usage defines the probability of an event as the ratio of the number of favorable cases to the total number of *equally likely* cases. This usage stems from a consideration of games of chance involving cards, dice, and coins. For example, on examining the structure of a die the assertion may be made that no basis exists for choosing one of the six alternatives in preference to another; consequently all six alternatives may be considered equally likely. The probability of throwing a particular result, say, a 3, in a single toss is then 1/6, there being one favorable case among six equally likely alternatives. This approach to probability involves a concept of equally likely cases, which has a degree of intuitive plausibility in relation to cards, dice, and coins. Difficulties present themselves, however, when we attempt to apply this approach in situations where it is impossible to delineate cases which can be construed to be equally likely. These difficulties have led to the argument that equally likely means the same as equally probable; therefore the definition is circular because it defines probability in terms of itself. Arguments have been advanced to escape this circularity. These need not detain us. The difficulty, however, is readily resolved by observing that the concept of equally likely in this definition of probability is a formal postulate and is not empirical. In effect we say, "Let us postulate that certain events are equally likely, and given this postulate let us deduce certain consequences." This means that a theory of probability employing this postulate is a formal mathematical model. It may or may not correspond to empirical events. It may be demonstrated, however, that this model does closely approximate certain empirical events and consequently is of value in dealing with practical problems.

The situation here is somewhat analogous to that in ordinary euclidian geometry. Euclidian geometry is a formal system comprised of a set of axioms, or primitive postulates, and their deductive consequences, called theorems. The proofs of the theorems hold regardless of questions of correspondence with the empirical world. We know, however, on the basis of lengthy experience, that these theorems can be shown to correspond closely to the world around us. In consequence, euclidian geometry provides a valuable model for dealing with problems in engineering, surveying, building construction, and many other fields. Both with euclidian geometry and probability it is useful to draw a clear distinction between the formal mathematical system and the empirical events for which the formal system may serve as a model.

A third approach to probability is through a consideration of relative frequencies. If a series of trials is made, say, N, and a given event occurs r times, then r/N is the relative frequency. This relative frequency may be considered an estimate of a value p. If a longer series of trials is made, the relative frequency will usually be closer to p. The difference between r/N and p may be made as small as we like by increasing the value of N. The probability p is defined as the limit approached by the relative frequency as the number of trials is increased. This approach to probability requires that a population of events be defined. Probability is the relative frequency in the population. It is a population parameter. The relative frequency in a sample of observations is an estimate of that parameter. To illustrate, consider a coin. The population of events may be regarded as an indefinitely large number of tosses which theoretically could be made. The proportion of heads p in this population is the probability of a head. This is often assumed to be 1/2. If the coin is tossed 100 times and a proportion .47 of heads is obtained, this may be taken as an estimate of p.

The ways of regarding probabilities described here—the subjective, the formal mathematical, and the empirical through the study of relative frequencies— are not incompatible, and, indeed, it may be argued that all three must of necessity coexist.

In recent years interest in subjective probability has increased. The reason for this resides in a growing interest in Bayesian statistics. The essential idca in thc Baycsian approach to statistics is that prior probabilities must be estimated and used in the final evaluation of any event or experimental outcome. In practice most experiments are conducted within a context of theory, related experimental evidence, accumulated knowledge, and experience. Clearly an investigator usually has anticipatory probability estimates, however vaguely conceptualized and defined, prior to the conduct of a particular experiment. If this were not so the experiment would possibly not be conducted in the first place. Bayesian statisticians argue that both the prior probability and the probability associated with the particular experimental result should be taken into account in reaching a final probabilistic evaluation. Although practical difficulties attach to the estimation of prior probabilities, progress has been made in the application of Bayesian statistics (Novick and Jackson, 1974). The student interested in further exploration of the concept of subjective probabilities is referred to the article by de Finetti (1978).

6.3 POSSIBLE OUTCOMES

Questions of probability frequently involve a consideration of the number of possible outcomes, sometimes spoken of as a set. For example, in tossing a single coin, there are two possible outcomes—either a head or a tail will occur. In tossing two coins, the four possible outcomes are HH, HT, TH, TT. This means that both coins may be heads, the first a head and thc second a tail, the first a tail and the second a head, or both tails. In tossing three coins, there are eight possible outcomes, which may be represented as HHH, HHT, HTH, THH, HTT, THT, TTH, TTT. In tossing 4 coins, the number of possible outcomes is 16. In throwing a single die, the number of possible outcomes is six; that is, either 1, 2, 3, 4, 5, or 6 may appear. In throwing two dice, the number of possible outcomes may be listed as follows:

11	21	31	41	51	61
12	22	32	42	52	62
13	23	33	43	53	63
14	24	34	44	54	64
15	25	35	45	55	65
16	26	36	46	56	66

These are the numbers which may come up in throwing two dicc. Both dice may come up 1, the first may come up 2 and the second 1, the first 3 and the second 1, and so on. Thus there are 36 possible outcomes to this experiment. In drawing a single card from a deck of 52 cards, the number of possible outcomes is 52. In drawing one card from a deck of 52 cards and another from a different deck, the number of possible outcomes is $52 \times 52 = 2{,}704$.

In the illustrative situations described above we may be willing to assume that the possible outcomes are equally likely. With most coins, dice, and cards this is a justifiable assumption, the validity of which may be readily checked by experiment. Under this assumption the probability of obtaining two heads, HH, in tossing two coins is 1/4. Two heads may occur once in four equally probable outcomes. We denote this probability by $p(\text{HH}) = 1/4$. The probability of obtaining three tails in tossing three coins is $p(\text{TTT}) = 1/8$, there being eight equally likely outcomes. The probability of obtaining two 6s in throwing two dice is $p(6,6) = 1/36$. This event is seen to occur once in the set of 36 possible, and equally likely, outcomes.

The illustrative applications above regard a probability as the ratio of the number of favorable cases to the total number of equally likely cases. This approach may be extended to situations where the number of possible outcomes are not equally likely, provided a method exists for either deducing or estimating the expected relative frequencies of the possible outcomes. In rolling two dice, we may decide simply to sum the two numbers which appear and to regard this sum as an outcome. The set of all possible outcomes, thus defined, consists of the 11 numbers 2, 3, 4, . . . , 11, 12. These outcomes are not, of course, equally likely. Their expected relative frequencies may be obtained, however, from a consideration of the 36 possible, equally likely, outcomes involved in rolling two dice. A 2 can occur only once in 36 equally likely outcomes; that is, $p(2) = 1/36$. A 3 can occur twice; that is, $p(3) = 2/36$. Likewise $p(4) = 3/36$, $p(5) = 4/36$, and so on. Thus the possible outcomes are not equally likely but have different relative frequencies or probabilities. In this illustrative example we were able to deduce the probabilities associated with the possible outcomes from more elementary considerations involving equally likely cases. In many practical situations this is not possible, and the required probabilities must be estimated empirically. For example, what is the probability that a person chosen at random from the population of the city of Boston is over 60 years of age? The estimation of this probability requires an empirical method. Presumably in this case the required probability can be estimated using census data.

6.4 JOINT AND CONDITIONAL PROBABILITIES

Consider a population of members classified with regard to two characteristics A and B, there being three classes, or strata, of A and two classes, or strata, of B, as follows:

	A_1	A_2	A_3	
B_1	.40	.15	.05	.60
B_2	.00	.10	.30	.40
	.40	.25	.35	1.00

The probability that a person is B_1 is .60; that is, the marginal probability $p(B_1) = .60$. The probability that he is A_1 is .40; that is, the marginal probability $p(A_1) = .40$. The probability that he is both A_1 and B_1 is .40; that is, $p(A_1 B_1) = .40$. This latter probability is a *joint* probability. It is the probability that a member will fall simultaneously within two classes. All probabilities in the above table are joint probabilities.

Now given the fact that a member is B_1, what is the probability that he is A_1, A_2, or A_3? To answer this question we divide the joint probabilities $p(A_1B_1) = .40$, $p(A_2B_1) = .15$, and $p(A_3B_1) = .05$ by the probability $p(B_1) = .60$ to obtain $p(A_1/B_1) = .67$, $p(A_2/B_1) = .25$, and $p(A_3/B_1) = .08$. These are *conditional* probabilities. They are the probabilities of A_1, A_2, and A_3 given B_1. Note that the sum of these three probabilities is 1.00. Likewise, if a member is B_2 what is the probability that he is A_1, A_2, or A_3? Here $p(A_1/B_2) = .00$, $p(A_2/B_2) = .25$, and $p(A_3/B_2) = .75$. These conditional probabilities may be written in tabular form as follows:

	A_1	A_2	A_3	
B_1	.67	.25	.08	1.00
B_2	.00	.25	.75	1.00

Likewise, the conditional probabilities of B given A are

	A_1	A_2	A_3
B_1	1.00	.60	.14
B_2	.00	.40	.86
	1.00	1.00	1.00

6.5 THE ADDITION AND MULTIPLICATION OF PROBABILITIES

In throwing a die, six possible events may occur. If we are prepared to assume that these six events are equally likely, then the probability of obtaining a 1, 2, 3, 4, 5, or 6 in a single throw is 1/6. What is the probability of obtaining either a 1, 2, or 3 in a single throw? The probability of obtaining a 1 is 1/6, the probability of a 2 is 1/6, and the probability of a 3 is also 1/6. Therefore, the probability of obtaining either a 1, a 2, or a 3 in a single throw is $1/6 + 1/6 + 1/6 = 1/2$. This example involves the addition of probabilities and illustrates what is known as the addition theorem of probability. This theorem states that the *probability that any one of a number of mutually exclusive events will occur is the sum of the probabilities of the separate events.* "Mutually exclusive" means that if one event occurs, the others cannot. Thus events that cannot happen at the same time are mutually exclusive.

The following are further examples of the addition of probabilities. In tossing two coins four possible events may occur: HH, TT, HT, TH. What is the probability of obtaining either two heads or two tails in tossing two coins? The probability of two heads is 1/4, and the probability of two tails is 1/4. Therefore, the probability of obtaining either two heads or two tails is $1/4 + 1/4 = 1/2$. In throwing two dice what is the probability that the number on the two dice will add up to 3? The probability that the first die is a 1 and the second a 2 is 1/36. The probability that the first die is a 2 and the second a 1 is 1/36. Therefore, the probability of obtaining either 1 and 2 or 2 and 1 is $1/36 + 1/36 = 1/18$. In rolling two dice what is the probability of obtaining either a 7 or an 11? If we consult the set of possible outcomes in rolling two dice in Section 5.3 above, we note that a 7 can occur in 6 different ways. Thus the probability of a 7 is 6/36. An 11 can occur in 2 different ways. Thus the probability of an 11 is 2/36. Therefore, the probability of either a 7 or an 11 is $6/36 + 2/36 = 8/36$. Consider a well-shuffled deck of 52 cards. In drawing a single card from the deck, what is the probability of obtaining either the ace, king, queen, or jack of spades? The probability that a single card is the ace of spades is 1/52. The probability that a single card is the king is 1/52, the queen, 1/52, and the jack, 1/52. Therefore, the probability of obtaining either the ace, king, queen, or jack of spades in drawing a single card is $1/52 + 1/52 + 1/52 + 1/52 = 4/52$.

The illustrative examples above involve the addition of probabilities. Probabilities can be added together when a set of mutually exclusive events can be specified, when the probabilities associated with these events can be stated, and when the probability associated with a subset of these events is required.

In tossing two coins the number of possible outcomes is 4, HH, HT, TH, TT, and the probability of any particular outcome, say, two heads, is 1/4. This is the product of the probabilities associated with the two independent events; that is, $1/2 \times 1/2 = 1/4$. The probability that the first coin is a head is 1/2 and that the second coin is a head is 1/2; therefore the probability that both are heads is 1/4. This is a simple application of the multiplication theorem of probability. This theorem states that the *probability of the joint occurrence of two or more independent events is the product of their separate probabilities.* Intuitively, by independent is meant that the occurrence of one event has nothing at all to do with the occurrence of the other event. In tossing two coins whether the first coin is a head or a tail does not affect in any way whether the second coin is a head or a tail. More precisely, two events are said to be independent when the probability of their joint occurrence is the product of the two separate probabilities.

The following examples illustrate further the multiplication of probabilities. What is the probability of obtaining four heads in four tosses of a coin? The probability that any one coin is a head is 1/2. The probability that all four coins are heads is $1/2 \times 1/2 \times 1/2 \times 1/2 = 1/16$. What is the probability of obtaining two 6s in rolling two dice? The probability that the first die is a 6 is 1/6. The probability that the second die is a 6 is 1/6. The probability that both dice are 6s is $1/6 \times 1/6 = 1/36$. What is the probability of drawing the ace, king, and queen of spades in that order, and without replacement, from a well-shuffled deck? The probability that the first card is the ace of spades is 1/52. Having drawn one card, 51 cards remain, and the probability that the second card is the king of spades is 1/51. Similarly the probability that the third card is the queen of spades is 1/50. The probability of the combined event is the product of the separate probabilities or $1/52 \times 1/51 \times 1/50$, or 1/132,600.

The above examples illustrate the multiplication of probabilities. Probabilities may be multiplied together when two or more independent events can be specified, when the probabilities associated with these events can be stated, and when the probability associated with the joint occurrence of these events is required.

6.6 PROBABILITY DISTRIBUTIONS

In Chapter 2 the concept of a frequency distribution was discussed. A frequency distribution was defined as an arrangement of the data that shows the frequency of occurrence of particular values of the variable or the frequency of occurrence of values falling within defined ranges of the variable. Consider now the situation in which three coins are tossed, the eight possible outcomes being HHH, HHT, HTH, THH, HTT, THT, TTH, TTT, these being regarded as equally probable. Let us now make a frequency distribution of the number of heads obtained in these eight possible outcomes. Note that three heads occur once, two heads occur three times, one head occurs three times, and no heads occurs once. Thus the frequency distribution of heads, with corresponding probabilities p, is as follows:

No. of heads	f	p
3	1	1/8
2	3	3/8
1	3	3/8
0	1	1/8
Total	8	1.00

The distribution to the right above is a probability distribution. It shows the probabilities of the outcomes, 3, 2, 1, 0 heads, on the assumption that the eight possible outcomes are equally probable. Thus in tossing three unbiased coins eight times, the probability of obtaining three heads is 1/8, two heads 3/8, and so on.

Consider a situation in which two dice are thrown. Here, as shown in Section 5.3 above, the number of possible outcomes is 36, these being assumed equally probable. Let us count the sum of the numbers that come up in throwing two dice. The lowest possible value is 2, the highest possible value is 12, and 11 values extending from 2 to 12 occur. The frequencies of these 11 values, and the associated probabilities, are as follows:

Number X	f	p
12	1	1/36
11	2	2/36
10	3	3/36
9	4	4/36
8	5	5/36
7	6	6/36
6	5	5/36
5	4	4/36
4	3	3/36
3	2	2/36
2	1	1/36
Total	36	1.00

The distribution to the right above is a probability distribution. A probability distribution specifies the probabilities associated with particular values of the variable or the probabilities associated, where class intervals are used, with defined ranges of the variable.

The distributions discussed above are theoretical probability distributions. In the examples involving coins and dice, a set of equally probable events is enumerated. The frequency and probability of certain other events (number of heads, sum of numbers on two dice) are then ascertained directly. No actual tossing of coins or rolling of dice is involved. The frequencies and probabilities are obtained directly from elementary theoretical considerations only. Of course it would be possible to actually toss coins and roll dice and obtain, thereby, the corresponding experimental frequency distributions. These could then be compared directly with those obtained from theoretical considerations. Probability distributions can be represented graphically by plotting probabilities along the vertical axis, instead of frequencies, as in the case of frequency distributions.

In the sections to follow, a particular type of theoretical probability distribution, known as the *binomial distribution* will be discussed. Before proceeding with a discussion of the binomial distribution, it will prove useful to review the readers' knowledge of permutations and combinations.

6.7 PERMUTATIONS AND COMBINATIONS

Consider questions of the following kind. In how many ways can four books be arranged on a shelf? In how many ways can six people be seated at a table? In how many ways can a dozen eggs be arranged in a carton of eggs? These questions are concerned with the number of possible arrangements. Any arrangement is called a *permutation*. Order is the essential idea here. A different order is a different arrangement.

Consider two objects labeled A and B. Two arrangements are possible, AB and BA. Here the number of permutations is 2. With three objects labeled A, B, and C, six arrangements are possible. These are ABC, ACB, BAC, BCA, CAB, CBA. The number of permutations in this case is 6. In this example 3 ways exist of choosing the first member. When the first member is chosen, 2 ways exist of choosing the second member. When the second member is chosen 1 way exists of choosing the third member. The number of permutations in this case is given by $3 \times 2 \times 1 = 6$. In general, if there are n distinguishable objects the number of permutations of these objects taken n at a time is given by $n!$, or n factorial, which is the product of all integers from n to 1, or $n(n-1)(n-2) \cdots 3 \times 2 \times 1$. For $n = 4$, $n! = 4 \times 3 \times 2 \times 1 = 24$; for $n = 5$, $n! = 5 \times 4 \times 3 \times 2 \times 1 = 120$. Consider the number of seating arrangements of 8 guests in 8 chairs at a dinner table. The first guest may sit in any one of 8 chairs. When the first guest is seated, the second guest may sit in any one of the remaining 7 chairs. Thus the number of possible arrangements for the first two guests is $8 \times 7 = 56$. When the first two guests are seated, the third guest may occupy any one of the remaining 6 chairs, and so on for the remaining guests. The number of possible seating arrangements for the 8 guests is 8!, or 40,320, a number which may help explain the indecision of many hostesses.

Instead of considering the number of ways of arranging n things n at a time, we may consider the number of ways of arranging n things r at a time, where r is less than n. Thus the possible arrangements of the objects A, B, and C taken two at a time are AB, AC, BA, BC, CA, and CB. Here we observe that there are 3 ways of selecting the first object and 2 ways of selecting the second. The number of arrangements is then $3 \times 2 = 6$. Similarly, on considering the number of arrangements of 10 objects taken 3 at a time, we observe that there are 10 ways of selecting the first, 9 ways of selecting the second, and 8 ways of selecting the third. The number of arrangements is then $10 \times 9 \times 8 = 720$. In general, the number of permutations of n things taken r at a time is

[6.1] $$P_r^n = n(n-1) \cdots (n-r+1) = \frac{n!}{(n-r)!}$$

The number of *different* ways of selecting objects from a set, ignoring the order in which they are arranged, is the number of *combinations*. Given the objects A, B, C, and D, the number of permutations of two from this set is $4 \times 3 = 12$. The arrangements are AB, BA, AC, CA, AD, DA, BC, CB, BD, DB, CD, and DC. Note that each pair occurs in two different orders. If we ignore the order in which each pair of objects is arranged, we have the number of combinations. In this example each pair occurs in two different orders. The number of combinations is then $(4 \times 3)/2 = 6$. In general, the number of different combinations of n things taken r at a time is

[6.2] $$C_r^n = \frac{n!}{r!(n-r)!}$$

The number of combinations of 10 things taken 3 at a time is $10!/(3!7!) = 120$.

The number of combinations of n things taken n at a time is clearly 1, because there is only one way of picking all n objects if we ignore the order of their arrangement.

6.8 THE BINOMIAL DISTRIBUTION

As indicated above in tossing three coins, eight possible, equally probable, outcomes exist. These are HHH, HHT, HTH, THH, TTH, THT, HTT, and TTT. Suppose now that we wish to count heads. The first outcome contains 3 heads, the second 2 heads, the third 2 heads, and so on. If now we wish to determine the probabilities of obtaining 3, 2, 1, and 0 heads, these probabilities are clearly 1/8, 3/8, 3/8, and 1/8, respectively. Denote the probability of a head by p and the probability of a tail by q; here $p + q = 1$. The eight possible outcomes with their associated probabilities may be represented as follows:

Outcome	Probability
HHH	p^3
HHT	p^2q
HTH	p^2q
THH	p^2q
TTH	pq^2
THT	pq^2
HTT	pq^2
TTT	q^3

The probabilities of obtaining 3, 2, 1, and 0 heads may now be represented as

No. of heads	Probability
3	p^3
2	$3p^2q$
1	$3pq^2$
0	q^3

Note that two heads can occur in three different ways, which is the number of combinations of 3 things taken 2 at a time; that is, $(3 \times 2)/(2 \times 1) = 3$. If the above probabilities are added together, we obtain

$$p^3 + 3p^2q + 3pq^2 + q^3 = 1$$

This is an example of what is called the binomial distribution. The probabilities associated with 3, 2, 1, and 0 heads in the situation described are given very simply by the successive terms of the expansion $(p + q)^3$. In this illustrative example $p = q = 1/2$, and the required probabilities are as follows:

$$\begin{aligned}(1/2 + 1/2)^3 &= (1/2)^3 + 3(1/2)^3 + 3(1/2)^3 + (1/2)^3 \\ &= 1/8 + 3/8 + 3/8 + 1/8\end{aligned}$$

Consider a more complex example involving 10 rather than 3 coins. In tossing 10 coins what is the probability of obtaining 0, 1, 2, . . . , 10 heads? We are required to determine the probability of obtaining 0 heads and 10 tails, 1 head and 9 tails, 2 heads and 8 tails, and so on. Let us designate the 10 coins by the letters A, B, C, D, E, F, G, H, I, and J. Let us assume that all 10 coins are unbiased and that the probability of throwing a head or a tail on a single toss of any coin is 1/2.

Let us attend first to the probability of throwing 0 heads and 10 tails in tossing all coins. The probability that coin A is not a head is 1/2, that B is not a head is 1/2, that C is not a head is 1/2, and so on. Therefore, from the multiplication theorem of probability, the probability that all 10 coins are not heads, or that they are tails, is obtained by multiplying 1/2 ten times; that is, $(1/2)^{10}$, or 1/1,024. Thus in tossing 10 coins there is 1 chance in 1,024 of obtaining 0 heads or 10 tails.

Now consider the problem of obtaining 1 head and 9 tails. The probability that coin A is a head is 1/2. The probability that all the remaining 9 coins are tails is $(1/2)^9$. Therefore the probability that A is a head and all other 9 coins are tails is $(1/2)^{10}$. It is readily observed, however, that 1 head can occur in 10 different ways. Coin A may be a head and all other coins tails, B may be a head and all the others tails, and so on. Since 1 head can occur in 10 different ways, the probability of obtaining 1 head and 9 tails is $10(1/2)^{10} = 10/1{,}024$. Thus in tossing 10 coins there are 10 chances in 1,024 of obtaining 1 head and 9 tails.

Determining the probability of obtaining 2 heads and 8 tails may be similarly approached. The probability that coins A and B are heads is $(1/2)^2$. The probability that all the remaining coins are tails is $(1/2)^8$. The probability that A and B are heads and all the remaining coins are tails is $(1/2)^{10}$. We readily observe, however, that two heads can occur in quite a number of different ways. This number is the number of combinations of 10 things taken 2 at a time, C_2^{10}, which is $10 \times 9/2 = 45$. Therefore the probability of obtaining 2 heads and 8 tails is $45(1/2)^{10}$, or 45/1,024. Similarly, the probability of obtaining 3 heads and 7 tails is $C_3^{10}(1/2)^{10} = 120/1{,}024$. Likewise, the probability of obtaining 4 heads and 6 tails is $C_4^{10}(1/2)^{10} = 210/1{,}024$; and so on. The probabilities of obtaining different numbers of heads in tossing 10 coins are then as follows:

No. of heads	Probability
10	1/1,024
9	10/1,024
8	45/1,024
7	120/1,024
6	210/1,024
5	252/1,024
4	210/1,024
3	120/1,024
2	45/1,024
1	10/1,024
0	1/1,024

The above probabilities are the successive terms in the expansion of the symmetrical binomial $(1/2 + 1/2)^{10}$. This expansion is

$$\left(\frac{1}{2}+\frac{1}{2}\right)^{10} = \left(\frac{1}{2}\right)^{10} + 10\left(\frac{1}{2}\right)^{10} + \frac{10 \times 9}{1 \times 2}\left(\frac{1}{2}\right)^{10} + \frac{10 \times 9 \times 8}{1 \times 2 \times 3}\left(\frac{1}{2}\right)^{10} + \cdots + \left(\frac{1}{2}\right)^{10}$$

This is a particular case of the symmetrical binomial $(1/2 + 1/2)^n$ whose terms are

[6.3] $$\left(\frac{1}{2}+\frac{1}{2}\right)^{n} = \left(\frac{1}{2}\right)^{n} + n\left(\frac{1}{2}\right)^{n} + \frac{n(n-1)}{1 \times 2}\left(\frac{1}{2}\right)^{n} + \frac{n(n-1)(n-2)}{1 \times 2 \times 3}\left(\frac{1}{2}\right)^{n} + \cdots + \left(\frac{1}{2}\right)^{n}$$

The symmetrical binomial is a particular case of the general form of the binomial $(p+q)^n$, where p is the probability that an event will occur and q is the probability that it will not occur; that is, $p + q = 1$. This binomial may be written as follows:

[6.4] $$(p+q)^n = p^n + np^{n-1}q + \frac{n(n-1)}{1 \times 2}p^{n-2}q^2 + \frac{n(n-1)(n-2)}{1 \times 2 \times 3}p^{n-3}q^3 + \cdots + q^n$$

The terms of this expansion for $n = 2$, $n = 3$, and $n = 4$ are as follows:

$$(p+q)^2 = p^2 + 2pq + q^2$$
$$(p+q)^3 = p^3 + 3p^2q + 3pq^2 + q^3$$
$$(p+q)^4 = p^4 + 4p^3q + 6p^2q^2 + 4pq^3 + q^4$$

The binomial $(p+q)^n$ can be readily illustrated by considering a problem involving the rolling of dice. What are the probabilities of obtaining five, four, three, two, one, and zero 6s in rolling five dice presumed to be unbiased? The probability of obtaining a 6 in rolling a single die is 1/6, whereas the probability of not obtaining a 6 is 5/6. The required probabilities are given by the six terms of the binomial $(1/6 + 5/6)^5$:

$$\left(\frac{1}{6}+\frac{5}{6}\right)^5 = \left(\frac{1}{6}\right)^5 + 5\left(\frac{1}{6}\right)^4\left(\frac{5}{6}\right) + \frac{5\times 4}{1\times 2}\left(\frac{1}{6}\right)^3\left(\frac{5}{6}\right)^2 + \frac{5\times 4}{1\times 2}\left(\frac{1}{6}\right)^2\left(\frac{5}{6}\right)^3 + 5\left(\frac{1}{6}\right)\left(\frac{5}{6}\right)^4 + \left(\frac{5}{6}\right)^5$$

Thus the probabilities of obtaining five, four, three, two, one, and zero 6s in rolling five dice are calculated as

No. of 6s	Probability
5	1/7,776
4	25/7,776
3	250/7,776
2	1,250/7,776
1	3,125/7,776
0	3,125/7,776

There is one chance in 7,776 of obtaining five 6s, 25 chances in 7,776 of obtaining four 6s, and so on. This distribution is seen to be asymmetrical.

Any term in the binomial expansion may be written as

[6.5] $$C_r^n p^r q^{n-r} = \frac{n!}{r!(n-r)!}\, p^r q^{n-r}$$

where C_r^n is the number of combinations of n things taken r at a time. Thus the probability of obtaining 3 heads in 10 tosses of a coin is

$$\frac{10!}{3!(10-3)!}\left(\frac{1}{2}\right)^3\left(\frac{1}{2}\right)^7 = \frac{120}{1{,}024}$$

The coefficients C_r^n in any expansion are

$$1,\ n,\ \frac{n(n-1)}{1\times 2},\ \frac{n(n-1)(n-2)}{1\times 2\times 3},\ \ldots$$

These coefficients may be rapidly obtained for different values of n from what is known as Pascal's triangle. The coefficients for different values of n are written in rows in the form of a triangle as shown in Table 6.1. The number in any row is the sum of the two numbers to the left and right on the row above. This device is very useful in generating expected frequencies and probabilities. For example, for $n = 10$, the entries in the triangle are the expected frequencies of heads, or tails, in tossing 10 coins 1,024 times. The required probabilities in this case are obtained by dividing the frequencies by 1,024.

Table 6.1
Pascal's triangle

n	
0	1
1	1 1
2	1 2 1
3	1 3 3 1
4	1 4 6 4 1
5	1 5 10 10 5 1
6	1 6 15 20 15 6 1
7	1 7 21 35 35 21 7 1
8	1 8 28 56 70 56 28 8 1
9	1 9 36 84 126 126 84 36 9 1
10	1 10 45 120 210 252 210 120 45 10 1

6.9 PROPERTIES OF THE BINOMIAL

For the symmetrical binomial, where $p = q = 1/2$, the mean, variance, skewness, and kurtosis are

[6.6]
$$\mu = \frac{n}{2}$$
$$\sigma^2 = \frac{n}{4}$$
$$\gamma_1 = 0$$
$$\gamma_2 = \frac{-2}{n}$$

Here we use the symbols μ, σ^2, γ_1, and γ_2 instead of $\bar{X}$, s^2, g_1, and g_2, because the binomial is a theoretical distribution. The symbols μ, σ^2, γ_1, and γ_2 may be viewed as population parameters rather than sample estimates. The above formulas may be illustrated by considering the tossing of 5 coins 32 times. The expected frequencies of 0, 1, 2, 3, 4, and 5 heads are 1, 5, 10, 10, 5, 1. These frequencies are the coefficients of the expansion $(1/2 + 1/2)^5$. Using the formula $\mu = n/2$, the mean is $\mu = 2.5$. This is the expected mean number of heads in tossing 5 unbiased coins 32 times. The variance of the distribution is $\sigma^2 = n/4 = 5/4 = 1.25$. Because the distribution is symmetrical, the measure of skewness $\gamma_1 = 0$. The measure of kurtosis $\gamma_2 = -2/n = -2/5 = -.40$. Note that as n increases in size, γ_2 becomes smaller. As n increases in size, γ_2 approaches zero as a limit.

The above values may be obtained by direct calculation. Denote the number of heads by X, the frequencies by f, and a deviation from the mean of X by x.

X	f	fX	x	fx^2	fx^3	fx^4
5	1	5	2.50	6.25	15.625	39.0625
4	5	20	1.50	11.25	16.875	25.3125
3	10	30	.50	2.50	1.250	.6250
2	10	20	−.50	2.50	−1.250	.6250
1	5	5	−1.50	11.25	−16.875	25.3125
0	1	0	−2.50	6.25	−15.625	39.0625
Total	32	80		40.00	.000	130.0000

The arithmetic mean, μ, of this distribution is $\Sigma fX/N = \frac{80}{32} = 2.50$. The variance, σ^2, is $\Sigma fx^2/N = \frac{40}{32} = 1.25$. Here, σ^2 is defined as $\Sigma fx^2/N$ and not $\Sigma fx^2/(N - 1)$, because we are dealing with a theoretical model, not a sample value. The skewness γ_1 is readily seen to be 0 because the third moment $\Sigma fx^3/N$ is zero. The fourth moment is $\Sigma fx^4/N = \frac{130}{32} = 4.0625$, and the kurtosis is $\gamma_2 = 4.0625/1.25^2 - 3 = -.40$. Here N denotes the number of tosses, or observations, whereas n denotes the number of coins.

In general the mean, variance, skewness, and kurtosis of any binomial distribution are given by

[6.7]
$$\mu = np$$
$$\sigma^2 = npq$$
$$\gamma_2 = \frac{q - p}{\sqrt{npq}}$$
$$\gamma_2 = \frac{1 - 6pq}{npq}$$

In tossing an unbiased die, the probability p of throwing a 6 is 1/6 and the probability q of not throwing a 6 is 5/6. The expected probability distribution of 6s in tossing 10 dice is given by the terms of the expansion $(1/6 + 5/6)^{10}$. The mean of this distribution is $\mu = np = 10/6 = 1.667$. The variance is $\sigma^2 = npq = 10 \times 1/6 \times 5/6 = 1.389$. The skewness $\gamma_1 = .566$ and the kurtosis $\gamma_2 = .12$.

BASIC TERMS AND CONCEPTS

Probability

Subjective probability

Probability as relative frequency

Set of possible outcomes

Joint probability

Conditional probability

Addition of probabilities

Multiplication of probabilities

Probability distribution

Permutation

Combination

Binomial distribution

Pascal's triangle

Mean and variance of the binomial

EXERCISES

1 In rolling a die, what is the probability of obtaining either a 5 or a 6?

2 In rolling two dice on one occasion, what is the probability of obtaining either a 7 or an 11? Note that there are 36 possible outcomes.

3 In rolling two dice, what is the probability that neither a 2 nor a 9 will appear?

4 In rolling two dice, what is the probability of obtaining a value less than 6?

5 On four consecutive rolls of a die a 6 is obtained. What is the probability of obtaining a 6 on the fifth roll?

6 In dealing 4 cards without replacement from a well-shuffled deck, what is the probability of obtaining 4 aces?

7 An urn contains 4 black and 3 white balls. If they are drawn without replacement, what is the probability of the order BWBWBWB?

8 In seating 8 people at a table with 8 chairs, what is the number of possible seating arrangements?

9 In how many ways can 2 people seat themselves at a table with 4 chairs?

10 In tossing 5 coins, what is the probability of obtaining fewer than 3 heads?

11 A multiple-choice test contains 100 questions. Each question has 5 alternatives. If a student guesses all questions, what score might he expect to obtain?

12 In how many ways can a committee of 3 be chosen from a group of 5 men?

13 Assume that intelligence and honesty are independent. If 10 percent of a population is intelligent and 60 percent is honest, what is the probability that an individual selected at random is both intelligent and honest?

14 A husband engages in random verbal behavior for 20 minutes in an hour. His wife is similarly engaged for 30 minutes in that hour. Neither listens to the other. Estimate the number of minutes of silence in the hour.

15 A coin is tossed 10 times. What is the probability that the third head will appear on the tenth toss?

16 The chances are 3 out of 4 that the weather tomorrow will be like the weather today. Today is Sunday and is a sunny day. What is the probability that it will be sunny all week? What is the probability that it will be sunny until Wednesday and not sunny on Thursday, Friday, and Saturday?

17 What is the expected distribution of heads in tossing 6 coins 64 times?

18 What is the expected distribution of 6s in rolling 6 dice 64 times?

19 What is the probability of obtaining either 9 or more heads or 3 or fewer heads in tossing 12 coins?

20 A man tosses 6 coins and rolls 6 dice simultaneously. What is the probability of 5 or more heads and 5 or more 6s?

ANSWERS TO EXERCISES

1 1/3

2 2/9

3 31/36

4 10/36

5 1/6

6 $4/52 \times 3/51 \times 2/50 \times 1/49$

7 $4/7 \times 3/6 \times 3/5 \times 2/4 \times 2/3 \times 1/2$

8 $8 \times 7 \times 6 \times 5 \times 4 \times 3 \times 2 \times 1 = 40{,}320$

9 12

10 1/2

11 20

12 10

13 6/100

14 20 minutes

15 36/1024

16 $(3/4)^6$, $1/4(3/4)^5$

17 The expected distribution of heads in tossing 6 coins 64 times is

H	0	1	2	3	4	5	6
f	1	6	15	20	15	6	1

18 The expected distribution of 6s in rolling 6 dice 64 times is:

6s	f
0	15,625/46,656 × 64
1	18,750/46,656 × 64
2	9,375/46,656 × 64
3	2,500/46,656 × 64
4	375/46,656 × 64
5	30/46,656 × 64
6	1/46,656 × 64

19 598/4,096

20 7/64 × 31/46,656

7

THE NORMAL CURVE

7.1 INTRODUCTION

In Chapter 6 the binomial distribution was discussed in detail. The symmetrical binomial $(\frac{1}{2} + \frac{1}{2})^{10}$ was used to illustrate the binomial expansion. Instead of considering $(\frac{1}{2} + \frac{1}{2})^{10}$ we might consider the more general form $(\frac{1}{2} + \frac{1}{2})^n$. As n increases in size this distribution will approach a continuous frequency curve. This frequency curve, which is bell-shaped, is the *normal curve* or normal distribution. The frequency distributions of many events in nature are found in practice to be approximated closely by the normal curve, and they are said to be normally distributed. Errors of measurement and errors made in estimating population values from sample values are often assumed to be normally distributed. The frequency distributions of many physical, biological, and psychological measurements are observed to approximate the normal form. Because the frequency of occurrence of many events in nature can be shown empirically to conform fairly closely to the normal curve, this curve can be used as a model in dealing with problems involving these events. The present chapter discusses the normal curve in detail. Before proceeding, however, with a detailed discussion of the normal curve it may be helpful to the reader to consider briefly the nature of functions and frequency curves in general.

7.2 FUNCTIONS AND FREQUENCY CURVES

When two variables are so related that the values of one depend on the values of the other they are said to be functions of each other. A function is descriptive of change in one variable with change in another. The area of a circle is a function of the radius, and the volume of a cube is a function of the length of the edge. Consider the equation $Y = bX + a$. This is a linear function. It is the equation for a straight line; Y and X are variables; b and a are constants. If b and a are known, different values of X can be substituted in the equation and the corresponding values of Y obtained. If the paired values of Y and X are plotted on graph paper, Y on the vertical and X on the horizontal axis, a straight line results. Variables Y and X bear a functional relation to each other, and this relation is linear. Sometimes Y is spoken of as the dependent and X the independent variable. A functional relation may be written in the general form $Y = f(X)$. This simply states that Y is some function of X. Here the nature of the function is not specified.

Consider now the binomial $(p + q)^n$. The terms in the binomial expansion are the expected relative frequencies or probabilities associated with particular events. Inspection of formula [6.4] indicates that any term in the binomial expansion is given by

[7.1] $$p_r = C_r{}^n p^r q^{n-r}$$

where p_r is the probability of the rth event. This expression is a function. For fixed values of n, p, and q, different values of r may be substituted on the right and the corresponding values of p_r obtained. Here p_r is the dependent and r the independent variable. The variable r is restricted to the $n + 1$ values $0, 1, 2, \ldots, n$; consequently p_r is also restricted to a fixed number of possible values. The paired values of p_r and r may be plotted on graph paper, p_r on the vertical and r on the horizontal axis. The resulting graph is a visual description of the functional relation between the event r and its relative frequency or probability p_r.

In the binomial the variable r is discrete and not continuous. In tossing 50 coins, for example, the number of heads or tails obtained is a discrete number. The value of p_r changes from r to $r + 1$ by discrete steps. We observe, however, that as n increases in size we obtain a larger and larger number of graduations of the distribution and by increasing the size of n we can make the graduations as fine as we like. By considering the situation where n becomes indefinitely large, that is, n approaches infinity, we arrive at the conception of a continuous frequency curve or function. This curve is the limiting form of the binomial.

Frequency curves are in certain instances conceptualized as extending along the X axis from minus infinity to plus infinity; that is, the curves taper off to 0 at the two extremities. Although this is so, the area between the curve and the horizontal axis is always finite. For convenience this area is often taken as unity.

On occasion it becomes necessary to find the proportion of the total area of the curve between ordinates erected at particular values of X, that is, between $X = a$ and $X = b$ as shown in Figure 7.1. This proportion is the probability that a particular value of X drawn at random from the popu-

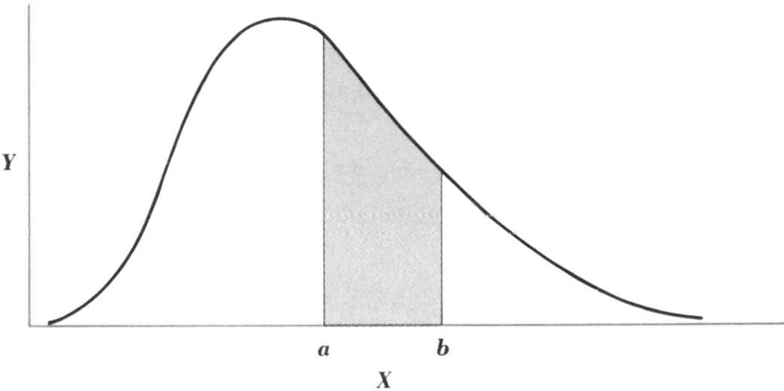

Figure 7.1 Frequency curve showing area between $X = a$ and $X = b$.

lation which the curve describes falls between a and b. Because of this, frequency curves are often referred to as probability curves or probability distributions. Statisticians use a variety of theoretical frequency curves as models. The normal curve is one of the more important of these.

7.3 THE NORMAL CURVE

In tossing n coins the frequency distribution of heads or tails is approximated more closely by the normal distribution as n increases in size. The normal curve is the limiting form of the symmetrical binomial. The equation for the normal curve is

[7.2]
$$Y = \frac{N}{\sigma\sqrt{2\pi}} e^{-(X-\mu)^2/2\sigma^2}$$

where Y = height of curve for particular values of X
π = a constant = 3.1416
e = base of napierian logarithms = 2.7183
N = number of cases, which means that the total area under the curve is N
μ and σ = mean and standard deviation of the distribution, respectively

We have used the notation μ and σ in this formula to represent the mean and standard deviation, instead of $\bar{X}$ and s, because the formula is a theoretical model. Presumably μ and σ may be regarded as population parameters. If N, μ, and σ are known, different values of X may be substituted in the equation and the corresponding values of Y obtained. If paired values of X and Y are plotted graphically, they will form a normal curve with mean μ, standard deviation σ, and area N.

The normal curve is usually written in standard-score form. Standard scores have a mean of 0 and a standard deviation of 1. Thus $\mu = 0$ and

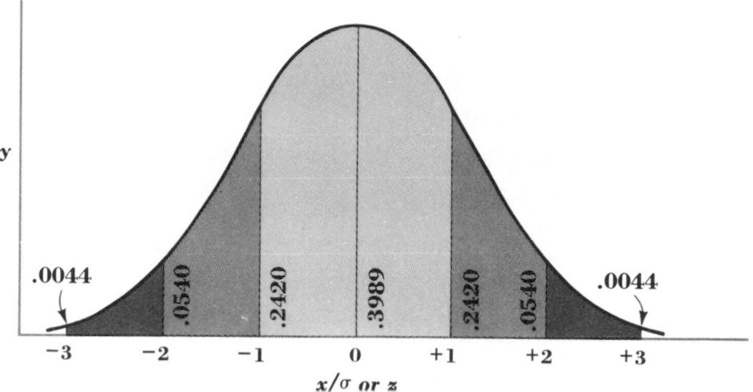

Figure 7.2 Normal curve showing height of the ordinate at different values of x/σ, or z.

$\sigma = 1$. The area under the curve is taken as unity; that is, $N = 1$. With these substitutions we may write

[7.3]
$$y = \frac{1}{\sqrt{2\pi}} e^{-z^2/2}$$

Here z is a standard score on X and is equal to $(X - \mu)/\sigma$. The score z is a deviation in standard deviation units measured along the base line of the curve from a mean of 0, deviations to the right of the mean being positive and those to the left negative. The curve has unit area and unit standard deviation. By substituting different values of z in the above formula, different values of y may be calculated. When $z = 0$, $y = 1/\sqrt{2\pi} = .3989$. This follows from the fact that $e^0 = 1$. Any term raised to the zero power is equal to 1. Thus the height of the ordinate at the mean of the normal curve in standard-score form is given by the number .3989. For $z = +1$, $y = .2420$, and for $z = +2$, $y = .0540$. Similarly, the height of the curve may be calculated for any value of z. In practice the student is not required to substitute different values of z in the normal-curve formula and solve for y to obtain the height of the required ordinate. These values may be obtained from Table A of the Appendix. This table shows different values of y corresponding to different values of z, and the area of the curve falling between the ordinates at the mean and different values of z.

The general shape of the normal curve can be observed by inspection of Figure 7.2. The curve is symmetrical. It is asymptotic at the extremities; that is, it approaches but never reaches the horizontal axis. It can be said to extend from minus infinity to plus infinity. The area under the curve is finite.

7.4 AREAS UNDER THE NORMAL CURVE

For many purposes it is necessary to ascertain the proportion of the area under the normal curve between ordinates at different points on the base line. We may wish to know (1) the proportion of the area under the curve between an ordinate at the mean and an ordinate at any specified point either above or below the mean, (2) the proportion of the total area above or below an ordinate at any point on the base line, (3) the proportion of the area falling between ordinates at any two points on the base line.

Table A of the Appendix shows the proportion of the area between the mean of the unit normal curve and ordinates extending from $z = 0$ to $z = 3$. Let us suppose that we wish to find the area under the curve between the ordinates at $z = 0$ and $z = +1$. We note from Table A that this area is .3413 of the total. Thus approximately 34 percent of the total area falls between the mean and one standard deviation unit above the mean. The proportion of the area of the curve between $z = 0$ and $z = 2$ is .4772. Thus about 47.7 percent of the area of the curve falls between the mean and two standard deviation units above the mean. The proportion of the area between $z = 0$ and $z = 3$ is .49865, or a little less than 49.9 percent.

The proportion of the area falling between $z = 0$ and $z = +1$ is .3413. Since the curve is symmetrical the proportion of the area falling between $z = 0$ and $z = -1$ is also .3413. The proportion of the area falling between the limits $z = \pm 1$ is therefore $.3413 + .3413 = .6826$, or roughly 68 percent. The proportion of the area falling between $z = \pm 2$ is $.4772 + .4772 = .9544$, or about 95 percent. The proportion between $z = \pm 3$ is $.49865 + .49865 = .99730$ or 99.73 percent. The area outside these latter limits is very small and is only .27 percent. For rough practical purposes the curve is sometimes taken as extending from $z = \pm 3$ (See Figure 7.3.)

Consider the determination of the proportion of the total area above or below any point on the base line of the curve. For illustration, let the point be $z = 1$. The proportion of the area between the mean and $z = 1$ is .3413. The proportion of the area below the mean is .5000. The proportion of the total area below $z = 1$ is $.5000 + .3413 = .8413$. The proportion above this point is $1.0000 - .8413 = .1587$. Similarly, the proportion of the area above or below any point on the base line can be readily ascertained.

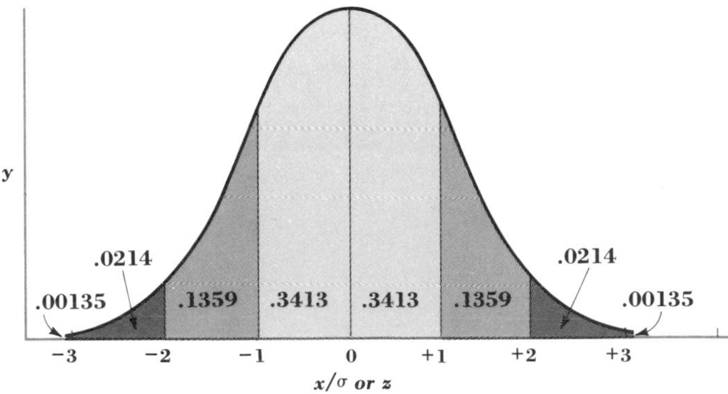

Figure 7.3 Normal curve showing areas between ordinates at different values of x/σ, or z.

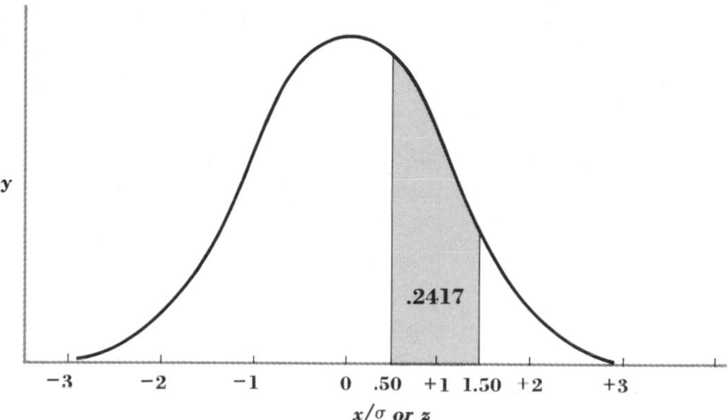

Figure 7.4 Normal curve showing area between ordinates of $z = .50$ and $z = 1.50$.

Consider the problem of finding the area between ordinates at any two points on the base line. Let us assume that we require the area between $z = .5$ and $z = 1.5$. From Table A of the Appendix we note that the proportion of the area between the mean and $z = .50$ is .1915. We note also that the area between the mean and $z = 1.5$ is .4332. The area between $z = .50$ and $z = 1.5$ is obtained by subtracting one area from the other and is $.4332 - .1915 = .2417$. The area for any other segment of the curve may be similarly obtained (see Figure 7.4).

On occasion we wish to find values of z which include some specified proportion of the total area. For example, the values of z above and below the mean which include a proportion .95 of the area may be required. We select a value of z above the mean which includes a proportion .475 of the total area and a value of z below the mean which also includes a proportion .475 of the total area. From Table A of the Appendix we observe that the proportion .475 of the area falls between $z = 0$ and $z = 1.96$. Since the curve is symmetrical the proportion .475 of the area falls between $z = 0$ and $z = -1.96$. Thus a proportion .95, or 95 percent, of the total area falls within the limits $z = \pm 1.96$. Also a proportion .05, or 5 percent, falls outside these limits. Similarly, it may be shown that 99 percent of the area of the curve falls within, and 1 percent outside, the limits $z = \pm 2.58$. Figure 7.5 is a normal curve showing values of z which include a proportion .95 of the total area.

7.5 AREAS UNDER THE NORMAL CURVE—ILLUSTRATIVE EXAMPLE

The distribution of IQs obtained by the application of a particular test is approximately normal with a mean of 100 and a standard deviation of 15. We are required to estimate what percent of individuals in the population

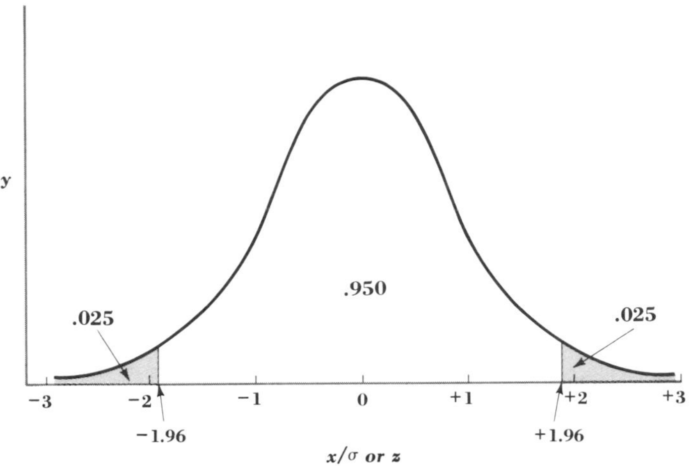

Figure 7.5 Normal curve showing values of z which include a proportion which is .95 of the total area.

have IQs of 120 and above. The IQ of 120 in standard-score form is $z = (120 - 100)/15 = 1.33$. Thus an IQ of 120 is 1.33 standard deviation units above the mean. Reference to a table of areas under the normal curve shows that the proportion of the area above a standard score of 1.33 is .092. Thus we estimate that on this particular test about 9.2 percent of the population have IQs equal to or greater than 120.

We are required to estimate for the same test the middle range of IQs which includes 50 percent of the population. A table of areas under the normal curve shows that 25 percent of the area under the curve falls between the mean and a standard score of $-.675$. Also 25 percent of the area falls between the mean and a standard score of $-.675$. Thus 50 percent of the area falls between the limits of $z = \pm.675$. The standard-score scale has a mean of 0 and a standard deviation of unity. Here we must transform standard scores to the original scale of IQs with a mean of 100 and a standard deviation of 15. To transform standard scores to IQs we multiply the standard score by 15 and add 100. Thus the standard score $-.675$ is transformed to $15 \times (-.675) + 100 = 89.88$ and $+.675$ to $15 \times .675 + 100 = 110.12$. Thus we estimate that about 50 percent of the population have IQs within a range of roughly 90 to 110.

7.6 THE NORMAL APPROXIMATION TO THE BINOMIAL

The observation has been made that as n increases in size the symmetrical binomial is more closely approximated by the normal distribution. This means that the normal distribution may be used to estimate binomial probabilities. If we are given a situation where 10 coins are tossed a large number of times, what is the probability of obtaining 7 or more heads? Here the mean of the binomial is $\mu = 10 \times \frac{1}{2} = 5.0$ and the standard deviation is $\sigma = \sqrt{\frac{10}{4}} = 1.58$. Because the normal distribution is continuous, and not discrete, we consider the value 7 as covering the exact limits 6.5 to 7.5. Thus we must ascertain the proportion of the area of the normal curve falling above an ordinate at 6.5, the mean of the curve being 5.0 and the standard deviation 1.58. In standard-score form the value 6.5 is equivalent to $z = (6.5 - 5.0)/1.58 = .949$. The proportion of the area of the normal curve falling above an ordinate at $z = .949$ can be readily ascertained from Table A of the Appendix and is .171. Thus using the normal-curve approximation to the binomial, we estimate the probability of obtaining 7 or more heads in tossing 10 coins as .171. We may compare this with the exact probabilities obtained directly from the binomial expansion shown in Table 7.1. This probability is .172. Here we note that the discrepancy between the estimate obtained from the normal curve and the exact binomial probability is trivial.

Table 7.1 compares the binomial and normal probabilities for $n = 10$ and $p = 1/2$. We note that in this instance the differences between the exact binomial probabilities and the corresponding normal approximations are small.

The accuracy of the approximation depends both on n and p; as n increases in size the accuracy of the approximation is improved. For any n as p departs from 1/2 the approximation becomes less accurate.

Table 7.1

Comparison of binomial probabilities with corresponding normal approximations for $n = 10$ and $p = 1/2$*

No. of heads	Exact binomial probability	Normal approximation
10	.001	.002
9	.010	.011
8	.044	.044
7	.117	.114
6	.205	.205
5	.246	.248
4	.205	.205
3	.117	.114
2	.044	.044
1	.010	.011
0	.001	.002
Total	1.000	1.000

7.7 SUMMARY OF PROPERTIES OF THE NORMAL CURVE

The following is a summary of properties of the normal curve.

1 The curve is symmetrical. The mean, median, and mode coincide.

2 The maximum ordinate of the curve occurs at the mean, that is, where $z = 0$, and in the unit normal curve is equal to .3989.

3 The curve is asymptotic. It approaches but does not meet the horizontal axis and extends from minus infinity to plus infinity.

4 The points of inflection of the curve occur at points plus or minus one standard deviation unit above and below the mean. Thus the curve changes from convex to concave in relation to the horizontal axis at these points.

5 Roughly 68 percent of the area of the curve falls within the limits plus or minus one standard deviation unit from the mean.

6 In the unit normal curve the limits $z = \pm 1.96$ include 95 percent and the limits $z = \pm 2.58$ include 99 percent of the total area of the curve, 5 percent and 1 percent of the area, respectively, falling beyond these limits.

BASIC TERMS AND CONCEPTS

Function

Normal curve

Ordinate

Areas under the normal curve

Area between $z = \pm 1.96$

Area between $z = \pm 2.58$

EXERCISES

1 Find the height of the ordinate of the normal curve at the following z values: -2.15, -1.53, $+.07$, $+.99$, $+2.76$.

2 Consider a normally distributed variable with $\bar{X} = 50$ and $s = 10$. For $N = 200$ find the height of the ordinates at the following values of X: 25, 35, 49, 57, and 63.

3 Find the proportion of the area of the normal curve (**a**) between the mean and $z = 1.49$, (**b**) between the mean and $z = 1.26$, (**c**) to the right

of $z = .25$, **(d)** to the right of $z = 1.50$, **(e)** to the left of $z = -1.26$, **(f)** to the left of $z = .95$, **(g)** between $z = \pm.50$, **(h)** between $z = -.75$ and $z = 1.50$, **(i)** between $z = 1.00$ and $z = 1.96$, **(j)** between $z = 1.00$ and $z = 1.01$.

4 Find a value of z such that the proportion of the area **(a)** to the right of z is .25, **(b)** to the left of z is .90, **(c)** between the mean and z is .40, **(d)** between $\pm z$ is .80.

5 On the assumption that IQs are normally distributed in the population with a mean of 100 and a standard deviation of 15, find the proportion of people with IQs **(a)** above 135, **(b)** above 120, **(c)** below 90, **(d)** between 75 and 125.

6 A teacher decides to fail 25 percent of the class. Examination marks are roughly normally distributed, with a mean of 72 and a standard deviation of 6. What mark must a student make to pass?

7 Error scores on a maze test for a particular strain of rats are known through prolonged experimentation to have an approximately normal distribution with a mean of 32 and a standard deviation of 8. In one experiment a control sample of six animals contains one animal with an error score of 66. What arguments may be advanced for discarding the results of this animal?

8 Scores on a particular psychological test are normally distributed, with a mean of 50 and a standard deviation of 10. The decision is made to use a letter-grade system A, B, C, D, and E with the proportions .10, .20, .40, .20, and .10 in the five grades, respectively. Find the score intervals for the five letter grades.

9 The following are data for test scores for two age groups:

	11-year group	14-year group
$\bar{X}$	48	56
s	8	12
N	500	800

Assuming normality, estimate how many of the 11-year-olds do better than the average 14-year-old and how many of the 14-year-olds do worse than the average 11-year-old.

ANSWERS TO EXERCISES

1 .0395, .1238, .3980, .2444, .0088

2 .350, 2.590, 7.940, 6.246, 3.428

3 **a** .4319 **b** .3962 **c** .4013 **d** .0668 **e** .1038
f .8289 **g** .3830 **h** .7066 **i** .1337 **j** .0025

4 **a** .675 **b** 1.281 **c** 1.281 **d** ±1.281

5 **a** .0099 **b** .0918 **c** .2514 **d** .9050

6 68

7 An error score of 66 is more than four standard deviations above the mean. The occurrence of a score equal to or greater than 66 is very improbable in random samples.

8

Grade	Score interval
A	62.8–
B	55.2–62.8
C	44.8–55.2
D	37.2–44.8
E	–37.2

9 **a** 79 **b** 201

8

CORRELATION

8.1 INTRODUCTION

In previous chapters we have considered the description of a single variable, the arrangement of a single variable in the form of a frequency distribution, and two theoretical models for single variable distributions, namely the binomial and the normal. In psychology and education many situations arise that involve two or more variables. This chapter is concerned with two variables only. Situations that involve more than two variables are discussed in later chapters.

The data under consideration here consist of pairs of observations or measurements; that is, we have two observations or measurements for each member of a group. Such data are called bivariate data. Examples are measures of height and weight for a group of schoolchildren, intelligence test scores and measures of scholastic performance for a group of university students, error scores in running two different mazes for a group of experimental animals, or years of schooling and income level for a group of adult males. The essential feature of such data is that one observation or measurement can be paired with another observation or measurement for each member of the group.

The study of data comprising paired measurements has two closely related aspects, correlation and prediction. We may concern ourselves with the problem of describing the degree or magnitude of the relation between the two variables, that is, the magnitude of concomitant variation. This is a problem in *correlation*. The statistic that describes the degree of relation between two variables is called a *correlation coefficient.* We may on the other hand concern ourselves with estimating or predicting one variable from a knowledge of another. This is a *prediction* problem. For example, consider height and weight. Tall persons tend in general to be heavier than short persons, and light persons shorter than heavy persons. We may attend to the problem of describing the magnitude of the relation between height and weight, or we may attend to the problem of predicting a person's height from a knowledge of weight, or conversely weight from height. Or our data may consist of scores obtained on a psychological test administered to students entering a university and the grade point averages of the same students at the end of the first year of university work. For such data we may concern ourselves with simply describing the degree of relation between psychological test scores and grade point averages. On the other hand we may focus attention on the prediction of grade point averages from a knowledge of psychological test scores, the purpose being to use psychological test scores when students enter a university to provide estimates of their subsequent scholastic performance.

Historically, the study of the prediction of one variable from a knowledge of another preceded the development of measures of correlation. In 1885 Francis Galton published a paper called "Regression towards Mediocrity in Hereditary Stature." Galton was interested in predicting the physical characteristics of offspring from a knowledge of the physical characteristics of their parents. He observed, for example, that the offspring of tall parents tended on the average to be shorter than their parents, whereas the offspring of short parents tended on the average to be taller than their parents. He used the term *regression* to refer to this effect. In modern statistics the term regression no longer has the biological interpretation assigned to it by Galton. In general, in statistics the term regression refers to the problem of predicting one variable from a knowledge of another or possibly several other variables. Karl Pearson extended Galton's ideas of regression and developed the methods of correlation that are used today with increasing frequency.

The most widely used measure of correlation is the Pearson *product-moment correlation coefficient.* This is a statistic of the interval-ratio type. Other varieties of correlation have been developed for use with nominal and ordinal variables, and for data with special characteristics. Many of the varieties of correlation, although not all, are particular cases of the product-moment correlation coefficient, and are discussed in later chapters of this book.

In this chapter we shall present a discussion of correlation and proceed in the following chapter to discuss prediction and its relation to correlation. The reader must bear in mind that correlation and prediction are two closely related topics. Certain matters pertaining to the interpretation of the correlation coefficient, and the assumptions underlying its use, can only be discussed following a consideration of prediction.

8.2 RELATIONS BETWEEN PAIRED OBSERVATIONS

Consider a group comprising N members. Denote these by A_1, A_2, A_3, . . . , A_N. Measurements are available on each member on two variables, X and Y. The data may be represented symbolically as follows:

	Measurement	
Members	X	Y
A_1	X_1	Y_1
A_2	X_2	Y_2
A_3	X_3	Y_3
. . .	. . .	. . .
A_N	X_N	Y_N

Let us assume that measurements have been arranged in order of magnitude on X extending from X_1, the highest, to X_N, the lowest, measurement. Given this arrangement on X, we may consider the possible arrangements of Y with respect to X. Consider an arrangement where the values of Y are in order of magnitude extending from the highest to the lowest. Thus the member who is highest on X is also highest on Y, the member who is next highest on X is next highest on Y, and so on, until the member who is lowest on X is also lowest on Y. This situation represents the maximum positive relation between the two variables. Consider now an arrangement where the values of Y are reversed so that Y_1 is the lowest and Y_N is the highest. The member who is highest on X is lowest on Y, the member who is next highest on X is next lowest on Y, and so on, until the member who is lowest on X is highest on Y. This situation represents the maximum negative relation between the variables. Consider a situation where the arrangement of Y is strictly random in relation to X. Values of Y may be inserted in a hat, shuffled, drawn at random, and paired with values of X. This is a situation of independence. The two sets of variate values bear a random relation to each other. Under this arrangement we may state that no relation exists between X and Y. Between the two extreme arrangements, representing the maximum positive and negative relations, we may consider arrangements which represent varying degrees of relation in either a positive or negative direction. To illustrate, let us assume that the values of X for the members A_1, A_2, A_3, A_4, and A_5 are the integers 5, 4, 3, 2, and 1. If the values of Y are the same integers and are also arranged in the order 5, 4, 3, 2, and 1, we have a maximum positive relation. If values of Y are arranged in the order 4, 5, 3, 2, and 1, we have clearly a high positive relation, although not the highest possible. If values of Y are arranged in the order 1, 2, 3, 4, and 5, we have a maximum negative relation. Again an arrangement on Y of the kind 1, 2, 4, 3, 5 would be high negative, although not the highest possible.

Relations of the kind described above may be examined by plotting the paired measurements on graph paper, each pair of observations being

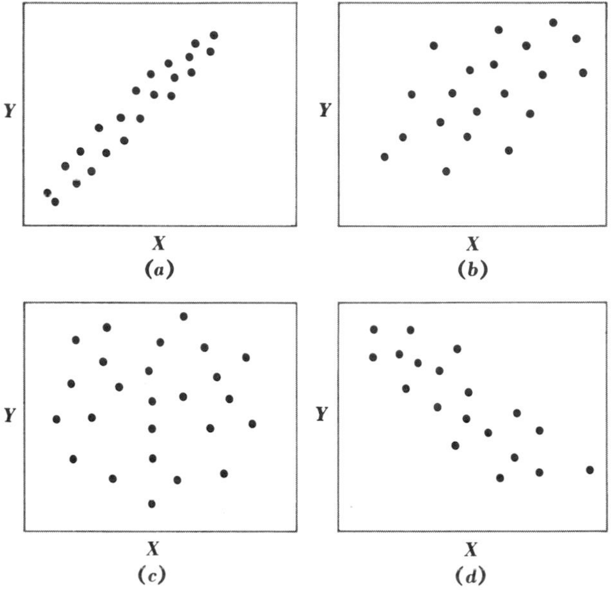

Figure 8.1 (*a*) High positive correlation, (*b*) low positive correlation, (*c*) zero correlation, (*d*) negative correlation.

represented by a point. Such a plotting of measurements is sometimes called a *scatter diagram*. Inspection of a scatter diagram yields an intuitive appreciation of the degree of relation between the two variables. Figure 8.1 shows four such diagrams.

Figure 8.1*a* is a graphical representation of a high positive relation. Note that the points fall very close to a straight line. If the points fall exactly on a straight line, a perfect positive relation exists between the variables. Figure 8.1*b* shows a low positive relation. Figure 8.1*c* shows a relation which is more or less random. No systematic tendency is observed for high values of X to be associated with high values of Y and low values of X to be associated with low values of Y, or vice versa. Figure 8.1*d* shows a fairly high negative relation. Again, if all the points fall exactly along a straight line, a perfect negative relation exists. It is obvious that between the two extremes of a perfect positive and a perfect negative relation an indefinitely large number of possible arrangements of points may occur, representing an indefinitely large number of possible relations between the two variables.

8.3 THE CORRELATION COEFFICIENT

Measures of correlation by common convention are defined to take values ranging from -1 to $+1$. A value of -1 describes a perfect negative relation. All points lie on a straight line, and X decreases as Y increases. A value of

+1 describes a perfect positive relation. All points lie on a straight line, and X increases as Y increases. A value of 0 means that X and Y are independent of each other or bear a random relation to each other.

As mentioned previously the most commonly used measure of correlation is the Pearson product-moment correlation coefficient, and many varieties of correlation are particular cases of this coefficient. In order to grasp the reasoning underlying the definition of this coefficient, readers may find it advantageous, as in previous discussions, to conceptualize the variables in standard-score form. Let X and Y represent paired observations with standard deviations s_x and s_y. To convert X to standard-score form we substract the mean $\bar{X}$ and divide this by the standard deviation, s_x, and similarly for Y. Thus

$$z_x = \frac{X - \bar{X}}{s_x}$$

$$z_y = \frac{Y - \bar{Y}}{s_y}$$

These standard scores have a mean of zero and a standard deviation of one. The product-moment correlation coefficient, denoted by the letter r, is defined as the sum of products of standard scores, divided by $N - 1$. The formula for r in standard-score form is

[8.1] $$r = \frac{\Sigma z_x z_y}{N - 1}$$

Thus the correlation coefficient may be obtained by converting the two variables to standard-score form, summing their product, and dividing by $N - 1$.

A brief and rather incomplete discussion of the rationale underlying the above coefficient is appropriate here. Consider a set of paired observations in standard-score form. The sum of products of standard scores $\Sigma z_x z_y$ is readily observed to be a measure of the degree of relationship between the two variables. Let us consider the *maximum* and *minimum* values of $\Sigma z_x z_y$. This sum of products is observed to take its maximum possible value when (1) the values of z_x and z_y are in the same order and (2) every value of z_x is equal to the value of z_y with which it is paired, the two sets of paired standard scores being identical. If the paired standard scores are plotted on graph paper, all points will fall exactly along a straight line with positive slope. Since all pairs of observations are such that $z_x = z_y$, we may write $z_x z_y = z_x^2 = z_y^2$ and $\Sigma z_x z_y = \Sigma z_x^2 = \Sigma z_y^2$. The quantity $\Sigma z_x^2 = \Sigma z_y^2 = N - 1$, as shown in Section 5.8. Thus we observe that the maximum possible value of $\Sigma z_x z_y$ is equal to $N - 1$. Similarly, $\Sigma z_x z_y$ will take its minimum possible value when (1) the values of z_x and z_y are in inverse order and (2) every value of z_x has the same absolute numerical value as the z_y with which it is paired, but differs in sign. This minimum value of $\Sigma z_x z_y$ is readily shown to be equal to $-(N - 1)$. Graphically, all points will fall exactly along a straight line with negative slope.

When z_x and z_y bear no systematic relation to each other, the expected value of $\Sigma z_x z_y$ will be 0. We may define a coefficient of correlation as the ratio of the observed value of $\Sigma z_x z_y$ to the maximum possible value of this quantity; that is, r is defined as $\Sigma z_x z_y/(N-1)$. Since $\Sigma z_x z_y$ has a range extending from $N-1$ to $-(N-1)$, the coefficient r will extend from $+1$ to -1.

The above discussion may be illustrated by a simple example. The following are raw scores on X and Y with the corresponding standard scores.

X	Y	z_x	z_y	$z_x z_y$
1	11	-1.2649	-1.2649	1.60
2	13	$-\ .6325$	$-\ .6325$	.40
3	15	.0000	.0000	.00
4	17	$+\ .6325$	$+\ .6325$	.40
5	19	$+1.2649$	$+1.2649$	1.60

$\Sigma z_x z_y = \Sigma z_x^2 = \Sigma z_y^2 = N - 1 = 4$

The variables X and Y are in the same order, and if plotted as points on a scatter diagram, the points will fall along a straight line. In this situation the pairs of standard scores z_x and z_y are identical, and $\Sigma z_x z_y = N - 1 = 4$ and $r = +1$. It is obvious by simple inspection that $N - 1$ is the maximum value that this sum of products can take. No arrangement of Y with respect to X can produce a larger value. If the variable Y is inverted with respect to X, the quantity $\Sigma z_x z_y = -(N-1) = -4$, and $r = -1$. This is the minimum value of $\Sigma z_x z_y$. Other arrangements of Y with respect to X will produce values of r between $+1$ and -1. Very simply a product-moment correlation coefficient is nothing other than the sum of products of standard scores divided by the maximum value the sum of products can assume. A variety of other descriptive measures are defined in a manner directly analogous to this.

The reader will note for any particular set of paired standard scores the maximum and minimum values of $\Sigma z_x z_y$, obtained by arranging the paired scores in direct and inverse orders, are not necessarily $N-1$ and $-(N-1)$. A maximum value equal to $N-1$ will occur only when the paired observations have the characteristic that every value of z_x is equal to the value of z_y with which it is paired. A minimum of $-(N-1)$ will occur only when every value of z_x is equal to z_y in absolute value, but differs in sign. When the data do not have these characteristics, the limits of the range of r, for the particular set of paired observations under consideration, will be less than $+1$ and greater than -1.

8.4 CALCULATION OF THE CORRELATION COEFFICIENT

The formula for the correlation coefficient in standard score form is $r = \Sigma z_x z_y/(N-1)$. In calculating a correlation coefficient in practice this formula is laborious, because it requires the conversion of all values to standard-score form. By substitution involving simple algebra it is possible to write a number of alternate formulas for the correlation coefficient. Formulas may be written in deviation-score form and in raw-score form.

Since $z_x = (X - \bar{X})/s_x$ and $z_y = (Y - \bar{Y})/s_y$, by simple substitution in $r = \Sigma z_x z_y/(N-1)$ we obtain

[8.2] $$r = \frac{\Sigma(X-\bar{X})(Y-\bar{Y})}{(N-1)s_x s_y} = \frac{\Sigma(X-\bar{X})(Y-\bar{Y})}{\sqrt{\Sigma(X-\bar{X})^2\Sigma(Y-\bar{Y})^2}} = \frac{\Sigma xy}{\sqrt{\Sigma x^2 \Sigma y^2}}$$

where x and y are deviations from the means $\bar{X}$ and $\bar{Y}$, respectively.

The above formula for the correlation coefficient may be used for computational purposes. The calculation is illustrated in Table 8.1. The first two columns contain the paired observations on X and Y. These columns are summed and divided by N to obtain the means $\bar{X}$ and $\bar{Y}$. Column 3 contains the deviations from the mean of X, and column 4 the deviations from the mean of Y. Columns 5 and 6 contain the squares of these deviations. These columns are summed to obtain Σx^2 and Σy^2. Column 7 contains the products of x and y, and this column is summed to obtain Σxy. The correlation coefficient in this example is +.58.

In some computational situations it is useful to use the formula for the correlation coefficient expressed in raw score form, that is, in terms of the original observations. This formula is as follows:

[8.3] $$r = \frac{N\Sigma XY - \Sigma X \Sigma Y}{\sqrt{[N\Sigma X^2 - (\Sigma X)^2]\,[N\Sigma Y^2 - (\Sigma Y)^2]}}$$

The reader should note that the application of this formula requires five

Table 8.1

Calculation of the correlation coefficient from ungrouped data using deviation scores

1	2	3	4	5	6	7
X	Y	x	y	x^2	y^2	xy
5	1	−1	−3	1	9	+3
10	6	+4	+2	16	4	+8
5	2	−1	−2	1	4	+2
11	8	+5	+4	25	16	+20
12	5	+6	+1	36	1	+6
4	1	−2	−3	4	9	+6
3	4	−3	0	9	0	0
2	6	−4	+2	16	4	−8
7	5	+1	+1	1	1	+1
1	2	−5	−2	25	4	+10
60	40	0	0	134	52	48
$\bar{X} = 6.0$	$\bar{Y} = 4.0$			Σx^2	Σy^2	Σxy

$$r = \frac{\Sigma xy}{\sqrt{\Sigma x^2 \Sigma y^2}} = \frac{48}{\sqrt{134 \times 52}} = +.58$$

Table 8.2
Calculation of the correlation coefficient from ungrouped data using raw scores

1	2	3	4	5
X	Y	X^2	Y^2	XY
5	1	25	1	5
10	6	100	36	60
5	2	25	4	10
11	8	121	64	88
12	5	144	25	60
4	1	16	1	4
3	4	9	16	12
2	6	4	36	12
7	5	49	25	35
1	2	1	4	2
60	40	494	212	288
ΣX	ΣY	ΣX^2	ΣY^2	ΣXY

$$r = \frac{N\Sigma XY - \Sigma X \Sigma Y}{\sqrt{[N\Sigma X^2 - (\Sigma X)^2][N\Sigma Y^2 - (\Sigma Y)^2]}}$$

$$= \frac{10 \times 288 - 60 \times 40}{\sqrt{(10 \times 494 - 60^2)(10 \times 212 - 40^2)}}$$

$$= \frac{480}{\sqrt{1{,}340 \times 520}} = +.58$$

terms in addition to N: ΣXY, ΣX^2, ΣY^2, ΣX, and ΣY. In practice, when an electric calculator is available, this formula, or some simple variant of it, is possibly the easiest formula to use in calculating correlation coefficients.

The application of the formula for computing the correlation coefficient from raw scores is illustrated in Table 8.2. The first two columns contain the paired observations on X and Y. These columns are summed to obtain ΣX and ΣY. Columns 3 and 4 contain the squares of the observations, and these are summed to obtain ΣX^2 and ΣY^2. Column 5 contains the product terms XY, and the sum of this column is ΣXY. The correlation is $+.58$, which checks with the value obtained by the previous method using deviation scores.

8.5 BIVARIATE FREQUENCY DISTRIBUTIONS

In Chapter 2 we discussed a frequency distribution for a single variable. A frequency distribution was defined as an arrangement of the data showing the frequency of occurrence of the values of the variable, or the frequency of occurrence of the values of the variable falling within defined ranges of the variable, the defined ranges being the class intervals. When one variable only is under consideration, the distribution is spoken of as *univariate*. The idea of a frequency distribution may readily be extended to two variables. A frequency distribution involving two variables is known as a *bivariate* distribution.

A bivariate frequency distribution is a table comprising a number of rows and columns. If class intervals are used, the columns correspond to class intervals of the X variable and the rows to class intervals of the Y variable. The paired observations on X and Y are entered in the appropriate cells of the bivariate table.

Table 8.3 shows hypothetical data for a simple bivariate distribution. Each pair of observations is entered in the appropriate cell. The numbers in the cells are the bivariate frequencies. For example, 4 values are observed to fall within the interval 10–14 on X and 15–19 on Y. By summing the rows the frequency distribution for the Y variable is obtained; by summing the columns the frequency distribution for the X variable is obtained. These are the univariate marginal distributions. Methods exist for the calculation of correlation coefficients from bivariate frequency distributions. These methods are now obsolete. The idea of a bivariate frequency distribution is, however, of some importance.

The reader will recall that in Chapter 2 the graphical representation of frequency distributions as histograms and frequency polygons was discussed. Such representation may be readily extended to bivariate distributions. It is possible to view the representation of a bivariate distribution as a three-dimensional graph. If bivariate frequencies in the cells are represented by three-dimensional blocks, or columns, the height of the block for each cell being proportional to the frequency in that cell, the result would be a three-dimensional histogram. Similarly by plotting points in three-dimensional space, a three-dimensional frequency polygon would result.

Table 8.3

Example of Bivariate Frequency Distribution

		X						
		0–4	5–9	10–14	15–19	20–24	25–29	f_y
	25–29					1	1	2
	20–24				2	4		6
	15–19			4	6		1	11
Y	10–14		1	8	3			12
	5–9	1	4	2	1			8
	0–4	2	1					3
	f_x	3	6	14	12	5	2	42

In Chapter 2 the observation was made also that with increase in the number of values of the variable and the number of class intervals, it was possible to arrive at the concept of a continuous frequency distribution. Similarly it is possible to imagine a continuous bivariate frequency distribution. One particular bivariate distribution is the *bivariate normal distribution.* The shape of this distribution is not independent of the size of the correlation. For zero correlation the distribution may be imagined as a circular mound; any slice, so to speak, parallel to the plane of X and Y is a circle. When the correlation is not zero this slice becomes an ellipse. As the correlation increases in absolute value, this ellipse becomes more elongated. For the bivariate normal distribution the two marginal distributions of X and Y are normal. Also, regardless of the size of the correlation, any slice of the ellipse parallel to either the Y or X axes results in a normal distribution, except, of course, in the case where either a perfect positive or negative correlation exists. In situations involving two variables the bivariate distribution is frequently assumed to be normal.

8.6 THE VARIANCE OF SUMS AND DIFFERENCES

A discussion of the variance of sums and differences is introduced here because of the relevance of this topic to subsequent work in statistics. Let X and Y be two sets of scores or measurements for the same group of individuals. For example, these may be marks on mathematics and history examinations for a group of university students or any set of paired measurements whatsoever. Given two sets of paired observations, X and Y, we may consider the variance of X, the variance of Y, and the correlation between X and Y. We may also consider the variance of $X+Y$, the variance of sums, and the variance of $X-Y$, the variance of differences.

To illustrate the situation in very simple terms consider the following data:

	X	Y	$X+Y$	$X-Y$
	5	2	7	3
	6	8	14	−2
	8	7	15	1
	12	20	32	−8
	15	6	21	9
Means	9.20	8.60	17.80	.60

Given a knowledge of X and Y, how can we write a simple expression for the variance of $X+Y$ and the variance of $X-Y$? This question has a simple and straightforward solution. Note that the mean of the sums of X and Y is $\bar{X}+\bar{Y}$, or the sum of the two means. Also the mean of the differences between X and Y is $\bar{X}-\bar{Y}$, or the difference between the two means.

The variance of the sum of X and Y may be written as follows:

[8.4]
$$s^2_{x+y} = \frac{\Sigma[(X+Y)-(\bar{X}+\bar{Y})]^2}{N-1}$$
$$= \frac{\Sigma[(X-\bar{X})+(Y-\bar{Y})]^2}{N-1}$$
$$= \frac{\Sigma(X-\bar{X})^2}{N-1} + \frac{\Sigma(Y-\bar{Y})^2}{N-1} + \frac{2\Sigma(X-\bar{X})(Y-\bar{Y})}{N-1}$$
$$= s_x{}^2 + s_y{}^2 + 2rs_xs_y$$

This shows that the variance of sums, s^2_{x+y}, is equal to the sum of the two variances plus the quantity $2\Sigma rs_xs_y$. Quantities of the kind rs_xs_y are called covariances. Note that the covariance is equal to $\Sigma(X-\bar{X})(Y-\bar{Y})/(N-1)$ or, using deviation score notation, $\Sigma xy/(N-1)$.

The formula for the variance of sums illustrates in a simple way a frequently recurring idea of fundamental importance in all statistics. It shows that a variance, in this particular case s^2_{x+y}, may be understood as comprising separate additive parts; or that a variance can be divided or partitioned into additive parts. Here the variance of sums is shown to comprise three additive parts, the variance of X, the variance of Y, and a part due to the covariation of X and Y. The simple notion involved here, which views a variance as made up of additive bits, is very general and applicable in many situations.

The discussion above is concerned with two variables only. The argument may be extended to any number of variables. Consider three variables X_1, X_2, and X_3. The variances and covariances may for convenience be written in the form of a small covariance table or matrix, as follows:

	1	2	3
1	$s_1{}^2$	$r_{12}s_1s_2$	$r_{13}s_1s_3$
2	$r_{12}s_1s_2$	$s_2{}^2$	$r_{23}s_2s_3$
3	$r_{13}s_1s_3$	$r_{23}s_2s_3$	$s_3{}^2$

The three variances appear along the main diagonal, and the covariances on either side of the main diagonal. The variance of sums is simply the sum of all the elements in the covariance table or matrix. Thus for three variables the variance of sums is

$$s^2_{1+2+3} = s_1{}^2 + s_2{}^2 + s_3{}^2 + 2r_{12}s_1s_2 + 2r_{13}s_1s_3 + 2r_{23}s_2s_3$$

The argument may be generalized to any number of variables. The variance of the sum of the variables is very simply the sum of all the elements in the covariance table or matrix with k rows and columns.

Consider now the variance of the differences between X and Y, that is, the variance of $X-Y$. This variance may be written as

$$s^2_{x-y} = \frac{\Sigma[(X-Y)-(\bar{X}-\bar{Y})]^2}{N-1}$$

By algebra similar to that used above for the variance of sums, the variance of differences is readily seen to be

[8.5] $$s^2_{x-y} = s^2_x + s^2_y - 2rs_xs_y$$

Thus the variance of differences is the sum of the two variances minus the covariance term $2rs_xs_y$.

An important observation is that if two variables are independent of each other, that is, if they are uncorrelated, $r = 0$, hence $2rs_xs_y = 0$, and the variance of sums is the sum of the separate variances. Thus $s^2_{x+y} = s^2_x + s^2_y$. This applies to any number of variables. The variance of the sum of k independent, or uncorrelated, variables is the sum of the separate variances. For two independent variables the variance of differences is also the sum of the separate variances, that is $s^2_x + s^2_y$. Thus we note that for two independent variables $s^2_{x+y} = s^2_{x-y}$.

BASIC TERMS AND CONCEPTS

Paired observations

Scatter diagram

Perfect positive relation

Perfect negative relation

$\Sigma z^2 = N - 1$

Correlation coefficient

Correlation coefficient: standard-score form; deviation-score form

Bivariate frequency distribution

Bivariate normal distribution

Variance of sums

Variance of differences

EXERCISES

1 Would you expect the correlation between the following to be positive, negative, or about 0? **(a)** The intelligence of parents and their offspring, **(b)** scholastic success and annual income 10 years after graduation, **(c)** age and mental ability, **(d)** marks on examinations in physics and mathematics, **(e)** wages and the cost of living, **(f)** birthrate and the numerosity of storks, **(g)** scores on a dominance-submission test for husbands and scores for their wives.

2 The following are paired measurements:

X	5	8	9	7	6	1
Y	3	7	8	8	5	9

Compute the correlation between X and Y.

3 Show that

$$\frac{\Sigma xy}{(N-1)s_x s_y} = \frac{\Sigma xy}{\sqrt{\Sigma x^2 \Sigma y^2}}$$

4 When $N = 2$, what are the possible values of the correlation coefficient?

5 The correlation coefficient is not necessarily equal to +1 when the paired measurements are in exactly the same rank order. Discuss.

6 Calculate the correlation coefficient for the following data using formula [7.3].

X	Y	X	Y	X	Y
22	18	19	25	11	17
15	16	7	36	5	6
9	31	6	27	26	45
7	8	46	45	19	30
4	2	11	18	8	18
45	36	27	18	1	3
19	12	19	37	9	7
26	16	36	42	18	28
35	47	25	20	46	21
49	22	10	12	9	25

7 Show that $s_{x-y}^2 = s_x{}^2 + s_y{}^2 - 2rs_x s_y$.

8 Under what conditions will the variance of the sum of two variables equal the variance of the difference between two variables?

9 Is the correlation between X and Y changed by adding a constant to X or by multiplying X by a constant?

10 The formulas found in this chapter assume that the variance is defined as $s^2 = \Sigma(X - \bar{X})^2/(N-1)$. What is the formula for r in standard-score form if the variance is defined as $s^2 = \Sigma(X - \bar{X})^2/N$?

ANSWERS TO EXERCISES

1 a The correlation between the intelligence of parents and that of their offspring has frequently been reported to be in the neighborhood of .50.

b Probably low positive.

c Fairly high positive in samples of individuals covering a broad age range up to about age of 17.
d Fairly high positive.
e Fairly high positive under relatively stable economic conditions.
f Probably about 0, although data have been reported to the contrary.
g Probably about 0, although arguments may be advanced for a possible negative correlation.

2 −.063

3 $s_x = \sqrt{\frac{\Sigma x^2}{N-1}}, \; s_y = \sqrt{\frac{\Sigma y^2}{N-1}}$

Hence by substitution

$$\frac{\Sigma xy}{(N-1)s_x s_y} = \frac{\Sigma xy}{\sqrt{\Sigma x^2 \Sigma y^2}}$$

4 +1 and −1

5 The paired measurements may be in exactly the same order, but the value of z_x may not be exactly equal to the value of z_y with which it is paired. Consequently $\Sigma z_x z_y$ will not necessarily be equal to $\Sigma z_x{}^2$ and $\Sigma z_y{}^2$. Under these circumstances $\Sigma z_x z_y$ will be less than either $\Sigma z_x{}^2$ or $\Sigma z_y{}^2$, and r will be less than 1.

6 .544

7 The mean of $X - Y$ is $\bar{X} - \bar{Y}$. Hence we may write

$$s_{X-Y}^2 = \frac{\Sigma[X-Y) - (\bar{X}-\bar{Y})]^2}{N-1} = \frac{\Sigma[(X-\bar{X}) - (Y-\bar{Y})]^2}{N-1}$$

$$= \frac{\Sigma(X-\bar{X})^2}{N-1} + \frac{\Sigma(Y-\bar{Y})^2}{N-1} - \frac{2\Sigma(X-\bar{X})(Y-\bar{Y})}{N-1}$$

$$= s_X{}^2 + s_Y{}^2 - 2rs_Xs_Y$$

8 The variance of sums is equal to the variance of the differences when the variables are uncorrelated.

9 No

10 $\frac{\Sigma z_x z_y}{N}$

9

PREDICTION AND THE INTERPRETATION OF CORRELATION

9.1 INTRODUCTION

This chapter is concerned with the problem of predicting one variable from a knowledge of another, and with the interpretation of the correlation coefficient. A variety of factors that must be kept in mind in the proper interpretation of a correlation coefficient are discussed.

Psychologists and educators are frequently concerned with problems of prediction. Educational psychologists are interested in predicting the scholastic performance of a child from a knowledge of intelligence-test scores. Industrial psychologists, in selecting an individual for a particular type of employment, make a prediction about the subsequent job performance of that individual from information available at the time of selection. Clinical psychologists may direct their attention to predicting the patient's receptivity to treatment from information obtained prior to treatment. In many areas of human endeavor predictions about the subsequent behavior of individuals are required. A somewhat elaborate statistical technology has evolved for dealing with the prediction problem. In this chapter we shall restrict attention to the simplest aspect of prediction, the prediction of one variable from a knowledge of another.

Prediction and correlation are closely related topics, and an understanding of one requires an understanding of the other. The presence of a zero correlation between two variables X and Y may usually be interpreted to mean that they bear no systematic relation to each other. A knowledge of X tells us nothing about Y, and a knowledge of Y tells us nothing about X. In predicting X from Y or Y from X, no prediction better than a random guess is possible. The presence of a nonzero correlation between X and Y

implies that if we know something about X we know something about Y, and vice versa. If knowing X implies some knowledge of Y, a prediction of Y from X is possible which is better than a random guess about Y made in the absence of a knowledge of X. The greater the absolute value of the correlation between X and Y, the more accurate the prediction of one variable from the other. If the correlation between X and Y is either -1 or $+1$, perfect prediction is possible.

9.2 THE EQUATION FOR A STRAIGHT LINE

In this chapter use is made of the equation for a straight line. The general equation of any straight line is given by

[9.1] $$Y = bX + a$$

The quantity a is a constant. It is the distance on the Y axis from the origin to the point where the line cuts the Y axis. It is the value of Y corresponding to $X = 0$. If we substitute $X = 0$ in the equation for a straight line, we observe that $Y = a$. The quantity b is the slope of the line. The slope of any line is simply the ratio of the distance in a vertical direction to the distance in a horizontal direction, as illustrated in Figure 9.1. The slope describes the rate of increase in Y with increase in X. If a and b are known, the location of the line is uniquely fixed, and for any given value of X we can compute a corresponding value of Y.

9.3 THE LINEAR REGRESSION OF Y ON X

Any set of paired observations may be plotted on graph paper, each pair of observations being represented by a point. Consider the data shown in Table 9.1, columns 2 and 3. These columns contain IQs and reading-test scores for a group of 18 school children. These data are plotted in graphic form in Figure 9.2. Although the arrangement of points when plotted graphically shows considerable irregularity, we observe a tendency for reading-test scores to increase as IQs increase.

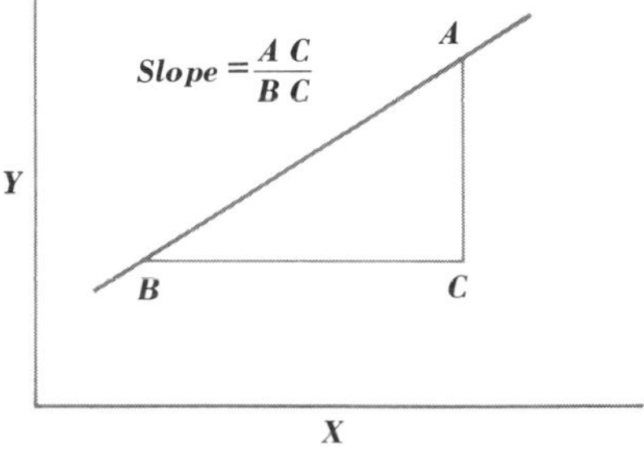

Figure 9.1 The slope of a line.

Table 9.1

Calculations for regression line of Y on X for ungrouped data

1 Pupil no.	2 IQ X	3 Reading score Y	4 X^2	5 XY	6 Expected reading score Y'
1	118	66	13,924	7,788	68
2	99	50	9,801	4,950	55
3	118	73	13,924	8,614	68
4	121	69	14,641	8,349	70
5	123	72	15,129	8,856	71
6	98	54	9,604	5,292	54
7	131	74	17,161	9,694	77
8	121	70	14,641	8,470	70
9	108	65	11,664	7,020	61
10	111	62	12,321	6,882	63
11	118	65	13,924	7,670	68
12	112	63	12,544	7,056	64
13	113	67	12,769	7,571	65
14	111	59	12,321	6,549	63
15	106	60	11,236	6,360	60
16	102	59	10,404	6,018	57
17	113	70	12,769	7,910	65
18	101	57	10,201	5,757	57
Sum	2,024	1,155	228,978	130,806	

SOURCE: R. W. B. Jackson and George A. Ferguson, "Manual of Educational Statistics," University of Toronto, Department of Educational Research, Toronto, 1942.

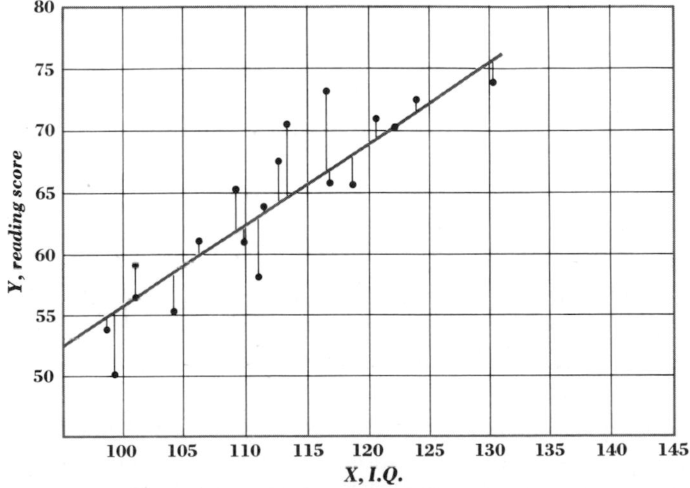

Figure 9.2 Scatter diagram for data of Table 9.1.

Let us suppose that we are given a child's IQ only and are required to predict his or her reading-test scores. How shall we proceed? Clearly, the data show considerable irregularity. An exact correspondence between the two sets of scores does not exist. In this situation we may proceed by fitting a straight line to the data. This straight line provides an average statement about the change in one variable with change in the other. It describes the trend in the data and is based on all the observations. If, then, we are given a child's IQ and are required to predict his or her reading-test score, we use the properties of the line. The method used in fitting a line to a set of points in a situation of this kind is the *method of least squares*. If our interest resides in predicting Y from X, the method of least squares locates the line in a position such that the sum of squares of distances from the points to the line taken parallel to the Y axis is a minimum. This line is known as the regression line of Y on X.

If values of X are given for which we do not have the corresponding values of Y, this regression line may be used to obtain estimates, or predictions, of Y corresponding to those values of X. A useful notational distinction may be made between an observed value Y and an estimated, or predicted, value which is denoted by Y'. In Figure 9.2 each value of X has a corresponding value Y, and also a value Y' corresponding to a point on the regression line. A departure of any point from the line, and parallel to the Y axis, is simply the difference $Y - Y'$. The method of least squares locates the regression line in such a position that the sum of squares of departures from the line parallel to the Y axis is a minimum; that is, the line is located in a position such that the quantity $\Sigma(Y - Y')^2$ is a minimum.

In this chapter the slope of the regression line for predicting Y from X will be denoted by b_{yx} and the point where the line cuts the Y axis by a_{yx}. The equation for such a regression line is then

[9.2] $$Y' = b_{yx}X + a_{yx}$$

The values of b_{yx} and a_{yx} may be calculated as follows:

[9.3] $$b_{yx} = \frac{N\,\Sigma XY - \Sigma X\,\Sigma Y}{N\,\Sigma X^2 - (\Sigma X)^2} = \frac{\Sigma XY - N\bar{X}\bar{Y}}{\Sigma X^2 - N\bar{X}^2}$$

[9.4] $$a_{yx} = \frac{\Sigma Y - b_{yx}\,\Sigma X}{N} = \bar{Y} - b_{yx}\bar{X}$$

where ΣX and ΣY = sums of X and Y, respectively
ΣXY = sum of the products of X and Y
ΣX^2 = sum of the squares of X
$\bar{X}$ and $\bar{Y}$ = means of X and Y, respectively

To illustrate, consider the data of Table 9.1. Columns 2 and 3 provide IQs and reading scores for the 18 schoolchildren. Column 4 provides the values X^2 and column 5 the products XY. Summing the columns, we get

$$\Sigma XY = 130{,}806$$
$$\Sigma X = 2{,}024$$
$$\Sigma Y = 1{,}155$$
$$\Sigma X^2 = 228{,}978$$
$$N = 18$$

Applying the left-hand form of formulas [9.3] and [9.4], we have

$$b_{yx} = \frac{18 \times 130{,}806 - 2{,}024 \times 1{,}155}{18 \times 228{,}978 - 2{,}024 \times 2{,}024} = .6708$$

$$a_{yx} = \frac{1{,}155 - .6708 \times 2{,}204}{18} = -11.25$$

The regression line for predicting Y from X is then described by the equation $Y' = .6708X - 11.25$. By substituting any value of X in this formula, we obtain Y', the estimated value of Y. Column 6 of Table 9.1 shows the estimated predicted reading-test scores obtained by applying this regression equation.

9.4 THE LINEAR REGRESSION OF X ON Y

Above we have considered the regression of Y on X. This regression line has been located in order to minimize the sum of squares of the distances from the points to the line parallel to the Y axis. The problem was to estimate, or predict, with minimum error, reading-test scores from IQs. If, however, we wish to estimate, or predict, IQs from reading-test scores, a different regression line is used. This is the regression line of X on Y. This line is located in a position such as to minimize the sum of squares of the distances from the point to the line parallel to the X axis. If X is an observed value, and X' is a value estimated or predicted from Y, this line is so located as to make the quantity $\Sigma(X - X')^2$ a minimum.

The formula for the regression line of X on Y is given by

[9.5] $$X' = b_{xy}Y + a_{xy}$$

where X' = estimated, or predicted, value of X
b_{xy} = slope of the regression line
a_{xy} = point where the line intercepts the X axis

The values of b_{xy} and a_{xy} may be calculated from the formulas

[9.6] $$b_{xy} = \frac{N\,\Sigma XY - \Sigma X\,\Sigma Y}{N\,\Sigma Y^2 - (\Sigma Y)^2} = \frac{\Sigma XY - N\bar{X}\bar{Y}}{\Sigma Y^2 - N\bar{Y}^2}$$

[9.7] $$a_{xy} = \frac{\Sigma X - b_{xy}\,\Sigma Y}{N} = \bar{X} - b_{xy}\bar{Y}$$

For the data of Table 9.1 the quantity $\Sigma Y^2 = 74{,}855$. The values

ΣXY, ΣX, and ΣY are as previously given in Section 9.3. Applying formulas [9.6] and [9.7], we have

$$b_{xy} = \frac{18 \times 130{,}806 - 2{,}204 \times 1{,}155}{18 \times 74{,}855 - (1{,}155)^2} = 1.207$$

$$a_{xy} = \frac{2{,}024 - 1.207 \times 1{,}155}{18} = 34.98$$

The regression line for predicting X from a knowledge of Y is then given by $X' = 1.207Y + 34.98$.

To summarize, for any scatter diagram two regression lines exist. The first line is used for predicting Y from a knowledge of X. The second line is used for predicting X from a knowledge of Y. The first line is located in order to minimize the sum of squares of deviations from the line parallel to the Y axis, that is, $\Sigma(Y - Y')^2$. The second line is located in order to minimize the sum of squares of deviations from the line parallel to the X axis, that is $\Sigma(X - X')^2$. If a perfect correlation exists between X and Y, these two lines coincide. If not it can be shown that $r = \sqrt{b_{yx}\, b_{xy}}$

9.5 REGRESSION LINES IN DEVIATION-SCORE FORM

Above we have considered the regression of Y on X and X on Y in raw-score form. Instead of considering raw scores we may consider regression lines for deviation scores, that is, for scores in the form $x = X - \bar{X}$ and $y = Y - \bar{Y}$. The slopes of the two regression lines expressed in deviation-score form are written very simply as

[9.8]
$$b_{yx} = \frac{\Sigma xy}{\Sigma x^2}$$
$$b_{xy} = \frac{\Sigma xy}{\Sigma y^2}$$

The slopes of the lines in the deviation-score model are, of course, exactly the same as in the raw-score model. The equations in [9.8] are the deviation-score forms of equations [9.3] and [9.6], respectively. The locations of the reference axes have, however, changed. The point of interception of the two regression lines with these reference axes is 0. The two lines pass through the origin; $a_{yx} = a_{xy} = 0$.

9.6 RELATION OF REGRESSION TO CORRELATION

If all points in a scatter diagram fall exactly along a straight line, the two regression lines coincide. Perfect prediction is possible. The correlation coefficient in this case is either -1 or $+1$. Where the correlation departs from either -1 or $+1$, the two regression lines have an angular separation. In general, as the degree of relationship between two variables decreases, the angular separation between the two regression lines increases. Where no systematic relationship exists at all, the two variables being independent, the two regression lines are at right angles to each other.

A simple relation exists between the correlation coefficient and the slopes of the two regression lines. As shown in equation [9.8], $b_{yx} = \Sigma xy/\Sigma x^2$ and $b_{xy} = \Sigma xy/\Sigma y^2$. Note, however, that $\Sigma xy = (N-1)rs_xs_y$, $\Sigma x^2 = (N-1)s_x^2$, and $\Sigma y^2 = (N-1)s_y^2$. By simple substitution we get

[9.9]
$$b_{yx} = r\frac{s_y}{s_x}$$
$$b_{xy} = r\frac{s_x}{s_y}$$

Consider a situation where the scores are in standard-score form, and $z_x = (X-\bar{X})/s_x$ and $z_y = (Y-\bar{Y})/s_y$. The standard deviation of standard scores is 1; that is, $s_{z_x} = s_{z_y} = 1$. Thus in the standard-score model the slope of each regression line is equal to the correlation coefficient. If pairs of standard scores are plotted graphically, and two regression lines are fitted to the data, the equation for these lines is

[9.10]
$$z'_y = rz_x$$
$$z'_x = rz_y$$

where z'_y and z'_x are the estimated or predicted standard scores. The regression model in standard-score form is shown in Figure 9.3.

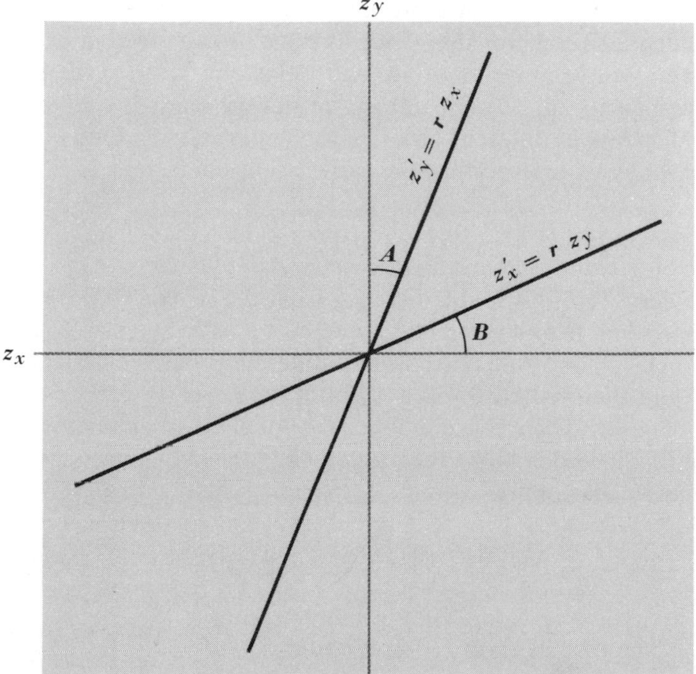

Figure 9.3 Regression lines in standard-score form. Angle A equals angle B.

In Figure 9.3 the slope of the line $z'_y = rz_x$ relative to the z_x axis is the same as the slope of the line $z'_x = rz_y$ relative to the z_y axis. For perfect correlation the lines coincide at a 45° angle to the z_x and z_y axes. The slope is obviously unity in this case. For a zero correlation the two regression lines have a 90° angular separation. One regression line will coincide exactly with the z_x axis and the other with the z_y axis.

9.7 INTERPRETING THE CORRELATION COEFFICIENT

Hitherto we have considered two related questions. First, how do we describe the magnitude of the relation between two variables? Second, how do we predict one variable from a knowledge of another? The questions now arise as to how correlation coefficients may be interpreted, and what conditions or assumptions underly their appropriate use. Correlation coefficients may be interpreted in a number of ways, depending on the objectives the investigator has in mind. Also a variety of circumstances influence the magnitude of the correlation coefficient. Some of these are discussed in this chapter, and others elsewhere in this book.

9.8 THE VARIANCE INTERPRETATION OF THE CORRELATION COEFFICIENT

A correlation coefficient is not a proportion. A coefficient of .60 does not represent a degree of relationship twice as great as a coefficient of .30. The difference between coefficients of .40 and .50 is not equal to the difference between coefficients of .50 and .60. The question arises as to how correlation coefficients of different sizes may be interpreted. One of the more informative ways of interpreting the correlation coefficient is in terms of variance.

In discussing the prediction of Y from a knowledge of X, it was noted that a score on Y could be viewed as comprised of two parts. One part is the predicted value Y'. This is the distance from the X axis to the regression line, corresponding to any particular value of X. All values of Y' lie on the regression line. The other part is the difference between the observed value of Y and the predicted value Y', that is, $Y - Y'$. Hence $Y = Y' + (Y - Y')$. These two parts are independent of each other; that is, they are uncorrelated. Since this is so, the variances of these two parts are directly additive, and we may write

[9.11] $$s_y^2 = s_{y'}^2 + s_{y \cdot x}^2$$

where s_y^2 = variance of Y
$s_{y'}^2$ = variance of values of Y predicted from X
$s_{y \cdot x}^2$ = variance of errors of prediction

The above equation illustrates clearly how a variance can be partitioned into additive parts, an event of very common occurrence in statistical work. The variance $s_{y'}^2$ is that part of the variance of Y that can be accounted for, predicted from, explained by, or attributed to the variance of X. It is a measure of the amount of information we have about Y from our knowledge of X. The variance $s_{y \cdot x}^2$ is the variance of the errors of prediction. It is that part of the variance of Y that cannot be attributed to the variance of X, but must be attributed to other influences. Because the variances are directly additive, if $s_y^2 = 400$, $s_{y'}^2 = 300$, and $s_{y \cdot x}^2 = 100$, it is appropriate to assert that the proportion $s_{y'}^2/s_y^2 = 300/400 = .75$, or 75 percent, of the variance of Y is predictable from X, and the proportion $s_{y \cdot x}^2/s_y^2 = 100/400 = .25$, or 25 percent, comprises errors of prediction.

The observations above lead to a simple interpretation of the correlation coefficient. It can be shown that the correlation coefficient squared, r^2, is the proportion of the variance of Y that can be predicted from X, that is

$$r^2 = \frac{s_{y'}^2}{s_y^2} \tag{9.12}$$

Thus r^2 is the ratio of two variances, and may, therefore, be viewed as a simple proportion. If $r = .80$, then $r^2 = .64$. We can state that 64 percent of the variance of the one variable is predictable from the variance of the other variable. In effect we know 64 percent of what we would have to know to make a perfect prediction of the one variable from the other. Thus r^2 can quite meaningfully be interpreted as a proportion, and $r^2 \times 100$ as a percent. In general, in attempting to conceptualize the degree of relationship represented by a correlation coefficient, it is more meaningful to think in terms of the square of the correlation coefficient instead of the correlation coefficient itself. The values of $r^2 \times 100$ for values of r from .10 to 1.00 are as follows:

r	$r^2 \times 100$
.10	1
.20	4
.30	9
.40	16
.50	25
.60	36
.70	49
.80	64
.90	81
1.00	100

Thus a correlation of .10 represents a 1 percent association, and a correlation of .50 represents a 25 percent association. A correlation of .7071 is required before we can state that 50 percent of the variance of one variable is predictable from the variance of the other. With a correlation as high as .90, the unexplained variance is 19 percent. Note also that equal differences between correlation coefficients may represent markedly different amounts of change in the degree of association. For example, the difference between correlation coefficients of .20 and .30 represents a 5 percent change in association, whereas the difference between coefficients of .80 and .90 represents a 17 percent change in association.

The variance interpretation of the correlation has been illustrated by discussing the prediction of Y from X. All discussion regarding the prediction of Y from X applies also to the prediction of X from Y. In predicting X from Y, $r^2 = s^2_{x'}/s^2_x$.

9.9 ACCURACY OF PREDICTION

If the investigator's concern is the prediction of one variable from a knowledge of another, he may choose to consider the relation between the correlation coefficient and the magnitude of errors of prediction. When Y is predicted from X, the variance of $(Y - Y')$, the difference between the actual and the predicted values, is denoted by $s^2_{y \cdot x}$. This variance, which is a measure of the accuracy of prediction, can be shown to be as follows:

[9.13] $$s^2_{y \cdot x} = \frac{\Sigma(Y - Y')^2}{N - 1} = s^2_y(1 - r^2)$$

The square root of this quantity is known as the *standard error of estimate* and may be written as

[9.14] $$s_{y \cdot x} = s_y \sqrt{1 - r^2}$$

The standard error of estimate varies from 0 to s_y. When $r = 1$, and all points fall exactly along a straight regression line, $s_{y \cdot x} = 0$; when $r = 0$ then $s_{y \cdot x} = s_y$. The standard error of estimate describes the degree of accuracy associated with predicting one variable from another. In a scatter diagram it predicts how closely the points cluster about the regression line.

Consider for illustrative purposes a variable Y with a standard deviation $s_y = 15$. For different values of r the corresponding values of $s_{y \cdot x}$ are as follows:

r	$\sqrt{1 - r^2}$	$s_{y \cdot x}$
.00	1.000	15.00
.10	.995	14.92
.20	.980	14.70
.30	.954	14.31
.40	.917	13.75
.50	.866	12.99
.60	.800	12.00
.70	.714	10.71
.80	.600	9.00
.90	.436	6.54
1.00	.000	.00

The above table shows that the errors of prediction, as described by $s_{y \cdot x}$, the standard error of estimate, must be viewed as quite substantial even for fairly large values of r. If the errors of prediction are assumed to be normally distributed with a standard deviation $s_{y \cdot x}$, interpretative statements can be made regarding the magnitude of error. The reader will recall that in the normal curve, 68 percent of the area falls between one standard deviation unit above and below the mean, and about 32 percent outside these limits. When $r = .00$, for example, $s_{y \cdot x} = 15$. Thus when the two variables are independent or uncorrelated, 68 percent of the errors of prediction will be less than 15 points of absolute magnitude, whereas about 32 percent will be greater than 15 points. When $r = .60$, 68 percent of errors will be less than 12 points whereas 32 percent will be greater than 12 points. When $r = .80$ the corresponding figure is 9 points of score. Considerations of this type illustrate clearly that even for fairly large values of the correlation coefficient the errors of prediction are substantial, and reduction in the size of the errors proceeds rather slowly with increase in the size of the correlation coefficient. This suggests that predicting one variable from a knowledge of another requires that considerable caution be attached to the predictions made. To illustrate, the correlation between the intelligence of parents and the intelligence of their offspring is commonly reported to be in the neighborhood of .50. Such data are used to support arguments for the role of genetic factors in intelligence. If we are prepared to argue that the correlation of .50 is about right and that genetic factors are possibly largely responsible for this, nonetheless, the variation that can be attributed to other than genetic factors, presumably environmental, is, indeed, very great. The standard deviation of many intelligence tests is about 15 points of I.Q. From a prediction point of view, if the parent-child correlation were 0 the standard error of estimate would be, of course, 15. If the correlation is about .50, as has frequently been reported, the standard error of estimate is about 13. Thus an $r = .50$ does not lead to a substantial reduction in the errors of prediction.

Note that the standard error of estimate makes no distinction between a positive and a negative relationship. From a prediction viewpoint a correlation of $-.70$, for example, provides the same accuracy of prediction as a correlation of $+.70$. Observe also that when standard scores are used, the standard error of estimate is simply $\sqrt{1-r^2}$. This quantity is sometimes called the *coefficient of alienation.* It is the obverse of the correlation coefficient. The correlation coefficient describes the degree of association between two variables. The quantity $\sqrt{1-r^2}$ describes the degree of lack of association, or alienation, between two variables.

The definition of $s^2_{y \cdot x}$, as shown in equation [9.13], is not an unbiased estimate of the population variance $\sigma^2_{y \cdot x}$. The number of degrees of freedom associated with the sum of squares $\Sigma(Y - Y')^2$ is not $N - 1$ but $N - 2$, there being $N - 2$ values of $(Y - Y')$ that are free to vary about the regression line. Here we have arbitrarily defined the variance of errors of estimate using $N - 1$, not $N - 2$, to simplify the explanation. For purposes of descriptive statistics it is algebraically more convenient to use $N - 1$ and not

$N - 2$. If, however, we wished to estimate $\sigma^2_{y \cdot x}$ as accurately as possible, the sum of squares should be divided by $N - 2$ rather than $N - 1$.

In the preceding paragraphs, we have discussed errors in predicting Y from X. The arguments involved in predicting X from Y are directly parallel.

9.10 THE GEOMETRY OF CORRELATION

For some purposes advantage attaches to the representation of correlation coefficients in geometrical terms. Such representation is commonly used in conceptualizing problems and their solutions where more than two, possibly many, variables are involved, as in the branch of statistics known as multivariate statistics.

The reader will recall that when two variables are expressed in standard-score form, a simple relation exists between correlation and regression. The slope of the regression line, related to its axis of reference, is the correlation coefficient, as illustrated in Figure 9.3. In the standard-score model a simple relation exists between the correlation coefficient and the angular separation between the two regression lines. The correlation coefficient is simply the cosine of the angle between the two regression lines. In a right-angle triangle the cosine of an angle is the length of the base line divided by the hypotenuse. When $r = 0$ the regression lines are at a 90° angle to each other. The cosine of a 90° angle is 0. When $r = 1$ the regression lines coincide. Their angular separation is 0°, and the cosine of this angle is 1.

Although the situation is more complex than suggested in the preceding discussion, the fundamental idea is that the correlation between two variables can be represented by the angular separation between two lines called *vectors,* a vector being simply a line with direction and length. Figure 9.4 shows the geometrical representation of correlation coefficients of different magnitudes. A correlation of 0 is represented by two vectors at right angles to each other, a correlation of .707 by two vectors at a 45° angle to each other, and a correlation of –.50 by two vectors with an angular separation of 120°. Note that a vector has both direction and length. In the preceding

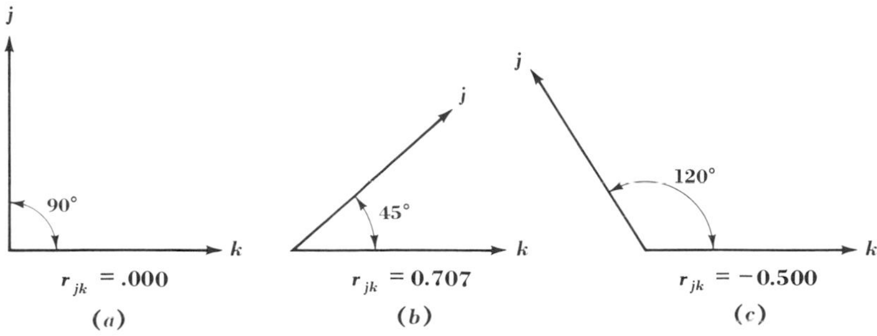

Figure 9.4 Geometrical representation of correlation coefficients.

discussion the simplifying assumption has been made that the vectors are of unit length. In some situations where this type of geometrical representation is used, the length of the vector may have a precise meaning and may be less than 1.

The following table shows the correlation coefficients, the cosines, corresponding to certain angular separations.

angle	*r*
90°	.000
80°	.174
70°	.342
60°	.500
50°	.642
40°	.766
30°	.866
20°	.940
10°	.985
0°	1.000

The geometrical representation of correlation coefficients may be readily extended to more than two variables. The following tables show correlations between three variables.

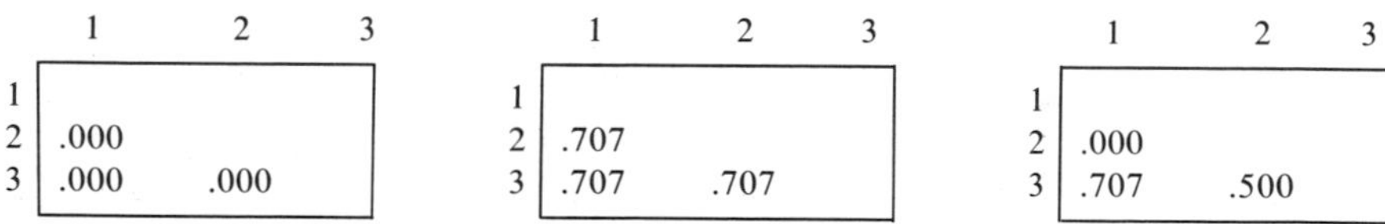

	1	2	3
1			
2	.000		
3	.000	.000	

	1	2	3
1			
2	.707		
3	.707	.707	

	1	2	3
1			
2	.000		
3	.707	.500	

The geometrical representations of these sets of correlations, assuming the vectors to be of unit length, are shown in Figure 9.5, which must be conceptualized in a space of three dimensions.

In Figure 9.5*a* all correlations are 0 and the three vectors are at right

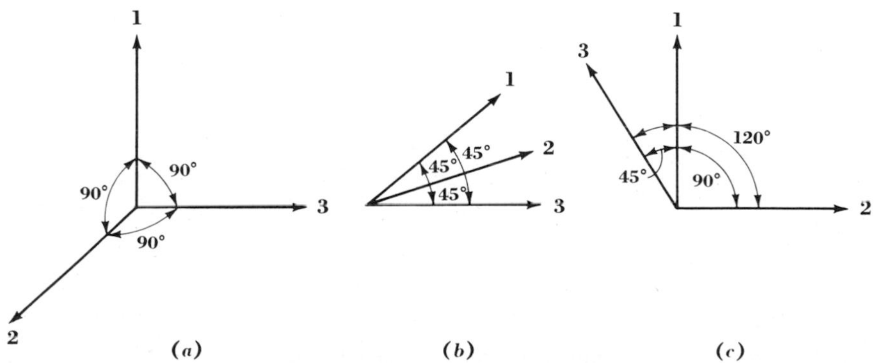

Figure 9.5 Geometrical representation of three tables of correlation coefficients.

angles to each other. In Figure 9.5*b* the three correlations are .707 and the three vectors are at a 45° angle to each other. In Figure 9.5*c* the three vectors have different angular separations. This type of vector model represents the geometrical image of the correlations between any set of variables and is useful in a variety of situations.

9.11 THE ASSUMPTION OF LINEARITY OF REGRESSION

In interpreting the correlation coefficient when r is neither 1 or 0, one should assume that the paired observations, when plotted as points on a scatter diagram, arrange themselves more or less in the shape of an ellipse, and that the fitting of two straight regression lines does not distort or conceal the functional relation between the two variables. Other types of arrangements of points that may occur are illustrated in Figure 9.6.

If the relation is nonlinear, low and possibly zero correlations may be obtained, and yet an orderly relation may exist between the two variables. Figure 9.6 shows a curvilinear relation between X and Y in the shape of an inverted U. If X is known, a fairly accurate prediction can be made of Y. If, however, two straight regression lines are fitted to the data, these lines will be about at right angles to each other, and r will be about 0. If a strictly random relation exists between X and Y, the correlation will be 0. The above example shows that the converse does not hold; that is, if the correlation is 0, it does not necessarily follow that X and Y bear a random relation to each other. This may mean that the linear regression model is a poor fit to the data, since it does not appropriately describe the nature of the relation among the data. In interpreting the correlation coefficient, one should ordinarily assume that the linear regression model is a good fit to the data and that a correlation of about 0 means a random relation. Consider a situation where $r = .80$. This means that 64 percent of the variance of the

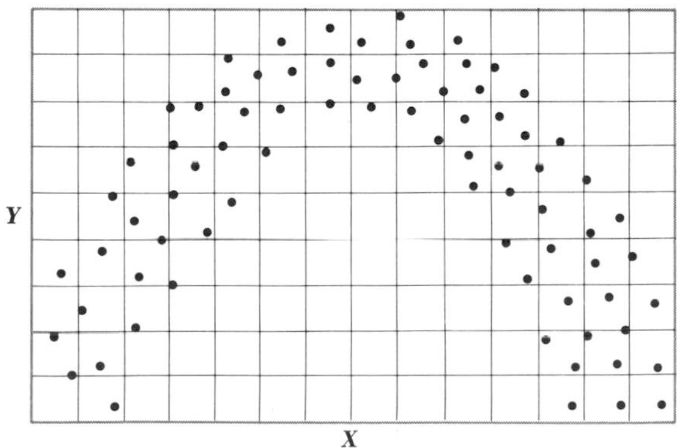

Figure 9.6 Scatter diagram showing curvilinear relation.

one variable is predictable from the other, and the residual 36 percent is due to other factors. The assumption is that these other factors do *not* result, at least to an appreciable extent, from a lack of goodness of fit of the linear regression lines to the data. If a large proportion of the residual 36 percent of the variance did result from nonlinearity, this would clearly affect the interpretation of the data. In interpreting a correlation coefficient, investigators should satisfy themselves that the linear regression lines are a good fit to the data. Any gross departure from the linear model can readily be detected by inspection of a scatter diagram or bivariate frequency table. For small values of N, nonlinear relations may be difficult to detect. In practice, for many of the variables used in psychology and education the assumption of linearity of regression is reasonably well satisfied.

9.12 UNIVARIATE SELECTION

The size of a correlation coefficient depends to some extent on the homogeneity or heterogeneity of the sample on which the correlation is calculated. Consider two variables, X and Y, with variances s_x^2 and s_y^2, and a nonzero correlation r in a random sample of a population. Consider now a second sample selected in such a way that its variance is less than s_x^2. Thus the variability of the sample is restricted or curtailed with respect to X. This restriction in variability on X will reduce the variance of Y and also the correlation between X and Y. The effect of univariate selection, sometimes referred to as restriction of range, on the correlation coefficient was first investigated by Karl Pearson and reported in 1902 in a paper called "On the Influence of Natural Selection on the Variability and Correlation of Organs." A discussion of this effect is given in McNemar (1969). A more detailed and thorough treatment of the topic will be found in Thomson (1950).

Formulas exist that enable the estimation of the correlation between two variables in an unselected sample from a knowledge of the correlation in the selected, or restricted, sample. The converse estimation also is possible. These formulas are not given here. In practice they are rarely used. These formulas, however, show clearly how selection with respect to one variable affects the correlation coefficient.

Some illustrative data may prove helpful. Consider two tests, a test of intelligence and a test of achievement in arithmetic. Assume that in an unselected sample the scores on the test of intelligence have a standard deviation of 16, and the correlation between the two tests is .80. Consider now a number of groups, selected on the basis of intelligence such that the standard deviations of intelligence test scores for these groups are 16, 14, 12, 10, 8, 6, 4, and 2. How will the correlations between the two tests for these groups be affected by the decreasing variability? The correlations can be estimated, and are as follows:

s_x	r
16	.800
14	.759
12	.707
10	.640
8	.555
6	.478
4	.318
2	.167

Clearly a high degree of selection, or restriction in range, sharply reduces the size of the correlation coefficient.

All users of the correlation coefficient should be aware of the effect discussed here. Data are frequently reported showing correlations between scores on scholastic aptitude or intelligence tests and performances on various subjects in college. Not uncommonly such correlations are found to be low. One reason for this is that students who gain admission to college are a highly selected group, selected on the basis of scholastic aptitude and possibly intelligence test scores.

This writer has recently encountered data where a group of psychological tests was administered to a number of different age groups. The oldest group consisted of quite elderly individuals. The correlation coefficients showed a systematic increase with age. The standard deviation of the tests increased also. If the increase in correlation is a real effect, and our abilities become more highly correlated as the years advance, this is a finding of considerable theoretical interest. If the increase in correlation can be attributed to the increase in variability, then the result may be a statistical artifact, subject to some other interpretation.

9.13 OTHER FACTORS AFFECTING THE CORRELATION COEFFICIENT

In interpreting or using the correlation coefficient, one need not assume that the distributions of the two variables are normal, although much mathematical work on correlation is based on the normality assumption. Correlations can be computed for rectangular and other types of distributions. Frequently correlations are calculated between variables with distributions of different shapes. This circumstance imposes constraints upon the correlation coefficient. If a positively skewed distribution is correlated with a negatively skewed distribution, the differences in the shapes of the distributions will affect the correlation coefficient. Some part of the departure of the correlation coefficient from unity will result because of the different shapes of the two distributions. In such a situation as this, the differences in the shapes of the distributions will in effect ensure that one or other or both of the regression lines are nonlinear. A rather extreme instance of this effect occurs when one variable is a dichotomy, that is, taking two values only, and the other takes many values and may in theory be continuous. An example of a two-valued variable may be pass-fail grades in college, and of a many-valued variable, scores on a scholastic aptitude test. With such data, correlations are frequently calculated by arbitrarily assigning numbers to the two categories of the two-valued variable, such as 1 for a pass and 0 for a failure. If this is done, and a product-moment correlation is calculated, the marked difference in the shapes of the two distributions will place restrictions on the range of possible values that the correlation may assume. A correlation of $+1$ or -1 would occur only under the most unusual and improbable circumstances. In psychological research, substantial differences in the shapes of the distributions under study occasionally are found. Under these circumstances it is a quite common practice to transform the variables to an approximately normal form.

Many other circumstances affect the correlation coefficient. Among these may be mentioned sampling error and errors of measurement. The effects of these on the correlation coefficient are discussed in later chapters.

9.14 CORRELATION AND CAUSALITY

The existence of a correlation does not necessarily imply a direct causal relation. By a direct causal relation is meant that if X and Y are correlated then X is at least in part the cause of Y, or Y at least in part the cause of X. If a direct causal relation exists between two variables, these two variables will be correlated, assuming of course that the linear regression model fits the data, or some form of nonlinear correlation is used, if appropriate. Examples of correlations that may be interpreted as implying direct causal connections are correlation between rainfall and crop yield, or food intake and weight in experimental animals. Of course the processes that relate these pairs of variables may be highly complex. Nonetheless, for practical purposes a direct causal connection may be assumed.

In psychology and education the presence of a correlation between two variables can rarely be interpreted as implying a direct causal relation. In many situations two variables are correlated, because both are correlated with an underlying variable or set of variables. Very simply X and Y may be correlated because both bear a direct causal relation to an underlying variable Z. For example, given a group of children with a substantial range of ages, a correlation may be found between a measure of intelligence and a measure of motor performance. Such a correlation may come about because measures of intelligence and motor ability are both correlated with age. If the effects of age are removed, the correlation may vanish. Another example is the correlation between scholastic aptitude test scores and performance in college. These variables are correlated not because of any direct causal connection but because of the presence of underlying individual abilities which are directly causally related to performance on aptitude tests and achievement in academic subjects within the university.

At times correlations are found where no line of plausible reasoning would suggest that any correlation at all should exist between the variables. An example here is the presence in the postwar years in Canada of a correlation between the birth rate and the consumption of alcohol. Clearly it cannot be argued that the prevalence of babies drives strong men to drink. Such correlations frequently involve data obtained over time and may result from underlying changes in some set of complex social or economic variables. Usually no very informative and straightforward interpretation is possible.

9.15 TYPES OF CORRELATION

Many types of correlation have been devised for various purposes. Most of these are described in this book. Many of them are particular cases of product-moment correlation, which is basic to all correlational work. Recall the distinction between nominal, ordinal, and interval-ratio variables. Methods exist for describing the relation between two nominal variables; also between a nominal variable and an interval-ratio variable. Ordinal, or rank correlation, methods are of much interest. Also the reader will encounter correlational methods that are used in situations involving more than two, perhaps many, variables. Some of these methods use the product-moment correlation between the weighted sums of sets of variables.

BASIC TERMS AND CONCEPTS

Prediction

Regression

Equation for straight line

Slope

Regression of Y on X

Regression of X on Y

Slope in standard-score form

r^2 as a proportion

Variance of errors of estimation

Standard error of estimate

Coefficient of alienation

Vector

Geometrical representation of correlation

Nonlinearity of regression

Univariate selection

Restriction of range

Correlation and causality

EXERCISES

1 Consider a straight line passing through the points

X	1	3
Y	4	5

What is the equation for this line?

2 A regression equation for predicting first-year university averages from high school leaving averages may be written in the form $Y' = .80X - 4.60$. Obtain predicted university averages corresponding to high school leaving averages of **(a)** 70, **(b)** 85, **(c)** 65.

3 The following are paired measurements:

X	1	5	6	6	2
Y	2	4	5	3	1

Compute **(a)** the correlation between X and Y, **(b)** the slope of the regression line for predicting Y from X, **(c)** the regression equation for predicting Y from X.

4 For the data of Exercise 3 above, obtain **(a)** the predicted values of Y corresponding to the given values of X, **(b)** the variance of the errors of estimation in predicting Y from X.

5 For the data of Exercise 3 above, write the regression equation for predicting a deviation score on Y from a knowledge of a deviation score on X.

6 The following are marks on a college entrance examination X and first-year averages Y for a sample of 20 students.

X	Y	X	Y	X	Y	X	Y
55	61	70	75	63	85	77	84
79	72	80	61	64	87	62	72
59	69	89	79	69	70	85	70
81	89	92	90	75	90	55	60
62	52	60	55	84	67	66	67

Compute **(a)** the correlation between entrance examination marks and first-year averages, **(b)** the regression equation for predicting first-year averages from examination marks, (**c**) the predicted first-year averages for the 20 students, (**d**) the variance of the errors of estimation.

7 Standard scores on variable X for four individuals are -2.0, -1.68, .19, 1.16. The correlation between X and Y is .50. (**a**) What are the estimated standard scores on Y? (**b**) What is the standard error in estimating standard scores on Y from standard scores on X?

8 The variance of errors of estimation $s_{y \cdot x}^2 = 200$ and the variance of Y is $s_y^2 = 600$. (**a**) What proportion of the variance of Y is under the control of X? (**b**) What is the correlation between X and Y?

9 A variance $s_y^2 = 400$ and the correlation between X and Y is .50. (**a**) What is the variance of the errors of estimation in predicting Y from X? (**b**) What is the variance of the predicted values?

10 Show that $r_{xy} = \sqrt{b_{yx} b_{xy}}$.

11 How many times is a difference in predictive capacity between correlations of .70 and .80 greater than between correlations of .20 and .30?

12 What correlation between X and Y is required in order to assert that 75 percent of the variance of X depends on the variance of Y?

13 The variance of differences is given by $s_{x-y}^2 = s_x^2 + s_y^2 - 2r_{xy}s_x s_y$. Write a formula for calculating the correlation coefficient from this formula.

14 Use the formula obtained in Exercise 13 above to calculate a correlation for the following data:

X	1	2	3	4	5
Y	2	1	4	5	3

15 Consider the paired observations

X	1	2	3	4	5	6	7
Y	4	3	2	1	2	3	4

What is the correlation between X and Y? Is this an appropriate use of the correlation coefficient?

16 The following are paired observations

X	1	2	3	4	5	6
Y	2	1	3	4	11	9

Variable X is symmetrical, variable Y is skewed. Calculate the correlation between X and Y. What is the maximum and minimum value that the correlation of X and Y can take?

17 If residuals $(Y - Y')$ are normally distributed with a standard deviation of $s_{y \cdot x}$, what limits will include 95 and 99 percent of $(Y - Y')$?

18 In a vector model, the vectors being of unit length, what angular separation corresponds to $r = -1$, $r = .174$, $r = .500$, $r = .866$?

ANSWERS TO EXERCISES

1 $Y = .50X + 3.5$

2 **a** 51.40 **b** 63.40 **c** 47.40

3 **a** $r = .809$ **b** $b_{yx} = .545$ **c** $Y = .545X + .820$

4 **a**

X	Y'
1	1.364
5	3.545
6	4.091
6	4.091
2	1.909

b .864

5 $y = .545x$

6 **a** .468 **b** $Y' = .482X + 38.34$ **c** 72.75 **d** 103.33

7 **a** $-1.0, -.84, .09, .58$
b .866

8 **a** .667 **b** .816

9 **a** 300 **b** 100

10 $b_{yx} = r\dfrac{s_y}{s_x}$ $b_{xy} = r\dfrac{s_x}{s_y}$

By multiplying, we obtain $b_{yx}b_{xy} = r^2$. Hence $r^2 = b_{yx}b_{xy}$.

11 3

12 .866

13 $r_{xy} = \dfrac{s_x{}^2 + s_y{}^2 - s_{x-y}^2}{2s_xs_y}$

14 .600

15 $r = 0$. This is not an appropriate use. The relation is clearly nonlinear.

16 $r = .871$, $r\,(\text{max}) = .950$, $r\,(\text{min}) = -.950$

17 **a** $\pm 1.96s_{y\cdot x}$ **b** $\pm 2.58s_{y\cdot x}$

18 **a** .00° **b** 80° **c** 60° **d** 30°

10

SAMPLING AND ESTIMATION

10.1 INTRODUCTION

In Chapter 1 the concepts of population and sample were discussed. A *population* is any defined aggregate of objects, persons, or events, the variables used as the basis for classification or measurement being specified. A *sample* is any subaggregate drawn from the population. Any statistic calculated on a sample of observations is an *estimate* of a corresponding population value or *parameter*. The symbol $\bar{X}$ is used to refer to the arithmetic mean of X calculated on a sample of size N. The symbol μ is used to refer to the mean of the population. Similarly, s^2 is used to refer to the variance in the sample, and σ^2 is the corresponding population parameter. $\bar{X}$ is an estimate of μ, and s^2 is an estimate of σ^2. Likewise, any other statistic calculated on a sample is an estimate of a corresponding population parameter. In most situations the parameters are unknown and must be estimated in some manner from the sample data.

Much statistical work in practice is concerned with the use of sample statistics as estimates of population parameters, and more particularly with describing the magnitude of error which attaches to such statistics. The body of statistical method concerned with the making of statements about population parameters from sample statistics is called *sampling statistics,* and the logical process involved is called *statistical inference,* this being a rigorous form of inductive inference. If inferences about population parameters are to be drawn from sample statistics, certain conditions must attach to the methods of sampling used.

10.2 METHODS OF SAMPLING

In drawing inferences about the characteristics of populations from sample statistics, the assertion is frequently made that the sample should be drawn *at random* from the population. The sample is spoken of as a *random sample*. The word "random" is used in at least three ways. It may refer to our subjective experience that certain events are haphazard or completely lacking in order. It may be used in a theoretical sense to refer to an assumption about the equiprobability of events. Thus a random sample is one such that every member of the population has an equal probability of being included in it. When the word is used in this way the meaning is assigned to it within the framework of probability theory. The word "random" is also used in an operational sense to describe certain operations or methods. Thus the drawing of numbers from a hat after they have been thoroughly mixed, or the drawing of cards from a deck after they have been well shuffled, or certain techniques used in sweepstakes, lotteries, and other games of chance are examples of random operations or methods. Sampling theory in statistics is based on the theoretical use of the word "random," that is, on the idea of the equiprobability of each population member being included in the sample. This equiprobability assumption underlies the derivation of many formulas used in sampling statistics. Practical operational methods of sampling are frequently such as to ensure that the theoretical assumption of equiprobability is closely approximated in practice. If methods of sampling ensure approximate equiprobability, then clearly the deductive consequences of a theory based on the idea of equiprobability can be used in dealing with practical sampling problems. The correspondence between the consequences of theory and what occurs in practice can, of course, always be checked by experiment.

For a finite population, if the numbers are listed or cataloged, a random sampling procedure may be applied. Consider, for example, a population of 1,000 members from which a sample of 100 is required. All names, or some identifying code number may be entered on slips of paper. These may be placed in a container from which 100 slips are drawn. A more convenient procedure is to use a table of random numbers. Tables of random numbers are comprised of digits so chosen that no systematic relation exists between any sequence of digits in the table, regardless of whether the table is read up, down, left, or right, or in any other way. To draw a sample by using a table of random numbers, a method must be used for assigning a number to each member of the population. If the population comprises 1,000 members, or less, each member must be assigned an identifying number ranging from 000 to 999. The identifying number may in some cases correspond to, or be associated with, the order in which the names appear on a list of names. To choose a sample of 100 from a population of 1,000, three-digit numbers, presumably nonrepeating, are read from the table either up, down, left, or right until 100 members are identified. The table may be entered arbitrarily in any way. If the population consisted of 100 members or less, two-digit numbers ranging from 00 to 99 would be used. Tables of random numbers are found in many texts in statistics, or in Fisher and Yates, "Statistical Tables for Biological, Agricultural and Medical Research" (1963). Tables of random numbers are commonly used to achieve randomization in experiments. Given a situation where 60 experimental animals must be assigned to three treatments at random, each animal may be assigned an identifying number and tables of random numbers used to assign animals to treatments.

If a list is arranged alphabetically, or in some other systematic fashion, every *n*th name may be chosen in the construction of the sample. This is sometimes spoken of as *systematic sampling.* Although in most practical situations such a sample may be viewed as random, the possibility exists that the variable under investigation may not be independent of the basis of ordering, and a biased sample may result. On occasion, samples are drawn from lists which do not provide a complete record of all members of the population, but are viewed, perhaps erroneously, as representative of the population. A telephone directory is such a list. Names chosen from a telephone directory may yield a biased sample of the population at large, because ownership or nonownership of a telephone may not be independent of the variable under investigation, e.g., how a person intends to vote in an election.

Forms of modified random sampling are sometimes used. One example is *stratified random sampling.* This procedure requires prior knowledge, perhaps obtained from census data, about the number or proportion of members in the population of various strata. Thus we may know the number of males and females, the number in various age groups, and the like. In constructing the sample, members are drawn at random from the various strata. If the members are drawn such that the proportions in the various strata in the sample are the same as the proportions in those strata in the population, the sample is a *proportional stratified sample.* For example, a university may have 10,000 students, of which 7,000 are males and 3,000 are females. A sample of 100 students is required. We may draw 70 males by a random method from the subpopulation, or stratum, of males, and 30 females from the subpopulation, or stratum, of females. Such a sample is a proportional stratified sample.

In much experimentation using human or animal subjects, the populations from which the samples are drawn may not be amenable to precise definition, and the methods of sampling described above may have little relevance. For example, a laboratory experiment may use a sample of experimental animals. Can such a sample be viewed meaningfully as a random sample from a known population of animals? In a study of the therapeutic effects of different operative procedures applied to a certain class of brain tumor, all cases admitted to a particular hospital during a specified time period may be used. This number may be small. In what sense may this group of cases be viewed as a random sample drawn from a larger group or population? Again, in an educational experiment the pupils in two or three classes in a particular school and grade may be used as subjects. Can such a group of subjects be viewed as a random sample or its equivalent? Such samples as these are clearly not random. The method by which they are selected is not a random method. Despite this the investigators may wish to draw inferences that transcend the particular samples under study. They may wish to argue that their findings are probably true for some large, although perhaps ill-defined, population of experimental animals; or that the therapeutic effects of their operative procedures may be extended to all patients suffering from the particular class of brain tumor; or that the results of the educational experiment may be generalized to a much larger group of school pupils in the same grade.

In situations of the above type, where strict random sampling procedures have not been used, does any basis exist for valid inference? A common practice is to investigate a posteriori a variety of characteristics of the sample. It may be possible to show that the sample does not differ appreciably in these characteristics from a larger group or population. Thus in the educational experiment the sample of students may be studied with respect to age, sex, IQ, socioeconomic level of the parents, and other characteristics. The sample may not differ much in these respects from a larger group or population in the same grade. Because the sample shows no bias on a number of known characteristics, that is, it may not differ from a random sample as far as these characteristics are concerned, the investigator may be prepared to regard it as representative of the larger group or population and treat it as if it were a random sample. Frequently precise knowledge is lacking about a larger reference group or population, and the investigator must rely on accumulated past experience and intuition in the attempt to detect possible bias. Clearly, where possible, random sampling is to be preferred to methods such as these. It must be recognized, however, that were we to insist on rigorous random sampling methods, much experimentation would not be possible.

In practice, experiments are clearly not always conducted in the way our statistical preconceptions suggest they should be conducted. Experienced experimentalists are frequently aware of this. Much of the art of the experimentalist is concerned with reaching conclusions from data which do not satisfy some of the conditions necessary for rigorous inference.

10.3 SAMPLING ERRORS

Let us now consider the nature of the errors associated with particular sample values. What precisely is a sampling error? A sampling error is a difference between a population value, or parameter, and a particular sample value. Thus, if μ is the population value of the mean and $\bar{X}_i$ is an estimate based on a random sample of size N, then the difference $\mu - \bar{X}_i = e_i$, where e_i is a sampling error. Let us suppose that we know that the mean scholastic aptitude test score for a population of 5,000 university students is $\mu = 562$. A sample of 100 provides an estimated mean of $\bar{X}_i = 566$. The sampling error in this case is $\mu - \bar{X}_i = 562 - 566 = -4$. Ordinarily, μ is not known, and we are unable to specify e_i exactly for any particular sample. Despite this, meaningful statements can be made about the magnitude of error which attaches to $\bar{X}_i$ as an estimate of the parameter μ. The reader should note that the concept of error in any context always implies a parametric, true, fixed, or standard value from which a given observed value may depart in greater or less degree. The idea that something in the nature of a parametric or true value can meaningfully be defined is essential to the concept of error. Without some appropriate definition of such a value, the concept of error has no meaning, and no theory of error is possible. Also, no science is possible.

How may the magnitude of error be estimated and described? Common sense suggests that in the measurement of any quantity some appreciation of the magnitude of error may be obtained by repeating the measurements a number of times, presumably under constant conditions, and observing how these repeated measurements vary from each other. Thus in the measurement of the length of a bar of metal a series of separate measurements may be made under constant conditions. Let us suppose that five such measurements are 55.95, 56.23, 56.25, 56.41, and 56.54 inches. In this case each measurement is an estimate of the same "true" length; hence the variation observed with repeated measurement is due to error. Let us suppose that five additional measurements are made using another measuring operation or procedure, these measurements being 54.80, 55.31, 56.44, 56.52, 57.29 inches. This latter set shows greater variation with repetition than the former. We may conclude that the magnitude of error associated with this latter set of measurements is greater.

The above example is concerned with errors associated with particular observations, namely, measurements of a bar of metal. In considering the magnitude of error associated with, say, a sample mean $\bar{X}_i$, as an estimate of a population mean μ, the situation is similar. The problem may be approached experimentally by considering how values of $\bar{X}_i$ vary in repeated samples of size N. Thus the mean scores on the scholastic aptitude test for five different samples of 100 students, drawn from a population of 5,000, may be 552, 558, 562, 568, and 569. These five sample means may be viewed as estimates of the same population mean μ, that is, the value that would have been obtained were information available on all 5,000 members of the population. The variation of these five means one from another may be attributed to sampling error.

In general, a number, say, k, of samples of size N may be drawn at random from the same population and a mean calculated for each sample. These means may be represented by the symbols $\bar{X}_1, \bar{X}_2, \bar{X}_3, \ldots, \bar{X}_k$. We may write

$$\mu - \bar{X}_1 = e_1$$
$$\mu - \bar{X}_2 = e_2$$

$$\mu - \bar{X}_3 = e_3$$

$$\cdots\cdots\cdots$$

$$\mu - \bar{X}_k = e_k$$

The variance and the standard deviation are ordinarily used to describe the magnitude of variation in any set of observations or values. In describing the magnitude of variation in sample means with repeated sampling, the variance and standard deviation are also used. These statistics describe the magnitude of sampling error, that is, the magnitude of error associated with $\bar{X}_i$ as an estimate of μ. Note that the variance and standard deviation of sample means $\bar{X}_i$ are the same as the variance and standard deviation of the sampling error e_i, because μ is a constant.

10.4 SAMPLING DISTRIBUTIONS

In the above discussion the problem of estimating error has been approached *experimentally;* that is, we considered the actual drawing of a number of samples and approached the experimental study of error through observed sample-to-sample fluctuation. Consider for illustrative purposes a small finite population of eight members. Let the members of the population be cards numbered from 1 to 8. These cards may be shuffled, a sample of four cards drawn without replacement, and a mean calculated for the sample. This procedure may be repeated 100 times, and a frequency distribution made of the 100 sample means. This distribution is an *experimental sampling distribution,* and its standard deviation is a measure of the fluctuation in means from sample to sample.

A *theoretical,* as distinct from an experimental, approach may be used. Given a finite population of eight members, a limited number of different samples of four cards exists. The number of such samples is the number of combinations of 8 things taken 4 at a time, or $C_4^8 = 70$. Each of these 70 samples may be considered equiprobable. The means for the 70 possible samples may be ascertained and a frequency distribution prepared. This frequency distribution is a *theoretical sampling distribution.* It is obtained by direct reference to probability considerations. No drawing of actual samples is involved. The standard deviation of the theoretical sampling distribution is a measure of fluctuation in means from sample to sample.

In the above example the population is small and finite. In practice, most of the populations with which we deal are indefinitely large, or if finite, they are so large that for all practical purposes they can be considered indefinitely large. In the study of sampling error the approach used in dealing with an indefinitely large population is a simple extension of that used with a small finite population. The distinction between an experimental and theoretical sampling distribution still applies. When the population is indefinitely large, the theoretical sampling distribution of, for example, the mean is the frequency distribution of means of the indefinitely large number of samples of size N which theoretically could be drawn.

The theoretical sampling distributions are known for all commonly used statistics. The standard deviation of the sampling distribution is called the *standard error*. Thus a standard error is always a standard deviation which describes the variability of a statistic over repeated sampling. The standard deviation of a theoretical sampling distribution is, in effect, a population parameter. It is descriptive of the variation of a statistic in a complete population of sample values. The standard deviation of the theoretical sampling distribution of the mean is represented by the symbol $\sigma_{\bar{x}}$. In practice this standard deviation must be estimated from sample data. This estimate in the case of the mean may be represented by the symbol $s_{\bar{x}}$. For most statistics fairly simple formulas are available for estimating the standard deviation of the theoretical sampling distribution.

The theoretical sampling distributions of some statistics are normal, or approximately so; others are not. For example, the theoretical sampling distribution of the mean $\bar{X}$ is normally distributed in sampling from an indefinitely large normally distributed population. The sampling distribution of the correlation coefficient presents a complicated problem. It is not normally distributed except under certain special circumstances. When the shape of the sampling distribution is known, certain kinds of statements can be made about a population value from a sample estimate. For example, it is possible to fix limits above and below a sample value and assert with a known degree of confidence that the population parameter falls within those limits. The fixing of such limits requires a knowledge of the shape of the sampling distribution.

10.5 SAMPLING DISTRIBUTION OF MEANS FROM A FINITE POPULATION

In practice, most samples are viewed as drawn from indefinitely large populations. The essential ideas of sampling may, however, be conveniently illustrated with reference to a small finite population. Suppose, as mentioned above, that we have a population of eight cards numbered from 1 to 8. These cards may be shuffled, and a sample of four cards drawn at random. After each card is drawn it is not returned; that is, the sampling is without replacement. A mean $\bar{X}$ may be calculated for this sample. The four cards may now be returned, the eight cards shuffled, another sample of four cards drawn, and another mean calculated. Let us continue this procedure until 100 samples of four cards have been drawn and their means calculated. Table 10.1, column 3, shows the frequency distribution of 100 such sample means. This distribution is an experimental sampling distribution of means. It shows experimentally how the means of samples of four drawn at random without replacement from a population of eight vary from sample to sample. The mean of the experimental sampling distribution $\bar{X}_{\bar{x}}$, that is, the mean of the 100 means based on samples of four,

Table 10.1

Experimental and theoretical sampling distributions of means of samples of four drawn from a population of eight members

		Experimental distribution		Theoretical distribution	
ΣX	$\bar{X}$	f	p	f	p
1	2	3	4	5	6
10	2.50	1	.010	1	.014
11	2.75	2	.020	1	.014
12	3.00	0	.000	2	.029
13	3.25	5	.050	3	.043
14	3.50	7	.070	5	.071
15	3.75	7	.070	5	.071
16	4.00	8	.080	7	.100
17	4.25	11	.110	7	.100
18	4.50	13	.130	8	.114
19	4.75	10	.100	7	.100
20	5.00	10	.100	7	.100
21	5.25	9	.090	5	.071
22	5.50	7	.070	5	.071
23	5.75	4	.040	3	.043
24	6.00	3	.030	2	.029
25	6.25	1	.010	1	.014
26	6.50	2	.020	1	.014
Total		100	1.000	70	.998

is found to be 4.56. The mean of the population from which the samples have been drawn is the mean of the integers from 1 to 8 and is 4.50. The standard deviation of the 100 means $s_{\bar{x}}$ is found to be .834.

The investigation of the fluctuation in sample means may be approached theoretically. The number of different samples of four in sampling without replacement from a population of eight members is the number of combinations of 8 things taken 4 at a time, or $C_4^8 = 70$. These 70 samples may be considered equiprobable. A listing of the 70 samples may readily be made, and the means calculated. The sample with the smallest mean will be 1, 2, 3, 4; the mean $\bar{X} = 2.50$. The sample with the largest mean will be 5, 6, 7, 8; here $\bar{X} = 6.50$. Thus $\bar{X}$ will range from 2.50 to 6.50. Table 9.1, column 5, shows the frequency distribution of the 70 sample means. This distribution is a theoretical sampling distribution of the mean of samples of four from a small finite population of eight members. It is based on the idea that there are 70 possible combinations of 8 things taken 4 at a time, all combinations being equiprobable.

The mean of the theoretical sampling distribution may be calculated. This mean $\mu_{\bar{x}}$ is found to be 4.50. The standard deviation $\sigma_{\bar{x}}$ is found to be .866. These values do not differ markedly from the mean and standard deviation of the experimental sampling distribution, these being $\bar{X}_{\bar{x}} = 4.56$ and $s_{\bar{x}} = .834$. Presumably, had a larger number of samples been drawn, say, 200 or 1,000, the experimental sampling distribution would be observed to approximate more closely to the theoretical distribution.

The mean and standard deviation of the theoretical sampling distribution of Table 10.1 were calculated directly from the 70 possible sample means. These values may, however, be readily obtained without using this time-consuming method. It can be shown that the mean of the theoretical sampling distribution is equal to the population mean; that is, $\mu_{\bar{x}} = \mu$. In our example the mean of the sampling distribution of samples of four from a population of eight members is observed to be 4.50. Likewise, the population mean, that is, the mean of the integers from 1 to 8, is also 4.50. The standard deviation of the theoretical sampling distribution is given by the formula

[10.1] $$\sigma_{\bar{x}} = \frac{\sigma}{\sqrt{N}} \sqrt{\frac{N_p - N}{N_p - 1}}$$

where σ = standard deviation in population
N_p = number of members in population
N = sample size

In our example σ is the standard deviation of the integers from 1 to 8 and is equal to 2.29. Population and sample size are, respectively, 8 and 4. Hence

$$\sigma_{\bar{x}} = \frac{2.29}{\sqrt{4}} \sqrt{\frac{8-4}{8-1}} = .866$$

If, then, the standard deviation σ of the population is known, we can readily obtain from the above formula the standard deviation of the theoretical sampling distribution and use this as a measure of fluctuation in means from sample to sample.

A knowledge of the standard deviation of a theoretical sampling distribution is of limited usefulness unless additional information is available on the shape of the distribution. In certain instances sampling distributions are normal or approximately normal in form. The theoretical sampling distribution of Table 10.1 departs appreciably from the normal form. If, however, both sample and population size were increased, the distribution would approximate more closely to the normal form. For $N = 30$ and $N_p = 100$, the normal distribution would be a good approximate fit. If the sampling distribution is approximately normal, we can, given its standard deviation, readily estimate the probability of obtaining values equal to or greater than any given size in random sampling from the population.

10.6 SAMPLING DISTRIBUTION OF MEANS FROM AN INDEFINITELY LARGE POPULATION

Many populations may be conceptualized as comprised of an indefinitely large number of members. Most applications of sampling theory encountered in psychology and education assume such populations. Sampling from an indefinitely large population is essentially the same as sampling from a finite population with replacement, that is, where each sample member is returned to the population prior to the drawing of the next member. In sampling without replacement from an indefinitely large population the probabilities remain unchanged regardless of the size of the sample drawn. Similarly, in sampling from a finite population with replacement, the population is not depleted and the probabilities are unchanged by the number of prior draws. It follows that problems of sampling from an indefinitely large population can be approached through the study of finite populations where samples are drawn with replacement.

To illustrate sampling from an indefinitely large population, an artificial population was constructed. This population was composed of 1,611 cards containing the numbers from 1 to 25. The distribution of numbers is approximately normally distributed. The distribution of this population is shown in Table 10.2. The mean μ of the population is 13, and the standard deviation σ is 3.56. The cards were inserted in a box, and samples of 10 cards drawn with replacement; that is, a card was drawn, its number noted,

Table 10.2

Population from which samples were drawn: frequency distribution of numbers

Number	Frequency	Number	Frequency
1	1	14	174
2	2	15	154
3	4	16	127
4	7	17	96
5	14	18	67
6	26	19	43
7	43	20	26
8	67	21	14
9	96	22	7
10	127	23	4
11	154	24	2
12	174	25	1
13	181		
Total			1,611

SOURCE: R. W. B. Jackson and George A. Ferguson, "Manual of Educational Statistics," University of Toronto, Department of Educational Research, Toronto, 1942.

Table 10.3

Means of samples of 10 drawn from the population in Table 10.2

10.9	13.5	11.7	13.3	13.8	12.5	15.0	12.7	14.3	12.7
13.0	13.2	14.0	13.1	13.2	12.7	12.6	11.5	13.2	12.9
12.4	13.9	14.1	12.2	13.1	11.7	11.5	14.6	12.6	12.9
13.9	14.0	11.7	12.1	13.2	13.6	14.4	14.0	12.2	13.7
12.6	11.6	11.8	12.1	13.1	13.2	12.5	14.0	16.4	12.2
12.6	13.7	13.6	14.0	12.1	13.2	14.8	13.6	12.5	14.5
14.4	13.9	13.8	15.1	14.2	14.4	13.5	12.7	14.5	14.4
12.9	11.3	14.5	13.0	12.0	13.3	12.7	14.8	11.3	11.0
12.7	14.6	15.2	14.1	16.1	14.7	12.3	11.2	14.3	14.7
12.9	12.3	11.9	14.0	14.5	12.4	11.9	12.3	12.4	12.6

SOURCE: R. W. B. Jackson and George A. Ferguson, "Manual of Educational Statistics," University of Toronto, Department of Educational Research, Toronto, 1942.

and the card then returned to the box before the next draw. Altogether 100 samples of 10 cards were drawn, and the 100 means calculated. Table 10.3 shows the means of the samples. Table 10.4, column 2, shows a frequency distribution of these means. This distribution is an experimental sampling distribution of means based on samples of size 10. The mean of the sampling distribution $\bar{X}_{\bar{x}}$ is 13.205, and the standard deviation $s_{\bar{x}}$ is 1.139. This standard deviation is a description based on experimental data of the sample-to-sample fluctuation of means of samples of size 10 drawn at random from this population.

The mean and standard deviation of the sampling distribution need not be estimated by the rather laborious experimental approach described above. It can be shown that the mean of the theoretical sampling distribu-

Table 10.4

Experimental and theoretical sampling distribution of 100 sample means for samples of size 10 drawn from the population of Table 10.2

	Frequency	
Class interval	Experimental	Theoretical
16.5–17.4	. . .	.1
15.5–16.4	2	1.4
14.5–15.4	13	8.4
13.5–14.4	27	24.6
12.5–13.4	31	34.3
11.5–12.4	22	22.8
10.5–11.4	5	7.2
9.5–10.4	. . .	1.1
Total	100	99.9

SOURCE: R. W. B. Jackson and George A. Ferguson, "Manual of Educational Statistics," University of Toronto, Department of Educational Research, Toronto, 1942.

tion of the mean in sampling from an indefinitely large population is equal to the population mean; that is, $\mu_{\bar{x}} = \mu$. It can also be shown that the standard deviation of the sampling distribution is given by

[10.2] $$\sigma_{\bar{x}} = \frac{\sigma}{\sqrt{N}}$$

where σ = standard deviation in population
N = size of sample

The reader will observe that the difference between this formula and the formula previously given for the standard deviation of the sampling distribution for the means of samples from a finite population resides in the absence here of the term $\sqrt{(N_p - N)/(N_p - 1)}$. As N_p increases, this term approaches 1 as a limit. It is equal to 1 when the population is indefinitely large. The standard deviation of the theoretical sampling distribution is $\sigma_{\bar{x}} = 3.56/\sqrt{10} = 1.126$. This is very close to the standard deviation of the experimental sampling distribution $s_{\bar{x}}$, which was found to be 1.139. In practice σ is usually unknown and is replaced by the sample standard deviation. The standard error of the mean is estimated by the formula

[10.3] $$s_{\bar{x}} = \frac{s}{\sqrt{N}}$$

This is the formula in common use for estimating the standard error of the mean.

The theoretical sampling distribution of the means of samples drawn from a normal population is normal. Thus if we know that the population distribution is normal, we know that the sampling distribution of means is normal. In many situations in psychology and education reasons may exist to suggest that the distribution of the variable in the population may depart from the normal form. Regardless of the shape of the population distribution, the sampling distribution of means, drawn from a population with variance σ^2 and mean μ, will approach a normal distribution with mean μ and variance σ^2/N as sample size N increases. This is a statement of what is called the *central-limit theorem.* This theorem simply asserts that the sampling distribution of means gets closer and closer to the normal form as sample size increases despite departures from normality in the population distributions.

In general, for practical purposes the sampling distribution of means can be viewed as approximately normal for samples of reasonable size, except perhaps in the case of gross departures in the population from normality. Table 10.4, columns 3 and 4, shows experimental frequencies and the theoretical normal frequencies for our illustrative example. The theoretical normal frequencies are the expected normal frequencies for a normal curve with a mean of 13.00 and a standard deviation of 1.126. The differences between the experimental and theoretical normal distribution are not very great.

Examination of the formula $\sigma_{\bar{x}} = \sigma/\sqrt{N}$ indicates that the standard error of the mean is directly related to the standard deviation of the population and inversely related to the size of the sample. Thus the greater the variation of the variable in the population, the greater the standard error; also the larger the size of N, the smaller the standard error. The standard error of means of samples of $N = 1$, the smallest sample size possible, is equal to the population standard deviation. For any fixed value of σ the standard error can be made as small as we like by increasing the size of the sample.

Useful insight into formula [10.2] results from the following argument. Consider a population with mean μ and variance σ^2. Draw samples of size 1. The long-range expected variance of such samples is σ^2. Draw samples of size 2 and consider the variance of sums. Since the values are paired at random and may be taken as independent, the expected variance of sums will be $2\sigma^2$. This argument may be extended to samples of size N. The expected variance of sums will be $N\sigma^2$. Concern here is with the variance of means, not sums. Thus the sums must be divided by N; that is multiplied by the constant $1/N$. If each observation in a set is multiplied by a constant, the variance is multiplied by the constant squared. The variance of means is then obtained by multiplying by the constant $1/N^2$. Hence the variance of means is $N\sigma^2/N^2 = \sigma^2/N$. The standard deviation of means, the standard error, is then $\sigma_{\bar{x}} = \sigma/\sqrt{N}$.

10.7 SAMPLING DISTRIBUTION OF PROPORTIONS

Many problems require the use of proportions. A study of the sampling distribution of a proportion may be approached either experimentally or theoretically. To illustrate, consider an urn containing a finite number N_p of black and white chips. Denote the proportion of black and white chips by θ and $1 - \theta$, respectively. Let us draw a large number of samples of size N at random, without replacement, from the urn, observe the proportion of black chips in each sample, and make a frequency distribution of these proportions. This frequency distribution is an experimental sample distribution of proportions for samples of size N. As with the arithmetic mean, we may use a theoretical, as distinct from an experimental, approach. In drawing samples of N from a finite population of N_p members, the number of different equiprobable samples is $C_N^{N_p}$. The proportion of black chips may be calculated from each of these samples and a frequency distribution made of the proportions. This distribution is a theoretical sampling distribution of proportions.

For illustrative purposes consider a hypothetical population of three black and three white chips. Denote the members of this population by B_1, B_2, B_3, W_4, W_5, W_6, the subscript identifying the particular population member. We may consider the set of equiprobable samples of three members which may be drawn without replacement from this population.

The number of such samples is $C_3^6 = 20$. The first sample is $B_1B_2B_3$, and the proportion of black chips is $p = 1.00$. The second sample is $B_1B_2W_4$ with $p = .67$; the third, $B_1B_2W_5$ with $p = .67$, and so on. A frequency distribution may be made of these proportions and is as follows:

p	f	f/N
0	1	.05
.33	9	.45
.67	9	.45
1.00	1	.05
Total	20	1.00

The standard deviation, or standard error, of this theoretical sampling distribution may be readily calculated and is $\sigma_p = .224$. This standard error may be obtained directly by the formula

[10.4] $$\sigma_p = \sqrt{\frac{\theta(1-\theta)}{N}\frac{N_p - N}{N_p - 1}}$$

In the above example $\theta = .50$ since 3 of the 6 chips are black. Population and sample size, respectively, are $N_p = 6$ and $N = 3$. Hence

$$\sigma_p = \sqrt{\frac{.5 \times .5}{3}\frac{6-3}{6-1}} = .224$$

which agrees with the value obtained by direct calculation.

The discussion above relates to sampling without replacement from a finite population. As with the arithmetic mean, the term $(N_p - N)/(N_p - 1)$ in formula [10.4] approaches unity for any finite value of N as N_p approaches infinity. Thus this term may be considered equal to unity for an indefinitely large population, and we obtain

[10.5] $$\sigma_p = \sqrt{\frac{\theta(1-\theta)}{N}}$$

as the standard error of a proportion in sampling from an indefinitely large population, or in sampling from a finite population with replacement.

Formula [10.5] may be obtained by reference to the binomial. If the proportion of black chips in a population is θ, the expected, or theoretical, sampling distribution of the *number* of black chips in samples of size N, as distinct from the proportion of black chips, is given by the terms of the binomial expansion $[\theta + (1-\theta)]^N$. The mean and standard deviation of this distribution are $N\theta$ and $\sqrt{N\theta(1-\theta)}$, respectively. We are interested in the distribution of the proportion, instead of the number, of black chips in the samples. To obtain the standard deviation of the distribution of the proportion, as distinct from the number, of black chips in samples of size N, we multiply $\sqrt{N\theta(1-\theta)}$ by $1/N$ to obtain $\sigma_p = \sqrt{\theta(1-\theta)/N}$, which is the same as formula [10.5] above. To illustrate, let $\theta = .25$ and $1 - \theta = .75$.

The expected distribution of the *number* of black chips in samples of size 10 is given by expanding the binomial $(.25 + .75)^{10}$. The mean in this example is $10 \times .25 = 2.5$, and the standard deviation is $\sqrt{10 \times .25 \times .75} = 1.37$. The standard deviation of the distribution of the *proportion* of black chips in samples of size 10 is obtained by dividing 1.37 by 10 and is .137.

Formulas [10.4] and [10.5] assume that θ is known. In practice θ is usually not known, and the sample value p is used as an estimate of θ. Also by definition $1 - p = q$. Hence the formula commonly used to estimate the standard error of a proportion, assuming an indefinitely large population, is

[10.6] $$s_p = \sqrt{\frac{pq}{N}}$$

Note that the formulas in this section imply a two-category nominal variable with a proportion θ of members in one category and $1 - \theta$ not in that category. Membership may be represented by, or coded, 1, and nonmembership by 0. The mean of such a variable is θ and the standard deviation is $\sqrt{\theta\ (1 - \theta)}$. Thus the formulas for the standard error of a proportion are particular cases of those for means; that is, formula [10.4] is a particular case of [10.1] and formula [10.5] is a particular case of [10.2].

10.8 SAMPLING DISTRIBUTION OF DIFFERENCES

For certain purposes a knowledge of the sampling distribution of the difference between two statistics, such as the difference between two arithmetic means or two proportions, is required. To conceptualize the sampling distribution of the difference between, say, two arithmetic means, let us consider two indefinitely large populations whose means are equal; that is, $\mu_1 = \mu_2$. Let $\bar{X}_1$ be the mean of a sample of N_1 cases drawn at random from the first population and $\bar{X}_2$ be the mean of a sample of N_2 cases drawn from the second population. The difference between means is $\bar{X}_1 - \bar{X}_2$. Since $\mu_1 = \mu_2$ this difference results from sampling error. A large number of pairs of samples may be drawn, and a frequency distribution made of the differences. It describes how the differences between means chosen at random from two populations, where $\mu_1 = \mu_2$, will vary with repeated sampling. From this distribution we may estimate the probability of obtaining a difference of any specified size in drawing samples at random from populations where $\mu_1 = \mu_2$. By considering an indefinitely large number of pairs of samples we arrive at the concept of a theoretical sampling distribution of differences between sample means. In this situation the individual measurements in the two populations are not paired with one another. The samples are independent. The means may be viewed as paired at random. No correlation exists between the pairs of means.

The variance of the sampling distribution of differences describes how the differences vary with repeated sampling. Consider the case of independent samples. If $\sigma_{\bar{x}_1}{}^2 = \sigma_1{}^2/N_1$ is the variance of the sampling distribution of means drawn from one population and $\sigma_{\bar{x}_2}{}^2 = \sigma_2{}^2/N_2$ is the corresponding variance from the other population, then the variance of the sampling distribution of differences between means is the sum of the two variances. Thus

[10.7] $$\sigma^2_{\bar{x}_1-\bar{x}_2} = \sigma_{\bar{x}_1}{}^2 + \sigma_{\bar{x}_2}{}^2 = \frac{\sigma_1{}^2}{N_1} + \frac{\sigma_2{}^2}{N_2}$$

When $\sigma_1{}^2 = \sigma_2{}^2 = \sigma^2$, the variances in the two populations being equal, we may write

[10.8] $$\sigma^2_{\bar{x}_1-\bar{x}_2} = \sigma^2\left(\frac{1}{N_1} + \frac{1}{N_2}\right)$$

Consider now a situation where measurements are paired with one another. Such data arise, for example, where measurements are made on the same group of subjects under both control and experimental conditions. The paired measurements may be correlated. In this instance, in approaching the sampling distribution of differences between means, we conceptualize two populations of paired measurements with equal means; thus $\mu_1 = \mu_2$. Denote the correlation between the paired measurements by the symbol ρ_{12}. Samples of size N are drawn at random, and the differences between means obtained. The distribution of differences between means for an indefinitely large number of samples is the sampling distribution of differences for correlated populations.

For correlated populations the variance of the sampling distribution of differences may be shown to be

[10.9] $$\sigma^2_{\bar{x}_1-\bar{x}_2} = \sigma_{\bar{x}_1}{}^2 + \sigma_{\bar{x}_2}{}^2 - 2\rho_{12}\sigma_{\bar{x}_1}\sigma_{\bar{x}_2}$$

where ρ_{12} is the correlation in the population. Note that the formula for independent samples is a particular case of the more general formula for correlated samples. It is the particular case which arises when $\rho_{12} = 0$. In the correlated case $N_1 = N_2 = N$.

Formulas [10.7] to [10.9] are simple applications of the formula for the variance of differences. (See Formula [8.5]). In practice the variances are estimated from the data. Formula [10.9] then becomes

[10.10] $$s^2_{\bar{x}_1-\bar{x}_2} = s^2_{\bar{x}_1} + s^2_{\bar{x}_2} - 2r_{12}s_{\bar{x}_1}s_{\bar{x}_2}$$

Obviously when $r_{12} = 0$, the sampling variance of the difference between means is the sum of the separate variances.

10.9 GENERALITY OF THE CONCEPTS

In the preceding sections above the reader was introduced to a number of important concepts including sampling error, sampling distribution, and standard error. These concepts were discussed using means and proportions. It should be clearly understood that these concepts are general and apply to all descriptive statistics including medians, variances, correlation coefficients, and so on. Further, it must not be assumed that all sampling distributions are normal. Some are, and some are not. A number of distributions, other than the normal distribution, are used in describing the sampling distributions of various statistics. Some of these distributions are discussed in later chapters in this book.

10.10 ESTIMATION

In practice population parameters are usually not known but must be estimated from the data. Thus $\bar{X}$ is used as an estimate of μ, and s^2 as an estimate of σ^2. A distinction is commonly made between two types of estimates, *point estimates* and *interval estimates.* A point estimate is a value obtained by direct calculation. If in a particular sample the mean $\bar{X} = 26.88$, this is a point estimate of the parameter μ. Another approach is to specify an interval within which we may assert with some known degree of confidence that the population mean lies. Thus, for example, instead of calculating the point estimate $\bar{X}$, we may perform a simple calculation, described later in this chapter, and assert with 95 percent confidence that the population mean falls within the limits 24.92 and 28.84. These values are *confidence limits,* and the interval they bound is called a *confidence interval.* Such an interval provides an interval estimate of the population value.

The formulas used in calculating $\bar{X}$, s^2, or any other statistic serve as methods of estimation. Different methods may provide different estimates of the same population value. For example, if the population distribution is normal, the population mean and median coincide. As a measure of central location the mean and the median are the same. For a sample drawn from this population the mean and median may be calculated. Why would one statistic be preferred to another as an estimate of the same population value? Again, $s^2 = \Sigma(X - \bar{X})^2/N$ and $s^2 = \Sigma(X - \bar{X})^2/(N - 1)$ both provide estimates of σ^2. Why is the one method of estimation preferred to the other? The answer to this question rests in the properties that attach to the estimates. Some estimates have properties that are more desirable than others, and these guide our choice of a method of estimation.

10.11 PROPERTIES OF ESTIMATES

Methods of estimation are sometimes said to yield *unbiased, consistent, efficient,* and *sufficient* estimates. These are desirable characteristics and serve as criteria for preferring one method of estimation to another.

A method of estimation provides an unbiased estimate when the mean of a large number of sample values, obtained by repeated sampling, approaches the population value in the limit as the number of samples increases. This simply means that a statistic is unbiased when it displays no systematic tendency to be either greater than or less than the population parameter; that is, it is not subject to a constant error. The arithmetic mean is an unbiased estimate. The sample mean $\bar{X}$ exhibits no systematic tendency to be either greater than or less than the parameter μ. Stated in somewhat different language, an estimate is unbiased when its *expected* value is equal to the parameter it purports to estimate.

The expected value of a statistic is the value we should expect to obtain upon averaging the values of the statistic over an indefinitely large number of repeated random samples. It is the value we should expect to obtain in the long run. The expected value of a variable is obtained by multiplying each value of the variable by its associated probability and summing the products. For example, consider the following frequency distribution with corresponding probabilities:

X	f	p
2	40	.40
3	50	.50
4	10	.10
Total	100	1.00

The expected value of X, that is, $E(X)$, is $2 \times .40 + 3 \times .50 + 4 \times .10 = 2.70$. Note that $E(X)$ is also the mean of X obtained by multiplying the values of X by the frequency of X and dividing by the total frequency. In general the expected value of the mean, $E(\bar{X})$, is the population mean μ. A statistic is a *biased* estimate when its expected value does not equal the population value but departs from it in systematic fashion. For example, it may be shown that the expected value of the variance defined as $s^2 = \Sigma(X - \bar{X})^2/N$ is not σ^2 but is $\sigma^2(N - 1)/N$. Thus the value of the variance, so defined, which we would expect to get in the long run is a biased estimate of σ^2.

A method of estimation is said to yield a consistent estimate if the estimate approaches the population parameter more closely as sample size increases. The arithmetic mean is a consistent estimate in that it tends to draw closer to the population parameter with increase in sample size.

The efficiency of a method of estimation is related to its sampling variance. The relative efficiency of two methods of estimation is the ratio of the two sampling variances, the larger variance being placed in the denominator. To illustrate, when the distribution of a variable in the population is symmetrical and unimodal, both the mean and the median are estimates of the same population parameter, μ. The sampling variance of the mean is $\sigma_{\bar{x}}^2 = \sigma^2/N$, and the median, $\sigma_{\text{mdn}}^2 = 1.57\sigma^2/N$. The relative efficiency is then

[10.11] $$\text{Relative efficiency} = \frac{\sigma^2/N}{1.57\sigma^2/N} = .64$$

Thus for this type of population the mean is more efficient than the median. Relative efficiency has meaning in terms of sample size. A median calculated on a sample of 100 cases has a sampling variance equal to that of a mean calculated on 64 cases. The mean is a more economical estimate and the saving achieved by its use in preference to the median is 36 cases in 100.

A method of estimation is sufficient if it is more efficient than any other possible method of estimation, that is, if its sampling variance is less. A sufficient method of estimation uses all the information in the sample. In this context, the concept of information is assigned a precise mathematical meaning.

10.12 THE DISTRIBUTION OF t

In discussing the sampling distribution of the arithmetic mean, we have stated that the distribution was normal, at least for samples of reasonable size. When means are expressed in standard-score form and the samples are not large, say $N < 30$, problems arise. Consideration of these problems leads to a discussion of an important and widely used theoretical distribution known as the distribution of t.

Consider a normal population with mean μ and variance σ^2. The variance of the sampling distribution of means drawn for this population is $\sigma_{\bar{x}}^2 = \sigma^2/N$. A particular mean $\bar{X}$ may be expressed in standard-score form as follows:

[10.12] $$z = \frac{\bar{X} - \mu}{\sigma_{\bar{x}}}$$

This ratio is simply a deviation of the sample mean $\bar{X}$ from the population mean μ divided by the standard deviation of the sampling distribution. This ratio is normally distributed with a mean of 0 and unit standard deviation.

In practice σ^2 is unknown and is estimated from the data using an unbiased estimator. We obtain, thereby, an estimate of $\sigma_{\bar{x}}$, denoted by $s_{\bar{x}}$. We may now consider the ratio

[10.13] $$t = \frac{\bar{X} - \mu}{s_{\bar{x}}}$$

This ratio contains the variable sample values $\bar{X}$ and $s_{\bar{x}}$ in the numerator and denominator, respectively. This is called a t ratio. It departs appreciably from the normal form for small N. Its theoretical sampling distribution is called the distribution of t. If 100 samples of, say, 5 members were drawn from a population, $\bar{X}$ and $s_{\bar{x}}$ calculated for each sample, and 100 values of t obtained, the frequency distribution of these 100 values of t would not be normal. It would be a symmetrical distribution with somewhat thicker tails than the normal distribution. The theoretical sampling distribution of t for small N is a symmetrical distribution and is thicker at the extremities than the corresponding normal curve. It tapers off to infinity at the two extremities. The distribution of t is not a single distribution but is a family of distributions. A different t distribution exists for each number of degrees of freedom. As the number of degrees of freedom increases, the t distribution approaches the normal form. Figure 10.1 compares the normal distribution with the t distribution for various degrees of freedom.

Hitherto we have considered two theoretical model frequency distributions, the binomial distribution and the normal distribution. The *t distribution* is a third theoretical model distribution with wide application to many sampling problems.

In sampling problems the t distribution is used in a manner directly analogous to the normal distribution. In the normal distribution 95 percent of the total area under the curve falls within plus and minus 1.96 standard deviation units from the mean and 5 percent of the area falls outside these limits. Likewise, 99 percent of the area under the normal curve falls within plus and minus 2.58 standard deviation units from the mean and 1 percent of the area falls outside these limits. In the t distribution, the distances along the base line of the curve that include 95 and 99 percent of the total area are different for different numbers of degrees of freedom. It is customary in tabulating areas under the t curve to use degrees of freedom, df, instead of N. While the df associated with the sample variance is $N - 1$, the df associated with other statistics may be $N - 2$, $N - 3$, and the like. Consequently, tables of t by degrees of freedom instead of N are more generally applicable. The distances from the mean, measured along

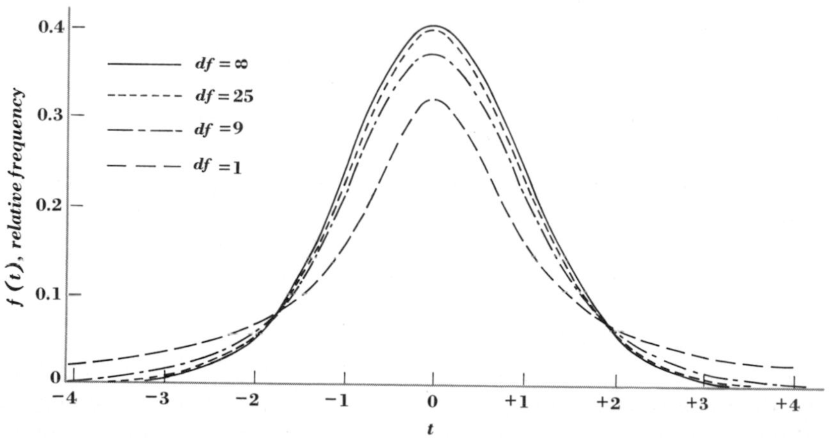

Figure 10.1 Distribution of t for various degrees of freedom. (*From D. Lewis, Quantitative methods in psychology, McGraw-Hill Book Company, New York, 1960.*)

the base line of the t distribution, that include 95 and 99 percent of the total area (analogous to the 1.96 and 2.58 of the normal distribution) for selected degrees of freedom are as follows:

df	95%	99%
1	12.71	63.66
2	4.30	9.93
3	3.18	5.84
4	2.78	4.60
5	2.57	4.03
10	2.23	3.17
15	2.13	2.95
20	2.09	2.85
30	2.04	2.75
120	1.98	2.62
∞	1.96	2.58

Note that as the number of degrees of freedom approaches infinity, t approaches the values 1.96 and 2.58. The difference between t for about 30 degrees of freedom and t for an indefinitely large number of degrees of freedom is sometimes interpreted for practical purposes as trivial. A more complete tabulation of t is given in Table B of the Appendix. A distinction is often made between *large* sample and *small* sample statistics. This distinction resides in the fact that the normal distribution is frequently found to be an appropriate model for use with sampling problems involving large samples. With small samples the distribution of t provides for many statistics a more appropriate model.

The distribution of t was developed originally in 1908 by W. S. Gossett (1876–1937) who wrote under the pen name "Student." The t distribution is frequently referred to as "Student's" distribution. Gossett spent his entire working life with the brewing firm of Guinness, first in Dublin and later in London. His interest in statistics originated in problems associated with the control of quality of materials used to make beer. He made important contributions to statistical methods and experimental design. He was not viewed as an outstanding mathematician, but brought so remarkable an intuitive capacity to the solution of statistical problems that he was once described as the Faraday of statistics.

10.13 DEGREES OF FREEDOM

In the above discussion on the distribution of t, mention is made of the number of degrees of freedom. This concept was discussed also in Chapter 5, where the sample variance was defined as the sum of squares of deviations about the arithmetic mean, divided by the number of degrees of freedom. The degree of freedom concept requires further elaboration.

As stated and illustrated in Chapter 5, the number of degrees of freedom is the number of values of the variable that are free to vary. The measurements 10, 14, 6, 5, and 5, when represented as deviations from a mean of 8, become +2, +6, −2, −3, −3. The sum of these deviations is 0. In consequence, if any four deviations are known, the remaining deviation is determined. The number of degrees of freedom is 4.

This type of situation may be represented in symbolic form. Let X_1, X_2, X_3 be three measurements with mean $\bar{X}$. The sum of deviations is $(X_1 - \bar{X}) + (X_2 - \bar{X}) + (X_3 - \bar{X}) = 0$. If $\bar{X}$ and any two of the values of X are known, the third value of X is determined. The number of degrees of freedom here is 2. The calculation of the variance and standard deviation requires the sum of squares of deviations about the mean, $\Sigma(X - \bar{X})^2$. There are $N - 1$ of the values that this sum of squares comprises which are free to vary independently. The number of degrees of freedom associated with the sum of squares is $N - 1$. Dividing this sum of squares by the number of degrees of freedom associated with it, as distinct from the number of observations, yields an unbiased estimate of the population variance σ^2. The symbol *df* is frequently used to represent degrees of freedom.

The number of degrees of freedom depends on the nature of the problem. In fitting a line to a series of points by the method of least squares the number of degrees of freedom associated with the sum of squares of deviations about the line is $N - 2$. If there are two points only, a straight line will fit the points exactly and the sum of squares of deviations about the line will, of course, be 0. No freedom of variation is possible. With three points, $df = 1$; with 15 points, $df = 13$. The equation of a straight line is $Y = bX + a$, where b is the slope of the line and a is the point where it cuts the Y axis. Both b and a are estimated from the data. It may be said that 2 degrees of freedom are lost in estimating b and a from the data. If b, a, and any $N - 2$ deviations from the line are known, the remaining two deviations are determined.

The concept of degrees of freedom has a geometric interpretation. A point on a line is free to move in one dimension only and has 1 degree of freedom. A point on a plane has freedom of movement in two dimensions and has 2 degrees of freedom. A point in a space of three dimensions has 3 degrees of freedom. Likewise, a point in a space of k dimensions has k degrees of freedom. It has freedom of movement in k dimensions.

The concept of degrees of freedom is widely used in statistical work and will be discussed subsequently in connection with contingency tables and the analysis of variance. The essence of the idea is simple. The number of degrees of freedom is always the number of values that are free to vary, given the number of restrictions imposed upon the data. It seems intuitively obvious that in the study of variation we should concern ourselves with the number of values that enjoy freedom to vary within the restrictions of the problem situation.

10.14 CONFIDENCE INTERVALS FOR MEANS OF LARGE SAMPLES

The calculation of confidence intervals for a mean based on a large sample is a relatively simple procedure. The sampling variance of the sampling distribution of the mean, as previously stated, is $\sigma_{\bar{x}}^2 = \sigma^2/N$. The population variance σ^2 is unknown. If we use an unbiased variance s^2 as an estimate of σ^2, our estimate of the sampling variance is $s_{\bar{x}}^2 = s^2/N$ and the estimated standard error is given by

[10.14] $$s_{\bar{x}} = \frac{s}{\sqrt{N}}$$

This is the commonly used formula for estimating the standard error of the arithmetic mean.

Consider now the ratio $z = (\bar{X} - \mu)/s_{\bar{x}}$. This ratio is a deviation of a sample mean from its population mean, divided by an estimate of the standard deviation of the sampling distribution. It is a standard score. Assuming the normality of z, it is correct to state that the probability is .95 that the following statement is true:

[10.15] $$-1.96 \leq \frac{\bar{X} - \mu}{s_{\bar{x}}} \leq 1.96$$

This inequality specifies the confidence interval in standard-score form. In effect it states that the chances are 95 in 100 that $(\bar{X} - \mu)/s_{\bar{x}}$ falls between ± 1.96. Ordinarily, however, we are not interested in the confidence interval in standard-score form, but in raw-score form. To convert the inequality to raw-score form, we multiply by $s_{\bar{x}}$ and add $\bar{X}$ to each term to obtain

[10.16] $$\bar{X} - 1.96s_{\bar{x}} \leq \mu \leq \bar{X} + 1.96s_{\bar{x}}$$

This states that the chances are 95 in 100 that μ falls between $\bar{X} \pm 1.96s_{\bar{x}}$; thus the upper limit is 1.96 standard error units above the sample mean, and the lower limit is 1.96 standard error units below the mean. The figure 1.96 derives, as the reader will recall, from the fact that 95 percent of the area of the normal curve falls within the limits ± 1.96 standard deviation units from the mean. To illustrate the fixing of confidence intervals, let the mean IQ of a random sample of 100 secondary school children be 114 and the standard deviation 17. The standard deviation here is the square root of the unbiased variance estimate. Our estimate of the standard error of the mean is $s_{\bar{x}} = 17/\sqrt{100} = 1.70$. The 95 percent confidence interval is then given by $114 \pm 1.96 \times 1.70$. The upper limit is 117.33 and the lower limit is 110.67. Thus we may assert with 95 percent confidence that the population mean falls within these limits. The 99 percent confidence limits are given by $\bar{X} \pm 2.58s_{\bar{x}}$. The figure 2.58 derives from the fact that 99 percent of the area of the normal curve falls within the limits ± 2.58 standard deviation units above and below the mean. In the above example the 99 percent confidence limits are given by $114 \pm 2.58 \times 1.70$. These limits are 109.61 and 118.39.

What meaning attaches to the statement that we are 95 percent confident that the actual population mean falls within certain specific limits? A particular sample may have a mean $\bar{X} = 26.88$ with 95 percent confidence intervals 24.92 and 28.84. Another sample of the same size may have a mean $\bar{X} = 25.68$ with 95 percent confidence intervals 23.72 and 27.64. Presumably we could draw a large number of samples, obtain a large number of upper and lower limits, and prepare frequency distributions of these upper and lower limits. These two distributions would be experimental sampling distributions for the 95 percent confidence limits. Without elaborating the details of this situation, we state that about 95 percent of the intervals so obtained would include the population mean and about 5 percent of the intervals would not include the population mean. Thus the statement that we are 95 percent confident implies that we expect about 95 percent of our assertions to be correct and the remaining 5 percent to be incorrect, or that the odds are 19 to 1 that the confidence interval includes the population value. The use of a 95 percent confidence interval is fairly common. If a greater degree of confidence is desired, a 99 percent interval may be used. This interval is, very roughly, 1.3 times as great as the 95 percent interval. Thus as we increase our level of confidence, the interval is increased. Likewise, of course, as we decrease the level of confidence, the interval is decreased. Any desired level of confidence can be obtained by varying the size of the confidence interval. As the confidence level is decreased and approaches 0, the confidence interval approaches 0 as a limit. As the confidence level is increased and approaches 100, the confidence interval approaches infinity as a limit. In practice, 95 and 99 percent confidence intervals are widely used.

Implicit in the above discussion is the assumption that the ratio $(\bar{X} - \mu)/s_{\bar{x}}$ is normally distributed. This ratio is not normally distributed when N is small, but approaches the normal form as N increases in size. It is a not uncommon statistical convention to consider a sample of 30 or more observations as large and a sample of less than 30 as small. This, of course, is highly arbitrary.

10.15 CONFIDENCE INTERVALS OF MEANS FOR SMALL SAMPLES

The line of reasoning used in determining confidence intervals for small samples is similar to that for large samples. With small samples, however, the distribution of t is used instead of the normal distribution in fixing the limits of the interval. For large samples the 95 and 99 percent confidence intervals for the mean are given, respectively, by $\bar{X} \pm 1.96s_{\bar{x}}$ and

$\bar{X} \pm 2.58 s_{\bar{x}}$. For small samples an unbiased estimate of σ^2 is used in estimating the standard error. The value of t used in fixing the limits of the 95 and 99 percent intervals will vary, depending on the number of degrees of freedom. Consider an example where $\bar{X} = 24.26$, $s^2 = 64$, $N = 16$, and $df = 16 - 1$. On reference to Table B of the Appendix we observe that for 15 degrees of freedom 95 percent of the area of the distribution falls within a t of ± 2.13 from the mean. The standard error using the unbiased variance estimate is $8/\sqrt{16}$. The 95 percent confidence limits are given by $24.26 \pm 2.13 \times 8/\sqrt{16}$. These limits are 20.00 and 28.52. We may assert with 95 percent confidence that the population mean falls within these limits. The 99 percent limits are given by $24.26 \pm 2.95 \times 8/\sqrt{16}$. These limits are 18.36 and 30.16.

10.16 STANDARD ERRORS AND CONFIDENCE INTERVALS OF OTHER STATISTICS

As stated previously the standard error of a proportion in sampling from an indefinitely large population may be estimated using the formula $s_p = \sqrt{pq/N}$. If it can be assumed that the sampling distribution of a proportion can be approximately represented by a normal distribution, then the 95 and 99 percent confidence limits for a proportion are given by $p \pm 1.96 s_p$ and $p \pm 2.58 s_p$, respectively. Whether or not the sampling distribution can be represented by a normal distribution depends both on the size of the sample and on the value of p. For any given value of N the sampling distribution of a proportion becomes increasingly skewed as p and q depart from .50. Quite clearly, the formula for the standard error of a proportion should not be used with reference to a normal curve for extreme values of p and q. It has been suggested that the formula for the standard error of a proportion should be used only when Np or Nq, whichever is the smaller, is equal to or greater than 5. Thus when $p = .10$ and $N = 20$, $Np = 2$. The use of the formula $s_p = \sqrt{pq/N}$ would be considered inappropriate here. When $p = .10$ and $N = 100$, $Np = 10$. Presumably, here the differences between the binomial and the normal distribution are quite small and can safely be ignored.

The standard error of the median may be estimated by

$$s_{\text{mdn}} = \frac{1.253 s}{\sqrt{N}} \tag{10.17}$$

where s is obtained from the unbiased estimate of σ^2. Confidence limits at the 95 and 99 percent levels may be located by taking $\pm 1.96 s_{\text{mdn}}$ and $\pm 2.58 s_{\text{mdn}}$ about the sample median. The above formulation assumes normality of the parent population and a large N. In many situations where the median is used, the distribution of the variable is not normal. This, indeed, is one of the reasons for using the median instead of the mean. In consequence the above formulation is of limited use.

The standard error of the standard deviation for large samples from a normal population is estimated by

[10.18] $$s_s = \frac{s}{\sqrt{2N}}$$

The 95 and 99 percent confidence limits can readily be obtained by taking $s \pm 1.96s_s$ and $s \pm 2.58s_s$, respectively. In using this formula a sample must be substantially greater than 30 to be regarded as large. The method of determining confidence limits for s based on small samples, and indeed the method which is perhaps most appropriate in all cases regardless of size of N, involves a knowledge of the distribution of chi square, or χ^2.

BASIC TERMS AND CONCEPTS

Population

Sample

Parameter

Estimate

Statistical inference

Random sample

Stratified random sample

Sampling error

Sampling distribution: experimental; theoretical

Standard error

Standard error of the mean

Standard error of a proportion

Sampling distribution of differences

Standard error of differences

Point estimate

Interval estimate

Confidence interval

Estimate: unbiased; consistent; efficient; sufficient

Relative efficiency

t ratio

t distribution

EXERCISES

1 Indicate the difference between **(a)** a random sample and a stratified random sample, **(b)** a stratified random sample and a proportional stratified sample, **(c)** an experimental and a theoretical sampling distribution.

2 How would you proceed to draw a random sample of 100 university students?

3 How would you proceed to draw a systematic stratified sample of 100 students from the students in a university?

4 Would a random sample of names selected from a telephone book be considered appropriate for the study of voting behavior?

5 Samples of three cards are drawn at random from a population of eight cards numbered from 1 to 8. Obtain **(a)** the theoretical sampling distribution of means, **(b)** the standard deviation of the sampling distribution, **(c)** the probability of obtaining a mean equal to or greater than 7.

6 The standard deviation of the sampling distribution of $\bar{X}$ is $\sigma/\sqrt{N}$. What is the standard deviation of the sampling distribution of $N\bar{X}$ or ΣX?

7 A university has a population of 1,000 students. The standard deviation of scholastic aptitude tests scores in this population is 80. Calculate the standard errors of mean scholastic aptitude test scores for a sample of 100 students drawn with and without replacement.

8 The variance of the sampling distribution of the mean for samples of 100 cases drawn from an indefinitely large population is 20. How large should the samples be to reduce this variance by one-half? How large should the samples be to reduce the standard deviation of the distribution by one-half?

9 A population consists of 6 black and 2 white chips. Obtain the frequency distribution of proportions of black chips in samples of 4 drawn from this population. What is the standard deviation of this distribution?

10 Will a negative correlation between paired observations increase or decrease the standard error of the difference between two means?

11 Using a large-sample procedure, obtain the 95 and 99 percent confidence intervals for a mean of 105, where $N = 100$ and $s = 10$. Obtain also the 75 and 85 percent confidence intervals.

12 A random sample of 400 observations has a mean of 50 and a standard deviation of 18. Estimate the 95 and 99 percent confidence limits for the mean.

13 How is the standard error of the mean affected by tripling sample size?

14 Using a large-sample procedure, estimate for the following data the 95 and 99 percent confidence intervals for means:

	$\bar{X}$	N	$\Sigma(X-\bar{X})^2$
a	26.2	7	77.0
b	58.3	11	249.0
c	46.3	25	1,525.0
d	8.4	16	444.7

15 Find the value of t for $df = 20$ such that the proportion of the area **(a)** to the right of t is .025, **(b)** to the left of t is .0005, **(c)** between the mean and t is .45, **(d)** between $\pm t$ is .90.

16 Obtain the values required in **(a)** to **(d)** of Exercise 5 above for $df = 5$.

17 What proportion of the area of the t distribution falls **(a)** above $t = 3.169$ where $df = 10$, **(b)** below $t = -1.725$ where $df = 20$, **(c)** between $t = \pm 3.659$ where $df = 29$, **(d)** between $t = 2.131$ and $t = 2.602$ where $df = 15$, **(e)** between $t = -4.541$ and $t = 3.182$ where $df = 3$?

18 Estimate the 95 and 99 percent confidence limits for $p = .75$ where $N = 169$.

ANSWERS TO EXERCISES

1 a A random sample is such that every member of the population is viewed as having an equal probability of being included in it. In a stratified random sample the members in each stratum are viewed as having an equal probability of being included in the subsample drawn from that stratum.

b If the proportions in the various strata in the sample are the same as the proportions in the population, the sample is a proportional stratified sample.

c An experimental sampling distribution is obtained by the actual drawing of samples from a population. It is the distribution of the values of the statistic thus obtained. A theoretical sampling distribution, on the other hand, is obtained using a theoretical method based on probability considerations. No samples are actually drawn.

2 A variety of methods suggest themselves. Students may be assigned numbers, using a directory. Numbers may be drawn using a lottery method, or a table of random numbers may be used.

3 Student population may be stratified by, for example, faculty. Let us suppose that a university has 5,000 students in four faculties, A, B, C, and D, the proportion in these four faculties being .50, .25, .15, and .10, respectively. In forming a sample of 100, 50 may be drawn from faculty A, 25 from faculty B, and so on. Faculty A contains 2,500 students. Using a directory, or faculty index, of students, every fiftieth name may be chosen. Faculty B contains 1,250 students. Here again every fiftieth name may be chosen, and so on. The resulting sample is a systematic proportional stratified sample.

4 A random sample of names selected from a telephone directory may not be considered an appropriate sample of voting behavior. Owning a telephone may be correlated with income, and possibly other variables, which may in turn be correlated with particular political predilections. Bias might result. Such a sample would clearly exhibit less bias in 1975 than, say, in 1931.

5 **a**

$\bar{X}$	f
2.00	1
2.33	1
2.67	2
3.00	3
3.33	4
3.67	5
4.00	6
4.33	6
4.67	6
5.00	6
5.33	5
5.67	4
6.00	3
6.33	2
6.67	1
7.00	1
Total	56

b 1.12 **c** .0179

6 $\sigma\sqrt{N}$

7 With replacement = 8.00. Without replacement = 7.53.

8. **a** 200 **b** 400

9 **a**

p	f
.50	15
.75	40
1.00	15

b $\sigma_p = .1637$

10 Increase

11 95: 103.04–106.96
99: 102.42–107.58
75: 103.85–106.15
85: 103.56–106.44

12 95: 48.24–51.76
99: 47.68–52.32

13 By tripling sample size the standard error is multiplied by .5774, or divided by $\sqrt{3}$.

14

	95%	99%
a	23.55–28.85	22.71–29.69
b	55.35–61.25	54.42–62.18
c	43.18–49.42	42.19–50.41
d	5.73–11.07	4.89–11.91

15 **a** 2.086 **b** −3.850 **c** 1.725 **d** ±1.725

16 **a** 2.571 **b** −6.859 **c** 2.015 **d** ±2.015

17 **a** .005 **b** .050 **c** .999 **d** .015 **e** .965

18 .685–.815, .664–.836

11

TESTS OF SIGNIFICANCE: MEANS

11.1 INTRODUCTION

In Chapter 10 we considered the sampling error associated with single sample values. Sampling distributions, standard errors of single sample values, and confidence intervals were discussed. In practical statistical work in psychology we are infrequently concerned with simply describing the magnitude of error associated with single sample values. Experimental data very often require a comparison and evaluation of two or more means, proportions, standard deviations, or other statistics obtained from separate samples or from the same sample for measurements obtained under two or more experimental conditions. To illustrate, we may wish to explore the effects of a tranquilizing drug on the estimation of time intervals as part of a study on time perception. We may administer a drug to an experimental group of subjects and a placebo, an inactive simulation of the drug, to a control group and measure the errors in time estimation made by the two groups. The mean error for the two groups may be calculated. The experiment requires an evaluation of the difference between these two means. Both means are subject to sampling error. May the difference between the two means be probably ascribed to sampling error, or may it be argued with confidence that the drug affects time perception? A decision is required between these alternatives. Statistical procedures which lead to decisions of this kind are known as *tests of significance*.

Tests of significance may be applied to the difference between statistics calculated on independent samples or between statistics obtained under different conditions for the same sample. Sometimes a test of significance is applied to test the difference between a single sample statistic and a fixed value. An example is the procedure used to test whether a correlation coefficient is significantly different from 0. In this case the fixed value is 0. While many tests of significance involve a comparison of two sample statistics, or a single sample statistic and a fixed value, such tests can readily be extended to cover situations where more than two sample statistics are involved. For example, in the experiment mentioned above on the effects of a drug on time perception, the experiment could be designed to include the administration of the drug in different dosages to different groups of subjects. Three or four or five different dosages might be used, resulting in three or four or five different means. The means could be compared two at a time to ascertain whether or not the differences between them could be attributed to sampling error. A more efficient form of analysis, the analysis of variance, provides a procedure for making an overall test in this type of situation.

11.2 THE NULL HYPOTHESIS

Consider an experiment using an experimental and a control group. A treatment is applied to the experimental group. The treatment is absent for the control group. Measurements are made on both groups. Presumably any significant difference between the two groups can be ascribed with confidence to the treatment and to no other cause. Let $\bar{X}_1$ and $\bar{X}_2$ be the means for the experimental and the control group, respectively. Both means are subject to sampling error. The means $\bar{X}_1$ and $\bar{X}_2$ are estimates of the population means μ_1 and μ_2. The trial hypothesis may be formulated that no difference exists between μ_1 and μ_2. This hypothesis is a null hypothesis and may be written

[11.1] $$H_0 : \mu_1 - \mu_2 = 0$$

The symbol H_0 represents the null hypothesis. Very simply, this hypothesis asserts that no difference exists between the two population means. Note that the statement $\mu_1 - \mu_2 = 0$ is the same as $\mu_1 = \mu_2$. Thus an alternative formulation of the hypothesis is to assert that the two samples are drawn from populations having the same mean. In general, regardless of the particular statistics used, the null hypothesis is a trial hypothesis asserting that no difference exists between population parameters. Thus a null hypothesis about two variances would take the form $H_0 : \sigma_1^2 - \sigma_2^2 = 0$, or $H_0 : \sigma_1^2 = \sigma_2^2$.

The logical steps used in applying a test of significance are these. *First,* assume the null hypothesis; that is, operate on the trial hypothesis that the treatment applied will have no effect. *Second,* examine the empirical data. Where the hypothesis pertains to two means, examine the difference between the two means, $\bar{X}_1 - \bar{X}_2$. *Third,* the question is asked, what is the probability of obtaining a difference equal to or greater than the one observed in drawing samples at random from populations where the null hypothesis is assumed to be true? In the case of two means, what is the probability of obtaining a difference equal to or greater than $\bar{X}_1 - \bar{X}_2$ in drawing random samples from populations where $\mu_1 - \mu_2 = 0$? *Fourth,* if this probability is *small,* the observed result being highly improbable on the basis of the null hypothesis, rejection of the null hypothesis is warranted. This means that the observed difference cannot reasonably be explained by sampling error and presumably may be attributed to the treatment applied. Thus the result may be said to be significant. If this probability cannot be considered small and the observed result is not highly improbable, then sampling error may account for the difference observed. Hence we cannot with confidence infer that the difference results from the treatment applied.

In the testing of any statistical hypothesis, it is necessary to specify an alternative hypothesis. This alternative is accepted if the initial hypothesis is rejected. Thus in the testing of the hypothesis $H_0 : \mu_1 - \mu_2 = 0$, the alternative may be $H_1 : \mu_1 - \mu_2 \neq 0$. Under certain circumstances, as described in detail in Section 11.5, some advantages may attach to a test of a null hypothesis against the alternative $\mu_1 - \mu_2 > 0$, or the alternative $\mu_1 - \mu_2 < 0$. The alternative hypothesis under consideration should be clearly recognized.

11.3 TWO TYPES OF ERROR

In making a decision about the null hypothesis H_0, two types of error may be made. An alternative H_1 may be accepted when the null hypothesis H_0 is true. This is called a Type I error. The null H_0 may be accepted when an alternative hypothesis H_1 is true. This is called a Type II error. The probabilities of committing Type I and Type II errors are represented, respectively, by α and β.

The situation may be represented as follows:

	H_0 is true	H_1 is true
Accept H_1	α	correct decision
Accept H_0	correct decision	β

When we accept H_1, and H_1 is true, this is a correct decision about nature. When we accept H_0, and H_0 is true, this is also a correct decision about nature. The acceptance of H_1 when H_0 is true and the acceptance of H_0 when H_1 is true are both errors. Both are incorrect decisions about nature.

The situation above is somewhat analogous to that which arises in the acceptance or rejection of applicants for employment or admission to a university. For example, a university admissions officer may accept applicants who subsequently are successful in the university, or she may reject applicants who would have failed if they had been accepted. In either case a correct decision may be said to have been made. On the other hand, the admissions officer may reject an applicant who would have been successful if he had been accepted, or she may accept an applicant who subsequently fails. In both instances an error has been made.

11.4 LEVELS OF SIGNIFICANCE

The probability of Type I, or α, error is called the level of significance of a test. Ordinarily the investigator adopts, perhaps rather arbitrarily, a particular level of significance. It is a common convention to adopt levels of significance of either .05 or .01. If the probability is equal to or less than .05 of asserting that there is a difference between two means, for example, when no such difference exists, then the difference is said to be significant at the .05, or 5 percent, level or less. Here the chances are 5 in 100, or less, that the difference could result when there is no difference in the population values. If the probability is .01, or less, the difference is said to be significant at the .01, or 1 percent, level. The .05 and .01 probability levels are descriptive of our degree of confidence that a real difference exists or that the observed difference is not due to the caprice of sampling. Usually in evaluating an experimental result, it is unnecessary to determine the probabilities with a high degree of accuracy. For most practical purposes it is sufficient to designate the probability as $p \leq .05$, or $p \leq .01$, or possibly $p \leq .001$ if the result is highly significant.

Decisions regarding the rejection, or otherwise, of the null hypothesis at a level of significance α are commonly made without reference to Type II, or β, error. A detailed description of the functioning of β error is beyond the scope of this book. A few comments, however, may be appropriate. Consider, for example, the difference between two means μ_1 and μ_2. For any particular value of α, say, .05, the value of β is a function of sample size N and the actual difference between μ_1 and μ_2. For specified α and N, the value of β, which is the probability of failing to recognize the existence of a difference when such a difference exists, will decrease with increase in the difference between μ_1 and μ_2. This means that the larger the difference between μ_1 and μ_2 the less likely we are to accept H_0. For a specified difference between μ_1 and μ_2 and a specified N the value of β will increase as α decreases. Accordingly, if too strict a level of significance is adopted, we may fail to reject the null hypothesis when in fact a fairly large difference between μ_1 and μ_2 exists. For any specified difference between μ_1 and μ_2, and any α, the Type II error β is a function of sample size N. The smaller the sample the greater the value of β. This means that although a large difference may exist between μ_1 and μ_2, it may be difficult to prove for small samples.

Although, quite clearly, failure to reject the null hypothesis does not imply that the null hypothesis is true, many investigators exhibit an inclination to conclude, even for quite small samples, that no difference, or a trivial difference, exists when a required level of significance is not achieved. Our discussion of Type II error clearly indicates that such conclusions are unwarranted. It would, of course, be possible to establish dual criteria such that if α were greater than, say, .05 and β were less than, say, .05, the investigator, in practice, might be allowed to conclude that the difference was nonexistent or inconsequential.

11.5 DIRECTIONAL AND NONDIRECTIONAL TESTS

An investigator may wish to test the null hypothesis $H_0: \mu_1 - \mu_2 = 0$ against the alternative $H_1: \mu_1 - \mu_2 \neq 0$. This means that if H_0 is rejected, the decision is that a difference exists between the two means. No assertion is made about the direction of the difference. Such a test is a *nondirectional* test. A test of this kind is sometimes called a *two-tailed* or *two-sided* test, because if the normal distribution, or the distribution of t, is used, the two tails, or the two sides, of the distribution are employed in the estimation of probabilities. Consider a 5 percent significance level. If the sampling distribution is normal, 2.5 percent of the area of the curve falls to the right of 1.96 standard deviation units above the mean, and 2.5 percent falls to the left of 1.96 standard deviation units below the mean. The area outside these limits is 5 percent of the total area under the curve. Under the null hypothesis the chances are 2.5 in 100 of getting a difference of 1.96 standard deviation units in one direction because of chance factors alone, and 2.5 in 100 in the other direction. Hence the chances in either direction are 5 in 100. Thus for significance at the 5 percent level for a nondirectional test, when the sampling distribution is normal, the observed difference must be equal to or greater than 1.96 times the standard deviation of the sampling distribution of differences. For significance at the 1 percent level a value of 2.58 is required for a nondirectional test. A nondirectional test is appropriate if concern is with the absolute magnitude of the difference, that is, with the difference regardless of sign.

Under certain circumstances we may wish to make a decision about the direction of the difference. It has been argued that few instances exist in scientific work where the traditional nondirectional test is of interest. If concern is with the direction of the difference we may test the hypothesis $H_0: \mu_1 - \mu_2 \leq 0$ against the alternative $H_1: \mu_1 - \mu_2 > 0$, or the hypothesis $H_0: \mu_1 - \mu_2 \geq 0$ against the alternative $H_2: \mu_1 - \mu_2 < 0$. Note that the symbol H_0 has been used to denote three different hypotheses: a hypothesis of no difference, a hypothesis of equal to or less than, and a hypothesis of equal to or greater than. Conventionally the term null hypothesis has been restricted to a hypothesis of no difference. It is not inappropriate, as pointed out by Kaiser (1960), to extend the meaning of the null hypothesis to include hypotheses of equal to or less than and equal to or greater than. Such tests are directional one-sided, or one-tailed, tests. If the normal, or t, distribution is used, one side or one tail only is employed to estimate the required probabilities. To reject $H_0 : \mu_1 - \mu_2 \leq 0$ and accept $H_1 : \mu_1 - \mu_2 > 0$, using the normal distribution, a normal deviate greater than $+1.64$ is required for significance at the .05 level. Likewise to reject $H_0 : \mu_1 - \mu_2 \geq 0$ and accept $H_2 : \mu_1 - \mu_2 < 0$, the corresponding normal deviate is less than -1.64. The figure 1.64 derives from the fact that for a normal distribution 5 percent of the area of the curve falls beyond $+1.64$ standard deviation units above the mean, and 5 percent beyond -1.64 standard deviation units below the mean. For significance at the .01 level for a directional test, normal deviates greater than $+2.33$ or less than -2.33 are required.

Kaiser (1960) discusses a directional two-sided test, which in the end result uses much the same procedure as the one described above. His procedure requires a decision between three hypotheses, $H_0 : \mu_1 - \mu_2 = 0$, $H_1 : \mu_1 - \mu_2 > 0$, and $H_2 : \mu_1 - \mu_2 < 0$. The rules here, at the 5 percent significance level, are that we decide upon H_0 when the normal deviate falls between -1.64 and $+1.64$. We decide upon H_1 when the normal deviate is greater than $+1.64$ and upon H_2 when the normal deviate is less than -1.64. This test amounts in practice to the same thing as making two directional one-sided tests. Involved in its development are errors of the third kind. These relate to a decision about a difference in the wrong direction.

When is it appropriate to use a directional as distinct from a nondirectional test? This question is open to some controversy. Clearly there are many instances in research where the direction of the differences is of substantial interest; indeed, it has been argued that there are few, if any, instances where the direction is not of interest. At any rate it is the opinion of this writer that directional tests should be used more frequently.

11.6 SIGNIFICANCE OF THE DIFFERENCE BETWEEN TWO MEANS FOR INDEPENDENT SAMPLES

Let $\bar{X}_1$ and $\bar{X}_2$ be the means of two samples of N_1 and N_2 cases, respectively, drawn from normally distributed populations with population means μ_1 and μ_2, and variances σ_1^2 and σ_2^2. The null hypothesis is $H_0 : \mu_1 - \mu_2 = 0$. The assumption is made that the samples are drawn from populations with equal variance; that is, that $\sigma_1^2 = \sigma_2^2 = \sigma^2$. This is known as the homogeneity of variance assumption. If this assumption is warranted, data from the two samples may be combined to obtain an unbiased estimate of the population variance σ^2. This estimate is obtained by adding together or pooling the two sums of squares of deviations about the two sample means and dividing this by the total number of degrees of freedom. This unbiased estimate of σ^2 may be written as

[11.2]
$$s^2 = \frac{\sum^{N_1}(X - \bar{X}_1)^2 + \sum^{N_2}(X - \bar{X}_2)^2}{N_1 + N_2 - 2}$$
$$= \frac{s_1{}^2(N_1 - 1) + s_2{}^2(N_2 - 1)}{N_1 + N_2 - 2}$$

The number of degrees of freedom associated with the two sums of squares is $N_1 - 1$ and $N_2 - 1$, respectively, and the total number of degrees of freedom is $N_1 + N_2 - 2$. The standard error of the difference between the two means is then given by

[11.3]
$$s_{\bar{x}_1 - \bar{x}_2} = \sqrt{\frac{s^2}{N_1} + \frac{s^2}{N_2}}$$
$$= \sqrt{s^2\left(\frac{N_1 + N_2}{N_1 N_2}\right)}$$

The difference between means, $\bar{X}_1 - \bar{X}_2$, is then divided by this estimate of the standard error to obtain the ratio

[11.4]
$$t = \frac{\bar{X}_1 - \bar{X}_2}{s_{\bar{x}_1 - \bar{x}_2}} = \frac{\bar{X}_1 - \bar{X}_2}{\sqrt{s^2/N_1 + s^2/N_2}}$$

This ratio has a distribution of t with $N_1 + N_2 - 2$ degrees of freedom. The values of t required for significance at the .05 and .01 levels will vary, depending on the number of degrees of freedom, and may be obtained by consulting Table B of the Appendix.

Another formula for calculating s^2 is

[11.5]
$$s^2 = \frac{\sum^{N_1} X^2 - \left(\sum^{N_1} X\right)^2/N_1 + \sum^{N_2} X^2 - \left(\sum^{N_2} X\right)^2/N_2}{N_1 + N_2 - 2}$$

This, if desired, may be further modified by writing

$$\frac{\left(\sum^{N_1} X\right)^2}{N_1} = N_1\bar{X}_1{}^2 \qquad \text{and} \qquad \frac{\left(\sum^{N_2} X\right)^2}{N_2} = N_2\bar{X}_2{}^2$$

Let the following be error scores obtained for two groups of experimental animals in running a maze under different experimental conditions.

Group *A*	16	9	4	23	19	10	5	2
Group *B*	20	5	1	16	2	4		

The following statistics are calculated from these data:

	Group *A*	Group *B*
N	8	6
ΣX	88	48
$\bar{X}$	11	8
ΣX^2	1,372	702

The unbiased estimate of variance is

$$s^2 = \frac{1{,}372 - 88^2/8 + 702 - 48^2/6}{8 + 6 - 2} = 60.17$$

The t ratio is then

$$t = \frac{11 - 8}{\sqrt{60.17/8 + 60.17/6}} = .72$$

The number of degrees of freedom in this example is $8 + 6 - 2 = 12$. For 12 degrees of freedom a t equal to 2.179 is required for significance at the .05 level. In this example the difference between means is not significant. No adequate grounds exist for rejecting the null hypothesis. We are not justified in drawing the inference from these data that the two experimental conditions are exerting a differential effect on the behavior of the animals.

The t test described here assumes that the distributions of the variables in the populations from which the samples are drawn are normal. It assumes also that these populations have equal variances. This latter condition is referred to as homogeneity of variance. The t test should be used only when there is reason to believe that the population distributions do not depart too grossly from the normal form and the population variances do not differ markedly from equality. Tests of normality and homogeneity of variance may be applied, but these tests are not very sensitive for small samples.

11.7 SIGNIFICANCE OF THE DIFFERENCE BETWEEN TWO MEANS FOR CORRELATED SAMPLES

Consider a situation where a single group of subjects is studied under two separate experimental conditions. The data may, for example, be autonomic response measures under stress and nonstress or measures of motor performance in the presence or absence of a drug. The data are composed of pairs of measurements. These may be correlated. This circumstance leads to a test of significance between means different from that for independent samples. A procedure for testing significance may be applied without actually computing the correlation coefficient between the paired observations. This method is sometimes called the difference method. Its nature is simply described. Given a set of N paired observations, the difference between each pair may be obtained. Denote any pair of observations by X_1 and X_2 and the difference between any pair $X_1 - X_2$ by D. The mean difference over all pairs is $(\Sigma D)/N = \bar{D}$. It is readily observed that the difference between the means of the two groups of observations is equal to the mean difference. The difference between any pair of observations is $X_1 - X_2 = D$. Summing over N pairs yields $\Sigma X_1 - \Sigma X_2 = \Sigma D$. Dividing by N, we obtain $\bar{X}_1 - \bar{X}_2 = \bar{D}$. Since the mean difference is the difference between the two means, we may test for the significance of the differences between means by testing whether or not $\bar{D}$ is significantly different from 0. Here in effect we treat the D's as a variable and test the difference between the mean of this variable and 0.

An unbiased estimate of the variance of the D's is given by

[11.6] $$s_D^2 = \frac{\Sigma(D - \bar{D})^2}{N - 1}$$

where N is the number of paired observations. Using this unbiased estimate, the sampling variance of $\bar{D}$ is given by

[11.7] $$s_{\bar{D}}^2 = \frac{s_D^2}{N}$$

To test whether $\bar{D}$ is significantly different from 0, we divide $\bar{D}$ by its standard error to obtain

[11.8] $$t = \frac{\bar{D}}{s_{\bar{D}}}$$

The number of degrees of freedom used in evaluating t is 1 less than the number of pairs of observations, or $N - 1$. The reader should note that the $\bar{D}$ in the numerator of the above formula is in effect $\bar{D} - 0$, which is of course $\bar{D}$. This test is concerned with the significance of $\bar{D}$ from 0.

Another formula for calculating t is

[11.9] $$t = \frac{\Sigma D}{\sqrt{[N\Sigma D^2 - (\Sigma D)^2]/(N - 1)}}$$

The data below are those obtained for a group of 10 subjects on a choice–reaction-time experiment under stress and nonstress conditions, the stress agent being electric shock. The figures are the number of false reactions over a series of trials. The problem here is to test whether the means

Subject	Stress X_1	Nonstress X_2	D	D^2
1	7	5	2	4
2	9	15	−6	36
3	4	7	−3	9
4	15	11	4	16
5	6	4	2	4
6	3	7	−4	16
7	9	8	1	1
8	5	10	−5	25
9	6	6	0	0
10	12	16	−4	16
Sum	76	89	−13	127
Mean	7.60	8.90	−1.30	

under the two conditions are significantly different. These means are 7.60

and 8.90. The difference between them is equal to the mean of the differences, or −1.30. The sum and sum of squares of D are, respectively, −13 and 127. Hence

$$t = \frac{-13}{\sqrt{[10 \times 127 - (-13)^2]/(10-1)}} = -1.18$$

We may ignore the negative sign of t and consider only its absolute magnitude. The number of degrees of freedom associated with this value of t is 9. For 9 degrees of freedom we require a t of 2.262 for significance at the 5 percent level. The observed value of t is well below this, and the difference between means is not significant. We cannot justifiably argue from these data that the mean number of false reactions under the two conditions is different.

The method described above takes into account the correlation between the paired measurements. This results because the variance of differences is related to the correlation between the paired measurements by the formula

[11.10] $$s_D^2 = s_1^2 + s_2^2 - 2r_{12}s_1s_2$$

When s_1, s_2, and r_{12} have been computed, as will frequently be the case, the variance of differences s_D^2 can be readily obtained from the above formula and need not be obtained by direct calculation on the differences themselves. A positive correlation between the paired measurements will reduce the size of s_D^2 and $s_{\bar{D}}^2$.

11.8 ROBUSTNESS OF THE t TEST

A statistical procedure is said to be *robust* when the probability statements resulting from its use are insensitive to, or not seriously affected by, violations of the assumptions made in its development. The t test for means of independent samples assumes (1) normality of the distributions of the variables in the populations from which the samples are drawn and (2) homogeneity of variance, that is, $\sigma_1^2 = \sigma_2^2$. When these assumptions are violated, how robust is the t test? This problem has been explored and discussed by a variety of investigators, more recently by Glass, Peckham, and Saunders (1972).

For large samples, say 25 or 30, nonnormality of the populations obviously will not seriously affect the estimation of probabilities, except possibly in cases of extreme skewness. Empirical evidence suggests that for quite small samples also, say 5 or 10, departures from normality, unless gross, will not seriously affect the estimation of probabilities for a nondirectional t test. A directional t test is apparently more seriously affected by nonnormality.

When the homogeneity of variance assumption is violated, that is, when $\sigma_1^2 \neq \sigma_2^2$, how robust is the t test? The answer to this question depends on both the size and the relative size of the two samples, and also on the difference between the two variances.

In testing the difference between means for independent samples, assuming homogeneity of variance, a single variance estimate, s^2, is used in forming the t ratio, $t = (\bar{X}_1 - \bar{X}_2)/\sqrt{s^2/N_1 + s^2/N_2}$. If the two population variances are different, two variance estimates are obtained; that is, s_1^2 and s_2^2 are estimates of σ_1^2 and σ_2^2, respectively. If the difference between means is divided by the standard error of the difference, the resulting ratio is

[11.11] $$t' = \frac{\bar{X}_1 - \bar{X}_2}{\sqrt{s_1^2/N_1 + s_2^2/N_2}}$$

This ratio, denoted by t', has neither a normal nor a t distribution. The distribution of t' is approximated by what is called the Behrens-Fisher distribution, tables of which are available in Fisher and Yates (1963).

An approximate method for estimating the required critical values of t' has been proposed by Cochran and Cox (1957). *First,* calculate t'. *Second,* if a nondirectional test at the .05 level is required, consult a t table and obtain the two values of t, t_1 and t_2, required for significance at the .05 level for $N_1 - 1$ and $N_2 - 1$ degrees of freedom, respectively. *Third,* the approximate value of t' required at the .05 level is given by

[11.12] $$t'_{.05} = \frac{(t_1 s_1^2)/N_1 + (t_2 s_2^2)/N_2}{s_1^2/N_1 + s_2^2/N_2}$$

Fourth, compare t' and $t'_{.05}$. If $t' > t'_{.05}$, the result is significant at the .05 level for a nondirectional test. Obviously other levels of significance, and a directional instead of nondirectional test, may be used.

The following is an example.

Group I	*Group II*
$\bar{X}_1 = 50.50$	$\bar{X}_2 = 40.50$
$N_1 = 10$	$N_2 = 5$
$s_1^2 = 100.00$	$s_2^2 = 25.00$
$t_1 = 2.228$	$t_2 = 2.776$

The critical value $t'_{.05}$ is

$$t'_{.05} = \frac{(2.228 \times 100.00)/10 + (2.776 \times 25.00)/5}{100.00/10 + 25.00/10} = 2.41.$$

The value of t' obtained using formula [11.11] is $t' = 2.58$. Because $t' > t_{.05}$, the result is significant at the .05 level for a nondirectional test.

If the usual t test had been applied to the data of the example above, ignoring the difference in variance, a $t = 2.08$ would result. For $df = 13$ this does not quite reach significance at the .05 level, the required value being $t_{.05} = 2.16$.

When $\sigma_1^2 \neq \sigma_2^2$ and $N_1 \neq N_2$, the t test is biased. When the larger sample has the larger variance, the ordinary t test will yield too few significant results; that is, it is too conservative. When the larger sample has the smaller variance, the t test will yield too many significant results; that is, it is insufficiently rigorous. For further discussion of the above points see Glass, Peckham, and Saunders (1972). Another approximate method which adjusts the number of degrees of freedom has been proposed by Welch (1938). Also, see fourth edition of this text.

BASIC TERMS AND CONCEPTS

Test of significance

Null hypothesis

Type I error

Type II error

Directional test

Independent samples

Correlated samples

t test for independent samples

t test for correlated samples

Robustness

Homogeneity of variance

EXERCISES

1 The following are data for two samples of subjects under two experimental conditions:

Sample A	2	5	7	9	6	7
Sample B	4	16	11	9	8	

Test the significance of the difference between means using a nondirectional test.

2 The following are data for two independent samples:

	Sample A	Sample B
$\bar{X}$	124	120
N	50	36
$\Sigma(X - \bar{X})^2$	5,512	5,184

Test whether the mean for sample A is equal to or greater than that for sample B.

3 The following are paired measurements obtained for a sample of eight subjects under two conditions:

Condition *A*	8	17	12	19	5	6	20	3
Condition *B*	12	31	17	17	8	14	25	4

Test the significance of the difference between means using a nondirectional test.

4 Calculate t for the following data:

	Sample *A*	Sample *B*
$\bar{X}$	20	25
N	25	10
ΣX^2	12,500	7,900

5 For a sample of 26 pairs of measurements, $\Sigma D = 52$ and $\Sigma D^2 = 400$. Calculate t.

6 What advantages attach to matched groups or paired observations in experimentation?

7 The means for two independent samples of 10 and 17 cases are 9.63 and 14.16, respectively. The unbiased variance estimates are 64.02 and 220.30. Use the method proposed by Cochran and Cox to test the significance of the difference between the two means.

ANSWERS TO EXERCISES

1 $t = 1.74$; $p > .05$ 2 $t = 1.62$; $p > .05$

3 $t = 2.81$; $p < .05$ 4 $t = 1.19$

5 $t = 2.96$

6 Groups are ordinarily matched to reduce sampling error. If matching leads to a positive correlation between the values of the independent variables for the matched pairs, the standard error of the difference between means, for example, will be less than that which would have been obtained for independent samples without matching.

7 Cochran and Cox: $t_{.05} = 2.17$; $p > .05$

12

TESTS OF SIGNIFICANCE: OTHER STATISTICS

12.1 INTRODUCTION

In Chapter 11, problems associated with the application of tests of significance to arithmetic means were discussed. Not infrequently, tests of significance for proportions, variances, correlation coefficients, or other statistics are required. The general rationale underlying the application of such tests of significance is precisely the same as that for arithmetic means, although the technical procedures used in estimating the required probabilities are different. The present chapter discusses procedures for applying tests of significance to proportions, variances, and correlation coefficients, for independent and correlated samples.

12.2 SIGNIFICANCE OF THE DIFFERENCE BETWEEN TWO INDEPENDENT PROPORTIONS

Questions arise in the interpretation of experimental results which require a test of significance of the difference between two independent proportions. The data comprise two samples drawn independently. Of the N_1 members in the first sample, f_1 have the attribute A. Of the N_2 members in the second sample, f_2 have the attribute A. The proportions having the attributes in the two samples are $f_1/N_1 = p_1$ and $f_2/N_2 = p_2$. Can the two samples be regarded as random samples drawn from the same population? Is p_1 significantly different from p_2? To illustrate, in a public opinion poll the proportion .65 in a sample or urban residents may express a favorable attitude toward a particular issue as against a proportion .55 in a sample of rural residents. May the difference between the proportions be interpreted as indicative of an actual urban-rural difference in opinion? To illustrate further, the proportion of failures in air-crew training in two training periods may be .42 and .50. Does this represent a significant change in the proportion of failures, or may the difference be attributed to sampling considerations?

The standard error of a single proportion is estimated by the formula

$$s_p = \sqrt{\frac{pq}{N}}$$

where p = sample value of a proportion
$q = 1 - p$

The standard error of the difference between two proportions based on independent samples is estimated by

[12.1] $$s_{p_1-p_2} = \sqrt{pq\left(\frac{1}{N_1} + \frac{1}{N_2}\right)}$$

where p is an estimate based on the two samples combined. The value p is obtained by adding together the frequency of occurrence of the attributes in the two samples and then dividing this by the total number in the two samples. Thus

$$p = \frac{f_1 + f_2}{N_1 + N_2}$$

where f_1 and f_2 are the two frequencies.

The justification for combining data from the two samples to obtain a single estimate of p resides in the fact that in all cases where the difference between two proportions is tested, the null hypothesis is assumed. This hypothesis states that no difference exists in the population proportions. Because the null hypothesis is assumed, we may use an estimate of p based on the data combined for the two samples. This procedure is analogous to that used in the t test for the difference between means for independent samples where the sums of squares for the two samples are combined to obtain a single variance estimate.

To test the difference between two proportions we divide the observed difference between the proportions by the estimate of the standard error of the difference to obtain

[12.2] $$z = \frac{p_1 - p_2}{s_{p_1-p_2}} = \frac{p_1 - p_2}{\sqrt{pq[(1/N_1) + (1/N_2)]}}$$

The value z may be interpreted as a deviate of the unit normal curve, provided N_1 and N_2 are reasonably large and p is neither very small nor very large. As usual for a two-tailed test, values of 1.96 and 2.58 are required for significance at the 5 and 1 percent levels.

How large should the N's be and how far should p depart from extreme values before this ratio can be interpreted as a deviate of the unit normal curve? An arbitrary rule may be used here. If the smaller value of p or q multiplied by the smaller value of N exceeds 5, then the ratio may be interpreted with reference to the normal curve. Thus if $p = .60$, $q = .40$, $N_1 = 20$, and $N_2 = 30$, the product $.40 \times 20 = 8$ and the normal curve may be used.

To illustrate, we refer to data obtained in a study of the attitudes of Canadians to immigrants and immigration policy. Independent samples of French- and English-speaking Canadians were used. Subjects were asked whether they agreed or disagreed with present government immigration practices. In the French-speaking sample of 300 subjects, 176 subjects indicated agreement. The proportion p_1 is $176/300 = .587$. In the English-speaking sample of 500 subjects, 384 indicated agreement. The proportion p_2 is $384/500 = .768$. By combining data for the two samples we obtain a value

$$p = \frac{176 + 384}{300 + 500} = .700$$

The value of q is $1 - .700 = .300$. The estimate of the standard error of the difference is $s_{p_1-p_2} = \sqrt{.700 \times .300(\frac{1}{300} + \frac{1}{500})} = .033$. The required z value is $z = (.768 - .587)/.033 = 5.48$.

Interpreting the value 5.48 as a unit-normal-curve deviate we observe immediately that the difference is highly significant. The chances are one in a great many millions that the observed difference could result from sampling. We may very safely conclude from these data that a real difference exists between French- and English-speaking Canadians on the question asked.

An alternative, but closely related, method exists for testing the significance of the difference between proportions for independent samples. This method uses chi square and is described in Chapter 13.

12.3 SIGNIFICANCE OF THE DIFFERENCE BETWEEN TWO CORRELATED PROPORTIONS

Frequently in psychological work we wish to test the significance of the difference between two proportions based on the same sample of individuals or on matched samples. The data consist of pairs of observations and are usually nominal in type. The paired observations may exhibit a correlation, which must be taken into consideration in testing the difference between proportions. To illustrate, a psychological test may be administered to a sample of N individuals. The proportions passing items 1 and 2 are p_1 and p_2. Paired observations are available for each individual. One individual may "pass" item 1 and also "pass" item 2. A second individual may "pass" item 1 and "fail" item 2. A third individual may "fail" both items. The paired observations may be tabulated in a 2×2 table. A tendency may exist for individuals who pass item 1 to also pass item 2 and for those who fail item 1 to also fail item 2. Thus the items are correlated. A further illustration arises where attitudes are measured with an attitude scale before and after a program designed to induce attitude change. On any particular attitude item a before response and an after response are available. Thus the data are comprised of a set of paired observations. To apply a test of significance to the difference between before and after proportions on any particular item requires that the correlation between responses be taken into account.

We proceed by tabulating the data in the form of a fourfold, or 2 × 2, table. A table with four cell frequencies is obtained. By way of illustration, assume that the data are "pass" and "fail" on two test items. The data may be represented schematically as follows:

Frequencies

		Item 2: Fail	Item 2: Pass	
Item 1	Pass	A	B	$A+B$
	Fail	C	D	$C+D$
		$A+C$	$B+D$	N

Proportions

		Item 2: Fail	Item 2: Pass	
Item 1	Pass	a	b	p_1
	Fail	c	d	q_1
		q_2	p_2	1.00

The capital letters represent frequencies. The lowercase letters are proportions obtained by dividing the frequencies by N. The proportions passing the two items are p_1 and p_2. We wish to test the significance of the difference between p_1 and p_2.

An estimate of the standard error of the difference between two correlated proportions is given by the formula

[12.3] $$s_{p_1-p_2} = \sqrt{\frac{a+d}{N}}$$

This formula is due to McNemar (1947). It takes into account the correlation between the paired observations. A normal deviate z is obtained by dividing the difference between the two proportions by the standard error of the difference.

Thus

[12.4] $$z = \frac{p_1 - p_2}{\sqrt{\frac{a+d}{N}}}$$

When the sum of the two cell frequencies, $A + D$, is reasonably large, this ratio can be interpreted as a unit-normal-curve deviate, values of 1.96 and 2.58 being required for significance at the 1 and 5 percent levels for a two-tailed test. In this context a reasonably large value of $A + D$ may be taken as about 20 or above.

It may be shown that the formula for the value of z given above reduces to

$$z = \frac{D - A}{\sqrt{A + D}} \tag{12.5}$$

where A and D are the cell frequencies. For computational purposes this is the more useful formulation.

To illustrate, consider the following fictitious data relating to attitude change. Let us assume an initial testing followed by a program intended to produce a change in attitude, and then a second testing with the same attitude scale. On a particular question let the data for the two testings be:

Frequencies

		2d Disagree	2d Agree	
1st	Agree	10	50	60
1st	Disagree	110	30	140
		120	80	200

Proportions

		2d Disagree	2d Agree	
1st	Agree	.05	.25	.30
1st	Disagree	.55	.15	.70
		.60	.40	1.00

Inspection of the above tables indicates a high correlation in response between the first and second testings. We wish to test the significance of the difference between .40 and .30. The standard error of the difference between the two proportions is

$$s_{p_1-p_2} = \sqrt{\frac{.05 + .15}{200}} = .0316$$

The value of z is

$$z = \frac{.40 - .30}{.0316} = 3.16$$

In this case the difference is significant. It exceeds the value of 2.58 required for significance at the 1 percent level for a two-tailed test. Arguments may be advanced for the use of a one-tailed test with the above data. It may be assumed that knowledge of a program intended to induce attitude change may warrant a hypothesis about the direction of the change. In either case the result is significant.

12.4 SIGNIFICANCE OF THE DIFFERENCE BETWEEN VARIANCES FOR INDEPENDENT SAMPLES

Occasions arise where a test of the significance of the difference between the variances of measurements for two independent samples is required.

In the conduct of a simple experiment using control and experimental groups, the effect of the experimental condition may reflect itself not only in a mean difference between the two groups but also in a variance difference. For example, in an experiment designed to study the effect of a distracting agent, such as noise, on motor performance, the effect of the distraction may be to greatly increase the variability of performance, in addition possibly to exerting some effect upon the mean. The variances obtained in any experiment should always be the object of scrutiny and comparison. A common situation, where a test of the significance of the difference between variances is required, is in relation to the t test for the significance of the difference between two means. This test assumes the equality of variances in the populations from which the samples are drawn; that is, it assumes that $\sigma_1^2 = \sigma_2^2 = \sigma^2$. This condition is usually spoken of as homogeneity of variance.

Let s_1^2 and s_2^2 be two variances based on independent samples. We may consider the difference $s_1^2 - s_2^2$. An alternative and more fruitful procedure is to consider the ratio s_1^2/s_2^2 or s_2^2/s_1^2. If the two variances are equal, this ratio will be unity. If they differ and $s_1^2 > s_2^2$, then $s_1^2/s_2^2 > 1$ and $s_2^2/s_1^2 < 1$. A departure of the variance ratio from unity indicates a difference between variances, the greater the departure the greater the difference. Quite clearly, a test of the significance of the departure of the ratio of two variances from unity will serve as a test of the significance of the difference between the two variances.

To apply such a test the sampling distribution of the ratio of two variances is required. To conceptualize such a sampling distribution, consider two normal populations A and B with the same variance σ^2. Draw samples of N_1 cases from A and N_2 cases from B, calculate unbiased variance estimates s_1^2 and s_2^2, and compute the ratio s_1^2/s_2^2. Continue this procedure until a large number of variance ratios is obtained. Always place the variance of the sample drawn from A in the numerator and the variance of the sample drawn from B in the denominator. Some of the variance ratios will be greater than unity; others will be less than unity. The frequency distribution of the variance ratios for a large number of pairs of variances is an experimental sampling distribution. The corresponding theoretical sampling distribution of variance ratios is known as the distribution of F. The variance ratio is known as an F ratio; that is, $F = s_1^2/s_2^2$, or $F = s_2^2/s_1^2$.

In the above illustration samples of N_1 are drawn from one population and samples of N_2 from another. Degrees of freedom $N_1 - 1$ and $N_2 - 1$ are associated with the two variance estimates. A separate sampling distribution of F exists for every combination of degrees of freedom. Table F of the Appendix shows values of F required for significance at the 5 and 1 percent levels for varying combinations of degrees of freedom. This table shows values of F equal to or greater than unity. It does not show values of F less than unity. The number of degrees of freedom associated with the variance estimates in the numerator and denominator are shown along the top and to the left, respectively, of Table F. The numbers in lightface type are the values for significance at the 5 percent level, and those in boldface type the values at the 1 percent level. These values cut off 5 and 1 percent of one tail of the distribution of F.

In testing the significance of the difference between two variances, the null hypothesis $H_0 : \sigma_1^2 = \sigma_2^2 = \sigma^2$ is assumed. We then find the ratio of the two unbiased variance estimates. These are

$$s_1^2 = \frac{\Sigma(X - \bar{X}_1)^2}{N_1 - 1}$$

and

$$s_2^2 = \frac{\Sigma(X - \bar{X}_2)^2}{N_2 - 1}$$

No prior grounds exist for deciding which variance estimate should be placed in the numerator and which in the denominator of the F ratio. In practice the larger of the two variance estimates is always placed in the numerator and the smaller in the denominator. In consequence the F ratio in this situation is always greater than unity. The F ratio is calculated, referred to Table F of the Appendix, and a significance level determined. At this point a slight complication arises. The obtained significance level must be doubled. Table F shows values required for significance at the 5 and 1 percent levels. In comparing the variances for two independent groups, these become the 10 and 2 percent levels. The reason for this complication resides in the fact that the larger of the two variances has been placed in the numerator of the F ratio. This means that we have considered one tail only of the F distribution. Not only must we consider the probability of obtaining s_1^2/s_2^2 but also the probability of s_2^2/s_1^2. Where interest is in the significance of the difference, regardless of direction, the required percent or probability levels are simply obtained by doubling those shown in Table F.

Table F has been prepared for use with the analysis of variance (Chapter 15) which makes extensive use of the F ratio. In the analysis of variance the decision as to which variance estimate should be put in the numerator and which in the denominator is made on grounds other than their relative size. Consequently, in the analysis of variance, F ratios less than unity can occur, and Table F provides the appropriate probabilities without any doubling procedure.

To illustrate, a psychological test is administered to a sample of 31 boys and 26 girls. The sum of squares of deviations, $\Sigma(X - \bar{X})^2$, is 1,926 for boys and 2,875 for girls. Unbiased variance estimates are obtained by dividing the sum of squares by the number of degrees of freedom. The df for boys is $31 - 1 = 30$ and for girls $26 - 1 = 25$. The variance estimate for boys is $1{,}926/30 = 64.20$ and for girls $2{,}875/25 = 115.00$.

Are boys significantly different from girls in the variability of their performance on this test? The F ratio is $115.00/64.20 = 1.79$. The df for the numerator is 25 and for the denominator 30. Referring this F to Table F we see that a value of F of about 1.88 is required for significance at the 5 percent level, and doubling this we obtain the 10 percent level. It is clear, therefore, that the difference between the variances for boys and girls cannot be considered statistically significant. The evidence is insufficient to warrant rejection of the null hypothesis.

12.5 SIGNIFICANCE OF THE DIFFERENCE BETWEEN CORRELATED VARIANCES

Given a set of paired observations, the two variances are not independent estimates. Data of this kind arise when the same subjects are tested under two experimental conditions, or matched samples are used. For example, in an experiment designed to study the effects of an educational program on attitude change, attitudes may be measured, an educational program applied, and attitudes remeasured. It may be hypothesized that some change in variance of attitude-test scores may result. An increase in variance may mean that the effect of the program is to reinforce existing attitudes, producing more extreme attitudes among individuals at both ends of the attitude continuum. A decrease in variance may mean that the effect of the program is to produce an attitudinal regression to greater uniformity.

If s_1^2 and s_2^2 are the two unbiased variance estimates and r_{12} is the correlation between the paired observations, the quantity

[12.6]
$$t = \frac{(s_1^2 - s_2^2)\sqrt{N-2}}{\sqrt{4s_1^2 s_2^2(1 - r_{12}^2)}}$$

has a t distribution with $N - 2$ degrees of freedom.

By way of illustration let s_1^2 and s_2^2 be unbiased variance estimates of attitude-scale scores before and after the administration of an educational program. Let $s_1^2 = 153.20$ and $s_2^2 = 102.51$ where $N = 38$. The correlation between the before-and-after attitude measures is .60. Are the two variances significantly different from each other? We obtain

$$t = \frac{(153.20 - 102.51)\sqrt{38-2}}{\sqrt{4 \times 153.20 \times 102.51(1 - .36)}} = 1.52$$

The number of degrees of freedom is $38 - 2 = 36$. For significance at the 5 percent level, a value of t equal to or greater than about 2.03 is required. The evidence is insufficient to warrant rejection of the null hypothesis. We cannot argue that the intervening educational program has changed the variability of attitudes.

12.6 SAMPLING DISTRIBUTION OF THE CORRELATION COEFFICIENT

We may draw a large number of samples from a population, compute a correlation coefficient for each sample, and prepare a frequency distribution of correlation coefficients. Such a frequency distribution is an experimental sampling distribution of the correlation coefficient. To illustrate, casual observation suggests that a positive correlation exists between height and weight. A number of samples of 25 cases may be drawn at random from a population of adult males and a correlation coefficient between height and weight computed for each sample. These coefficients will display variation one from another. By arranging them in the form of a frequency distribution an experimental sampling distribution of the correlation coefficient is obtained. The mean of this distribution will tend to approach the population value of the correlation coefficient with increase in the number of samples. Its standard deviation will describe the variability of the coefficients from sample to sample. A further illustration may prove helpful. By throwing a pair of dice, say, a white one and a red one, a number of times, a set of paired observations is obtained. A correlation coefficient may be calculated for the paired observations. Since the two dice are independent, the expected value of this correlation coefficient is zero. However, for any particular sample of N throws, a positive or a negative correlation may result. A large number of samples of N throws may be obtained, a correlation coefficient computed for each sample, and a frequency distribution of the coefficients prepared. The mean of this experimental sampling distribution will tend to approach 0, the population value of the correlation coefficient, and its standard deviation will be descriptive of the variability of the correlation in drawing samples of size N from this particular kind of population. Note that here, as in all sampling problems, a distinction is drawn between a population value and an estimate of that value based on a sample. The symbol ρ, the Greek letter "rho," is used to refer to the population value of the correlation coefficient, and r is the sample value.

The shape of the sampling distribution of the correlation coefficient depends on the population value ρ. As ρ departs from 0, the sampling distribution becomes increasingly skewed. When ρ is high positive, say, $\rho = .80$, the sampling distribution has extreme negative skewness. Similarly, when ρ is high negative, say, $\rho = -.80$, the distribution has extreme positive skewness. When $\rho = 0$, the sampling distribution is symmetrical and for large values of N, say, 30 or above, is approximately normal. The reason for the increase in skewness in the sampling distribution as ρ departs from 0 is intuitively plausible. In sampling, for example, from a population where $\rho = .90$, values greater than 1.00 cannot occur, whereas values extending from .90 to -1.00 are theoretically possible. The range of possible variation below .90 is far greater than the range above .90. This suggests that the sample values may exhibit greater variability below then above .90, a circumstance which leads to negative skewness.

The standard deviation of the theoretical sampling distribution of ρ, the standard error, is given by the formula

[12.7] $$\sigma_r = \frac{1 - \rho^2}{\sqrt{N - 1}}$$

When ρ departs appreciably from 0, this formula is of little use, because the departures of the sampling distribution from normality make interpretation difficult.

Difficulties resulting from the skewness of the sampling distribution of the correlation coefficient are resolved by a method developed by R. A. Fisher. Values of r are converted to values of z_r, using the transformation

[12.8] $$z_r = \tfrac{1}{2} \log_e (1 + r) - \tfrac{1}{2} \log_e (1 - r)$$

Values of z_r corresponding to particular values of r need not be computed directly from the above formula but may be simply obtained from Table E in the Appendix. For $r = .50$, the corresponding $z_r = .549$, for $r = .90$, $z_r = 1.472$, and so on. For negative values of r the corresponding z_r values may be given a negative sign. In a number of sampling problems involving correlation, r's are converted to z_r's, and a test of significance is applied to the z_r's instead of to the original r's.

One advantage of this transformation resides in the fact that the sampling distribution of z_r is for all practical purposes independent of ρ. The distribution has the same variability for a given N regardless of the size of ρ. Another advantage is that the sampling distribution of z_r is approximately normal. Values of z_r can be interpreted in relation to the normal curve. The standard error of z_r is given by

[12.9] $$s_{z_r} = \frac{1}{\sqrt{N - 3}}$$

The standard error is seen to depend entirely on the sample size.

The z_r transformation may be used to obtain confidence limits for r. Let $r = .82$ for $N = 147$. The corresponding $z_r = 1.157$. The standard error of z_r, given by $1/\sqrt{N - 3}$, is .083. The 95 percent confidence limits are obtained by taking 1.96 times the standard error above and below the observed value of z_r, or $z_r \pm 1.96 s_{z_r}$. These are $1.157 + 1.96 \times .083 = 1.320$ and $1.157 - 1.96 \times .083 = .994$. These two z_r's may now be converted back to r's, where $z_r = 1.320$, $r = .867$ and where $z_r = .994$, $r = .759$. Thus we may assert with 95 percent confidence that the population value of the correlation coefficient falls within the limits .759 and .867.

In practice we are infrequently concerned with fixing confidence intervals for correlation coefficients. The most frequently occurring problems are testing the significance of a correlation coefficient from 0 and testing the significance of the difference between two correlation coefficients.

12.7 SIGNIFICANCE OF A CORRELATION COEFFICIENT

Testing the significance of the correlation between a set of paired observations is a frequent problem in psychological research. We begin by assuming the null hypothesis that the value of the correlation coefficient is equal to 0, or $H_0 : \rho = 0$. A test of significance may then be applied using the distribution of t. The t value required is given by the formula

[12.10] $$t = r\sqrt{\frac{N-2}{1-r^2}}$$

The number of degrees of freedom associated with this value of t is $N - 2$. The loss of 2 degrees of freedom results because testing the significance of r from 0 is equivalent to testing the significance of the slope of a regression line from 0. The reader will recall that the correlation coefficient is the slope of a regression line in standard-score form. The number of degrees of freedom associated with the variability about a straight line fitted to a set of points is 2 less than the number of observations. A straight line will always fit two points exactly, and no freedom to vary is possible. With three points there is 1 degree of freedom, with four points 2 degrees of freedom, and so on.

Consider an example where $r = .50$ and $N = 20$. We obtain

$$t = .50\sqrt{\frac{20-2}{1-.50^2}} = 2.45$$

The $df = 20 - 2 = 18$. Referring to the table of t, Table B in the Appendix, we find that for this df a t of 2.10 is required for significance at the 5 percent level and a t of 2.88 at the 1 percent level. The sample value of r falls between these two values. It may be said to be significant at the 5 percent level.

Table D of the Appendix presents a tabulation of the values of r required for significance at different levels. We note that where the number of degrees of freedom is small, a large value of r is required for significance. For example, where $df = 5$, a value of $r \gtreqless .754$ is required before we can argue at the 5 percent level that the r is significant. Even for $df = 20$, a value of $r \gtreqless .423$ is required for significance at the 5 percent level. This means that little importance can be attached to correlation coefficients calculated on small samples unless these coefficients are fairly substantial in size.

12.8 SIGNIFICANCE OF THE DIFFERENCE BETWEEN TWO CORRELATION COEFFICIENTS FOR INDEPENDENT SAMPLES

Consider a situation where two correlation coefficients r_1 and r_2 are obtained on two independent samples. The correlation coefficients may, for example, be correlations between intelligence-test scores and mathematics-examination marks for two different freshman classes. We wish to test whether r_1 is significantly different from r_2, that is, whether the two samples can be considered random samples from a common population. The null hypothesis is $H_0 : \rho_1 = \rho_2$ or $H_0 : \rho_1 - \rho_2 = 0$.

The significance of the difference between r_1 and r_2 can be readily tested using Fisher's z_r transformation. Convert r_1 and r_2 to z_r's, using Table E of the Appendix. As stated previously, the sampling distribution of z_r is approximately normal with a standard error given by $s_{z_r} = 1\sqrt{N-3}$. The standard error of the difference between two values of z_r is given by

[12.11] $$s_{z_{r1}-z_{r2}} = \sqrt{s^2_{z_{r1}} + s^2_{z_{r2}}} = \sqrt{\frac{1}{N_1 - 3} + \frac{1}{N_2 - 3}}$$

By dividing the difference between the two values of z_r by the standard error of the difference, we obtain the ratio

[12.12] $$z = \frac{z_{r1} - z_{r2}}{\sqrt{1/(N_1 - 3) + 1/(N_2 - 3)}}$$

This is a unit-normal-curve deviate and may be so interpreted. Values of 1.96 and 2.58 are required for significance at the 1 and 5 percent levels.

To illustrate, let the correlations between intelligence scores and mathematics-examination marks for two freshman classes be .320 and .720. Let the number of students in the first class be 53 and in the second 23. Are the two coefficients significantly different? The corresponding z_r values obtained from Table E of the Appendix are .332 and .908. The required normal deviate is

$$z = \frac{.908 - .332}{\sqrt{1/(53 - 3) + 1/(23 - 3)}} = 2.18$$

The difference between the two correlations is significant at the 5 percent level.

The application of a test of significance in a situation of this kind is simple. The interpretation of what the difference in correlation means may be difficult.

BASIC TERMS AND CONCEPTS

Significance test: independent proportions; correlated proportions

F distribution

Sampling distribution of correlation coefficient

z_r transformation

Significance test: independent correlation coefficients; correlated correlation coefficients

EXERCISES

1 Given two random samples of size 100 with sample values $p_1 = .80$ and $p_2 = .60$, test the significance of the difference between p_1 and p_2.

2 Consider two test items *A* and *B*. In a sample of 100 people, 30 pass item *A* and fail item *B*, whereas 20 fail item *A* and pass item *B*. Are the proportions passing the two items significantly different from each other?

3 In a market survey 24 out of 96 males and 63 out of 180 females indicate a preference for a particular brand of cigarettes. Do the data warrant the conclusion that a sex difference exists in brand preference?

4 On an attitude scale, 63 and 39 individuals from a sample of 140 indicate agreement to items *A* and *B*, respectively, and 29 individuals indicate agreement to both items. Is there a significant difference in the response elicited by the two items?

5 Given two independent samples of size 20 with $s_1^2 = 400$ and $s_2^2 = 625$, test the hypothesis that the variances are significantly different from each other.

6 Given two correlated samples of size 20 with $s_1^2 = 400$, $s_2^2 = 625$, and $r_{12} = .7071$, test the hypothesis that the variances are significantly different from each other.

7 Calculate values of z_r for $r = .70$, $r = .05$, $r = -.60$, and $r = -.99$.

8 Calculate, using formula [12.10], values of t for the following values of r and N:

r	.20	.30	.40	.50
N	50	40	30	20

9 The correlation between psychological-test scores and academic achievement for a sample of 147 freshmen is .40. The corresponding correlation for a sample of 125 sophomores is .59. Do these correlations differ significantly?

ANSWERS TO EXERCISES

1 $z = 3.086$; $p < .01$

2 $z = 1.414$; $p > .05$

3 $z = 1.704$; $p > .05$

4 $z = 3.620$; $p < .001$

5 $F = 1.563$; $p > .10$

6 $t = 1.350$; $p > .05$

7 **a** .867 **b** .050 **c** $-.693$ **d** -2.647

8 **a** $t = 1.414$ **b** $t = 1.939$ **c** $t = 2.309$ **d** $t = 2.450$

9 $z = 2.064$; $p < .05$

13

THE ANALYSIS OF FREQUENCIES USING CHI SQUARE

13.1 INTRODUCTION

We have previously discussed the application of the binomial, normal, t, and F distributions. Another distribution of considerable theoretical and practical importance is the distribution of chi square, or χ^2. The distribution of χ^2, just as the binomial, normal, t, and F distributions, is a theoretical model.

In many experimental situations we wish to compare a set of observed frequencies with a set of theoretical frequencies. The observed frequencies are those obtained empirically by direct observation or experiment. The theoretical frequencies are generated on the basis of some hypothesis or line of theoretical speculation which is independent of the data at hand. The question arises as to whether the differences between the observed and theoretical frequencies are significant. In this context the null hypothesis is that no difference exists between the observed and theoretical frequencies. If the observed frequencies depart significantly from the theoretical frequencies, this constitutes evidence for the rejection of the hypothesis or theory that gave rise to the theoretical frequencies. The theoretical frequencies are usually called the expected frequencies. They are the frequencies the investigator would expect to get if the particular theory in question were true.

To illustrate, consider a coin. The hypothesis may be formulated that the coin is unbiased. Let us now toss the coin 100 times with the following results:

	O	E
Heads	45	50
Tails	55	50

The observed frequencies, denoted by O, are 45 heads and 55 tails. The expected frequencies, denoted by E, are those the investigator would expect to get, if the coin were unbiased, and are 50 heads and 50 tails. How may these two sets of frequencies be compared? What constitutes evidence for the rejection of the hypothesis that the coin is unbiased? The statistic χ^2 is used to answer these questions.

Consider a further illustrative example involving a die. Let us conduct an experiment in which the die is thrown 300 times. The observed frequencies are as shown below:

X	O	E
1	43	50
2	55	50
3	39	50
4	56	50
5	63	50
6	44	50
Total	300	300

In a series of 300 throws the expected or theoretical frequencies of 1, 2, 3, 4, 5, and 6 are 50, 50, 50, 50, 50, and 50. These are the frequencies the investigator would expect to obtain if the die were unbiased. How may the observed and expected frequencies be compared? What constitutes evidence for the rejection of the null hypothesis? The statistic χ^2 is used to answer these questions.

Consider an example involving the binomial distribution. Let us formulate the hypothesis that in litters of rabbits the probability of any birth being either male or female is 1/2. Using the binomial distribution, we ascertain that the expected or theoretical frequencies of 0, 1, 2, 3, 4, 5, and 6 males in 64 litters of six rabbits are 1, 6, 15, 20, 15, 6, and 1. By counting the number of males in 64 litters of six rabbits, the corresponding observed frequencies are 0, 3, 14, 19, 20, 6, and 2. Do the observed and theoretical frequencies differ significantly from each other?

Consider a market research project where two varieties of soap, A and B, are distributed to a random sample of 200 households. After a period of use the households are asked which they prefer. The results show that 115 prefer A and 85 prefer B. The hypothesis may be formulated that no difference exists in consumer preference for the two varieties of soap, that a 50:50 split exists. Do the observed frequencies constitute evidence for the rejection of this hypothesis?

13.2 DEFINITION OF CHI SQUARE

In situations where sets of observed and theoretical frequencies are to be compared, χ^2 is defined by

[13.1] $$\chi^2 = \sum \frac{(O-E)^2}{E}$$

where O and E denote the observed and expected, or theoretical, frequencies respectively. Inspection of this definition shows that χ^2 is a descriptive measure of the magnitude of the discrepancies between the observed and expected frequencies. The larger these discrepancies the larger χ^2 will tend to be. If no discrepancies exist, and the observed and expected frequencies are the same, χ^2 will be 0. The reader should note also that χ^2 in this definition is always 0 or a positive number. Negative values cannot occur.

Table 13.1 shows the calculation of χ^2 in comparing heads and tails for 100 tosses of a coin with the frequencies expected if the coin were unbiased. Here the value of $\chi^2 = 1$. Does this value of χ^2 constitute evidence that the coin is unbiased? How may this value of χ^2 be evaluated?

Table 13.2 illustrates the calculation of χ^2 in comparing the observed and expected frequencies for 300 rolls of a die. The value of χ^2 obtained in Table 13.2 is 8.72. Does a value of $\chi^2 = 8.72$ constitute evidence at an accepted level of significance for rejecting the null hypothesis? How may a value of $\chi^2 = 8.72$ be evaluated?

The interpretation of a particular value of χ^2 requires a knowledge of the sampling distribution of χ^2. This will be considered in the section to follow.

Table 13.1

Calculation of χ^2 in comparing observed and expected frequencies for 100 tosses of a coin

	O	E	$O-E$	$(O-E)^2$	$\frac{(O-E)^2}{E}$
H	45	50	−5	25	.50
T	55	50	+5	25	.50
					$\chi^2 = 1.00$

Table 13.2

Calculation of χ^2 in comparing observed and expected frequencies for 300 throws of a die

Value of die X	Observed frequency O	Expected frequency E	$O-E$	$(O-E)^2$	$\frac{(O-E)^2}{E}$
1	43	50	−7	49	.98
2	55	50	5	25	.50
3	39	50	−11	121	2.42
4	56	50	6	36	.72
5	63	50	13	169	3.38
6	44	50	−6	36	.72
Total	300	300	0		$\chi^2 = 8.72$

13.3 THE SAMPLING DISTRIBUTION OF CHI SQUARE

The sampling distribution of χ^2 may be illustrated with reference to the tossing of coins. Let us assume that in tossing 100 *unbiased* coins 46 heads and 54 tails result. The expected frequencies are 50 heads and 50 tails. A value of χ^2 may be calculated as follows:

	O	E	$O-E$	$(O-E)^2$	$\frac{(O-E)^2}{E}$
H	46	50	−4	16	.32
T	54	50	+4	16	.32
					$\chi^2 = .64$

In the tossing of 100 coins two frequencies are obtained, one for heads and one for tails. These frequencies are not independent. If the frequency of heads is 46, the frequency of tails is $100 - 46 = 54$. If the frequency of heads is 62, the frequency of tails is $100 - 62 = 38$. Quite clearly, given either frequency, the other is determined. One frequency only is free to vary. In this situation 1 degree of freedom is associated with the value of χ^2.

Let us toss the 100 coins a second time, a third time, and so on, to obtain different values of χ^2. A large number of trials may be made and a large number of values of χ^2 obtained. The frequency distribution of these values is an experimental sampling distribution of χ^2 for 1 degree of freedom. It describes the variation in χ^2 with repeated sampling. By inspecting this experimental sampling distribution, estimates may be made of the proportion of times, or the probability, that values of χ^2 equal to or greater than any given value will occur due to sampling fluctuation for 1 degree of freedom. In the present illustration this assumes, of course, that the coins are unbiased.

Instead of tossing 100 coins, let us throw an unbiased die 100 times, obtain observed and expected frequencies, and calculate a value of χ^2. In this situation if any five frequencies are known, the sixth is determined. Five degrees of freedom are associated with the value of χ^2 obtained. The 100 dice may be tossed a great many times, a value of χ^2 calculated for each trial, and a frequency distribution made. This frequency distribution is an experimental sampling distribution of χ^2 for 5 degrees of freedom.

The theoretical sampling distribution of χ^2 is known, and probabilities may be estimated from it without using the elaborate experimental approach described for illustrative purposes above. The equation for χ^2 is complex and is not given here. It contains the number of degrees of freedom as a variable. This means that a different sampling distribution of χ^2 exists for each value of *df*. Figure 13.1 shows different chi-square distributions for different values of *df*. The value χ^2 is always positive, a circumstance which results from squaring the difference between the observed and expected values. Values of χ^2 range from 0 to infinity. The right-hand tail of the curve is asymptotic to the abscissa. For 1 degree of freedom the curve is asymptotic to the ordinate as well as to the abscissa.

The χ^2 distribution is used in tests of significance in much the same way that the normal, *t*, or the *F* distributions are used. The null hypothesis is assumed. This hypothesis states that no actual differences exist between the observed and expected frequencies. A value of χ^2 is calculated. If this value is equal to or greater than the critical value required for significance at an accepted significance level for the appropriate *df*, the null hypothesis is rejected. We may state that the differences between the ob-

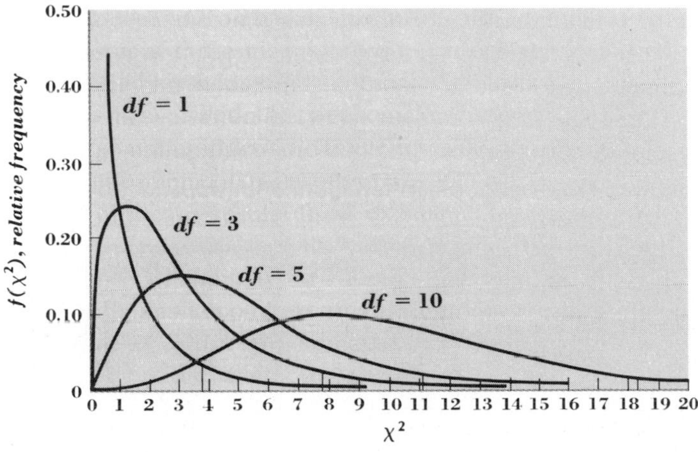

Figure 13.1 Chi-square distribution and 5 percent critical regions for various degrees of freedom. (*From Francis G. Cornell, "The Essentials of Educational Statistics," John Wiley & Sons, Inc., New York,* 1956.)

served and expected frequencies are significant and cannot reasonably be explained by sampling fluctuation. Table C in the Appendix shows values of χ^2 required for significance at various probability levels for different values of df. The critical values at the 5 and 1 percent levels for $df = 1$ are, respectively, 3.84 and 6.64. This means that 5 and 1 percent of the area of the curve fall to the right of ordinates erected at distances 3.84 and 6.64 measured along the base line from a zero origin. For $df = 5$, the corresponding 5 and 1 percent critical values are 11.07 and 15.09.

Table C of the Appendix provides the 5 and 1 percent critical values for $df = 1$ to $df = 30$. This covers the great majority of situations ordinarily encountered in practice. Situations where a χ^2 is calculated based on a $df > 30$ are infrequent. Where $df > 30$ the expression $\sqrt{2\chi^2} - \sqrt{2df - 1}$ has a sampling distribution which is approximately normal. Values of this expression required for significance at the 5 and 1 percent levels are 1.64 and 2.33.

The above discussion of χ^2 has undoubtedly given the impression that χ^2 is a statistic which is exclusively concerned with the comparison of observed and expected frequencies. The distribution of χ^2 is, in fact, a more general statistical distribution, and its use in the study of frequencies is one particular application. The more general approach defines χ^2 as the sum of squared normal deviates. Consider a population with mean μ, variance σ^2, and a normal distribution of scores Y. A standard score drawn from this population is $z = (Y - \mu)/\sigma$. Such standard scores have, of course, a normal distribution. We may, however, consider squared standard scores drawn from this population. Such a squared standard score is $z^2 = (Y - \mu)^2/\sigma^2$. If members are drawn one at a time from this population, the frequency distribution of z^2 will be a χ^2 distribution with 1 degree of freedom. Thus, very simply, if z is a normally distributed standard score, z^2 will have a χ^2 distribution with 1 degree of freedom. This means that for 1 degree of freedom $z = \sqrt{\chi^2}$. Critical values of χ^2 for significance at the .05 and .01 levels for $df = 1$ are 3.84 and 6.64. Corresponding critical values of z obtained from the normal distribution are 1.96 and 2.58. Note that $1.96 = \sqrt{3.84}$ and $2.58 = \sqrt{6.64}$. This simple relation means that for $df = 1$ the use of either the normal or χ^2 distribution will lead to equivalent results. The above discussion relates to situations where $df = 1$. If members are drawn two at a time from a population, the distribution of $z_1^2 + z_2^2$ will be a χ^2 distribution with 2 degrees of freedom. If members are drawn three at a time from a population, the distribution of $z_1^2 + z_2^2 + z_3^2$ will be a χ^2 distribution with 3 degrees of freedom. In general for samples of size N the quantity Σz_i^2 has a χ^2 distribution with N degrees of freedom.

13.4 GOODNESS OF FIT

In the analysis of frequencies using χ^2 a distinction is made between *tests of goodness of fit* and *tests of independence*. For both types of tests ob-

Table 13.3

Comparison of observed and expected frequencies in shape and color of peas in experiment by Mendel

	O	E	$O-E$	$(O-E)^2$	$\frac{(O-E)^2}{E}$
Round yellow	315	312.75	2.25	5.06	.016
Round green	108	104.25	3.75	14.06	.135
Angular yellow	101	104.25	−3.25	10.56	.101
Angular green	32	34.75	−2.75	7.56	.218
Total	556	556	0.00		$\chi^2 = .470$

served and expected frequencies are compared. Usually, in tests of goodness of fit, a set of observed frequencies on a single variable is compared with a corresponding set of expected, or theoretical, frequencies. In tests of independence two variables are involved, and observed and expected frequencies compared. Here, the expected frequencies are those the investigator would expect to get if the two variables were independent of each other. It may justifiably be argued that a test of independence is really a particular case of a test of goodness of fit.

Numerous examples may be found to illustrate the goodness of fit of a theoretical to an observed frequency distribution. In one experiment Abbé Mendel observed the shape and color of peas in a sample of plants. The distribution he obtained is shown in Table 13.3. According to his genetic theory the expected frequencies should follow the ratio 9:3:3:1. The correspondence between observed and expected frequencies is close. The value of $\chi^2 = .470$, and no grounds exist for rejecting the null hypothesis. The data lend confirmation to the theory. The value of χ^2 is smaller than we should ordinarily expect, the probability associated with it being between .90 and .95. Assuming the null hypothesis, a fit as good as or better than the one observed may be expected to occur in between 5 and 10 percent of samples of the same size.

In testing goodness of fit the hypothesis may be entertained that the distribution of a variable conforms to some widely known distribution such as the binomial or normal distribution. To illustrate the goodness of fit of the theoretical binomial distribution to an observed distribution, P. Johnson tossed 10 coins 512 times and recorded the proportion of tails. His data are shown in Table 13.4, together with the corresponding theoretical binomial frequencies. The mean and standard deviation of the observed distribution are $\bar{X} = 0.5$ and $s = .162$. The mean and standard deviation of the theoretical binomial are $\bar{X} = 0.5$ and $s = .156$.

Note that in the calculation of χ^2 for these data, the small frequencies at the tails of the distributions are combined, a procedure that is generally advisable with data of this type. Problems in the application of chi square resulting from the presence of small frequencies are discussed later in this

Table 13.4

Goodness of fit of binomial distribution to observed distribution of proportion of tails from 512 tosses of 10 coins

Proportion of tails	O	E	$O-E$	$\frac{(O-E)^2}{E}$
1.0	2 } 7	0.5 } 5.5	1.5	0.409
0.9	5	5.0		
0.8	15	22.5	−7.5	2.500
0.7	68	60.0	8.0	1.067
0.6	105	105.0	0.0	0.000
0.5	134	126.0	8.0	0.508
0.4	95	105.0	−10.0	0.952
0.3	55	60.0	−5.0	0.417
0.2	23	22.5	0.5	0.011
0.1	8 } 10	5.0 } 5.5	4.5	3.682
0.0	2	0.5		
Total	512	512		$\chi^2 = 9.546$

SOURCE: Palmer Johnson, "Statistical Methods in Research," Prentice-Hall, Inc., Englewood Cliffs, N.J., 1949.

chapter. With the present data, combining small frequencies reduces the number of frequencies from 11 to 9 and the number of degrees of freedom from 10 to 8. The value of χ^2 for these data is 9.55. The value required for significance at the 5 percent level for 8 degrees of freedom is 15.51. The conclusion is that the evidence is insufficient to justify rejection of the null hypothesis. Reference to a table of χ^2 shows that a value of χ^2 equal to or greater than the one observed might be expected to occur in about 30 percent of samples due to sampling fluctuation alone.

Chi square may be used to test the representativeness of a sample where certain population values are known. This in effect is a test of goodness of fit. To illustrate, in a study of attitudes toward immigrants, a sample of 200 cases is drawn from the city of Montreal. The observed frequencies and percentages by national origin are shown in Table 13.5.

Table 13.5

Application of χ^2 in comparing sample frequencies of national origin with population frequencies

National origin	O	Sample, percent	Population, percent	E	$O-E$	$\frac{(O-E)^2}{E}$
French	95	47.5	62.5	125	−30	7.20
English	67	33.5	19.4	39	28	20.10
Other	38	19.0	18.1	36	2	.11
Total	200	100.0	100.0	200	0	$\chi^2 = 27.41$

The population percentages obtained from census returns are also shown. These population percentages are used to obtain the expected, or theoretical, frequencies. The value of χ^2 is 27.41. For $df = 2$ this is highly significant, the value required for significance at the 1 percent level being 9.21. We may conclude that the sample is biased and cannot be considered a random sample with respect to national origin. Since attitudes toward immigrants may be linked to national origin, results obtained on this sample may be highly questionable unless a correction is applied to adjust for the sample bias.

13.5 TESTS OF INDEPENDENCE

In tests of independence two variables are involved. These are usually nominal variables. The question arises as to whether the two variables are independent of each other. The data are arranged in the form of a table called a *contingency* table. Contingency tables may be composed of any number of rows, R, and any number of columns, C. For example, consider a contingency table of two rows and two columns. Such tables are known as 2×2 tables. Here is an illustrative example of such a table, where the two variables are denoted by A and B.

Observed frequencies

	B_1	B_2	
A_1	50	10	60
A_2	20	20	40
	70	30	100

The frequencies shown are observed frequencies. The question is whether the variable A is independent of the variable B, or whether it may be argued that an association exists between A and B.

First, expected frequencies must be obtained. What frequencies would we expect to find in the four cells of the table if A were independent of B? These frequencies are readily obtained using the multiplication theorem of probability. The argument runs as follows. The probability of any member chosen at random having the attribute A_1 is $60/100 = .60$. The probability of any member chosen at random having the attribute B_1 is $70/100 = .70$. Therefore, using the multiplication theorem of probability, if A is independent of B, the probability of any member chosen at random being both A_1 and B_1 is the product of the separate probabilities or $.60 \times .70 = .42$. Thus, if A is independent of B, the expected proportion in the top left-hand cell is .42, and the expected frequency is $.42 \times 100 = 42$, there being 100 observations in all in this example. Similarly, expected frequencies for the other cells of the table may be obtained.

The expected frequencies for the complete table are as follows:

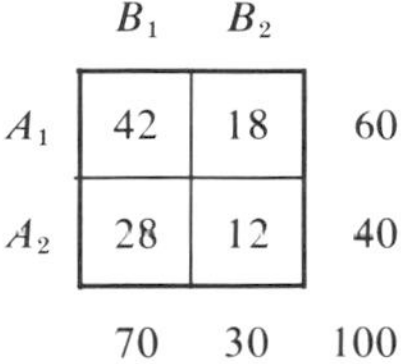

Expected frequencies

	B_1	B_2	
A_1	42	18	60
A_2	28	12	40
	70	30	100

In practical work the expected frequencies are usually not obtained by using the probabilistic argument described above. The expected frequency for the A_1B_1 cell is obtained directly by multiplying the A_1 row total by the B_1 column total and dividing by the total number of cases N. Thus the expected frequency for the A_1B_1 cell is given simply by $60 \times 70/100 = 42$. Similarly, the A_1B_2 expected frequency is $60 \times 30/100 = 18$, the A_2B_1 expected frequency is $40 \times 70/100 = 28$, and the A_2B_2 expected frequency is $40 \times 30/100 = 12$. In general, for any contingency tables of R rows and C columns, the expected frequencies are obtained by multiplying the appropriate row and column marginal totals, and dividing by the total numbers of observations N.

Given a set of observed and expected frequencies, χ^2 may be calculated in the usual way as follows:

O	E	$O - E$	$(O - E)^2$	$\frac{(O - E)^2}{E}$
50	42	8	64	1.52
10	18	−8	64	3.56
20	28	−8	64	2.29
20	12	8	64	5.33
100	100			$\chi^2 = 12.70$

Here the value of χ^2 is 12.70. How many degrees of freedom are associated with this value of χ^2? In any 2×2 table, given the restrictions of the marginal totals, only one cell frequency is free to vary. In the illustrative 2×2 table above, the observed cell frequency in the A_1B_1 cell is 50. If this frequency is known, and the marginal totals are fixed, all other frequencies are determined. The A_1B_2, A_2B_1, and A_2B_2 frequencies are obtained by simple subtraction. Thus only one cell frequency is free to vary. The value of χ^2 required for significance at the .05 level for $df = 1$ is 3.84 and at the .01 level is 6.64. The obtained value $\chi^2 = 12.70$ exceeds the value required for significance at the .01 level. Clearly A is not independent of B. An association can be said to exist between A and B.

For any 2×2 table, given the marginal totals, the number of degrees of freedom associated with the value of χ^2 is 1. In general for any table of

Table 13.6

Contingency table showing relationship between eye and hand laterality for 413 subjects and calculation of expected values

	Left-eyed	Ambiocular	Right-eyed	Total
Left-handed	34	62	28	124
	(35.4)	(58.5)	(30.0)	
Ambidextrous	27	28	20	75
	(21.4)	(35.4)	(18.2)	
Right-handed	57	105	52	214
	(61.1)	(101.0)	(51.8)	
Total	118	195	100	413

Calculation of expected values:

$$\frac{124 \times 118}{413} = 35.4 \qquad \frac{124 \times 195}{413} = 58.5 \qquad \frac{124 \times 100}{413} = 30.0$$

$$\frac{75 \times 118}{413} = 21.4 \qquad \frac{75 \times 195}{413} = 35.4 \qquad \frac{75 \times 100}{413} = 18.2$$

$$\frac{214 \times 118}{413} = 61.1 \qquad \frac{214 \times 195}{413} = 101.0 \qquad \frac{214 \times 100}{413} = 51.8$$

R rows and C columns the number of degrees of freedom associated with the value of χ^2, given the marginal totals, is $(R - 1)(C - 1)$. Thus for a 3×2 table the number of degrees of freedom is $(3 - 1)(2 - 1) = 2$. The reader can readily ascertain that for any 3×2 table, given the marginal totals, if two frequencies are known the remaining four frequencies are fixed and can exhibit no freedom of variation.

Consider a further illustration. Table 13.6 presents data collected by Woo (1928) on the relation between eyedness and handedness in a sample of 413 subjects. Subjects were tested for eyedness and handedness and grouped in one of three categories on both variables. Paired observations were available for each subject. Table 13.6 presents the contingency table for 413 paired observations. The expected cell frequencies have been calculated by multiplying the appropriate row and column totals and dividing by the number of cases. The expected cell frequencies are shown in parentheses in Table 13.6.

If eye and hand laterality are independent of each other, the 124 observations in the first row of Table 13.6 will be distributed in the three cells in that row in a manner proportional to the column sums. The expected values of 35.4, 58.5, and 30.0 are proportional to the column sums 118, 195, and 100. Likewise, the 118 cases in the first column will be distributed in the three cells in that column in a manner proportional to the row sums. The expected values 35.4, 21.4, and 61.1 are proportional to the row sums 124, 75, and 214. A similar proportionality exists throughout the table.

Table 13.7
Calculation of χ^2 for data of Table 13.6

O	E	$O-E$	$(O-E)^2$	$\frac{(O-E)^2}{E}$
34	35.4	−1.4	1.96	.055
62	58.5	3.5	12.25	.209
28	30.0	−2.0	4.00	.133
27	21.4	5.6	31.36	1.465
28	35.4	−7.4	54.76	1.547
20	18.2	1.8	3.24	.178
57	61.1	−4.1	16.81	.275
105	101.0	4.0	16.00	.158
52	51.8	.2	.04	.001
413	412.8			$\chi^2 = 4.021$

The expected cell frequencies in the rows and columns of any contingency table are proportional to the marginal totals.

Table 13.7 shows the calculation of χ^2 for the data of Table 13.6. The value of χ^2 obtained is 4.021. The number of degrees of freedom associated with this table is $(R-1)(C-1) = (3-1)(3-1) = 4$. The value of χ^2 required for significance at the 5 percent level is 9.488. We have, therefore, no grounds for rejecting the hypothesis of independence between eye and hand laterality. Apparently there is no relationship between the two variables.

13.6 CALCULATION OF χ^2 FOR 2 × 2 TABLES

A frequently occurring type of contingency table is the 2 × 2 or fourfold contingency table. A χ^2 test for independence can be readily obtained for such a table without calculating the expected values. Let us represent the cell and marginal frequencies by the following notation:

A	B	$A+B$
C	D	$C+D$
$A+C$	$B+D$	N

Chi square may then be calculated by the formula

[13.2]
$$\chi^2 = \frac{N(AD-BC)^2}{(A+B)(C+D)(A+C)(B+D)}$$

Note that the term in the numerator, $AD - BC$, is simply the difference between the two cross products and the term in the denominator is the product of the four marginal totals.

Consider the following 2 × 2 table showing the relationship between ratings of successful or unsuccessful on a job and pass or fail on an ability-test item.

	Test item Fail	Pass	
Successful	20	40	60
Unsuccessful	25	15	40
	45	55	100

Is there an association between performance on the job and performance on the test item? Does the item differentiate significantly between the successful and unsuccessful individuals? Chi square is as follows:

$$\chi^2 = \frac{100(20 \times 15 - 40 \times 25)^2}{60 \times 40 \times 45 \times 55} = 8.25$$

For $df = 1$, a $\chi^2 = 8.25$ is significant at better than the 1 percent level. The data provide fairly conclusive evidence that the test item differentiates between individuals on the basis of their job performance.

Formula 13.2 applies only to 2 × 2 tables. Analogous formulas can be obtained for tables of more than two rows and two columns. These are usually cumbersome and have no computational advantage over the direct calculation of χ^2.

13.7 THE APPLICATION OF χ^2 IN TESTING THE SIGNIFICANCE OF THE DIFFERENCE BETWEEN PROPORTIONS

In Chapter 12, procedures were described for testing the significance of the difference between both *independent* and *correlated* proportions. These procedures involved dividing the difference between two proportions by the standard error of the difference to obtain a normal deviate which could be referred to a table of areas under the normal curve. Because of a simple relationship for 1 degree of freedom between χ^2 and the normal deviate, χ^2 provides an alternative but equivalent procedure for testing the significance of the difference between proportions. For 1 degree of freedom it may be shown that χ^2 is equal to the normal deviate squared. Thus $\chi^2 = (x/s)^2 = z^2$ or $\sqrt{\chi^2} = z$.

We shall now consider the use of χ^2 in testing the significance of the difference between proportions for *independent* samples. Let the following be data obtained in response to an attitude-test statement for a group of males and females:

Frequency	Agree	Disagree	
Males	70	70	140
Females	20	40	60
	90	110	200

Proportion	Agree	Disagree	
Males	.500	.500	1.00
Females	.333	.667	1.00
	.450	.550	1.00

The number of males and females are $N_1 = 140$ and $N_2 = 60$, respectively. The proportions of males and females indicating agreement to the attitude statement are $p_1 = 70/140 = .500$ and $p_2 = 20/60 = .333$. Is there a significant difference in the attitudes of males and females? To apply the method previously described we calculate a proportion p based on a combination of data for the two samples. With the above data

$$p = \frac{70 + 20}{140 + 60} = \frac{90}{200} = .450$$

$$q = 1 - p = 1 - .450 = .550$$

The required normal deviate is then

$$z = \frac{p_1 - p_2}{\sqrt{pq[(1/N_1) + (1/N_2)]}}$$

$$= \frac{.500 - .333}{\sqrt{.450 \times .550(\frac{1}{140} + \frac{1}{60})}} = 2.172$$

The difference between the two proportions falls between the 5 and 1 percent levels. Reference to a table of areas under the normal curve shows that the proportion of the area falling beyond plus and minus 2.172 standard deviation units from the mean is close to .03. The difference may be taken as significant at about the 3 percent level. Let us now apply the formula for calculating χ^2 for a 2×2 contingency table to the same data:

$$\chi^2 = \frac{N(AD - BC)^2}{(A + B)(C + D)(A + C)(B + D)}$$

$$= \frac{200(70 \times 40 - 70 \times 20)^2}{140 \times 60 \times 90 \times 110} - 4.717$$

Consulting a table of χ^2 with 1 degree of freedom, we observe that the proportion of the area in the tail of the distribution of χ^2 is about .03 and the difference between proportions may be said to be significant at about the 3 percent level. We observe also that $\chi^2 = (2.172)^2 = 4.717$. The two procedures for testing the significance of the difference between proportions for independent samples lead to identical results. From a computational viewpoint the χ^2 test is the more convenient. Considerations pertaining to small frequencies apply also to the application of χ^2 in testing the significance of the difference between proportions (Section 13.8).

Where the data are correlated and are composed of paired observations, the normal deviate for testing the significance of the differences between proportions is given (Section 12.3) by the formula

$$z = \frac{D - A}{\sqrt{A + D}}$$

where D and A are cell frequencies in the bottom right and top left cells, respectively, of a 2×2 table. Instead of calculating a critical ratio and referring this to the normal curve, we may calculate χ^2 by the formula

[13.3] $$\chi^2 = \frac{(D - A)^2}{A + D}$$

For the data shown in Section 12.3, where we wish to test the significance of the differences between proportions of agreements to an attitude question for the same individuals tested on two occasions, we obtain a $z = 3.16$. The difference is significant at better than the 1 percent level. The value of the probability is .0016. The value of χ^2 calculated on the same data is $(3.16)^2$, or 9.986. The probability is the same as before.

13.8 ALTERNATIVE MODELS FOR 2 × 2 TABLES

For a given value of N the marginal frequencies for a 2×2 table may arise in three different ways. *First,* both sets of marginal frequencies may vary with repeated sampling. This means that if the study were repeated, and one or more additional samples of size N drawn, the marginal frequencies would be expected to vary with repetition. This is known as a random model. By far the most commonly occurring instances of 2×2 tables are of the random type. *Second,* one set of marginal frequencies may be fixed, and the other may vary at random. For example, a sample of 50 males and 50 females may be classified as smokers versus nonsmokers. Here sex is fixed, and smokers versus nonsmokers is random. If the study were repeated, another 50 males and 50 females would be used. This is known as a mixed model. *Third,* both sets of marginal frequencies may be fixed. If the experiment were repeated, the same marginal frequencies would arise. This is a fixed model. Good examples are rare. An example might arise where a subject received N presentations of two stimuli and is required to judge whether these are the same or different. The subjects are told that half are the same and half different, and consequently they adopt a strategy of judging half the same and half different. The χ^2 test for 2×2 tables may be used for data conforming to all three models – random, mixed, and fixed – given certain restrictions and qualifications discussed in the section to follow.

13.9 SMALL EXPECTED FREQUENCIES

The distribution of χ^2 used in determining critical significance values is a continuous theoretical frequency curve. Where the expected frequencies are small, the actual sampling distribution of χ^2 may exhibit marked discontinuity. The situation here is analogous to that found in using the normal curve as a fit to the binomial. For small values of N the continuous normal curve is a poor fit to the discrete binomial.

A correction, commonly recommended for use with 2×2 tables, is known as *Yates' correction for continuity*. To apply this correction we reduce by .5 the obtained frequencies that are greater than expectation and increase by .5 the obtained frequencies that are less than expectation. This brings the observed and expected frequencies closer together and decreases χ^2. Many texts recommend that the correction should be used when any of the expected frequencies are less than 5. The formula used in computing χ^2 from a 2×2 table may be written to incorporate Yates' correction for continuity as follows:

[13.4]
$$\chi^2 = \frac{N(|AD - BC| - N/2)^2}{(A+B)(C+D)(A+C)(B+D)}$$

The term $|AD - BC|$ is the absolute difference, that is, the difference taken regardless of sign. The correction amounts to subtracting $N/2$ from this absolute difference.

Work by Camilli and Hopkins (1978, 1979) and others indicates that Yates' correction for continuity is unduly conservative for data conforming to the random and the mixed models, these being by far the great majority of instances encountered in practice. By conservative is meant that the use of the correction will lead to too few statistically significant results. It detracts from the accuracy of the Type I probability estimates. The results of Camilli and Hopkins indicate that the χ^2 test without a continuity correction, when applied to 2×2 tables for both the random and mixed models, provides reasonably accurate estimates of Type I error for $N \geq 8$. They recommend that the correction should not be used for data of the type described. Presumably the correction should be used in the fixed model case or an exact test applied.

13.10 THE CONTINGENCY COEFFICIENT

On occasion a descriptive statistic, analogous to a correlation coefficient, is required to describe the degree of association in a contingency table. One such statistic is the contingency coefficient defined by the following equation:

[13.5]
$$C = \sqrt{\frac{\chi^2}{N + \chi^2}}$$

In Table 13.7 a 3 × 3 table shows the relation between eye and hand laterality. The contingency coefficient for these data is

$$C = \sqrt{\frac{4.021}{413 + 4.021}} = .031$$

This coefficient describes the almost complete absence of association between eye and hand laterality.

The minimum value of C is 0. The maximum value of the contingency coefficient depends on the number of categories of the variables. When the number of rows is equal to the number of columns, the maximum value of C is given by $\sqrt{(k-1)/k}$. Thus for a 2 × 2 table the maximum upper limit of C is $\sqrt{\frac{1}{2}} = .707$. For a 3 × 3 table the maximum upper limit is $\sqrt{\frac{2}{3}} = .816$. Contingency coefficients are not directly comparable unless calculated on tables containing the same number of rows and columns.

13.11 MISCELLANEOUS OBSERVATIONS ON CHI SQUARE

In this section we shall consider a number of miscellaneous points about χ^2 not hitherto discussed.

One-tailed and two-tailed tests Tables of χ^2 used for tests of significance are based on one tail only, the tail to the right, of the sampling distribution of χ^2. Table C of the Appendix shows that for 1 degree of freedom 5 percent of the area of the distribution falls to the right of $\chi^2 = 3.84$ and 1 percent to the right of $\chi^2 = 6.64$. These are *not* critical values for directional, or one-tailed, tests as described in Chapter 11. Although one tail only of the sampling distribution of χ^2 is used, the table values are those required for testing the significance of a difference regardless of direction, that is, for two-tailed tests. The critical ratio or normal deviate required for significance at the 5 percent level for a two-tailed test is 1.96. If this value is squared, we obtain 3.84, the χ^2 value at the 5 percent level for 1 degree of freedom. For 1 degree of freedom the square root of χ^2 is a normal deviate and may be used with reference to the normal curve in applying two-tailed tests. In effect, because χ^2 is the square of the normal deviate for 1 degree of freedom, both tails of the normal curve are incorporated in the right tail of the χ^2 curve. In many situations where χ^2 is applied, the idea of a directional, or one-tailed, test has little meaning. In tests of goodness of fit and in most tests of independence we are usually not concerned with the direction of the difference observed. If a one-tailed test is required, the proportionate areas in the chi-square tables should be halved. The value of χ^2 required for significance at the 5 percent level for a one-tailed test is 2.71 for $df = 1$. The corresponding value at the 1 percent level is 5.41. These are the squares of the normal deviates 1.64 and 2.33 required for significance for a one-tailed test at the 5 and 1 percent levels, respectively.

Chi square and sample size The value of χ^2 is related to the size of the sample. If an actual difference exists between observed and expected values, this difference will tend to increase as sample size increases. Chi square will also increase, and the associated probability value will decrease. Consider the following tables:

1

6	4	10
4	11	15
10	15	25

$\chi^2 = 2.78$

2

12	8	20
8	22	30
20	30	50

$\chi^2 = 5.56$

3

24	16	40
16	44	60
40	60	100

$\chi^2 = 11.12$

As the samples are doubled in size from 25 to 50 to 100, the differences between the observed and expected values, $O - E$, are doubled and the χ^2 values are doubled. If no actual difference exists between observed and expected values, χ^2 will tend to remain unchanged as sample size increases. For a constant difference between observed and expected values χ^2 will decrease as sample size increases. If we double sample size and hold the difference between observed and expected values fixed, the value of χ^2 will be reduced by one-half.

Alternative formula for chi square We can readily demonstrate that

[13.6]
$$\chi^2 = \sum \frac{(O - E)^2}{E} = \sum \frac{O^2}{E} - N$$

This alternative way of writing χ^2 is sometimes useful for computational purposes.

Reduction of an $R \times C$ table to a 2 × 2 table A table with R rows and C columns may be reduced to a 2×2 table in order to facilitate a rapid test of association with χ^2. This procedure is legitimate enough provided the points of dichotomy of the two variables are made without reference to the cell frequencies. The investigator may decide a priori to dichotomize about the two medians, or something of the sort. Data are found where the points of dichotomy have been located in order to maximize the association in the data and obtain thereby a significant χ^2. This practice is spurious and should be enthusiastically discouraged.

BASIC TERMS AND CONCEPTS

Theoretical frequency

χ^2

Sampling distribution of χ^2

Test of goodness of fit

Test of independence

Contingency table

Correction for continuity

Contingency coefficient

EXERCISES

1 On a true-false test of 100 items a student obtains a score of 65. Does this differ from what the student would expect to obtain if he or she had guessed all the items?

2 In tossing a coin 200 times the following results are obtained:

Heads	90
Tails	110

Can it be argued that the coin is biased?

3 In 180 throws of a die the observed frequencies of the values from 1 to 6 are 34, 27, 41, 25, 18, and 35. Test the hypothesis that the die is unbiased.

4 Six coins are tossed 64 times and the number of heads counted. The following results were obtained:

No. of heads	f
0	1
1	9
2	16
3	18
4	16
5	3
6	1
Total	64

Do these frequencies differ significantly from those obtained from the binomial $(1/2 + 1/2)^6$?

5 In the game "acey-deucey," which is a variant of backgammon, special privilege attaches to rolling an ace and a deuce, that is, a one and a two, in rolling two dice. In 360 rolls of two dice an ace and a deuce occur 25 times. Does this differ from chance expectation?

6 How many cell frequencies are free to vary in tables with **(a)** two rows and two columns, **(b)** two rows and three columns, **(c)** three rows and five columns? Assume fixed marginal totals.

7 The following data relate to patients in a mental hospital:

	Rating: Improvement	No improvement	
Therapy *A*	16	28	44
Therapy *B*	9	37	46
	25	65	90

Test the hypothesis that method of therapy is independent of rating assigned.

8 The following contingency table describes the relation between pass and fail on an examination and ratings of job performance for 100 employees.

	Rating: Below average	Average	Above average	
Pass	11	25	35	71
Fail	15	7	7	29
	26	32	42	100

Test the hypothesis that job performance is independent of examination results.

9 A sample used in a market survey contains 100 males and 100 females. Of these, 33 males and 18 females state a preference for brand *A*. Use χ^2 to test the hypothesis that no sex differences exist in consumer preference.

ANSWERS TO EXERCISES

1 $\chi^2 = 9$, $df = 1$; $p < .01$

2 $\chi^2 = 2.00$, $df = 1$; $p > .05$

3 $\chi^2 = 11.33$, $df = 5$; $p < .05 > .02$

4 $\chi^2 = 3.332$, $df = 6$; $p > .05$

5 $\chi^2 = 1.323$, $df = 1$; $p > .05$

6 **a** 1 **b** 2 **c** 8

7 $\chi^2 = 3.163$, $df = 1$; $p > .05$

8 $\chi^2 = 14.287$, $df = 2$; $p < .001$

9 $\chi^2 = 5.922$, $df = 1$; $p < .02$

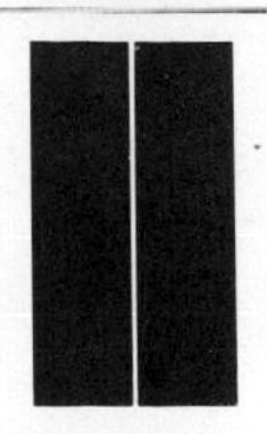

THE DESIGN OF EXPERIMENTS

14

THE STRUCTURE AND PLANNING OF EXPERIMENTS

14.1 INTRODUCTION

Several subsequent chapters of this book are concerned with the analysis of variance and covariance. These procedures are used in the analysis of the data of experiments. A very brief, and elementary, discussion of the structure and planning of experiments should serve as a useful preliminary to a detailed study of these methods of analysis. The study of the structure and planning of experiments is a field of investigation commonly called the design of experiments. This subject has many aspects, some quite complex. The present discussion will deal in a nonmathematical way with only a few of the simplest aspects of experimental design.

All experiments are concerned with the relations between variables. In the simplest type of experiment two variables only are involved, an independent variable and a dependent variable. To illustrate, an experiment may be initiated to compare three different methods of teaching French, designated A, B, and C. Each method may be applied to a different group of experimental subjects. Following a period of instruction, performance may be measured using an achievement test. In this experiment the different methods of teaching French constitute the levels or categories of the independent variable. The investigator decides which methods will be used and the size of the groups to which they will be applied; that is, he or she controls the values or categories of the independent variable and the frequency of occurrence of those values or categories. The measures of achievement constitute the dependent variable. The experiment is concerned with the way in which achievement in French depends on the method of instruction used. The essence of the idea of an experiment lies in the simple fact that the investigator selects the values or categories of the treatment variable and the frequency of their occurrence. This enables him or her to study an indefinitely large number of relations which are not amenable to study by observational or correlational methods, and may not in fact have any existence in nature at all prior to the conduct of the experiment. New knowledge is thereby produced. The tremendous proliferation of human knowledge in the past 20 years is in large measure a result of experimentation. It is clearly desirable, therefore, that some of the principles underlying such experimentation should be understood.

In developing the design for an experiment, the investigator must (1) select the values or categories of the independent variable, or variables, to be compared; (2) select the subjects for the experiment; (3) apply rules or procedures whereby subjects are assigned to the particular values or categories of the independent variable; (4) specify the observations or measurements to be made on each subject. In the experiment mentioned above, on the comparison of different methods of teaching French, the investigator must select the different methods of teaching to be compared. He or she must choose the experimental subjects to which these methods are applied and must allocate subjects to methods. Also decisions must be made regarding appropriate measures of achievement which will yield valid comparisons of the methods used.

14.2 TERMINOLOGY

Comment on terminology is appropriate here. An independent variable used in an experiment may be either a *treatment* variable or a *classification* variable. A treatment variable involves a modification in the experimental subjects, a modification which is controlled by the experimenter. Different dosages of a drug or different methods of learning are administered to different groups of subjects. In effect, the subjects are treated in some way by the experimenter. Experimental subjects may, however, be classified on a characteristic which was present prior to, and quite apart from, the experiment, and does not result from the manipulations of the experimenter. Such a variable is a classification variable. Examples are sex, age, disease entity, IQ level, socioeconomic status, and so on. Although the values of a classification variable are not, as it were, created by the experimenter, as is the case with a treatment variable, the investigator nonetheless selects the classification variables which are included in the experiment or are the objects of attention.

Any independent variable, whether of the treatment or classification type, is spoken of as a *factor*. Experiments which investigate simultaneously the effects of two independent variables are spoken of as two-factor, or two-way classification, experiments. When three factors are involved, the experiment is said to be a three-factor, or three-way classification, experiment, and so on. The different values or categories of the independent variable are spoken of as *levels;* thus we may have two, or three, or more levels of a factor.

In the literature on experimental design the unit to which a treatment is applied is frequently spoken of as a *plot,* a term which derives from agricultural experimentation. In experimental work in psychology and education the plot is usually a human subject or an experimental animal. The term "plot" will be used infrequently in this text. Measurements obtained from a plot are sometimes spoken of as the *yield,* a term which stems also from agricultural usage. In psychology and education the measurements or observations made on a group of subjects or animals correspond to the yield for a number of plots. A grouping of experimental units which is homogeneous with regard to some basis of classification is spoken of as a *block.* An experimenter may select 50 subjects, of which 25 are male and 25 female. The two groupings by sex are blocks.

14.3 THE CLASSIFICATION OF VARIABLES IN RELATION TO EXPERIMENTS

As indicated above, experiments are concerned with the relations between variables. In Chapter 1, variables were classified as nominal, ordinal, or interval-ratio types. In a simple two-variable experiment the treatment variables may be nominal, ordinal, or interval-ratio; the dependent variable may also be nominal, ordinal, or interval-ratio. The methods to be applied in the analysis of the data of experiments and the type of questions which experiments can answer are determined by the nature of the variables. Thus, in effect, the nature of the information which an experiment can yield, and the analytic methods which communicate that information to our understanding, depend on the nature of the variables.

To illustrate, consider an experiment in which three types of therapy are applied to three groups of depressed patients. After a time period, observations are made indicating those subjects that show evidence of recovery and those that do not. In this experiment both variables are nominal. The investigator may calculate the proportions in the three groups that show recovery and compare these proportions one with another. The magnitude of the differences between proportions is a measure of the difference between treatments. No further analysis of these data is possible. Consider, on the other hand, an experiment in which both the treatment variable and the dependent variable are of the interval or ratio type. The treatment variable may consist of five equally spaced dosages of a drug, and the dependent variable may be reaction time. Here the investigator may not only compare the mean reaction times for each dosage with every other dosage, but he or she may explore the nature of the functional relation between the two variables. Reaction time may increase or decrease in a linear fashion with change in the treatment variable; it may increase and then decrease; or some other type of relation may exist. Here we have considered two experiments. In one experiment both variables were nominal; in the other, both were either interval or ratio. In many experiments the treatment variables may be nominal, and the dependent variable may be of the interval or ratio type; or the treatment variable may be ordinal, and the dependent variable may be nominal, and so on. The nature of the variables determines the method of analysis employed and the nature of the conclusions drawn.

14.4 SINGLE-FACTOR EXPERIMENTS

Many experiments involve a single treatment, or classification, variable with two or more levels. Let us initially consider experiments in which the single factor is a treatment variable with k levels or categories, and not a classification variable. Such experiments are of a variety of types, some of which are discussed here. *First,* a group of experimental subjects may be divided into k independent groups, using a random method. A different treatment may then be applied to each group. One group may be a control group, that is, a group to which no treatment is applied. A meaningful interpretation of the experiment may require a comparison of results obtained under treatment with results obtained in the absence of treatment. Comparisons may be made between treatments and a control, between treatments, or both. *Second,* some single-factor experiments involve a single group of subjects. Each subject receives all k treatments. Repeated observations or measurements are made under k conditions, one of which may be a control condition, on the same subjects. In such an experiment as this the measurements made under the k treatments will not be independent. Positive correlations will usually exist between the paired measurements obtained under any two treatments. These correlations will reduce the magnitude of the error in the comparisons of the separate treatment means. *Third,* a single-factor experiment may consist of groups that are matched on one or more variables which are known to be correlated with the dependent variable. In an experiment on three different methods of teaching French, IQ may be known to be correlated with achievement in French. Three groups of subjects, paired or matched subject by subject by IQ, may be used. The rationale here is that because IQ is correlated with French achievement, a correlation between French achievement scores for paired subjects may result. The error term would be reduced thereby. In some practical experimental situations the gains made by matching are trivial in relation to the work involved. An interesting variant of the matched-group experiment is one in which subjects are not matched subject by subject but are matched with regard to distribution on one or more control variables. This results in the same error reduction as the individual matching of subjects (see McNemar, 1969).

14.5 RANDOMIZATION

In the design of experiments frequent use is made of randomization. In general the purpose of randomization is to ensure that extraneous variables which are concomitant with the dependent variable, and may be correlated with it, will not introduce systematic bias in the experimental results. Consider again the experiment in which three methods of teaching French are compared. If subjects are assigned to the three groups using a random method, IQ, which may be correlated with French achievement, will not vary in any systematic way from group to group. Also, an indefinitely large number of other variables, the specifications of many of which may not have entered the purview of the investigator, will be rendered powerless to introduce systematic bias. Although methods of randomization differ from one type of experiment to another, the general purpose of randomization is to protect the validity of the experiment by controlling the biasing influence of extraneous variables. A quotation from Cochran and Cox (1957) is relevant here.

> Randomization is somewhat analogous to insurance, in that it is a precaution against disturbances that may or may not occur and that may or may not be serious if they do occur. It is generally advisable to take the trouble to randomize even when it is not expected that there will be any serious bias from failure to randomize. The experimenter is thus protected against unusual events that upset his expectations.

In experimental design the term *randomization* refers not to any subjective impression of haphazard arrangement but to clearly stated operational procedures. These procedures may involve the tossing of coins, the drawing of numbered cards from a well-shuffled deck, or the use of tables of random numbers. Such tables are composed of series of digits from 0 to 9. Each digit occurs with approximately equal frequency, and adjacent digits are independent of each other. Tables of random numbers are available in Fisher and Yates (1963) and elsewhere. To illustrate, suppose we wish to select a sample of 12 experimental subjects from a sample of 40 subjects. Identify the subjects by the numbers 1 to 40. Enter a row, column, or diagonal of two-digit numbers in a table of random numbers, and select the first 12 different numbers between 1 and 40 as they occur. This procedure will yield a random set of 12 subjects. Other procedures may, of course, be used.

In a single-factor experiment with k independent groups of n_1, n_2, . . . , n_k subjects, an appropriate procedure is to choose n_1 subjects at random for the first group, n_2 at random for the second, and so on. This procedure is continued until n_k subjects remain for the kth group. In an experiment in which k experimental treatments are administered to the same group of n subjects and repeated measurements are made on the same subjects, an order effect may exist; that is, the result observed may not be independent of the order of treatment for reason of fatigue, practice, or some other cause. For example, measures of reaction time may be made on the same group of subjects under four different dosages of a drug. Here quite clearly an order effect may exist. In such an experiment as this the order of treatments may be randomized for each subject, thereby eliminating any systematic influence of order on the treatment means. When there are two treatments only, A and B, a common practice is to use the order AB for half the subjects and the order BA for the other half, assigning subjects to orders at random. In experiments involving matched groups, the members of each matched pair, triplet, or quadruplet may be allocated to treatments by a random method.

14.6 FACTORIAL EXPERIMENTS

The experiments discussed hitherto in this chapter have involved a single independent variable or factor. Experiments may, however, be designed to study simultaneously the effects of two or more independent variables. For example, the experiment in which three methods of teaching French are compared may be extended to include a comparison of spaced versus massed learning. Under the spaced conditions, subjects receive short periods of intensive instruction separated by time intervals. Under the massed learning conditions, subjects receive intensive instruction for a prolonged time period. The effect on achievement of the six possible combinations of three methods of teaching and two learning conditions may be investigated, each combination being applied to a different group of experimental subjects. Such an experiment is called a *factorial experiment*. Experiments in which the treatments are combinations of levels of two or more factors are said to be factorial. If all possible treatment combinations are studied, the experiment is said to be a *complete factorial experiment*. In some experiments the factors have two levels only. We may speak of a 2×2 experiment, a $2 \times 2 \times 2$ experiment, or a 2^n experiment, where n is the number of factors.

It is informative to examine in more detail the structural features of a factorial experiment. Let us suppose that in our 3×2 experiment on teaching French in relation to spaced versus massed learning, four subjects were used in each of the six groups and the following achievement test scores were obtained.

Learning condition	Teaching method					
	A		*B*		*C*	
Spaced	72	34	78	29	16	29
	96	55	25	20	24	41
Massed	81	85	75	45	19	19
	99	62	36	26	41	46

Three variables are involved in this experiment. Two of these, method and learning condition, are independent variables, and one, achievement-test score, is the dependent variable. Three variate values are available for each subject—the method used, the learning condition, and achievement-test scores. Method and learning condition are nominal variables. These data could readily be written in columnar fashion as follows:

Learning condition	Teaching method	Test score	Learning condition	Teaching method	Test score
S	*A*	72	*M*	*A*	81
S	*A*	96	*M*	*A*	99
S	*A*	34	*M*	*A*	85
S	*A*	55	*M*	*A*	62
S	*B*	78	*M*	*B*	75
S	*B*	25	*M*	*B*	36
S	*B*	29	*M*	*B*	45
S	*B*	20	*M*	*B*	26
S	*C*	16	*M*	*C*	19
S	*C*	24	*M*	*C*	41
S	*C*	29	*M*	*C*	19
S	*C*	41	*M*	*C*	46

The purpose of this experiment is to explore the relation between the learning condition and test scores and the teaching method and test scores. The relations between the six combinations of learning condition and teaching method may also be studied. An important feature of the structure of this experiment is that the two nominal variables, learning condition and teaching method, are independent of each other. By choosing six groups of equal size, the independence of these two variables is assured. If the numbers of cases in the six groups, when written in the form of a 2×3 contingency table, are proportional to the marginal totals, and χ^2 for the table is 0, the independence of the two treatment variables will also be assured. In the design of factorial experiments the groups should be either of equal size or proportional. Departures from equality or proportionality should be avoided.

One advantage of the factorial experiment is that information is obtained about the interaction between factors. For example, in the experiment on teaching French one method of teaching may interact with a condition of learning and render that combination either better or worse than any other combination. The concept of interaction is discussed more explicitly, and in detail, in Section 16.5. One disadvantage of the factorial experiment is that the number of combinations may become quite unwieldy, and, from a practical point of view, the experiment may be very difficult to conduct. Also, the meaningful interpretation of the interactions may prove difficult. In general, in psychology and education, it is usually advisable to avoid factorial experiments with more than a few factors.

The reader should note that factorial experiments may involve repeated measurement on the same subjects. For example, in a 3×2 design, repeated measurements may be made on the same subjects under each of the six treatment combinations. The result is an $N \times 3 \times 2$ arrangement of numbers.

14.7 OTHER EXPERIMENTAL DESIGNS

For a complete factorial experiment all possible treatment combinations are used. Each combination is applied to a different group of n subjects. Obviously experiments with a number of factors, each with several levels, may involve a large number of treatment combinations. A $3 \times 4 \times 5$ complete factorial experiment would involve 60 different treatment combinations. Clearly such experiments may prove difficult, costly, and perhaps impossible to properly complete. Under such circumstances a reduced number, a subset, of systematically selected treatment combinations may be used. Such experiments are known as *incomplete factorial experiments*. To illustrate, consider an experiment with independent groups with three factors, R, C, and A, with three levels of each factor. A complete factorial experiment would require 27 treatment combinations and groups of subjects. The experiment may be conducted using nine treatment combinations, according to the following plan:

	C_1	C_2	C_3
R_1	A_2	A_1	A_3
R_2	A_3	A_2	A_1
R_3	A_1	A_3	A_2

Here nine groups of n subjects are used. The first n subjects receive the treatment combination R_1, C_1, A_2, the second the combination R_1, C_2, A_1, and so on. If a single measurement X is made on each subject, four variables are involved in the experiment, the dependent variable X, and three independent variables R, C, and A. Note that in the arrangement above the three levels of the A variable occur once in each row and column. This arrangement is called a *Latin square*. A Latin square is simply an arrangement of k letters, or symbols, in the form of a square in such a way that each letter, or symbol, occurs once in each row and column. Incomplete factorial experiments involve assumptions, and problems of interpretation, that are not involved in complete factorial experiments.

In a simple complete factorial experiment the design may be represented as a matrix. For a two-factor experiment with three levels of each factor and n subjects in each group, such a design may be represented as:

	C_1	C_2	C_3
R_1	n	n	n
R_2	n	n	n
R_3	n	n	n

Here, R_1, R_2, and R_3 may, for example, be three methods of teaching French, and C_1, C_2, and C_3 may be three instructors. In this design each instructor is required to use each method of instruction. All levels of the one variable occur at all levels of the other. The two variables are said to be *crossed*. This arrangement at times may prove both impractical and undesirable.

Another type of experiment employs what is called a *nested design*. For example, an alternate procedure is to use not three but nine instructors with three instructors using each of the three methods. This design may be represented as follows:

R_1			R_2			R_3		
C_1	C_2	C_3	C_4	C_5	C_6	C_7	C_8	C_9
n	n	n	n	n	n	n	n	n

In this design the C variable (instructors) is said to be *nested* under the R variable (methods). In general in such designs several levels of one variable are restricted to a single level of the other variable. This is the essence of the idea of nesting. Experiments may be designed with more than one layer or level of nesting.

Situations arise where a basis exists for the classification of experimental subjects into subgroups or blocks. The blocks may exhibit some degree of homogeneity with respect to a variable which may be correlated with the dependent variable under study in the experiment. Consider an experiment involving four different methods of teaching arithmetic. Subjects may be classified into six groups according to performance on an intelligence test, which is the blocking variable. These groupings constitute six blocks. Subjects within blocks may be allocated at random to the four methods. Such an experiment is a *randomized block experiment*. The purpose underlying the use of this design is the reduction of error. To illustrate further consider an animal experiment with four experimental treatments. A sample of 24 animals is used, comprised of 6 sets of 4 litter mates. Each set of litter mates is a block. The blocking variable is membership in a litter. Treatments are allocated to the animals in each block, or vice versa, on a random basis. The result is a 6×4 design with one animal in each cell.

14.8 CLASSIFICATION VARIABLES

Hitherto we have discussed experiments in which the independent variables are treatment variables. In many investigations in psychology and education the independent variables are not treatment variables but are, in fact, classification variables; that is, subjects are classified according to a characteristic which was present prior to the conduct of the experiment and did not result from the manipulations of the investigator. For example, normals, neurotics, and psychotics may be compared on flicker-fusion rates, or some other variable. Here the independent variable is a classification, and not a treatment, variable. No treatment by the investigator is involved. Quite clearly in such an experiment it is not possible to assign subjects to experimental groups at random. Randomization is not possible, because the attribute which determines membership in a particular group is not under the control of the investigator.

Three types of situations in which classification variables are involved may be recognized. *First,* the sample of subjects used may be a random sample drawn from a defined population, and the proportions in the various classes, or strata, may, within the limits of sampling error, correspond to the population proportions. The object of the investigation is to describe relations between the classification variable and some other variable. An investigation of this type is, in effect, not an experiment at all, but a correlational study. *Second,* samples may be drawn for comparison at random from two or more subpopulations, but the proportions in these samples may not correspond to the population proportions. For example, an investigator may choose to compare 50 normals with 50 psychotics on a specified variable or variables. Clearly the proportions in the two samples do not correspond to the proportions of normals and psychotics in the population. This, in effect, is a form of disproportional stratified sampling. Also we note that certain classes which exist in the population at large may be excluded. Thus, for example, not all nonpsychotics are necessarily normal. This again is, in effect, a correlational study, the object of which is simply to identify differences between groups. Frequently there is a presupposition that a knowledge of such differences may ultimately prove useful in the construction of causal arguments. *Third,* investigations may be conducted involving classification variables in which the attempt is made to control the influence of certain variables which in a straightforward correlational study would not be controlled. In comparing hospitalized normals with hospitalized psychotics, the two groups may be equated for age, IQ, sex, length of hospitalization, socioeconomic status, and other variables which might be construed to be correlated with the dependent variable. In studies such as this the investigator excludes certain variables from the group of causal influences affecting the results observed. Although no direct causal argument linking the classification variable and the dependent variable can be advanced, the influence of certain variables on the results observed can be excluded. Such investigations narrow the range of possible causal influences. Factorial investigations involving classification variables may perhaps be construed legitimately to be experiments in that by design they involve the experimental control of certain variables which otherwise might be uncontrolled.

14.9 CONCLUDING OBSERVATION

Discussion in this chapter has touched briefly on only a few aspects of the structure and planning of experiments. This subject can be elaborated at great length and complexity. For a more comprehensive discussion the reader is referred to Cochran and Cox (1957), Winer (1971), Keppel (1973), and Myers (1979).

BASIC TERMS AND CONCEPTS

Treatment variable

Classification variable

Randomization

Single factor experiment

Repeated measurement design

Complete factorial experiment

Incomplete factorial experiment

Latin square

Nested design

Randomized block experiment

Matching in experimental design

EXERCISES

1 Distinguish, with examples, between **(a)** an independent and a dependent variable and **(b)** a treatment and a classification variable.

2 Discuss the rationale underlying the use of matched samples in the planning of experiments.

3 Why does randomization eliminate systematic bias in experimental results?

4 Outline a randomization procedure for allocating 100 experimental subjects to five experimental groups.

5 What is the rationale underlying the use of equal or proportional groups in factorial experiments?

6 What is meant by a randomized block experiment?

7 Give an example of a Latin square containing five rows and five columns.

ANSWERS TO EXERCISES

1 a In the simplest type of experiment two variables are involved, an independent and a dependent variable. The value of the dependent variable is thought to depend in some way on the value of the independent variable. To illustrate, an experiment might be conducted to investigate the effects of different amounts of practice on intelligence test items on intelligence test performance. Here intelligence test performance is the dependent variable and amount of practice constitutes the independent variable. In an experiment designed to investigate the relation between size of lesion in a particular brain locus and maze performance, measures of maze performance constitute the dependent variable and size of lesion, the independent variable. The investigator wishes to ascertain how maze performance depends on lesion size.

b A treatment variable is in effect the creation of the investigator, who decides the particular values which the variable will assume and the frequency of occurrence of these values. Different methods of learning a language, different dosages of a drug, and different amounts of practice in learning a task are examples of treatment variables. A classification variable, on the other hand, is one whose values, and the frequency of their occurrence, are not within the control of the investigator but exist, as it were, in nature. Height, weight, College Board test scores, and the like, are examples of classification variables.

2 Consider a simple experimental situation involving an experimental and a control group. Subjects in the two groups are matched on a variable X; consequently the two groups are equivalent with respect to X. If X is positively correlated with Y, a possibility exists that the paired values of Y for the experimental and control groups will also be positively correlated. If this is so, the sampling variance of the difference between means, and other statistics, will be less than had the two groups been independent.

3 In the above question matching of experimental subjects on a single variable X was discussed. Clearly the subjects do not have a single attribute which may be correlated with Y, but possibly a large number of such attributes. Randomization tends to ensure that groups of subjects will not differ significantly on a large number of such attributes.

Randomization also tends to ensure that groups of subjects will not differ significantly on a large number of attributes, some of which may be correlated with Y.

4 One very simple procedure is to number subjects from 1 to 100, write the numbers on slips of paper, and draw the slips of paper from a well-shuffled container, allocating a subject in turn to each of the five groups. Tables of random numbers could be used.

5 Unless either equal or proportional groups are used, the factors are not independent of each other. When the groups are either equal or proportional, conclusions can be reached about the effect of each of the factors on the dependent variable, quite apart from, and independent of, the other factors used in the experiment.

6 A randomized block experiment is one in which a basis exists for arranging subjects into subgroups or blocks.

7

$$\begin{matrix} A & B & C & D & E \\ B & C & D & E & A \\ C & D & E & A & B \\ D & E & A & B & C \\ E & A & B & C & D \end{matrix}$$

15

ANALYSIS OF VARIANCE: ONE-WAY CLASSIFICATION

15.1 ITS NATURE AND PURPOSE

The analysis of variance is a method for dividing the variation observed in experimental data into different parts, each part assignable to a known source, cause, or factor. We may assess the relative magnitude of variation resulting from different sources and ascertain whether a particular part of the variation is greater than expectation under the null hypothesis. The analysis of variance is inextricably associated with the design of experiments. Obviously, if we are to relate different parts of the variation to particular causal circumstances, experiments must be designed to permit this to occur in a logically rigorous fashion.

The partitioning of variance is a common occurrence in statistics. The particular body of technology known as the analysis of variance was developed by R. A. Fisher and reported by him in 1923. Since that time it has found wide application in many areas of experimentation. Its early applications were in the field of agriculture. If the variance is understood as the square of the standard deviation of a variable X, $s_x{}^2$, the analysis of variance does not in fact divide this variance into additive parts. The method divides the sum of squares $\Sigma(X - \bar{X})^2$ into additive parts. These are used in the application of tests of significance to the data.

In its simplest form the analysis of variance is used to test the significance of the differences between the means of a number of different populations. We may wish to test the effects of k treatments. These may be different methods of memorizing nonsense syllables, different methods of instruction, or different dosages of a drug. A different treatment is applied to each of the k samples, each sample being comprised of n members.

Members are assigned to treatments at random. The means of the k samples are calculated. The null hypothesis is formulated that the samples are drawn from populations having the same mean. Assuming that the treatments applied are having no effect, some variation due to sampling fluctuation is expected between means. If the variation cannot reasonably be attributed to sampling error, we reject the null hypothesis and accept the alternative hypothesis that the treatments applied are having an effect. With only two means, $k = 2$, this approach leads to the same result as that obtained from the t test for the significance of the difference between means for independent samples.

Consider an agricultural experiment undertaken to compare yields of four varieties of wheat. Thirty-two experimental plots are prepared, and each of the 4 varieties grown in 8 plots. Thus $k = 4$ and $n = 8$. Assume that appropriate precautions have been exercised to randomize uncontrolled factors such as variation in soil fertility from plot to plot. The yield for each plot is obtained, and the mean yield for each variety on the eight plots calculated. Differences in yield reflect themselves in the variation in the four means. If this variation is small and can be explained by sampling error, the investigator has no grounds for rejecting the null hypothesis that no difference exists between the yields of the four varieties. If the variation between means is not small and of such magnitude that it could arise in random sampling in less than 1 or 5 percent of cases, then the evidence is sufficient to warrant rejection of the null hypothesis and acceptance of the alternative hypothesis that the varieties differ in yield.

In the above agricultural experiment the sampling unit is the plot. In psychological experimentation the analog of the plot is usually either a human subject or an experimental animal. In an experiment on the relative efficacy of four different methods of memorizing nonsense syllables, four groups of subjects may be selected, a different method used on each group, and means on a measure of recall obtained for the four groups. A comparison of these means provides information on the relative efficacy of the different methods, and the analysis of variance may be used to decide whether the variation between means is greater than that expected from random sampling fluctuation.

The problem of testing the significance of the differences between a number of means results from experiments designed to study the variation in a dependent variable with variation in an independent variable. The independent variable may be varieties of wheat, methods of memorizing nonsense syllables, or different environmental conditions. The dependent variable may be crop yield, number of nonsense syllables recalled, or number of errors made by an animal in running a maze. Experiments which employ one independent variable are said to involve one basis of classification. The analysis of variance may be used in the analysis of data resulting from experiments which involve more than one basis of classification. For example, an experiment may be designed to permit the study both of varieties of wheat and types of fertilizer on crop yield. This experiment employs two independent variables. We wish to discover how crop yield depends on these two variables. The analysis of variance may be used to extract a part of the total variation resulting from the differences in varieties of wheat and another part resulting from differences in fertilizers, in addition to interaction and error components. A further example is a psychological experiment designed to permit the study of the effects of both free-versus-restricted environment and early-versus-late blindness on maze performance in the rat. Here we have two independent variables. Each variable has two categories. There are four combinations of conditions: free environment and early blindness, free environment and late blindness, restricted environment and early blindness, restricted environment and late blindness. Four groups of experimental animals may be used, and one of the four conditions applied to each group. The analysis of variance may be applied to identify parts of the variation in maze performance assignable to the different environmental and blindness conditions, and other parts as well. Experiments may be designed to permit the simultaneous study of any number of experimental variables within practical limits.

Let us proceed by considering in detail the simple case of a one-way-classification problem where the analysis of variance provides a composite test of the significance of the difference between a set of means.

15.2 NOTATION FOR ONE-WAY ANALYSIS OF VARIANCE

Consider an experiment involving k experimental treatments. The treatments may be different dosages of a drug, different methods of memorizing nonsense syllables, or different environmental variations in the rearing of experimental animals. Each treatment is applied to a different experimental group. Denote the number of members in the k groups by $n_1, n_2, \ldots, n_k$. The number of members in the jth group is n_j. The total number of members in all groups combined is $n_1 + n_2 + \cdots + n_k = N$. When the groups are of equal size we may write $n_1 = n_2 = \cdots = n_k = n = N/k$. The data may be represented as follows:

Group 1	Group 2	Group k
X_{11}	X_{12}	X_{1k}
X_{21}	X_{22}	X_{2k}
X_{31}	X_{32}	X_{3k}
.	.	.
$X_{n_1 1}$	$X_{n_2 2}$	$X_{n_k k}$
$\sum_{i=1}^{n_1} X_{i1}$	$\sum_{i=1}^{n_2} X_{i2}$	$\sum_{i=1}^{n_k} X_{ik}$

Here a system of double subscripts is used. The first subscript identifies the member of the group; the second identifies the group. Thus X_{21} represents the measurement for the second member of the first group, X_{32} represents the measurement for the third member of the second group, and so on. In general, the symbol X_{ij} means the ith member of the jth group. Where the data for each group are tabulated in a separate column, the first subscript identifies the row and the second the column. The sum of measurements in the k groups is represented by

$$\sum_{i=1}^{n_1} X_{i1},\ \sum_{i=1}^{n_2} X_{i2},\ \ldots,\ \sum_{i=1}^{n_k} X_{ik}$$

We may denote the group means by $\bar{X}_{.1}, \bar{X}_{.2}, \ldots, \bar{X}_{.k}$. The symbol $\bar{X}_{.1}$ refers to the mean of the first column, $\bar{X}_{.2}$ the mean of the second column, and $\bar{X}_{.j}$ the mean of the jth column. The convention is to use a dot to indicate the variable subscript over which the summation extends. The mean of all the observations taken together may be represented by the symbol $\bar{X}_{..}$, sometimes called the grand mean. In a one-way classification the meaning associated with the various symbols is quite clear without the use of the dot notation. In discussion of one-way classification we shall therefore simplify the notation and represent the group means by $\bar{X}_1, \bar{X}_2, \ldots, \bar{X}_k$ and the grand mean by $\bar{X}$. The dot notation is necessary in the more complex applications of the analysis of variance, and we shall return to it in Chapter 16.

The total variation in the data is represented by the sum of squares of deviations of all the observations from the grand mean. The sum of squares of deviations of the n_1 observations in the first group from the grand mean is $\sum_{i=1}^{n_1} (X_{i1} - \bar{X})^2$ and the sum of squares of the n_j observations in the jth group from the grand mean is $\sum_{i=1}^{n_j} (X_{ij} - \bar{X})^2$. For k groups each comprising n_j observations the total sum of squares of deviations about $\bar{X}$ is $\sum_{j=1}^{k} \sum_{i=1}^{n_j} (X_{ij} - \bar{X})^2$. When the meaning is clearly understood from the context, it is common practice to represent this total sum of squares by $\Sigma(X_{ij} - \bar{X})^2$, or more simply by $\Sigma(X - \bar{X})^2$.

15.3 PARTITIONING THE SUM OF SQUARES

Simple algebra may be used to demonstrate that the total sum of squares may be divided into two additive and independent parts, a *within-group* sum of squares and a *between-group* sum of squares. We proceed by writing the identity $(X_{ij} - \bar{X}) = (X_{ij} - \bar{X}_j) + (\bar{X}_j - \bar{X})$. This identity states that the deviation of a particular score from the grand mean is comprised of two parts, a deviation from the mean of the group to which the score belongs

$(X_{ij} - \bar{X}_j)$ and a deviation of the group mean from the grand mean $(\bar{X}_j - \bar{X})$. We square this identity and sum over the n_j cases in the jth group to obtain

$$\sum_{i=1}^{n_j} (X_{ij} - \bar{X})^2 = \sum_{i=1}^{n_j} (X_{ij} - \bar{X}_j)^2 + \sum_{i=1}^{n_j} (\bar{X}_j - \bar{X})^2 + 2(\bar{X}_j - \bar{X}) \sum_{i=1}^{n_j} (X_{ij} - \bar{X}_j)$$

The second term to the right requires the summation of a constant $(\bar{X}_j - \bar{X})^2$ over all n_j values of the jth group and may be written $n_j(\bar{X}_j - \bar{X})^2$. The third term to the right disappears because the sum of deviations about the mean $\bar{X}_j$ is 0. We obtain thereby

$$\sum_{i=1}^{n_j} (X_{ij} - \bar{X})^2 = \sum_{i=1}^{n_j} (X_{ij} - \bar{X}_j)^2 + n_j(\bar{X}_j - \bar{X})^2$$

This expression says that the sum of the squares of the deviations of the n_j observations in the jth group from the grand mean $\bar{X}$ is equal to the sum of squares of deviations of the observations from the group mean plus n_j times the square of the difference between the group mean and the grand mean. We now sum over the k groups to obtain

$$\sum_{j=1}^{k} \sum_{i=1}^{n_j} (X_{ij} - \bar{X})^2 = \sum_{j=1}^{k} \sum_{i=1}^{n_j} (X_{ij} - \bar{X}_j)^2 + \sum_{j=1}^{k} n_j(\bar{X}_j - \bar{X})^2 \qquad [15.1]$$

The term to the left is the total sum of squares: the sum of squares of all the observations from the grand mean $\bar{X}$. The first term to the right is the sum of squares within groups: the sum of squares of deviations from the respective group means. The second and last terms to the right are the sums of squares between groups: the sum of squares of deviations of the group means from the grand mean, each term $(\bar{X}_j - \bar{X})^2$ being weighted by n_j, the number of cases in the group. Thus the total sum of squares is partitioned into two additive parts, a sum of squares within groups, and a sum of squares between groups. These two parts are independent.

15.4 THE VARIANCE ESTIMATES OR MEAN SQUARES

Each sum of squares has an associated number of degrees of freedom. The total number of observations is $n_1 + n_2 + \cdots + n_k = \Sigma n_j = N$. The total sum of squares has $N - 1$ degrees of freedom. One degree of freedom is lost by taking deviations about the grand mean. Of these deviations $N - 1$ are free to vary. The number of degrees of freedom associated with the within-groups sum of squares is

$$(n_1 - 1) + (n_2 - 1) + \cdots + (n_k - 1) = \sum_{j=1}^{k} n_j - k = N - k$$

The number of degrees of freedom for each group is $n_j - 1$. Hence the number of degrees of freedom for k groups is $\Sigma n_j - k$, or $N - k$. The number of degrees of freedom associated with the between-groups sum of squares is $k-1$. We have k means, and 1 degree of freedom is lost by expressing the group means as deviations from the grand mean. The degrees of freedom are additive:

$$\underset{\text{Total}}{N-1} = \underset{\text{within}}{(N-k)} + \underset{\text{between}}{(k-1)}$$

The within- and between-groups sums of squares are divided by their associated degrees of freedom to obtain a within-groups variance estimate s_w^2 and a between-group variance estimate s_b^2. Thus

[15.2] $$s_w^2 = \frac{\sum_{j=1}^{k}\sum_{i=1}^{n_j}(X_{ij}-\bar{X}_j)^2}{N-k}$$

[15.3] $$s_b^2 = \frac{\sum_{j=1}^{k} n_j(\bar{X}_j-\bar{X})^2}{k-1}$$

The sums of squares and degrees of freedom are additive. The variance estimates are not additive. The variance estimate is sometimes spoken of as the *mean square*.

15.5 THE MEANING OF THE VARIANCE ESTIMATES

What meaning attaches to the variance estimates s_w^2 and s_b^2? Let us assume that the k samples are drawn from populations having the same variance. The assumption is that $\sigma_1^2 = \sigma_2^2 = \cdots = \sigma_k^2 = \sigma^2$. If this assumption is tenable, the expected value of s_w^2 is σ^2; that is, $E(s_w^2) = \sigma^2$. Thus s_w^2 is an unbiased estimate of the population variance. It is an estimate obtained by combining the data for the k samples. It may be written in the form

[15.4] $$s_w^2 = \frac{\sum_{i=1}^{n_1}(X_{i1}-\bar{X}_1)^2 + \sum_{i=1}^{n_2}(X_{i2}-\bar{X}_2)^2 + \cdots + \sum_{i=1}^{n_k}(X_{ik}-\bar{X}_k)^2}{n_1+n_2+\cdots+n_k-k}$$

The reader will recall that in applying the t test to determine the significance of the differences between two means for independent samples, an unbiased estimate of the population variance was obtained by combining the sums of squares about the means of the two samples and dividing this by the total number of degrees of freedom. The within-group variance s_w^2 is an estimate of precisely the same type. It is obtained by adding together the sums of squares about the k sample means and dividing this by the total number of degrees of freedom. The variance estimate used in the t test is the particular case of s_w^2 which occurs when $k=2$.

The expected value of s_b^2 may be shown to be

[15.5] $$E(s_b^2) = \sigma^2 + \frac{\sum_{j=1}^{k} (\mu_j - \mu)^2}{k-1} \left(\frac{N - \sum_{j=1}^{k} n_j^2/N}{k-1} \right)$$

where μ_j and μ are population means. Under the null hypothesis $\mu_1 = \mu_2 = \cdots = \mu_k = \mu$, and the second term to the right of the above expression is equal to 0. Hence under the null hypothesis both s_w^2 and s_b^2 are estimates of the population variance σ^2.

That s_b^2 is an estimate of σ^2 under the null hypothesis may be illustrated by considering the particular situation where $n_1 = n_2 = \cdots = n_k = n$. The between-group variance estimate may then be written as

$$\frac{n \sum_{j=1}^{k} (\bar{X}_j - \bar{X})^2}{k-1}$$

This is n times the variance of the k means, or $ns_{\bar{x}}^2$. The error variance of the sampling distribution of the arithmetic mean for samples of size n is given by $\sigma_{\bar{x}}^2 = \sigma^2/n$. Hence $n\sigma_{\bar{x}}^2 = \sigma^2$. The quantity $ns_{\bar{x}}^2$ is an estimate of $n\sigma_{\bar{x}}^2$, hence also of σ^2. Thus s_b^2 is an estimate of σ^2.

Where the null hypothesis is false and the means of the populations from which the k samples are drawn differ one from another, the second term to the right of the expression for $E(s_b^2)$ is not equal to 0. It is a measure of the variation of the separate population means μ_j from the grand mean μ.

To test the hypothesis $H_0: \mu_1 = \mu_2 = \cdots = \mu_k$, consider the ratio s_b^2/s_w^2. This is an F ratio. If the population means differ from each other, $E(s_b^2/s_w^2)$ will be greater than unity. If s_b^2/s_w^2 is found to be significantly greater than unity, this may be construed to be evidence for the rejection of the null hypothesis and for the acceptance of the alternative hypothesis that differences exist between the population means. The significance of the F ratio s_b^2/s_w^2 may be assessed with reference to the table of F (Table F of the Appendix) with $k - 1$ degrees of freedom associated with the numerator and $N - k$ degrees of freedom associated with the denominator.

15.6 COMPUTATION FORMULAS

The calculation of the required sums of squares may be simplified by the use of computation formulas. To simplify the notation, denote the sum of all the observations in the jth group by T_j. Thus

$$\sum_{i=1}^{n_j} X_{ij} = T_j$$

Denote the sum of all observations in the k groups by T. Thus

$$\sum_{j=1}^{k} \sum_{i=1}^{n_j} X_{ij} = T$$

The computation formulas are readily obtained. The formula for the *total* sum of squares is

[15.6] $$\sum_{j=1}^{k} \sum_{i=1}^{n_j} (X_{ij} - \bar{X})^2 = \sum_{j=1}^{k} \sum_{i=1}^{n_j} X_{ij}^2 - \frac{T^2}{N}$$

Thus we find the sum of squares of all observations and subtract T^2/N. The *within-groups* sum of squares is

[15.7] $$\sum_{j=1}^{k} \sum_{i=1}^{n_j} (X_{ij} - \bar{X}_j)^2 = \sum_{j=1}^{k} \sum_{i=1}^{n_j} X_{ij}^2 - \sum_{j=1}^{k} \left(\frac{T_j^2}{n_j}\right)$$

The quantity T_j^2/n_j is the square of the sum of the jth group divided by the number of cases in that group. These values are calculated and summed over the k groups. The *between-groups* sum of squares is

[15.8] $$\sum_{j=1}^{k} n_j(\bar{X}_j - \bar{X})^2 = \sum_{j=1}^{k} \left(\frac{T_j^2}{n_j}\right) - \frac{T^2}{N}$$

The above formulas are generally applicable to groups of unequal or equal size. In the particular case where the groups are of equal size and $n_1 = n_2 = \cdots = n_k = n$, the within-groups sum of squares may be written as

[15.9] $$\sum_{j=1}^{k} \sum_{i=1}^{n} X_{ij}^2 - \frac{\sum_{j=1}^{k} T_j^2}{n}$$

and the between-groups sum of squares becomes

[15.10] $$\frac{\sum_{j=1}^{k} T_j^2}{n} - \frac{T^2}{N}$$

15.7 SUMMARY

Table 15.1 presents in summary form the formulas hitherto discussed.

In summary, to test the significance of the difference between k means using the analysis of variance, the following steps are involved:

1 Partition the total sum of squares into two components, a within-groups and a between-groups sum of squares, using the appropriate computation formulas.

2 Divide these sums of squares by the associated number of degrees of

Table 15.1

Analysis of variance: one-way classification summary of formulas

	Source of variation		
	Between	Within	Total
Sum of squares	$\sum_{j=1}^{k} n_j(\bar{X}_j - \bar{X})^2$	$\sum_{j=1}^{k}\sum_{i=1}^{n_j} (X_{ij} - \bar{X}_j)^2$	$\sum_{j=1}^{k}\sum_{i=1}^{n_j} (X_{ij} - \bar{X})^2$
Degrees of freedom	$k - 1$	$N - k$	$N - 1$
Variance estimate; mean square	$s_b^2 = \frac{\sum_{j=1}^{k} n_j(\bar{X}_j - \bar{X})^2}{k - 1}$	$s_w^2 = \frac{\sum_{j=1}^{k}\sum_{i=1}^{n_j} (X_{ij} - \bar{X}_j)^2}{N - k}$	
Expectation	$E(s_b^2) = \sigma^2 + \frac{\sum_{j=1}^{k} (\mu_j - \mu)^2}{k - 1}\left(\frac{N - \sum_{j=1}^{k} n_j^2/N}{k - 1}\right)$	$E(s_w^2) = \sigma^2$	
Computation formulas	$\sum_{j=1}^{k}\left(\frac{T_j^2}{n_j}\right) - \frac{T^2}{N}$	$\sum_{j=1}^{k}\sum_{i=1}^{n_j} X_{ij}^2 - \sum_{j=1}^{k}\left(\frac{T_j^2}{n_j}\right)$	$\sum_{j=1}^{k}\sum_{i=1}^{n_j} X_{ij}^2 - \frac{T^2}{N}$
Computation formulas: equal groups	$\frac{\sum_{j=1}^{k} T_j^2}{n} - \frac{T^2}{N}$	$\sum_{j=1}^{k}\sum_{i=1}^{n} X_{ij}^2 - \frac{\sum_{j=1}^{k} T_j^2}{n}$	$\sum_{j=1}^{k}\sum_{i=1}^{n} X_{ij}^2 - \frac{T^2}{N}$

freedom to obtain s_w^2 and s_b^2, the within- and between-groups variance estimates.

3 Calculate the F ratio s_b^2/s_w^2 and refer this to the table of F (Table F of the Appendix).

4 If the probability of obtaining the observed F value is small, say, less than .05 or .01, under the null hypothesis, reject that hypothesis.

15.8 ILLUSTRATIVE EXAMPLE: ONE-WAY CLASSIFICATION

Table 15.2 shows the number of nonsense syllables recalled by four groups of subjects using four different methods of presentation. Fictitious data are used here for simplicity of illustration. The sums of squares have been calculated using the computation formulas. The data are presented in summary form in Table 15.3. The number of groups is 4. The number of degrees of freedom associated with the between-groups sum of squares is

Table 15.2

Computation for the analysis of variance; one-way classification. Number of nonsense syllables correctly recalled under four methods of presentation

	Method				
	I	II	III	IV	
	5	9	8	1	
	7	11	6	3	
	6	8	9	4	
	3	7	5	5	
	9	7	7	1	
	7		4	4	
	4		4		
	2				
n_j	8	5	7	6	$N = 26$
T_j	43	42	43	18	$T = 146$
$\bar{X}_j$	5.38	8.40	6.14	3.00	$T^2/N = 819.85$
$\sum_{i=1}^{n_j} X_{ij}^2$	269	364	287	68	$\sum_{j=1}^{k}\sum_{i=1}^{n_j} X_{ij}^2 = 988$
$\frac{T_j^2}{n_j}$	231.13	352.80	264.14	54.00	$\sum_{j=1}^{k} \frac{T_j^2}{n_j} = 902.07$

Sum of squares	
Between	902.07 − 819.85 = 82.22
Within	988 − 902.07 = 85.93
Total	988 − 819.85 = 168.15

Table 15.3
Analysis of variance for data of Table 15.2

Source of variation	Sum of squares	Degrees of freedom	Variance estimate
Between	82.22	3	$27.41 = s_b^2$
Within	85.93	22	$3.91 = s_w^2$
Total	168.15	25	$F = 7.01$

$k - 1 = 4 - 1 = 3$. The number of degrees of freedom associated with the within-groups sum of squares is $N - k = 26 - 4 = 22$. The number of degrees of freedom associated with the total is $N - 1 = 26 - 1 = 25$. The between and within sums of squares are divided by the associated degrees of freedom to obtain the variance estimates s_b^2 and s_w^2.

The F ratio is $s_b^2/s_w^2 = 27.41/3.91 = 7.01$. Consulting a table of F with $df = 3$ associated with the numerator and $df = 22$ with the denominator, we find that the value of F required for significance at the .01 level is 4.82.

15.9 THE ANALYSIS OF VARIANCE WITH TWO GROUPS

With two groups only, the significance of the differences between means may be tested using either a t test or the analysis of variance. These procedures lead to the same result. Where $k = 2$ it may be readily shown that $\sqrt{F} = t$.

Consider a situation where $k = 2$ and $n_1 = n_2 = n$. Under these circumstances the between-groups variance estimate s_b^2 is

$$s_b^2 = \frac{n(\bar{X}_1 - \bar{X})^2 + n(\bar{X}_2 - \bar{X})^2}{2 - 1}$$

For groups of equal size the grand mean $\bar{X}$ is halfway between the two group means $\bar{X}_1$ and $\bar{X}_2$. Thus $(\bar{X}_1 - \bar{X}) = (\bar{X}_2 - \bar{X}) = \frac{1}{2}(\bar{X}_1 - \bar{X}_2)$ and $(\bar{X}_1 - \bar{X})^2 = (\bar{X}_2 - \bar{X})^2 = \frac{1}{4}(\bar{X}_1 - \bar{X}_2)^2$. We may therefore write

$$s_b^2 = \frac{n}{2}(\bar{X}_1 - \bar{X}_2)^2$$

When $k = 2$ the within-groups variance estimate s_w^2 is the unbiased variance estimate s^2, obtained by adding the two sums of squares about the means of the two samples and dividing by the total number of degrees of freedom (Section 15.5). Hence

$$F = \frac{(\bar{X}_1 - \bar{X}_2)^2}{s^2(2/n)} \qquad [15.11]$$

and
$$\sqrt{F} = \frac{X_1 - X_2}{s\sqrt{1/n + 1/n}} = t$$

Thus $\sqrt{F} = t$ and $F = t^2$. To illustrate, let $n_1 = n_2 = 8$. In applying the analysis of variance with $df = 1$ associated with the numerator and $df = 14$ associated with the denominator of the F ratio, an F of 4.60 is required for significance at the .05 level. The corresponding t for $df = 14$ required for significance at the .05 level is $\sqrt{4.60} = 2.145$. The t test may be considered a particular case of the F test. It is a particular case which arises when $k = 2$.

In the above discussion we have considered two groups of equal size. The result $\sqrt{F} = t$ is, however, quite general and holds when n_1 and n_2 are unequal. For unequal groups the algebraic development is a bit more cumbersome than that given here. The grand mean does not fall midway between the two group means.

15.10 ASSUMPTIONS UNDERLYING THE ANALYSIS OF VARIANCE

In the mathematical development of the analysis of variance a number of assumptions are made. Questions may be raised about the nature of these assumptions and the extent to which the failure of the data to satisfy them leads to the drawing of invalid inferences.

One assumption is that the distribution of the dependent variable in the population from which the samples are drawn is normal. For large samples the normality of the distribution may be tested using a test of goodness of fit, although in practice this is rarely done. When the samples are fairly small, it is usually not possible to rigorously demonstrate lack of normality in the data. Unless there is reason to suspect a fairly extreme departure from normality, it is probable that the conclusions drawn from the data using an F test will not be seriously affected. The effect of a departure from normality is to make the results appear somewhat more significant than they are. Consequently, where a fairly gross departure from normality occurs, a somewhat more rigorous level of confidence than usual may be employed.

A further assumption in the application of the analysis of variance is that the variances in the populations from which the samples are drawn are equal. This is known as homogeneity of variance. A variety of tests of homogeneity of variance may be applied. These are discussed in more advanced texts (Winer, 1971). Moderate departures from homogeneity should not seriously affect the inferences drawn from the data. Gross departures from homogeneity may lead to results which are seriously in error. Under certain circumstances a transformation of the variable, which leads to greater uniformity of variance, may be used. Under other circumstances it may be possible to use a nonparametric procedure.

A further assumption is that the effects of various factors on the total variation are additive, as distinct from, say, multiplicative. The basic model underlying the analysis of variance is that a given observation may be partitioned into independent and additive bits, each bit resulting from an identifiable source. In most situations there are no grounds to suspect the validity of this model.

With most sets of real data the assumptions underlying the analysis of variance are, at best, only roughly satisfied. The raw data of experiments frequently do not exhibit the characteristics which the mathematical models require. One advantage of the analysis of variance is that reasonable departures from the assumptions of normality and homogeneity may occur without seriously affecting the validity of the inferences drawn from the data.

15.11 TRANSFORMATION OF DATA

As indicated in Section 15.10 the appropriate use of the analysis of variance involves assumptions of homogeneity of variance, normality, and additivity. With some data these assumptions are obviously not satisfied. Sometimes it is possible to use a simple transformation, resulting in a set of transformed values which conform more closely to one or more of the assumptions which the appropriate use of the analysis requires. Most transformations are intended to bring the variances closer to equality. In many cases the transformed values will approximate the normal form more closely than do the original observations.

Commonly used transformations are the square root transformation ($\sqrt{X}$), the logarithmic transformation ($\log X$), the reciprocal transformation ($1/X$), and the arc sine transformation ($\arcsin \sqrt{X}$). A decision regarding which transformation is appropriate is made by exploring the relation between the variances and the treatment means. The nature of this relation determines which transformation to use.

The square root transformation replaces each of the original observations by its square root. This transformation is appropriate when the variances are proportional to the means, that is, when $s_i^2/\bar{X}_i = c$, where c is a constant. Of course with real data this relation will be only roughly approximated.

Table 15.4 shows an example of a square root transformation. The original data are shown on the left. Note that the variances differ markedly one from another. Note also that the variances are roughly proportional to the means. The ratio s_i^2/X_i is in the neighborhood of 1.25. Transformed values are shown at the right in Table 15.4. Note that the effect of the transformation is to reduce the disparity between the variances. The variances for the four groups are now .35, .28, .32, .35, and the assumption of homogeneity of variance is approximately satisfied in the transformed data.

Table 15.4
Comparison of original and transformed data using a square root transformation

	Original data, X				Transformed data, $\sqrt{X}$			
	I	II	III	IV	I	II	III	IV
	1	5	14	2	1.00	2.24	3.74	1.41
	5	9	18	7	2.24	3.00	4.24	2.65
	2	5	9	3	1.41	2.24	3.00	1.73
	6	11	21	10	2.45	3.32	4.58	3.16
	7	12	20	7	2.65	3.46	4.47	2.65
	3	5	12	5	1.73	2.24	3.46	2.24
	3	8	16	6	1.73	2.83	4.00	2.45
Total	27	55	110	40	13.21	19.33	27.49	16.29
$\bar{X}$	3.86	7.86	15.71	5.71	1.89	2.76	3.93	2.33
s^2	4.81	8.81	18.90	7.24	.35	.28	.32	.35

With data containing small numbers, and with 0s, it is advisable to add .5 to each value before taking the square root. The transformation becomes $\sqrt{X + .5}$.

When the variances are proportional not to the means but to the square of the means, a logarithmic transformation, log X, is appropriate. Here again, of course, with real data the relation $s_i^2/\bar{X}_i^2 = c$ may be only roughly satisfied. If some of the observations are 0 and small numbers, the suggestion is that the transformation log $(X + 1)$ be used. When the standard deviations, and not the variances, are proportional to the square of the treatment means, that is, when $s_i/\bar{X}_i^2 = c$, a reciprocal transformation, $1/X$, may prove useful.

In some experiments the observations obtained are proportions or percentages, such as the proportion of successful responses in a given number of trials, or the percentage of correct answers in performing a series of tasks. When the data are proportions, a relation between means and variances roughly of the kind $s_i^2 = \bar{X}_i\,(1 - \bar{X}_i)$ may be observed. Under this circumstance an arcsine transformation, arcsin $\sqrt{X}$, may be used. Each of the original observations is replaced by an angle whose sine is the square root of the original observation.

15.12 THE CORRELATION RATIO

In an analysis of variance for one-way classification two variables are involved, an independent variable which is frequently nominal in type and a dependent variable which is usually an interval-ratio variable. The total sum of squares is partitioned into two parts, a between-group and a within-group sum of squares. The between-group sum of squares is that part of the variation attributed to the independent variable. The within-group sum of squares is that part of the variation attributed to other factors. The strength of the relation between the two variables can be expressed in the form of a ratio as follows:

[15.12] $$\eta_{y.x}^2 = \frac{\text{between SS}}{\text{total SS}} = 1 - \frac{\text{within SS}}{\text{total SS}}$$

The symbol η is the Greek letter eta, and SS denotes sum of squares. This statistic is known as the correlation ratio. It may be interpreted as a simple proportion in the same way that r^2 is interpreted as a proportion. It is a measure of the strength of association between the dependent and independent variables involved in the experiment.

The correlation ratio has been described in many texts as a measure appropriate for use in describing the relation between two variables when the regression lines are nonlinear. If one variable, say, the independent variable, is a nominal variable and the other an interval-ratio variable, the idea of linearity or nonlinearity of regression is quite meaningless. Under these circumstances the correlation ratio is a measure of the strength of association between a nominal variable and an interval-ratio variable. If both variables are of the interval-ratio type, questions of linearity and nonlinearity can meaningfully be raised.

To test whether a correlation ratio is significantly different from 0, we may use the F ratio

[15.13] $$F = \frac{\eta_{y.x}^2/(k-1)}{(1-\eta_{y.x}^2)(N-k)}$$

where k = number of categories of nominal variable
N = total number of observations

To illustrate for the data of Table 15.2 the correlation ratio $\eta_{y.x}^2 = 82.22/168.15 = .489$. Thus 48.9 percent of the variation in the data can be attributed to the independent variable.

The F ratio used in testing the significance of this correlation ratio is

$$F = \frac{.489/(4-1)}{(1-.489)/(26-4)} = 7.02$$

This is the same F ratio, within the rounding of decimals, as that previously obtained in Table 15.3. The analyses of variance and the correlation-ratio approach are equivalent procedures and lead to the same result.

BASIC TERMS AND CONCEPTS

Sum of squares

Between-group sum of squares

Within-group sum of squares

Variance estimate

Between group variance estimate, s_b^2

Within group variance estimate, s_w^2

F ratio, s_b^2/s_w^2

Square root transformation

Logarithmic transformation

Correlation ratio

EXERCISES

1 How many degrees of freedom are associated with the variation in the data for **(a)** a comparison of two means for independent samples, each containing 20 cases, **(b)** a comparison of four means for independent samples, each containing 14 cases, **(c)** a comparison of four means for independent samples of size 10, 16, 18, and 11, respectively?

2 The following are measurements obtained for five equal groups of subjects:

Group	Measurements					$\bar{X}_j$
I	4	7	9	9	14	8.60
II	5	6	12	12	7	8.40
III	15	18	21	26	20	20.00
IV	35	27	29	30	25	29.20
V	17	26	17	20	12	18.40

Apply the analysis of variance to test the null hypothesis

$$H_0: \mu_1 = \mu_2 = \mu_3 = \mu_4 = \mu_5$$

3 The following are error scores on a psychomotor test for four groups of subjects tested under four experimental conditions:

Group	Error scores							$\bar{X}_j$
I	16	7	19	24	31			19.40
II	24	6	15	25	32	24	29	22.14
III	16	15	18	19	6	13	18	15.00
IV	25	19	16	17	42	45		27.33

Apply the analysis of variance to test the null hypothesis

$$H_0: \mu_1 = \mu_2 = \mu_3 = \mu_4$$

4 Apply the analysis of variance to test the significance of the difference between means for the following data:

	I	II	III
n	10	10	10
$\bar{X}_j$	7.40	8.30	10.56
$\sum_{i=1}^{n} X_{ij}^2$	649	755	1,263

5 The following are test scores for two groups:

Group	Scores							$\bar{X}_j$
I	12	26	31	12	14	16	10	17.29
II	8	14	29	7	14	6		13.00

Calculate s_w^2, s_b^2, and F. Test the null hypothesis $H_0: \mu_1 = \mu_2$.

6 What assumptions underly the analysis of variance?

7 The following are experimental data for three independent groups:

Group	Data					$\bar{X}_j$
I	4	16	49	64	81	42.80
II	49	121	144	169	196	135.80
III	16	36	81	100	121	70.80

What would be an appropriate transformation on these data? Transform the data and apply an analysis of variance to the transformed values, calculating the required F ratio.

ANSWERS TO EXERCISES

1 **a** $df_1 = 1,\ df_2 = 38$ **b** $df_1 = 3,\ df_2 = 52$ **c** $df_1 = 3,\ df_2 = 51$

2 $F = 23.31,\ df_1 = 4,\ df_2 = 20;\ p < .01$

3 $F = 2.06,\ df_1 = 3,\ df_2 = 21;\ p > .05$

4 $F = 2.27,\ df_1 = 2,\ df_2 = 27;\ p > .05$

5 $s_b^2 = 59.34,\ s_w^2 = 68.49,\ F = 0.87,\ df_1 = 1,\ df_2 = 11;\ p > .05$

6 The assumption of normality, homogeneity of variance, and additivity

7 $s^2/\bar{X}$ is roughly 25 for the three groups, and a square root transformation is appropriate. $F = 4.60,\ df_1 = 2,\ df_2 = 12;\ p < .05$.

16

ANALYSIS OF VARIANCE: TWO-WAY CLASSIFICATION

16.1 INTRODUCTION

Experiments may be designed to permit the simultaneous investigation of two experimental variables. Such experiments involve two bases of classification. To illustrate, assume that an investigator wishes to study the effects of two methods of presenting nonsense syllables on recall after 5 minutes, 1 hour, and 24 hours. One experimental variable is method of presentation, the other the interval between presentation and recall. There are six combinations of experimental conditions. One method of conducting such an experiment is to select a group of subjects and allocate these at random to the experimental conditions, an equal number being assigned to each condition. With, say, 10 subjects allocated to each experimental condition, the total number of subjects will be $2 \times 3 \times 10 = 60$. The data may be arranged in a table containing two rows and three columns. The rows correspond to methods, the columns to time intervals. The 10 observations for each group may be entered in each cell of the table. Differences in the means of the rows result from differences in recall under the two methods of presentation. Differences in the means of the columns result from differences in recall after the three time intervals.

Experiments with two-way classification may be conducted with only one sampling unit, and measurement, for each experimental condition. With one measurement for each experimental condition the total sum of squares is partitioned into three parts, a between-rows, a between-columns, and an interaction sum of squares. With more than one measurement for each experimental condition, the total sum of squares is partitioned into four parts, a between-rows, a between-columns, an interaction, and a within-cells sum of squares. Each sum of squares has an associated number of degrees of freedom. By dividing the sums of squares by the associated degrees of freedom, four variance estimates are obtained. These variance estimates are used to test the significance of the differences between row means, column means, and, with more than one measurement per cell, the interaction effect.

16.2 NOTATION FOR TWO-WAY ANALYSIS OF VARIANCE

Consider an experiment involving R experimental treatments of one variable and C experimental treatments of another variable. The number of treatment combinations is RC. Let us consider the particular case where we have one sampling unit, and one measurement, for each of the RC treatment combinations. The total number of measurements is $RC = N$. The data may be represented as follows:

	1	2	3	$\cdots$	C	Row mean
1	X_{11}	X_{12}	X_{13}	$\cdots$	X_{1C}	$\bar{X}_{1.}$
2	X_{21}	X_{22}	X_{23}	$\cdots$	X_{2C}	$\bar{X}_{2.}$
3	X_{31}	X_{32}	X_{33}	$\cdots$	X_{3C}	$\bar{X}_{3.}$
.	.	.	.	$\cdots$	.	.
.	.	.	.	$\cdots$		.
.	.	.	.	$\cdots$		.
R	X_{R1}	X_{R2}	X_{R3}	$\cdots$	X_{RC}	$\bar{X}_{R.}$
Column mean	$\bar{X}_{.1}$	$\bar{X}_{.2}$	$\bar{X}_{.3}$	$\cdots$	$\bar{X}_{.C}$	$\bar{X}_{..}$

Double subscripts are used. The first subscript identifies the *row;* the second subscript identifies the *column.* Thus X_{32} is the measurement in the third row and the second column. Usually X_{rc} denotes the measurement in the rth row and cth column, where $r = 1, 2, \ldots, R$ and $c = 1, 2, \ldots, C$. A dot notation is used to identify means. The symbol $\bar{X}_{1.}$ refers to the mean of the first row, $\bar{X}_{2.}$ the mean of the second row, and $\bar{X}_{r.}$ the mean of the rth row. Similarly, $\bar{X}_{.1}$ refers to the mean of the first column, $\bar{X}_{.2}$ the mean of the second column, and $\bar{X}_{.c}$ the mean of the cth column. The grand mean, the mean of all N observations, is $\bar{X}_{..}$. The total sum of squares of deviations about the grand mean is given by

$$\sum_{r=1}^{R} \sum_{c=1}^{C} (X_{rc} - \bar{X}_{..})^2$$

Consider now a situation where we have n sampling units and n measurements for each of the RC treatment combinations. The total number of measurements is $nRC = N$. Where $R = 2$, $C = 3$, and $n = 3$, the data may be represented as follows:

	1	2	3	Row mean
1	X_{111} X_{112} X_{113}	X_{121} X_{122} X_{123}	X_{131} X_{132} X_{133}	$\bar{X}_{1..}$
2	X_{211} X_{212} X_{213}	X_{221} X_{222} X_{223}	X_{231} X_{232} X_{233}	$\bar{X}_{2..}$
Column mean	$\bar{X}_{.1.}$	$\bar{X}_{.2.}$	$\bar{X}_{.3.}$	

Triple subscripts are used. The first subscript identifies the *row*, the second the *column*, and the third the measurement *within* the cell. Thus X_{231} means the measurement for the first individual in the second row and the third column. In general X_{rci} denotes the measurement for the ith individual in the rth row and cth column, where $i = 1, 2, \ldots, n$. Row, column, and cell means are identified by a dot notation. The mean of all the observations in the first row is $\bar{X}_{1..}$. The mean of all the observations in the rth row is $\bar{X}_{r..}$. Similarly, the mean of the first column is $\bar{X}_{.1.}$ and of the cth column $\bar{X}_{.c.}$. The mean of all the observations in the cell corresponding to the rth row and cth column is $\bar{X}_{rc.}$. The mean of all nRC observations, the grand mean, is $\bar{X}_{...}$. The total sum of squares of all observations about the grand mean is

$$\sum_{r=1}^{R}\sum_{c=1}^{C}\sum_{i=1}^{n}(X_{rci} - \bar{X}_{...})^2$$

The sum of squares of deviations about the grand mean, both where $n = 1$ and $n > 1$, is partitioned into additive components.

16.3 PARTITIONING THE SUM OF SQUARES

With one measurement only for each of the RC treatment combinations, the total sum of squares may be partitioned into three additive components, a between-rows, a between-columns, and an interaction sum of squares. We proceed by writing the identity

$$(X_{rc} - \bar{X}_{..}) = (\bar{X}_{r.} - \bar{X}_{..}) + (\bar{X}_{.c} - \bar{X}_{..}) + (X_{rc} - \bar{X}_{r.} - \bar{X}_{.c} + \bar{X}_{..})$$

This identity states that the deviation of an observation from the grand mean may be viewed as composed of three parts, a deviation of the row mean from the grand mean, a deviation of the column mean from the grand mean, and a remainder, or residual term, known as an interaction term. By squaring both sides of the above identity, an expression is obtained containing six terms. This may be summed over R rows and C columns. Three of these terms conveniently vanish, because they contain a sum of deviations about a mean, which, of course, is 0. The resulting total sum of squares may be written as

$$\sum_{r=1}^{R}\sum_{c=1}^{C}(X_{rc}-\bar{X}_{..})^2 = C\sum_{r=1}^{R}(\bar{X}_{r.}-\bar{X}_{..})^2 + R\sum_{c=1}^{C}(\bar{X}_{.c}-\bar{X}_{..})^2 + \sum_{r=1}^{R}\sum_{c=1}^{C}(X_{rc}-\bar{X}_{r.}-\bar{X}_{.c}+\bar{X}_{..})^2$$

The first term to the right is C times the sum of squares of deviations of row means from the grand mean. This is the between-rows sum of squares. It describes the variation in row means. The second term is R times the sum of squares of deviations of column means from the grand mean. This is the between-columns sum of squares. It describes variation in column means. The third term is a residual, or interaction, sum of squares. The meaning of the interaction term is discussed in detail in Section 16.5.

With n measurements for each of the RC treatment combinations the total sum of squares may be partitioned into four additive components. These are a between-rows, a between-columns, an interaction, and a within-cells sum of squares. In this situation we write the identity

$$(X_{rci}-\bar{X}_{...}) = (\bar{X}_{r..}-\bar{X}_{...}) + (\bar{X}_{.c.}-\bar{X}_{...}) + (\bar{X}_{rc.}-\bar{X}_{r..}-\bar{X}_{.c.}+\bar{X}_{...}) + (X_{rci}-\bar{X}_{rc.})$$

This expression may be squared and summed over rows, columns, and within cells. All but four terms vanish, and the resulting total sum of squares may be written as

$$[16.1]\quad \sum_{r=1}^{R}\sum_{c=1}^{C}\sum_{i=1}^{n}(X_{rci}-\bar{X}_{...})^2 = nC\sum_{r=1}^{R}(\bar{X}_{r..}-\bar{X}_{...})^2 + nR\sum_{c=1}^{C}(\bar{X}_{.c.}-\bar{X}_{...})^2 + n\sum_{r=1}^{R}\sum_{c=1}^{C}(\bar{X}_{rc.}-\bar{X}_{r..}-X_{.c.}+\bar{X}_{...})^2 + \sum_{r=1}^{R}\sum_{c=1}^{C}\sum_{i=1}^{n}(X_{rci}-\bar{X}_{rc.})^2$$

The first term to the right is descriptive of the variation of row means, the second of column means, and the third of interaction. The fourth term is the within-cells sum of squares. It is the sum of squares of the deviations of observations from the means of the cells to which they belong.

16.4 VARIANCE ESTIMATES OR MEAN SQUARES

With a single entry in each cell, $n = 1$ and $RC = N$. The number of degrees of freedom associated with the total sum of squares is $RC - 1 =$

$N - 1$. The numbers of degrees of freedom associated with row and column sums are $R - 1$ and $C - 1$, respectively. The number of degrees of freedom associated with the interaction sum of squares is $(R - 1)(C - 1)$. The degrees of freedom are additive, and

$$\underset{\text{Total}}{N - 1} = \underset{\text{row}}{(R - 1)} + \underset{\text{column}}{(C - 1)} + \underset{\text{interaction}}{(R - 1)(C - 1)}$$

The sums of squares are divided by the associated degrees of freedom to obtain three variance estimates, or mean squares. The between-rows, between-columns, and interaction variance estimates are s_r^2, s_c^2, and s_{rc}^2, respectively.

With n entries in each cell, where $n > 1$, the total number of observations is $nRC = N$. The number of degrees of freedom associated with the total sum of squares is $nRC - 1 = N - 1$. The numbers of degrees of freedom associated with row, column, and interaction sums of squares are $R - 1$, $C - 1$, and $(R - 1)(C - 1)$, respectively. The number of degrees of freedom associated with the within-cells sum of squares is $nRC - RC = RC(n - 1)$. Because the deviations are taken about the cell means, 1 degree of freedom is lost for each cell. In each cell $n - 1$ deviations are free to vary. The number of degrees of freedom for RC cells, therefore, is $RC(n - 1)$. The degrees of freedom are additive. The sums of squares are divided by the associated degrees of freedom to obtain the variance estimates, or mean squares.

Table 16.1 shows in summary form the sum of squares, degrees of freedom, and variance estimates for a two-way classification with n entries per cell.

F ratios are formed from the variance estimates and used to test the significance of row, column, and, where $n > 1$, interaction effects. The

Table 16.1

Analysis of variance for two-way classification with n entries per cell: $n > 1$

Source	Sums of squares	*df*	Variance estimate
Rows	$nC \sum_{r=1}^{R} (\bar{X}_{r..} - \bar{X}_{...})^2$	$R - 1$	s_r^2
Columns	$nR \sum_{c=1}^{C} (\bar{X}_{.c.} - \bar{X}_{...})^2$	$C - 1$	s_c^2
Interaction	$n \sum_{r=1}^{R} \sum_{c=1}^{C} (\bar{X}_{rc.} - \bar{X}_{r..} - \bar{X}_{.c.} + \bar{X}_{...})^2$	$(R - 1)(C - 1)$	s_{rc}^2
Within cells	$\sum_{r=1}^{R} \sum_{c=1}^{C} \sum_{i=1}^{n} (X_{rci} - \bar{X}_{rc.})^2$	$RC(n - 1)$	s_w^2
Total	$\sum_{r=1}^{R} \sum_{c=1}^{C} \sum_{i=1}^{n} (X_{rci} - \bar{X}_{...})^2$	$nRC - 1$	

correct procedure here, and the interpretation of the variance estimates, depends on the statistical model appropriate for the experiment. Three models may be identified: fixed, random, and mixed. The investigator must decide which model fits the experiment. This decision determines how the variance estimates are used in the application of tests of significance to the data. Before proceeding with a discussion of these models (Section 16.6), the meaning of the interaction term is discussed.

16.5 THE NATURE OF INTERACTION

The algebraic partitioning of sums of squares in a two-way classification leads to the interaction term

$$n \sum_{r=1}^{R} \sum_{c=1}^{C} (\bar{X}_{rc.} - \bar{X}_{r..} - \bar{X}_{.c.} + \bar{X}_{...})^2$$

The nature of interaction may be illustrated by example. Consider a simple agricultural experiment with two varieties of wheat and two types of fertilizer. Assume that one variety of wheat has a higher yield than the other. If the yield is uniformly higher regardless of which fertilizer is used, then there is no interaction between the two experimental varieties. If, however, one variety produces a relatively higher yield with one type of fertilizer than with the other, then the two variables may be said to interact. To illustrate further, assume that we have two methods of teaching arithmetic and two teachers. Each teacher uses the two methods on separate groups of pupils. The achievement of the pupils is measured. If one method of instruction is uniformly superior or inferior regardless of which teacher uses it, then there is no interaction between methods and teacher. If, however, one teacher obtains better results with one method than the other, and the opposite holds for the other teacher, then teachers and methods may be said to interact.

Table 16.2 shows observed cell means for a two-way classification with three categories for each of the experimental variables. The observed cell entries are means based on an equal number of cases. What are the expected cell means on the assumption of zero interaction? This situation is somewhat analogous to the calculation of expected values for contingency tables. For a contingency table we calculate expected cell frequencies. Here we are required to calculate the expected cell means on the assumption that the two experimental variables function independently.

Assuming zero interaction, certain constant differences will be maintained between cell means. In Table 16.2 the observed row mean for A is 10 points less than the row mean for B. If the interaction were 0, we should expect a constant 10-point difference to occur between means for A and B under treatments I, II, and III. A similar relationship would be expected on comparing all other rows and columns of this table. Obviously,

Table 16.2

Comparison of observed cell means and means expected under zero interaction

Observed, $\bar{X}_{rc.}$

	I	II	III	
A	2	12	16	10
B	6	20	34	20
C	64	76	40	60
	24	36	30	30

Expected, $E(\bar{X}_{rc.})$

	I	II	III	
A	4	16	10	10
B	14	26	20	20
C	54	66	60	60
	24	36	30	30

the observed values in Table 16.2 do not exhibit this characteristic. The interaction is not 0.

Where the interaction is 0, a deviation of a cell mean from the mean of the row (or column) to which it belongs will be equal to the deviation of its column (or row) mean from the grand mean. If $\bar{X}_{rc.}$ is a cell mean and $\bar{X}_{r..}$ and $\bar{X}_{.c.}$ are its corresponding row and column means, then under zero interaction, $\bar{X}_{rc.} - \bar{X}_{r..} = \bar{X}_{.c.} - \bar{X}_{...}$. Thus the expected value of $\bar{X}_{rc.}$ under zero interaction is given by $E(\bar{X}_{rc.}) = \bar{X}_{r..} + \bar{X}_{.c.} - \bar{X}_{...}$. These expected values have been calculated for the observed data of Table 16.2 and are shown to the right of the table. On comparing the expected values in any two rows or columns, note the constant increment or decrement. If $\bar{X}_{rc.}$ is an observed and $E(\bar{X}_{rc.})$ an expected value, the deviation of an observed from an expected value is $\bar{X}_{rc.} - \bar{X}_{r..} - \bar{X}_{.c.} + \bar{X}_{...}$. The interaction term in the analysis of variance is n times the sum of squares of deviations of the observed cell means from the expected cell means.

16.6 FINITE, RANDOM, FIXED, AND MIXED MODELS

Difficulties arise in the selection of procedures for testing row, column, and interaction effects in a two-way analysis of variance. These difficulties are resolved by the recognition of a general statistical model underlying the analysis of variance. This model is referred to here as the *finite* model. Three particular cases of the finite model may be identified. These are the *random, fixed,* and *mixed* models. The models appropriate for different experiments differ. Investigators must decide which model best represents their experiment. The choice of model determines the procedure for testing row, column, and interaction effects. The choice of model depends on the nature of the variables used as the basis of classification in the experimental design.

The general finite model makes the linearity assumption that a deviation of an observation X_{rci} from the population value of the grand mean μ may be expressed in the form

[16.2] $$X_{rci} - \mu = a_r + b_c + (ab)_{rc} + e_{rci}$$

The four quantities to the right are in deviation form. Thus $a_r = \mu_{r..} - \mu$, a deviation of the population value of the row mean from the grand mean μ. Similarly, $b_c = \mu_{.c.} - \mu$, a deviation of a column mean from the grand mean. The interaction term $(ab)_{rc} = (\mu_{rc.} - \mu_{r..} - \mu_{.c.} + \mu)$, and the error term $e_{rci} = X_{rci} - \mu_{rc.}$. Where this model is used to represent experimental data the implicit assumption is made that treatment effects can meaningfully be partitioned into additive components for each sampling unit. Because a_r, b_c, $(ab)_{rc}$, and e_{rci} are in deviation form, they each sum to 0. The population variances of the four components are σ_a^2, σ_b^2, σ_{ab}^2, and σ_e^2.

The null hypothesis under test, for example, for row effects is $H_r : \mu_{1..} = \mu_{2..} = \cdot \cdot \cdot = \mu_{R..}$. This hypothesis may be stated in the form $H_r : \sigma_a^2 = 0$. Similarly, the null hypotheses for column and interaction effects may be stated as $H_c : \sigma_b^2 = 0$ and $H_{rc} : \sigma_{ab}^2 = 0$. We wish to obtain from the experimental data information which will provide a valid test of these hypotheses.

We now consider an actual experiment involving R levels of one variable and C levels of another. The R and C levels may be regarded as samples drawn at random from two populations of levels comprising R_p and C_p members, respectively. Thus we conceptualize two populations of levels. The levels used in a particular experiment are construed to be drawn at random from these two populations. R_p, R, C_p, and C may take any integral values, provided, of course, that $R \leqq R_p$ and $C \leqq C_p$. The RC treatment combinations are assigned at random to the nRC sampling units or individuals. Under these conditions, and given the basic linearity assumption, Wilk and Kempthorne (1955) have shown that the expectations of the mean squares for the general finite model are as shown in Table 16.3.

Table 16.3

Expectation of mean squares for general finite model: two-way analysis of variance with n entries in each cell: $n > 1$

Mean square	Expectation of mean square
Row, s_r^2	$\sigma_e^2 + \frac{(C_p - C)}{C_p} n\sigma_{ab}^2 + nC\sigma_a^2$
Column, s_c^2	$\sigma_e^2 + \frac{(R_p - R)}{R_p} n\sigma_{ab}^2 + nR\sigma_b^2$
Interaction, s_{rc}^2	$\sigma_e^2 + n\sigma_{ab}^2$
Within cells, s_w^2	σ_e^2

Thus the mean squares provide estimates of variance components, and these are used to test the significance of row, column, and interaction effects. How these are used depends on a consideration of three particular instances of the general finite model.

Consider an experiment involving R levels of one variable and C of another, these being regarded as random samples of levels from populations comprising R_p and C_p members. We may consider a case where R_p and C_p are very large, so that $R_p >> R$ and $C_p >> C$, where $>>$ denotes much greater than. Under these circumstances such terms as $(R_p - R)/R_p$ and $(C_p - C)/C_p$ approach unity. When this is so we have what is referred to as a *random-model* situation. The expectations for the random model are obtained by substituting $(R_p - R)/R_p = 1$ and $(C_p - C)/C_p = 1$ in the expectations of the mean squares for the general finite model given in Table 16.3. Thus the random model is a particular case of the finite model.

In psychological research, experiments where the random model is appropriate are not numerous. Satisfactory examples are not readily found. One example is an experiment where each member of a sample of R job applicants is assigned a rating by each member of a sample of C interviewers. Here both job applicants and interviewers may be viewed as samples drawn at random from populations such that $R_p >> R$ and $C_p >> C$.

In many experiments the R levels of one variable and the C levels of the other are not conceptualized as random samples. In agricultural experiments where R varieties of wheat and C varieties of fertilizer are used, the investigator is usually concerned with the yield of particular wheat varieties and with the effect of particular fertilizers on yield. He is not concerned with drawing inferences about hypothetical populations of wheat and fertilizer varieties. Both variables or factors are *fixed.* Any factor is fixed if the investigator on repeating the experiment would use the same levels of it. Under the fixed model $R = R_p$ and $C = C_p$. By substituting $(R_p - R)/R_p = 0$ and $(C_p - C)/C_p = 0$ in the expectations of the mean squares for the finite model given in Table 16.3, the expectations for the fixed model are obtained.

In psychological experiments different methods of learning, environmental conditions, methods of inducing stress, and the like, are examples of fixed factors or variables. In many experiments different levels of the experimental variable are introduced, i.e., levels of illumination, time intervals, size of brain lesion, and dosages of a drug. While the levels may be thought to constitute a representative set and interpolation between levels may be possible, such variables are usually regarded as fixed. Of course it is possible to conceptualize a study where, for example, levels of illumination or dosages of a drug are sampled at random from a population of levels or dosages. Ordinarily, however, experiments are not designed in this way.

In many experiments one basis of classification is a random factor or variable and the other is fixed. Measurements may be obtained for a sample of R individuals for each of C treatments or experimental conditions. Here one basis of classification is random and the other is fixed. This is a *mixed* model. In the mixed-model situation either $R_p = R$ and $C_p >> C$ or $R_p >> R$ and $C_p = C$. By substituting $(R_p - R)/R_p = 1$ and $(C_p - C)/C_p = 0$, or vice versa, in the expectations for the finite model of Table 16.3, we obtain the expectations for the mixed model.

Table 16.3 may be used to provide the required expectations for a two-way classification where $n = 1$. Under this circumstance no within-cells variance estimate s_w^2 is available. The expectations for row, column, and interaction effects for the random, fixed, and mixed models are obtained by writing $n = 1$ and substituting the appropriate values of $(R_p - R)/R_p$ and $(C_p - C)/C_p$.

16.7 CHOICE OF ERROR TERM

By choice of error term is meant the selection of the appropriate variance estimate for the denominator of the F ratio in testing row, column, and interaction effects. In general, in forming an F ratio, the expectation of the variance estimate in the numerator should contain one term more than the expectation of the variance estimate in the denominator, the additional term involving the effect under test. On applying this principle to the expectations of Table 16.3, the following rules may be formulated:

1 **Random model: $n > 1$** The proper error term for testing the interaction effect is s_w^2.

$$F_{rc} = \frac{s_{rc}^2}{s_w^2}$$

The correct error term for testing row and column effects is s_{rc}^2.

$$F_r = \frac{s_r^2}{s_{rc}^2} \quad \text{and} \quad F_c = \frac{s_c^2}{s_{rc}^2}$$

2 **Fixed model: $n > 1$** The proper error term is s_w^2 for interaction, row, and column effects. The three F ratios are

$$F_{rc} = \frac{s_{rc}^2}{s_w^2} \qquad F_r = \frac{s_r^2}{s_w^2} \quad \text{and} \quad F_c = \frac{s_c^2}{s_w^2}$$

3 **Mixed model: $n > 1$** The proper error term for testing the interaction effect is s_w^2.

$$F_{rc} = \frac{s_{rc}^2}{s_w^2}$$

When R is random and C is fixed, the proper error term for testing row effects is s_w^2.

$$F_r = \frac{s_r^2}{s_w^2}$$

The proper term for testing column effects is s_{rc}^2.

$$F_c = \frac{s_c^2}{s_{rc}^2}$$

When R is fixed and C is random, the converse procedure applies.

$$F_r = \frac{s_r^2}{s_{rc}^2} \qquad \text{and} \qquad F_c = \frac{s_c^2}{s_w^2}$$

4 **Random model: $n = 1$** No s_w^2 is available. The correct error term for testing both row and column effects is s_{rc}^2.

$$F_r = \frac{s_r^2}{s_{rc}^2} \qquad \text{and} \qquad F_c = \frac{s_c^2}{s_{rc}^2}$$

5 **Fixed model: $n = 1$** The point of view may be adopted that no test of either row or column effects can be made. This point of view requires some modification. The ratio s_r^2/s_{rc}^2 is an estimate of $(\sigma_e^2 + C\sigma_a^2)/(\sigma_e^2 + \sigma_{ab}^2)$ and will, where $\sigma_{ab}^2 > 0$, be an underestimate of $(\sigma_e^2 + C\sigma_a^2)/\sigma_e^2$. This means that if a significant result is obtained, the investigator knows *a fortiori* that the effect tested is significant. If the result is not significant, the probability of accepting the null hypothesis, $H_0 : \sigma_a^2 = 0$, when it is false may be high. Thus in the absence of significance no conclusions should be drawn from the data.

6 **Mixed model: $n = 1$** When R is random and C is fixed, the situation pertaining to the testing of row effects is as described above for the fixed model, $n = 1$. The proper error term for the column effect is s_{rc}^2.

$$F_c = \frac{s_c^2}{s_{rc}^2}$$

When C is random and R is fixed, the argument relating to the fixed model, $n = 1$, again applies. The proper error term for the row effect is s_{rc}^2.

$$F_r = \frac{s_r^2}{s_{rc}^2}$$

The above rules, excluding the modifications of rules 5 and 6 above, can be very simply obtained by using the following schema for the proper choice of error term:

ROW	$\frac{C_p - C}{C_p} s_{rc}^2 + \frac{C}{C_p} s_w^2$
COLUMN	$\frac{R_p - R}{R_p} s_{rc}^2 + \frac{R}{R_p} s_w^2$
INTERACTION	s_w^2

For the random model, $(C_p - C)/C_p = 1$ and $C/C_p = 0$. The proper error term for row and column effects is s_{rc}^2. Similarly, the proper error term for the fixed and mixed models may be obtained. When $n = 1$, all terms containing s_w^2 vanish. For the random model, s_{rc}^2 becomes the correct error term for row and column effects. For the fixed model, no tests are possible. When rows are random and the columns are fixed, the column effect may be tested, but not the row.

16.8 POOLING SUMS OF SQUARES: $n > 1$

Under certain circumstances the within-cells and interaction sums of squares may be added together and divided by the combined degrees of freedom to obtain an estimate of variance based on a larger number of degrees of freedom. Caution should be exercised in applying this procedure.

For the fixed model, the within-cells variance estimate is the proper error term for testing interaction, row, and column effects. For the random model, the interaction variance estimate is the proper error term for testing row and column effects. These procedures are always correct. For both models, when the interaction is quite clearly *not* significant, the within-cells and interaction sums of squares may be pooled to obtain a variance estimate for the denominator of the F ratio based on a larger number of degrees of freedom. Of course, when row and column effects are clearly significant, when tested without pooling, the pooling procedure is unnecessary.

When doubt exists as to the significance of the interaction, the investigator may or may not choose to pool the sums of squares. If the interaction effect in fact exists, σ_{ab}^2 being greater than 0, and terms are pooled, the pooling may be said to be erroneous.

For the fixed model, erroneous pooling will increase the size of the error term. For the random model, erroneous pooling will decrease the size of the error term. In both instances the number of degrees of freedom is increased. Erroneous pooling will for the fixed model usually lead to too few significant results and for the random model to too many significant results.

For the mixed model, when rows are random and columns are fixed, pooling may be applied with nonsignificant interaction. In this situation erroneous pooling will tend to make the error term too large for testing row effects and too small for testing column effects, leading to too few significant effects for rows and too many for columns.

In general, it is advisable not to pool unless the investigator is quite confident that the interaction is not significant. For a detailed discussion of this rather troublesome problem, see Keppel (1973).

16.9 COMPUTATION FORMULAS FOR SUMS OF SQUARES

Computation formulas are used to calculate the required sums of squares. A simplified notation is used. Denote the sum of all observations in the rth row by $T_{r.}$, the sum of all observations in the cth column by $T_{.c}$, the sum of all observations in the cell corresponding to the rth row and cth column by T_{rc}, and the sum of all N observations by T.

With *one* entry in each cell, the computation formulas for sums of squares are as follows:

[16.3] ROWS
$$\frac{1}{C}\sum_{r=1}^{R} T_{r.}^{2} - \frac{T^2}{N}$$

[16.4] COLUMNS
$$\frac{1}{R}\sum_{c=1}^{C} T_{.c}^{2} - \frac{T^2}{N}$$

[16.5] INTERACTION
$$\sum_{r=1}^{R}\sum_{c=1}^{C} X_{rc}^{2} - \frac{1}{C}\sum_{r=1}^{R} T_{r.}^{2} - \frac{1}{R}\sum_{c=1}^{C} T_{.c}^{2} + \frac{T^2}{N}$$

[16.6] TOTAL
$$\sum_{r=1}^{R}\sum_{c=1}^{C} X_{rc}^{2} - \frac{T^2}{N}$$

The interaction sum of squares may be obtained by adding row and column sums and subtracting this from the total sum of squares. This provides no check on the accuracy of the calculation; consequently it is preferable to compute the interaction term directly.

Computation formulas for sums of squares with n entries in each cell are as follows:

[16.7] ROWS
$$\frac{1}{nC}\sum_{r=1}^{R} T_{r.}^{2} - \frac{T^2}{N}$$

[16.8] COLUMNS
$$\frac{1}{nR}\sum_{c=1}^{C} T_{.c}^{2} - \frac{T^2}{N}$$

[16.9] INTERACTION
$$\frac{1}{n}\sum_{r=1}^{R}\sum_{c=1}^{C} T_{rc}^{2} - \frac{1}{nC}\sum_{r=1}^{R} T_{r.}^{2} - \frac{1}{nR}\sum_{c=1}^{C} T_{.c}^{2} + \frac{T^2}{N}$$

[16.10] WITHIN CELLS
$$\sum_{r=1}^{R}\sum_{c=1}^{C}\sum_{i=1}^{n} X_{rci}^{2} - \frac{1}{n}\sum_{r=1}^{R}\sum_{c=1}^{C} T_{rc}^{2}$$

[16.11] TOTAL
$$\sum_{r=1}^{R}\sum_{c=1}^{C}\sum_{i=1}^{n} X_{rci}^{2} - \frac{T^2}{N}$$

Here again the interaction sum of squares may be obtained by subtracting the row, column, and within-cells sums of squares from the total, although direct calculation of the interaction term is preferable.

The reader should note that the analysis of variance for two-way classification with a single entry in each cell is a particular case of the more general case with more than one entry in each cell. When $n = 1$, formulas for the latter case become the formulas for the former.

16.10 ILLUSTRATIVE EXAMPLE OF TWO-WAY CLASSIFICATIONS: $n > 1$

Table 16.4 shows data obtained in an animal experiment designed to study the effects of two variables on measures of performance of rats in a maze test. Three strains of rats were used: bright, mixed, and dull. A group from each strain was reared under free or restricted environmental conditions. Thus there are 6 groups of experimental animals with 8 animals in each group. The total N is 48. The data are arranged in a 2×3 table with eight observations in each of the six cells. The row means permit a comparison of environments, and the column means a comparison of strains. Table 16.5 shows the sums, means, and sums of squares of row, column, and cell totals. The sum of squares for all the observations is also given.

Applying the computation formulas, the calculations are as follows:

ROWS

$$\frac{1}{nC}\sum_{r=1}^{R} T_{r.}^2 - \frac{T^2}{N} = \frac{5{,}944{,}837}{24} - \frac{(3{,}343)^2}{48} = 14{,}875.52$$

COLUMNS

$$\frac{1}{nR}\sum_{c=1}^{C} T_{.c}^2 - \frac{T^2}{N} = \frac{4{,}015{,}617}{16} - \frac{(3{,}343)^2}{48} = 18{,}150.04$$

WITHIN CELLS

$$\sum_{r=1}^{R}\sum_{c=1}^{C}\sum_{i=1}^{n} X_{rci}^2 - \frac{1}{n}\sum_{r=1}^{R}\sum_{c=1}^{C} T_{rc}^2 = 309{,}851 - \frac{2{,}137{,}469}{8}$$
$$= 42{,}667.38$$

INTERACTION

$$\frac{1}{n}\sum_{r=1}^{R}\sum_{c=1}^{C} T_{rc}^2 - \frac{1}{nC}\sum_{r=1}^{R} T_{r.}^2 - \frac{1}{nR}\sum_{c=1}^{C} T_{.c}^2 + \frac{T^2}{N}$$
$$= \frac{2{,}137{,}469}{8} - \frac{5{,}944{,}837}{24} - \frac{4{,}015{,}617}{16} + \frac{(3{,}343)^2}{48}$$
$$= 1{,}332.04$$

TOTAL

$$\sum_{r=1}^{R}\sum_{c=1}^{C}\sum_{i=1}^{n} X_{rci}^2 - \frac{T^2}{N} = 309{,}851 - \frac{(3{,}343)^2}{48} = 77{,}024.98$$

Table 16.4

Data for the analysis of variance with two-way classification: $n > 1$ error scores for three strains of rats reared under two environmental conditions

Environment	Strain: Bright		Mixed		Dull	
Free	26	14	41	82	36	87
	41	16	26	86	39	99
	28	29	19	45	59	126
	92	31	59	37	27	104
Restricted	51	35	39	114	42	133
	96	36	104	92	92	124
	97	28	130	87	156	68
	22	76	122	64	144	142

Table 16.5

Computation for data of Table 16.4

Environment	Strain: Bright	Mixed	Dull	Total
Free	$T_{11} = 277$ $\bar{X}_{11} = 34.63$	$T_{12} = 395$ $\bar{X}_{12} = 49.38$	$T_{13} = 577$ $\bar{X}_{13} = 72.13$	$T_{1.} = 1{,}249$ $\bar{X}_{1.} = 52.04$
Restricted	$T_{21} = 441$ $\bar{X}_{21} = 55.13$	$T_{22} = 752$ $\bar{X}_{22} = 94.00$	$T_{23} = 901$ $\bar{X}_{23} = 112.63$	$T_{2.} = 2{,}094$ $\bar{X}_{2.} = 87.25$
Total	$T_{.1} = 718$ $\bar{X}_{.1} = 44.88$	$T_{.2} = 1{,}147$ $\bar{X}_{.2} = 71.69$	$T_{.3} = 1{,}478$ $\bar{X}_{.3} = 92.38$	$T = 3{,}343$ $\bar{X}_{...} = 69.65$

$\sum_{r=1}^{R} T_{r.}^2 \cdots 5{,}944{,}837$ $\qquad \sum_{c=1}^{C} T_{.c}^2 = 4{,}015{,}617$

$\sum_{r=1}^{R}\sum_{c=1}^{C} T_{rc}^2 = 2{,}137{,}469$ $\qquad \sum_{r=1}^{R}\sum_{c=1}^{C}\sum_{i=1}^{n} X_{rci}^2 = 309{,}851$

The analysis-of-variance table for these data is given in Table 16.6. The degrees of freedom for rows is $R - 1 = 2 - 1 = 1$, for columns $C - 1 = 3 - 1 = 2$, for interaction $(R - 1)(C - 1) = (2 - 1)(3 - 1) = 2$, and for within cells $RC(n - 1) = 2 \times 3(8 - 1) = 42$. These sum to the total degrees of freedom $RCn - 1 = 2 \times 3 \times 8 - 1 = 47$. For these data a fixed model is appropriate and s_w^2 is the proper error term for testing row,

Table 16.6

Analysis of variance for data of Table 16.4

Source of variation	Sum of squares	Degrees of freedom	Variance estimate
Rows (environments)	14,875.52	1	$14,875.52 = s_r^2$
Columns (strains)	18,150.04	2	$9,075.02 = s_c^2$
Interaction	1,332.04	2	$666.02 = s_{rc}^2$
Within cells	42,667.38	42	$1,015.89 = s_w^2$
Total	77,024.98	47	

$F_{rc} = \frac{s_{rc}^2}{s_w^2} = .656 \qquad F_r = \frac{s_r^2}{s_w^2} = 14.64 \qquad F_c = \frac{s_c^2}{s_w^2} = 8.93$

column, and interaction effects. For interaction we have

$$F_{rc} = \frac{s_{rc}^2}{s_w^2} = \frac{666.02}{1,015.89} = .656$$

This is less than unity. The expectation on the basis of the null hypothesis is unity. The interaction is somewhat less than we would ordinarily expect under the null hypothesis. We may safely conclude that there is no significant interaction between the two experimental variables. For differences in environments we have

$$F_r = \frac{s_r^2}{s_w^2} = \frac{14,875.52}{1,015.89} = 14.64$$

with 1 degree of freedom associated with the numerator and 42 degrees of freedom with the denominator. For these degrees of freedom the values required for significance at the 5 and 1 percent levels are 4.07 and 7.27. We conclude that the different environments have affected the maze performance of the animals. For strains the required ratio is $F_c = s_c^2/s_w^2 = 9,075.02/1,015.89 = 8.93$ with 2 degrees of freedom associated with the numerator and 42 degrees of freedom with the denominator. Again, this difference is significant at well beyond the 1 percent level.

16.11 UNEQUAL AND DISPROPORTIONATE NUMBERS IN THE SUBCLASSES

Situations arise in educational and psychological research where the number of observations in the subclasses, the cell frequencies, in a two-way analysis of variance is unequal. In animal experimentation in psychology this situation may result from loss by death or accident of a number of animals during the conduct of the experiment. Situations also arise where the cell frequencies are unequal but proportional to the marginal totals. Consider a 2 × 3 factorial experiment with the following numbers in the cells:

	C_1	C_2	C_3	
R_1	2	4	6	12
R_2	4	8	12	24
	6	12	18	

Here the cell frequencies are proportional to the marginal totals. Situations also arise where the cell frequencies are not only unequal but are also disproportional.

No difficulties arise when the cell frequencies are unequal but are proportional to the marginal totals. Disproportionality creates difficulties, however, because the partitioning of the sums of squares into independent components as described in Section 16.3 cannot be completed in a simple way. Certain terms are not independent of each other but are correlated.

A number of methods exist for making adjustments to the data when the cell frequencies are unequal and disproportional. Some of these methods are approximate but are of considerable practical value in data analysis. Exact methods exist which involve the fitting of constants using the method of least squares. A description of these exact methods is beyond the scope of this book but will be found in Winer (1971) and Bancroft (1968).

16.12 THE METHOD OF UNWEIGHTED MEANS

A commonly used method for adjusting data for unequal numbers in the subclasses is called the *method of unweighted means*. This method is appropriate when the cell frequencies do not depart in any substantial way from equality. It is appropriate when the investigator originally planned the experiment with an equal number of observations in the cells, but for one reason or another some data are missing. This method is in effect an analysis of variance applied to the means of the subclasses. The sums of squares for rows, columns, and interaction are then adjusted, using the harmonic mean of the cell frequencies. The argument underlying the use of the harmonic mean, and not the arithmetic mean, is based on the observation that the square of the standard error of the mean is proportional to $1/n$ instead of n.

Consider a two-factor experiment with R levels of one factor and C levels of the other. Denote the cell frequencies by n_{rc}. The following steps are involved in applying the method of unweighted means.

1 Calculate the harmonic mean of the cell frequencies as follows:

$$\bar{n}_h = \frac{RC}{1/n_{11} + 1/n_{22} + \cdot \cdot \cdot + 1/n_{rc}}$$

2 Calculate the cell means, and also the row and column *totals* and *means*. To illustrate, for a 2 × 3 factorial experiment these means may be denoted as follows:

	C_1	C_2	C_3		
R_1	$\bar{X}_{11}$	$\bar{X}_{12}$	$\bar{X}_{13}$	$T_{1.}$	$\bar{X}_{1.}$
R_2	$\bar{X}_{21}$	$\bar{X}_{22}$	$\bar{X}_{23}$	$T_{2.}$	$\bar{X}_{2.}$
	$T_{.1}$	$T_{.2}$	$T_{.3}$	T	
	$\bar{X}_{.1}$	$\bar{X}_{.2}$	$\bar{X}_{.3}$		$\bar{X}_{..}$

Here the reader should note that the row and column means are the means of the means in the rows and columns. They are *not* the means of all the observations in the rows and columns. Likewise, the totals $T_{1.}$, $T_{2.}$ and also $T_{.1}$, $T_{.2}$, and $T_{.3}$ are the sums of the row and column means. They are *not* the sums of all the observations in the rows and columns. The quantity $\bar{X}_{..}$ is the mean of the six means, and T is the total of the six means. In calculating row, column, and interaction effects, the data are treated as if there were a single observation in each cell. The sums of squares are then adjusted to estimate what the sums of squares would have been, if there had been $\bar{n}_h$ observations in each cell.

3 The following sums of squares are used:

[16.12] ROWS $$\bar{n}_h C \sum^{R} (\bar{X}_{r.} - \bar{X}_{..})^2$$

[16.13] COLUMNS $$\bar{n}_h R \sum^{C} (\bar{X}_{.c} - \bar{X}_{..})^2$$

[16.14] INTERACTION $$\bar{n}_h \sum^{R} \sum^{C} (\bar{X}_{rc} - \bar{X}_{r.} - \bar{X}_{.c} + \bar{X}_{..})^2$$

WITHIN

[16.15] $$\sum^{R} \sum^{C} \sum^{n_{rc}} (X_{rci} - \bar{X}_{rc})^2$$

These sums of squares may be obtained using the following computation formulas:

ROWS

$$\bar{n}_h \left[\frac{1}{C} \sum^{R} T_{r.}^2 - \frac{T^2}{RC} \right] \tag{16.16}$$

COLUMNS

$$\bar{n}_h \left[\frac{1}{R} \sum^{C} T_{.c}^2 - \frac{T^2}{RC} \right] \tag{16.17}$$

INTERACTION

$$\bar{n}_h \left[\sum^{R} \sum^{C} \bar{X}_{rc}^2 - \frac{1}{C} \sum^{R} T_{r.}^2 - \frac{1}{R} \sum^{C} T_{.c}^2 + \frac{T^2}{RC} \right] \tag{16.18}$$

WITHIN CELLS

$$\sum^{R} \sum^{C} \sum^{n_{rc}} X_{rci}^2 - \sum^{R} \sum^{C} \left(\frac{T_{rc}^2}{n_{rc}} \right) \tag{16.19}$$

In the within-cells sum of squares T_{rc} is the sum or total of all the observations in the cell corresponding to the rth row and the cth column.

4 The total number of degrees of freedom associated with row, column, and interaction effects is $R - 1$, $C - 1$, and $(R - 1)(C - 1)$. The degrees of freedom associated with the within group sum of squares are $\Sigma n_{rc} - RC$ or $N - RC$.

5 Proceed with the analysis of variance in the usual way.

An alternative to the method of unweighted means using the harmonic mean is to use a least-squares procedure described in more advanced texts such as Winer (1971). This is a more complex method. Rankin (1974) has investigated some of the properties of the method of unweighted means in relation to the least-squares procedure. His finding is that it is a reasonable alternative to the least-squares procedure. The general conclusion is that errors in the probability estimates resulting from this method do not become appreciable until the ratio of the smallest to largest sample size is about one to three. In many practical situations the method might be expected to work quite well.

16.13 ILLUSTRATIVE EXAMPLE OF THE METHOD OF UNWEIGHTED MEANS

Table 16.7 shows fictitious illustrative data for a two-way classification experiment with two levels of one factor and three levels of the other. The number of observations in each cell and the total for each cell are also shown at the bottom of Table 16.7. Table 16.8 shows cell means, row and column totals and means, and the quantities used in the computation formulas.

Table 16.7
Illustrative data for the method of unweighted means

	C_1	C_2	C_3
R_1	7 6 6 2 4 3	8 17 12 19 16 21 24 22	16 8 14 15 17
R_2	23 18 14 22 9 26	11 26 15 14 26 13 31	9 16 27 17 31 18 42 20

	C_1	C_2	C_3
R_1	$n_{11} = 6$ $T_{11} = 28$	$n_{12} = 8$ $T_{12} = 139$	$n_{13} = 5$ $T_{13} = 70$
R_2	$n_{21} = 6$ $T_{21} = 112$	$n_{22} = 7$ $T_{22} = 136$	$n_{23} = 8$ $T_{23} = 180$

Table 16.8
Means and other computations for the data of Table 16.7

	C_1	C_2	C_3	
R_1	$\bar{X}_{11} = 4.67$	$\bar{X}_{12} = 17.38$	$\bar{X}_{13} = 14.00$	$T_{1.} = 36.05$ $\bar{X}_{1.} = 12.02$
R_2	$\bar{X}_{21} = 18.67$	$\bar{X}_{22} = 19.43$	$\bar{X}_{23} = 22.50$	$T_{2.} = 60.60$ $\bar{X}_{2.} = 20.20$
	$T_{.1} = 23.34$ $\bar{X}_{.1} = 11.67$	$T_{.2} = 36.81$ $\bar{X}_{.2} = 18.41$	$T_{.3} = 36.50$ $\bar{X}_{.3} = 18.25$	$T = 96.65$ $\bar{X}_{..} = 16.11$

$\frac{1}{C}\sum^{R} T_{r.}^2 = 1{,}657.32$ $\frac{1}{R}\sum T_{.c}^2 = 1{,}615.99$

$\sum^{R}\sum^{C} \bar{X}_{rc}^2 = 1{,}752.22$ $\sum^{R}\sum^{C}\sum^{n_{rc}} X_{rci}^2 = 13{,}913$

$\frac{T^2}{RC} = 1{,}556.87$

The harmonic mean of the cell frequencies is as follows:

$$\bar{n}_h = \frac{6}{\frac{1}{6}+\frac{1}{8}+\frac{1}{5}+\frac{1}{6}+\frac{1}{7}+\frac{1}{8}} = 6.48$$

Applying the computation formulas, the following are obtained:

ROWS

$$\tilde{n}_h\left(\frac{1}{C}\sum^{R} T_{r.}^2 - \frac{T^2}{RC}\right) = 6.48(1657.32 - 1556.87) = 650.92$$

COLUMNS

$$\tilde{n}_h\left(\frac{1}{R}\sum^{R} T_{.c}^2 - \frac{T^2}{RC}\right) = 6.48(1615.99 - 1556.87) = 383.10$$

INTERACTION

$$\tilde{n}_h\left(\sum^{R}\sum^{C} \bar{X}_{rc}^2 - \frac{1}{C}\sum^{R} T_{r.}^2 - \frac{1}{R}\sum^{C} T_{.c}^2 + \frac{T^2}{RC}\right)$$

$$= 6.48(1752.22 - 1657.32 - 1615.99 + 1556.87) = 231.85$$

WITHIN CELLS

$$\sum^{R}\sum^{C}\sum^{n_{rc}} X_{rci}^2 - \sum^{R}\sum^{C}\left(\frac{T_{rc}^2}{n_{rc}}\right)$$

$$= 13{,}913 - \left[\frac{(28)^2}{6} + \frac{(139)^2}{8} + \cdots + \frac{(180)^2}{8}\right] = 1604.26$$

The analysis-of-variance table for these data is given in Table 16.9. The degrees of freedom for rows is $R - 1 = 2 - 1 = 1$, for columns $C - 1 = 3 - 1 = 2$, for interaction $(R - 1)(C - 1) = (2 + 1)(3 - 1) = 2$, and for within cells $\Sigma n_{rc} - RC = 40 - 6 = 34$. In this illustrative example the row effect is significant at better than the .01 level, the column effect is significant at the .05 level, and the interaction effect is not significant.

Table 16.9

Analysis of variance for data of Table 16.8

Source of variation	Sum of squares	Degrees of freedom	Variance estimate
Rows	650.92	1	$650.92 = s_r^2$
Columns	383.10	2	$191.55 = s_c^2$
Interaction	231.85	2	$115.93 = s_{rc}^2$
Within cells	1604.26	34	$47.18 = s_w^2$

$F_r = \frac{s_r^2}{s_w^2} = 13.80$ $p < .01$

$F_c = \frac{s_c^2}{s_w^2} = 4.06$ $p < .05$

$F_{rc} = \frac{s_{rc}^2}{s_w^2} = 2.46$ $p > .05$

BASIC TERMS AND CONCEPTS

Two-way classification

Sum of squares: between rows, s_r^2; between columns, s_c^2; within cells, s_w^2; interaction, s_{rc}^2

Interaction

Random model

Fixed model

Mixed model

Choice of error term

Pooling sums of squares

Unequal numbers in the subclasses

Disproportionality

Method of unweighted means

EXERCISES

1 In an experiment involving double classification within 10 observations in each cell, the following cell and marginal means were obtained:

	C_1	C_2	C_3	
R_1	8.3	3.2	17.4	9.6
R_2	12.5	4.6	12.6	9.9
	10.4	3.9	15.0	9.8

Compute (**a**) the cell means expected under zero interaction and (**b**) the interaction sum of squares.

2 The following are data for a double-classification experiment involving two fixed variables:

	C_1		C_2		C_3	
R_1	29	31	23	62	17	32
	26	50	31	60	18	49
	42	25	18	20	50	58
R_2	17	62	35	83	17	28
	27	62	50	42	14	58
	50	29	62	19	49	62

Apply the analysis of variance to test the significance of row, column, and interaction effects.

3 The following are data with unequal numbers in the subclasses:

	C_1	C_2
R_1	8 9 20 5 16 11 23 4 2	6 11 6 2 4 1 3
R_2	8 20 14 16 12 15	11 15 6 12 18 3 6 2

Apply the analysis of variance by using the method of unweighted means to test row, column, and interaction effects, on the assumption that the two experimental variables are fixed.

ANSWERS TO EXERCISES

1 **a**

	C_1	C_2	C_3	
R_1	10.2	3.7	14.8	9.6
R_2	10.5	4.0	15.1	9.9
	10.4	3.9	15.0	9.8

b 212.13

2 $F_r = 1.23$, $p > .05$; $F_c = .23$, $p > .05$; $F_{rc} = .31$, $p > .05$

3 $F_r = 3.48$, $p > .05$; $F_c = 7.40$, $p < .05$; $F_{rc} = .08$, $p > .05$

17

ANALYSIS OF VARIANCE: THREE-WAY CLASSIFICATION

17.1 INTRODUCTION

Many experiments involve the simultaneous study of more than two independent variables or factors. For example, an experiment may involve 2 levels or categories of one factor, 3 of a second factor, and 5 of a third factor, with n subjects assigned at random to each of the $2 \times 3 \times 5$ groups. Such an experiment may be spoken of as a "$2 \times 3 \times 5$ factorial experiment." The data from such an experiment may be conceptualized as a three-dimensional cube of numbers containing 2 rows, 3 columns, and 5 layers, with n observations in each of the 30 different cells of which the cube may be thought to be comprised.

The analysis and interpretation of data resulting from such experiments are a direct extension of the analysis and interpretation of data for two-way classification. In a two-factor experiment with n observations in each cell, the total sum of squares is divided into four parts, a between-rows, a between-columns, an interaction, and a within-cells sum of squares. In a three-way classification, or three-factor, experiment with n observations per cell, the total sum of squares is partitioned into eight parts, three sums of squares for main effects, four interaction sums of squares, and a within-cells sum of squares. Each sum of squares has an associated number of degrees of freedom. Sums of squares are, as previously, divided by their associated degrees of freedom to obtain variance estimates, or mean squares, which are used to test the significance of main effects and interactions. Although such analysis and interpretation are clearly more complex than for two-way classification, the essential ideas are the same.

17.2 NOTATION FOR THREE-WAY ANALYSIS OF VARIANCE

Consider an experiment involving R levels of one factor, C levels of a second factor, and L levels of a third factor. The number of treatment combinations is RCL. Consider the particular case where we have one measurement, or observation, for each of the RCL combinations, the total number of measurements being N. The data for the first layer of numbers may be represented as follows:

Layer 1	1	2	3	$\cdots$	C	Row mean
1	X_{111}	X_{121}	X_{131}	$\cdots$	X_{1C1}	$\bar{X}_{1.1}$
2	X_{211}	X_{221}	X_{231}	$\cdots$	X_{2C1}	$\bar{X}_{2.1}$
3	X_{311}	X_{321}	X_{331}	$\cdots$	X_{3C1}	$\bar{X}_{3.1}$
.	.	.	.	.	.	.
.	.	.	.	.	.	.
.	.	.	.	.	.	.
R	X_{R11}	X_{R21}	X_{R31}	$\cdots$	X_{RC1}	$\bar{X}_{R.1}$
Column mean	$\bar{X}_{.11}$	$\bar{X}_{.21}$	$\bar{X}_{.31}$	$\cdots$	$\bar{X}_{.C1}$	$\bar{X}_{..1}$

Here the first subscript identifies the row, the second the column, and the third the layer. Thus, for example, X_{321} denotes the observation in the third row and second column of the first layer. The mean $\bar{X}_{1.1}$ is the mean of the first row of the first layer, $\bar{X}_{.11}$ is the mean of the first column of the first layer, and $\bar{X}_{..1}$ is the mean of all the observations in the first layer.

Notation for the second layer is as follows:

Layer 2	1	2	3	$\cdots$	C	Row mean
1	X_{112}	X_{122}	X_{132}	$\cdots$	X_{1C2}	$\bar{X}_{1.2}$
2	X_{212}	X_{222}	X_{232}	$\cdots$	X_{2C2}	$\bar{X}_{2.2}$
3	X_{312}	X_{322}	X_{332}	$\cdots$	X_{3C2}	$\bar{X}_{3.2}$
.	.	.	.	.	.	.
.	.	.	.	.	.	.
.	.	.	.	.	.	.
R	X_{R12}	X_{R22}	X_{R22}	$\cdots$	X_{RC2}	$\bar{X}_{R32}$
Column mean	$\bar{X}_{.12}$	$\bar{X}_{.22}$	$\bar{X}_{.32}$	$\cdots$	$\bar{X}_{.C2}$	$\bar{X}_{..2}$

Similarly the third, fourth, and Lth layers may be considered. In general X_{rcl} denotes a measurement for the rth row, the cth column, and the lth layer. The reader should note that R denotes the number of rows, C the number of columns, and L the number of layers. The sum r denotes the rth row, where r may take the values $1, 2, \ldots, R$. Similarly c and l denote the cth column and the lth layer, respectively.

The grand mean of all the $RCL = N$ observations is $\bar{X}_{...}$. The total sum of squares of deviations about the grand mean is given by

$$\sum_{r=1}^{R} \sum_{c=1}^{C} \sum_{l=1}^{L} (X_{rcl} - \bar{X}_{...})^2$$

In many experiments there are n sampling units and measurements in each of the RCL treatment combinations. The total number of measurements is then $nRCL = N$. The notation for such data in the particular case where $R = 2$, $C = 3$, $L = 2$, and $n = 3$ may be represented as follows:

Layer 1

	1	2	3	
1	X_{1111} X_{1112} X_{1113}	X_{1211} X_{1212} X_{1213}	X_{1311} X_{1312} X_{1313}	$\bar{X}_{1.1.}$
2	X_{2111} X_{2112} X_{2113}	X_{2211} X_{2212} X_{2213}	X_{2311} X_{2312} X_{2313}	$\bar{X}_{2.1.}$
	$\bar{X}_{.11.}$	$\bar{X}_{.21.}$	$\bar{X}_{.31.}$	$\bar{X}_{..1.}$

Layer 2

	1	2	3	
1	X_{1121} X_{1122} X_{1123}	X_{1221} X_{1222} X_{1223}	X_{1321} X_{1322} X_{1323}	$\bar{X}_{1.2.}$
2	X_{2121} X_{2122} X_{2123}	X_{2221} X_{2222} X_{2223}	X_{2321} X_{2322} X_{2323}	$\bar{X}_{2.2.}$
	$\bar{X}_{.12.}$	$\bar{X}_{.22.}$	$\bar{X}_{.32.}$	$\bar{X}_{..2.}$

Quadruple subscripts are used. The first identifies the row, the second the column, the third the layer, and the fourth the measurement within the cell. Thus, for example, the symbol X_{1323} indicates the third measurement in the cell, corresponding to the first row, the third column, and the second layer. In general X_{rcli} denotes the ith measurement in the rth row and cth column of the lth layer, where $i = 1, 2, \ldots, n$. Row and column means for each layer are shown. Thus $\bar{X}_{1.1.}$ is the mean for the first row of the first layer, $X_{.22.}$ is the mean for the second column of the second layer, and so on. The grand mean, the mean of all $nRCL$ observations, is $\bar{X}_{....}$. The total sum of squares for a triple-classification experiment with n observations per cell may be denoted by

$$\sum_{r=1}^{R} \sum_{c=1}^{C} \sum_{l=1}^{L} \sum_{i=1}^{n} (X_{rcli} - \bar{X}_{....})^2$$

The sum of squares, both in the case where $n = 1$ and where $n > 1$, may be partitioned into additive sums of squares.

17.3 PARTITIONING THE SUM OF SQUARES

With a single measurement in each of the RCL-treatment combinations, the total sum of squares may be partitioned into seven additive parts: three main effects for rows, columns, and layers, three first-order interaction terms, and one second-order interaction term. The first-order interaction terms are row by column, row by layer, and column by layer. The second-order interaction term is row by column by layer. The precise meaning which attaches to each of these terms is described in detail in Section 17.5. With more than one observation per cell, $n > 1$, the total sum of squares is partitioned into eight additive parts: three main effects, three first-order interaction terms, one second-order interaction term, and one within-cells sum of squares.

The procedure used in partitioning the sum of squares is directly analogous to that used in the two-way classification case, although somewhat more complex. For $n > 1$ we begin with the rather involved identity

$$\begin{aligned}(X_{rcli} - \bar{X}_{....}) = {} & (\bar{X}_{r...} - \bar{X}_{....}) + (\bar{X}_{.c..} - \bar{X}_{....}) + (\bar{X}_{..l.} - \bar{X}_{....}) \\ & + (\bar{X}_{rc..} - \bar{X}_{r...} - \bar{X}_{.c..} + \bar{X}_{....}) + (\bar{X}_{r.l.} - \bar{X}_{r...} - \bar{X}_{..l.} + \bar{X}_{....}) \\ & + (\bar{X}_{.cl.} - \bar{X}_{.c..} - \bar{X}_{..l.} + \bar{X}_{....}) \\ & + (\bar{X}_{rcl.} - \bar{X}_{rc..} - \bar{X}_{r.l.} - \bar{X}_{.cl.} + \bar{X}_{r...} + \bar{X}_{.c..} + \bar{X}_{..l.} - \bar{X}_{....}) \\ & + (X_{rcli} - \bar{X}_{rcl.})\end{aligned}$$

As in the two-way classification case both sides are squared, summed over R rows, C columns, and L layers, and over the n observations in each cell. Certain terms contain a sum of deviations about a mean. These vanish, and the eight sums of squares as shown in Table 17.1 result. Note that three sums of squares for main effects are obtained, three first-order interaction sums of squares, one second-order interaction sum of squares, and one within-cells sum of squares. With one observation in each cell the required sums of squares are obtained from Table 17.1 simply by writing $n = 1$. No within-cells sum of squares exists, and in this case the total sum of squares is partitioned into seven, and not eight, additive parts.

17.4 DEGREES OF FREEDOM AND MEAN SQUARES

As shown in Table 17.1 the number of degrees of freedom for rows is $R - 1$, for columns $C - 1$, and for layers $L - 1$. The first-order interaction for row-by-column has $(R - 1)(C - 1)$ degrees of freedom. Similarly the row-by-layer and column-by-layer interactions have $(R - 1)(L - 1)$ and $(C - 1)(L - 1)$ degrees of freedom, respectively. The second-order interaction term has $(R - 1)(C - 1)(L - 1)$ degrees of freedom. Each cell has associated with it $n - 1$ degrees of freedom. Since there are RCL cells, the total number of degrees of freedom associated with the within-cells sum of squares is $RCL(n - 1)$. The degrees of freedom are directly

Table 17.1
Analysis of variance for three-way classification with n entries per cell: $n > 1$

Source	Variance Sum of squares	df	Variance estimate
Rows	$nCL \sum^{R} (\bar{X}_{r...} - \bar{X}_{....})^2$	$R - 1$	s_r^2
Columns	$nRL \sum^{C} (\bar{X}_{.c..} - \bar{X}_{....})^2$	$C - 1$	s_c^2
Layers	$nCR \sum^{L} (\bar{X}_{..l.} - \bar{X}_{....})^2$	$L - 1$	s_l^2
$R \times C$	$nL \sum^{R} \sum^{C} (\bar{X}_{rc..} - \bar{X}_{r...} - \bar{X}_{.c..} + \bar{X}_{....})^2$	$(R-1)(C-1)$	s_{rc}^2
$R \times L$	$nC \sum^{R} \sum^{L} (\bar{X}_{r.l.} - \bar{X}_{r...} - \bar{X}_{..l.} + \bar{X}_{....})^2$	$(R-1)(L-1)$	s_{rl}^2
$C \times L$	$nR \sum^{C} \sum^{L} (X_{.cl.} - \bar{X}_{.c..} - \bar{X}_{..l.} + \bar{X}_{....})^2$	$(C-1)(L-1)$	s_{cl}^2
$R \times C \times L$	$n \sum^{R} \sum^{C} \sum^{L} (\bar{X}_{rcl.} - \bar{X}_{rc..} - \bar{X}_{r.l.} - \bar{X}_{.cl.} + \bar{X}_{r...} + \bar{X}_{.c..} + \bar{X}_{..l.} - \bar{X}_{....})^2$	$(R-1)(C-1)(L-1)$	s_{rcl}^2
Within cells	$\sum^{n} \sum^{R} \sum^{C} \sum^{L} (X_{rcli} - \bar{X}_{rcl.})^2$	$RCL(n-1)$	s_w^2
Total	$\sum^{n} \sum^{R} \sum^{C} \sum^{L} (X_{rcli} - \bar{X}_{....})^2$	$nRCL - 1$	

additive; that is, it may be shown that

$$\begin{aligned} N - 1 &= nRCL - 1 \\ &= (R-1) + (C-1) + (L-1) + (R-1)(C-1) \\ &\quad + (R-1)(L-1) + (C-1)(L-1) \\ &\quad + (R-1)(C-1)(L-1) + RCL(n-1) \end{aligned}$$

For $n = 1$ the degrees of freedom are the same as for $n > 1$ except that no within-cells sum of squares exists.

Sums of squares are divided by their associated number of degrees of freedom to obtain variance estimates, or mean squares. As previously, F ratios are formed using these mean squares to obtain tests of significance for main effects and interaction effects. As in the two-way classification case, the correct choice of error term, the appropriate mean square to insert in the denominator of the F ratio, depends on whether the model appropriate to the experiment is fixed, random, or mixed.

17.5 INTERACTION IN THREE-WAY ANALYSIS OF VARIANCE

As stated above, partitioning a sum of squares in a three-way analysis of variance results in four interaction sums of squares, $R \times C$, $R \times L$, $C \times L$, and $R \times C \times L$. What meaning may be attached to these sums of squares? How may they be interpreted? The answers to these questions may be clarified by a hypothetical example. The following are cell means for a $3 \times 3 \times 2$ factorial design with n observations per cell.

Layer 1 — Columns

Rows			
	10	20	6
	5	15	1
	15	25	11

Layer 2 — Columns

Rows			
	16	26	12
	11	21	7
	21	31	17

The data to the left are means for the first layer, and those to the right are means for the second layer. These data may be visualized as a $3 \times 3 \times 2$ cube of numbers with the second layer superimposed on the first as shown in Figure 17.1. The means shown in Figure 17.1 are means on the surface of the cube obtained by averaging cell means over rows, columns, and layers, a procedure which is, of course, correct only when the cell means are based on equal n's. Thus in Figure 17.1 the mean of 13 for the first row and column is obtained by averaging the means for the first row and column over the two layers. Thus the mean of 13 is the average of the two cell means 10 and 16. Likewise the mean of 10 for the first column and the first layer is obtained by averaging over rows. Thus 10 is the average of the three cell means for the first layer, 10, 5, 15. All other means in Figure 17.1 are similarly obtained. Figure 17.1 also shows certain marginal means, or means along the edge of the cube. These means are obtained by averaging over rows, columns, and layers. The grand mean, the mean of all the observations taken together, is also shown, which in this case is 15.

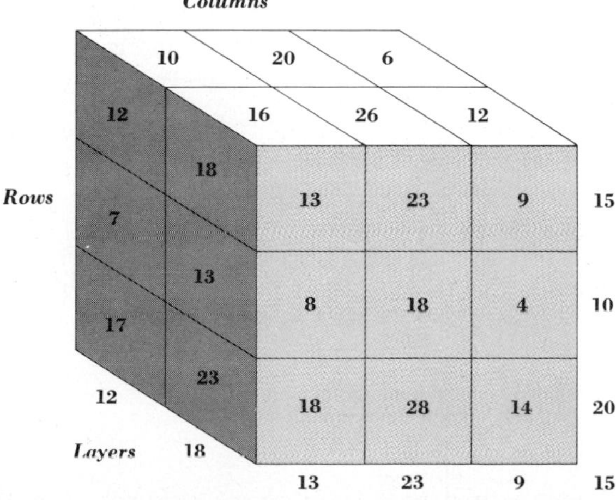

Figure 17.1 Representation of data for three-way classification as a cube of numbers.

The first-order interactions in a three-way analysis of variance, that is, the $R \times C$, $R \times L$, and $C \times L$ interactions, are concerned with the parallelism, or its absence, of the means on the surface of the cube. The rationale is essentially the same as that used in the double-classification case. Consider, for example, the means obtained by averaging over layers in Figure 17.1. These means are

		Columns I	II	III	
	I	13	23	9	15
Rows	II	8	18	4	10
	III	18	28	14	20
		13	23	9	15

Here we note that all sets of means for both rows and columns are parallel. For example, means for the first row, 13, 23, and 9, are uniformly 5 points higher than the means for the second row, 8, 18, 4. Likewise, the means for the first column, 13, 8, 18, are uniformly 10 points less than the means for the second column, 23, 18, and 28, and so on. In Figure 17.1 the $R \times C$ interaction is 0.

As in the two-way classification case the values of the cell means for rows and columns expected in the case of zero interaction are given by $E(\bar{X}_{rc..}) = \bar{X}_{r...} + \bar{X}_{.c..} - \bar{X}_{....}$. The observed means are $\bar{X}_{rc..}$. The differences between the observed and expected means are $\bar{X}_{rc..} - \bar{X}_{r...} - \bar{X}_{.c..} + \bar{X}_{....}$. The $R \times C$ interaction sum of squares is given by

$$nL \sum^{R} \sum^{C} (\bar{X}_{rc..} - \bar{X}_{r...} - \bar{X}_{.c..} + \bar{X}_{....})^2$$

This is seen to be nL times the sum of squares of these differences summed over rows and columns and is simply an overall measure of the departures of the means from parallelism. Note that the $R \times C$ interaction is concerned with parallelism among means obtained by averaging over layers. The $R \times L$ and $L \times C$ interactions involve means obtained by averaging over columns and rows, respectively.

The $R \times C \times L$ interaction in a three-way analysis of variance is concerned with the similarity, or otherwise, of the interaction, whether it be 0 or not, between each pair of variables at different levels of the third variable. Consider the following means for a $2 \times 2 \times 2$ factorial experiment:

L_1	C_1	C_2
R_1	10	30
R_2	10	20

L_2	C_1	C_2
R_1	20	40
R_2	20	30

Inspection of the means for L_1 indicate that the means are not parallel and an interaction is present. Likewise inspection of the means for L_2 indicates again that the means are not parallel and an interaction is present. We note, however, that for L_1 the difference between R_1 and R_2 for C_1 is 0. The corresponding difference for L_2 is also 0. Note also that for L_1 the difference between R_1 and R_2 for C_2 is 10. The corresponding difference for L_2 is also 10. It is appropriate to say here that the interaction between rows and columns is the same, or geometrically similar, for the two layers. In this case the triple interaction will be 0. In this illustrative example it may be ascertained that the $R \times C$ interaction is nonzero whereas the $R \times L$, the $C \times L$, and the $R \times C \times L$ interactions are 0.

The $R \times C \times L$ interaction is a measure of the differences between a set of observed values and a set of expected values. The expected values are those we would obtain if the interactions were the same between all pairs of variables at different levels of the third variable.

17.6 EXPECTATIONS OF MEAN SQUARES IN THREE-WAY ANALYSIS OF VARIANCE

As in the two-way analysis of variance the correct choice of error term in a three-way analysis of variance depends on the nature of the variables which serve as the basis of classification in the experimental design. All three variables may be *fixed,* all three may be *random,* any two may be fixed and the third random, or any one may be fixed and the other two random. Thus we may have a fixed, random, or mixed model.

The expectations of mean squares for the general finite model are shown in Table 17.2. The reader should note that in Table 17.2 the quantity σ_a^2 is the variance in the population for row means. The null hypothesis under test for rows is $H_r : \mu_{1...} = \mu_{2...} = \cdot \cdot \cdot = \mu_{R...}$. The variance of these means is σ_a^2, and the null hypothesis for rows may be stated as $H_r : \sigma_a^2 = 0$. Likewise σ_b^2 and σ_c^2 are the variances in the population for column and layer means, respectively.

From the general model of Table 17.2 we can deduce the expectations of the mean squares for the fixed model, the random model, or any of the mixed models. For the fixed model $R = R_p$, $C = C_p$, and $L = L_p$. By substituting $(R_p - R)/R_p = 0$, $(C_p - C)/C_p = 0$, and $(L_p - L)/L_p = 0$ in the general finite model, the expectations for the fixed model are obtained. These expectations are shown in Table 17.3.

For the random model the expectations of the mean squares are obtained by substituting $(R_p - R)/R_p = 1$, $(C_p - C)/C_p = 1$, and $(L_p - L)/L_p = 1$ in the expectation for the general finite model. These expectations are shown in Table 17.4 (page 283).

For the mixed model where rows are random and columns and layers are fixed, the expectations of the mean squares are obtained by substituting $(R_p - R)/R_p = 1$, $(C_p - C)/C_p = 0$, and $(L_p - L)/L_p = 0$ in the expectation

Table 17.2

Expectations of mean squares for general finite model: three-way analysis with n entries in each cell

Mean square	Expectation of mean square
Rows, s_r^2	$\sigma_e^2 + \frac{L_p - L}{L_p}\frac{C_p - C}{C_p} n\sigma_{abc}^2 + \frac{C_p - C}{C_p} Ln\sigma_{ab}^2 + \frac{L_p - L}{L_p} Cn\sigma_{ac}^2 + LCn\sigma_a^2$
Columns, s_c^2	$\sigma_e^2 + \frac{R_p - R}{R_p}\frac{L_p - L}{L_p} n\sigma_{abc}^2 + \frac{R_p - R}{R_p} Ln\sigma_{ab}^2 + \frac{L_p - L}{L_p} Rn\sigma_{bc}^2 + RLn\sigma_b^2$
Layers, s_l^2	$\sigma_e^2 + \frac{R_p - R}{R_p}\frac{C_p - C}{C_p} n\sigma_{abc}^2 + \frac{R_p - R}{R_p} Cn\sigma_{ac}^2 + \frac{C_p - C}{C_p} Rn\sigma_{bc}^2 + RCn\sigma_c^2$
$R \times C$, s_{rc}^2	$\sigma_e^2 + \frac{L_p - L}{L_p} n\sigma_{abc}^2 + Ln\sigma_{ab}^2$
$R \times L$, s_{rl}^2	$\sigma_e^2 + \frac{C_p - C}{C_p} n\sigma_{abc}^2 + Cn\sigma_{ac}^2$
$C \times L$, s_{cl}^2	$\sigma_e^2 + \frac{R_p - R}{R_p} n\sigma_{abc}^2 + Rn\sigma_{bc}^2$
$R \times C \times L$, s_{rcl}^2	$\sigma_e^2 + n\sigma_{abc}^2$
Within cells, s_w^2	σ_e^2

for the general finite model given in Table 17.2. The expectations for this particular mixed model are shown in Table 17.5 (page 284). The expectations for other varieties of the mixed-model case can be similarly obtained.

17.7 CHOICE OF ERROR TERM

In forming F ratios the expectation of the mean square in the numerator should contain one more term than the expectation of the mean square in the denominator, all other terms in both the numerator and the denominator being the same. Applying this rule to the expectations shown in Tables 17.3 to 17.5 indicates how we should proceed in the forming of F ratios.

1 **Fixed model:** $n > 1$ The correct error term for testing all main effects and all interaction effects is s_w^2. All F ratios are formed with s_w^2 in the denominator.

2 **Random model:** $n > 1$ The correct error term for testing the $R \times C \times L$ interaction is s_w^2. The correct error term for testing the

Table 17.3

Expectation of mean squares for fixed model: three-way analysis of variance with n entries in each cell

Mean square	Expectation of mean squares
Rows, s_r^2	$\sigma_e^2 + LCn\sigma_a^2$
Columns, s_c^2	$\sigma_e^2 + RLn\sigma_b^2$
Layers, s_l^2	$\sigma_e^2 + RCn\sigma_c^2$
$R \times C$, s_{rc}^2	$\sigma_e^2 + Ln\sigma_{ab}^2$
$R \times L$, s_{rl}^2	$\sigma_e^2 + Cn\sigma_{ac}^2$
$C \times L$, s_{cl}^2	$\sigma_e^2 + Rn\sigma_{bc}^2$
$R \times C \times L$, s_{rcl}^2	$\sigma_e^2 + n\sigma_{abc}^2$
Within cells, s_w^2	σ_e^2

Table 17.4

Expectation of mean squares for random model: three-way analysis of variance with n entries in each cell

Mean square	Expectation of mean squares
Rows, s_r^2	$\sigma_e^2 + n\sigma_{abc}^2 + Ln\sigma_{ab}^2 + Cn\sigma_{ac}^2 + LCn\sigma_a^2$
Columns, s_c^2	$\sigma_e^2 + n\sigma_{abc}^2 + Ln\sigma_{ab}^2 + RN\sigma_{bc}^2 + RLn\sigma_b^2$
Layers, s_l^2	$\sigma_e^2 + n\sigma_{abc}^2 + Cn\sigma_{ac}^2 + Rn\sigma_{bc}^2 + RCn\sigma_c^2$
$R \times C$, s_{rc}^2	$\sigma_e^2 + n\sigma_{abc}^2 + Ln\sigma_{ab}^2$
$R \times L$, s_{rl}^2	$\sigma_e^2 + n\sigma_{abc}^2 + Cn\sigma_{ac}^2$
$C \times L$, s_{cl}^2	$\sigma_e^2 + n\sigma_{abc}^2 + Rn\sigma_{bc}^2$
$R \times C \times L$, s_{rcl}^2	$\sigma_e^2 + n\sigma_{abc}^2$
Within cells, s_w^2	σ_e^2

Table 17.5

Expectation of mean squares for fixed model: rows are random, columns and layers are fixed

Mean square	Expectation of mean squares
Rows, s_r^2	$\sigma_e^2 + LCn\sigma_a^2$
Columns, s_c^2	$\sigma_e^2 + Ln\sigma_{ab}^2 + RLn\sigma_b^2$
Layers, s_l^2	$\sigma_e^2 + Cn\sigma_{ac}^2 + RCn\sigma_c^2$
$R \times C$, s_{rc}^2	$\sigma_e^2 + Ln\sigma_{ab}^2$
$R \times L$, s_{rl}^2	$\sigma_e^2 + Cn\sigma_{ac}^2$
$C \times L$, s_{cl}^2	$\sigma_e^2 + n\sigma_{abc}^2 + Rn\sigma_{bc}^2$
$R \times C \times L$, s_{rcl}^2	$\sigma_e^2 + n\sigma_{abc}^2$
Within cells, s_w^2	σ_e^2

$R \times C$, $R \times L$, and $C \times L$ interactions is s_{rcl}^2. No exact tests of the main effects can be made, although approximate F tests can be used. A discussion of such approximate tests can be found in Scheffé (1959).

3 **Mixed model: $n > 1$** Inspectation of the expectations in the particular mixed-model case where rows are random and columns and layers are fixed indicates that the correct error term for testing rows, $R \times C$, $R \times L$, and $R \times C \times L$ interactions is the within-cells variance estimate s_w^2. The correct error term for testing the column effect is the $R \times C$ interaction mean square s_{rc}^2, and for the layer effect the $R \times L$ interaction mean square s_{rl}^2. The correct error term for testing the $C \times L$ interaction is the $R \times C \times L$ interaction mean square s_{rcl}^2. The correct error term for other examples of the mixed-model case will, of course, differ from the above, but may be deduced from the general finite model.

17.8 COMPUTATION FORMULAS FOR SUMS OF SQUARES

As previously, computation formulas are used to calculate the required sums of squares. The sum of all N observations is denoted by T. The sum of all the observations for rows, summed within cells and then over columns and layers, is $T_{r...}$. If we conceptualize the data as comprising a cube of observations, $T_{r...}$ are totals along the edge of the cube. Likewise $T_{.c..}$ are totals summed over rows and layers, and $T_{..l.}$ are totals summed over rows and columns.

The quantities $T_{rc..}$, $T_{r.l.}$, and $T_{.cl.}$ are totals summed within cells, and then over layers, columns, and rows, respectively. These are totals on the surface of the cube. The quantity $T_{rcl.}$ is an individual cell total corresponding to the rth row, the cth column, and the lth layer.

With n entries in each cell the computation formulas for the sums of squares are as follows:

ROWS

[17.1] $$\frac{1}{nCL}\sum^{R} T^2_{r...} - \frac{T^2}{N}$$

COLUMNS

[17.2] $$\frac{1}{nRL}\sum^{C} T^2_{.c..} - \frac{T^2}{N}$$

LAYERS

[17.3] $$\frac{1}{nRC}\sum^{L} T^2_{..l.} - \frac{T^2}{N}$$

$R \times C$ INTERACTION

[17.4] $$\frac{1}{nL}\sum^{R}\sum^{C} T^2_{rc..} - \frac{1}{nLC}\sum^{R} T^2_{r...} - \frac{1}{nRL}\sum^{C} T^2_{.c..} + \frac{T^2}{N}$$

$R \times L$ INTERACTION

[17.5] $$\frac{1}{nC}\sum^{R}\sum^{L} T^2_{r.l.} - \frac{1}{nLC}\sum^{R} T^2_{r...} - \frac{1}{nRC}\sum^{L} T^2_{..l.} + \frac{T^2}{N}$$

$C \times L$ INTERACTION

[17.6] $$\frac{1}{nR}\sum^{C}\sum^{L} T^2_{.cl.} - \frac{1}{nLR}\sum^{C} T^2_{.c..} - \frac{1}{nRC}\sum^{L} T^2_{..l.} + \frac{T^2}{N}$$

$R \times C \times L$ INTERACTION

[17.7] $$\frac{1}{n}\sum^{R}\sum^{C}\sum^{L} T^2_{rcl.} - \frac{1}{nL}\sum^{R}\sum^{C} T^2_{rc..} - \frac{1}{nC}\sum^{R}\sum^{L} T^2_{r.l.} - \frac{1}{nR}\sum^{C}\sum^{L} T^2_{.cl.}$$
$$+ \frac{1}{nCL}\sum^{R} T^2_{r...} + \frac{1}{nRL}\sum^{C} T^2_{.c..} + \frac{1}{nRC}\sum^{L} T^2_{..l.} - \frac{T^2}{N}$$

WITHIN CELLS

[17.8] $$\sum^{R}\sum^{C}\sum^{L}\sum^{n} X^2_{rcli} - \frac{1}{n}\sum^{R}\sum^{C}\sum^{L} T^2_{rcl.}$$

TOTAL

[17.9] $$\sum^{R}\sum^{C}\sum^{L}\sum^{n} X^2_{rcli} - \frac{T^2}{N}$$

Although some of the above formulas are cumbersome in appearance, their application is really quite simple and involves the calculation of only nine different terms from the data, as can be seen in the illustrative example in the following section. The most cumbersome term of all is the triple interaction term, which is sometimes obtained very simply by subtracting from the total sum of squares the sum of all the other sums of squares. The reader should notc also that if computational formulas for triple classification for $n = 1$ are required, these are directly obtained by writing $n = 1$ in the computation formulas given above.

17.9 ILLUSTRATIVE EXAMPLE OF THREE-WAY CLASSIFICATION

Table 17.6 shows data for a $2 \times 3 \times 2$ factorial experiment. Early- and late-blinded animals belonging to three strains, bright, mixed, and dull,

Table 17.6

Data for the analysis of variance with three-way classification: $n > 1$, error scores for early- and late-blinded rats for three strains reared under two environmental conditions: cell totals shown in parentheses

Early blinded

Environment	Strain: Bright		Strain: Mixed		Strain: Dull	
Free	27	22	31	37	55	62
	45	18	52	45	76	85
	76	33	86	66	104	126
	(221)		(317)		(508)	
Restricted	55	40	77	76	132	104
	81	50	98	68	96	70
	36	70	42	104	89	142
	(332)		(465)		(633)	

Late blinded

Environment	Strain: Bright		Strain: Mixed		Strain: Dull	
Free	61	39	61	71	140	122
	76	60	82	92	99	92
	46	59	103	105	68	101
	(341)		(514)		(622)	
Restricted	88	92	100	120	142	150
	95	103	120	131	96	105
	51	73	89	76	80	125
	(502)		(636)		(698)	

were reared under two environmental conditions, free and restricted. Twelve groups of animals were used, each group comprising six animals. In this example rows are environments, columns are strains, and layers are early- versus late-blinded. In Table 17.6 the cell totals are shown in parentheses.

For computational purposes it is necessary to write down the totals for rows by columns summed over layers, rows by layers summed over columns, and layers by columns summed over rows. If these data are visualized as a cube of numbers, these are the totals on the surface of the cube. The totals for rows by columns summed over layers, $T_{rc..}$, are as follows:

$T_{rc..}$

	Columns			$T_{r...}$
Rows	562	831	1,130	2,523
	834	1,101	1,331	3,266
$T_{.c..}$	1,396	1,932	2,461	$5{,}789 = T_{....}$

The totals above are obtained very simply by adding the cell totals over layers, that is, for the early- and late-blinded animals. Thus $221 + 341 = 562$, $317 + 514 = 831$, and so on. Totals for rows, $T_{r...}$, and for columns, $T_{.c..}$, are also shown, together with the grand total, $T_{....}$. Totals for rows by layers summed over columns, $T_{r.l.}$, are as follows:

$T_{r.l.}$

	Layers		$T_{r...}$
Rows	1,046	1,477	2,523
	1,430	1,836	3,266
$T_{..l.}$	2,476	3,313	$5{,}789 = T_{....}$

Here 1,046 is obtained by summing the cell totals for the first layer over columns; thus, $221 + 317 + 508 = 1{,}046$. Likewise $332 + 465 + 633 = 1{,}430$, and so on. Totals for layers by columns summed over rows, $T_{.cl.}$, are as follows:

$T_{.cl.}$

	Columns			$T_{..l.}$
Layers	553	782	1,141	2,476
	843	1,150	1,320	3,313
$T_{.c..}$	1,396	1,932	2,461	$5{,}789 = T_{....}$

Here 553 is obtained by summing the cell totals for the first column over rows; thus, $221 + 332 = 553$, $317 + 465 = 782$, and so on.

From the data as arranged above *nine* quantities are calculated which are the different terms in the computation formulas. These quantities in this illustrative example are as follows:

$$\frac{1}{nCL}\sum^{R} T^2_{r...} = \frac{1}{6\times 3\times 2}\times 17{,}032{,}285 = 473{,}119$$

$$\frac{1}{nRL}\sum^{C} T^2_{.c..} = \frac{1}{6\times 2\times 2}\times 11{,}737{,}961 = 489{,}082$$

$$\frac{1}{nRC}\sum^{L} T^2_{..l.} = \frac{1}{6\times 2\times 3}\times 17{,}106{,}545 = 475{,}182$$

$$\frac{1}{nL}\sum^{R}\sum^{C} T^2_{rc..} = \frac{1}{6\times 2}\times 5{,}962{,}623 = 496{,}885$$

$$\frac{1}{nC}\sum^{R}\sum^{L} T^2_{r.l.} = \frac{1}{6\times 3}\times 8{,}691{,}441 = 482{,}858$$

$$\frac{1}{nR}\sum^{C}\sum^{L} T^2_{.cl.} = \frac{1}{6\times 2}\times 5{,}994{,}763 = 499{,}564$$

$$\frac{1}{n}\sum^{R}\sum^{C}\sum^{L} T^2_{rcl.} = \frac{1}{6}\times 3{,}045{,}597 = 507{,}600$$

$$\sum^{R}\sum^{C}\sum^{L}\sum^{n} X^2_{rcli} = 536{,}369$$

$$\frac{T^2}{N} = \frac{1}{72}\times 33{,}512{,}521 = 465{,}452$$

In the above computation the quantity $\sum^{R} T^2_{r...} = 2{,}523^2 + 3{,}266^2 = 17{,}032{,}285$, $\sum^{C} T^2_{.c..} = 1{,}396^2 + 1{,}932^2 + 2{,}461^2 = 11{,}737{,}961$, and so on. Applying the computation formulas the required sums of squares are then as follows:

ROWS

$$\frac{1}{nLC}\sum^{R} T^2_{r...} - \frac{T^2}{N} = 473{,}119 - 465{,}452 = 7{,}667$$

COLUMNS

$$\frac{1}{nRL}\sum^{C} T^2_{.c..} - \frac{T^2}{N} = 489{,}082 - 465{,}452 = 23{,}630$$

LAYERS

$$\frac{1}{nRC}\sum^{L} T^2_{..l.} - \frac{T^2}{N} = 475{,}182 - 465{,}452 = 9{,}730$$

$R \times C$ INTERACTION

$$\frac{1}{nL}\sum^{R}\sum^{C} T^2_{rc..} - \frac{1}{nLC}\sum^{R} T^2_{r...} - \frac{1}{nRL}\sum^{C} T^2_{.c..} + \frac{T^2}{N}$$

$$= 496{,}885 - 473{,}119 - 489{,}082 + 465{,}452 = 136$$

$R \times L$ INTERACTION

$$\frac{1}{nC}\sum^{R}\sum^{L} T^2_{r.l.} - \frac{1}{nLC}\sum^{R} T^2_{r...} - \frac{1}{nRC}\sum^{L} T^2_{..l.} + \frac{T^2}{N}$$

$$= 482{,}858 - 473{,}119 - 475{,}182 + 465{,}452 = 9$$

$C \times L$ INTERACTION

$$\frac{1}{nR}\sum^{C}\sum^{L} T^2_{.cl.} - \frac{1}{nLR}\sum^{C} T^2_{.c..} - \frac{1}{nRC}\sum^{L} T^2_{..l.} + \frac{T^2}{N}$$

$$= 499{,}564 - 489{,}082 - 475{,}182 + 465{,}452 = 752$$

$R \times C \times L$ INTERACTION

$$\frac{1}{n}\sum^{R}\sum^{C}\sum^{L} T^2_{rcl.} - \frac{1}{nL}\sum^{R}\sum^{C} T^2_{rc..} - \frac{1}{nC}\sum^{R}\sum^{L} T^2_{r.l.} - \frac{1}{nR}\sum^{C}\sum^{L} T^2_{.cl.}$$

$$+ \frac{1}{nCL}\sum^{R} T^2_{r...} + \frac{1}{nRL}\sum^{C} T^2_{.c..} + \frac{1}{nRC}\sum^{L} T^2_{..l.} - \frac{T^2}{N}$$

$$= 507{,}600 - 496{,}885 - 482{,}858 - 499{,}564 + 473{,}119$$

$$+ 489{,}082 + 475{,}182 - 465{,}452 = 224$$

WITHIN CELLS

$$\sum^{R}\sum^{C}\sum^{L}\sum^{n} X^2_{rcli} - \frac{1}{n}\sum^{R}\sum^{C}\sum^{L} T^2_{rcl.} = 536{,}369 - 507{,}600 = 28{,}769$$

TOTAL

$$\sum^{R}\sum^{C}\sum^{L}\sum^{n} X^2_{rcli} - \frac{T^2}{N} = 536{,}369 - 465{,}452 = 70{,}917$$

The analysis-of-variance table for these data is given in Table 17.7. The degrees of freedom for

rows, $R - 1 = 2 - 1 = 1$

columns, $C - 1 = 3 - 1 = 2$

layers, $L - 1 = 2 - 1 = 1$

$R \times C$ interactions, $(R-1)(C-1) = (2-1)(3-1) = 2$

Table 17.7
Analysis of variance for the data of Table 17.6

Source of variation	Sum of squares	Degrees of freedom	Variance estimate
Rows (environments)	7,667	1	$7{,}667 = s_r^2$
Columns (strains)	23,630	2	$11{,}815 = s_c^2$
Layers (early vs. late)	9,730	1	$9{,}730 = s_l^2$
$R \times C$	136	2	$68 = s_{rc}^2$
$R \times L$	9	1	$9 = s_{rl}^2$
$C \times L$	752	2	$376 = s_{cl}^2$
$R \times C \times L$	224	2	$112 = s_{rcl}^2$
Within cells	28,769	60	$479 = s_w^2$
Total	70,917	71	

$R \times L$ interactions, $\quad (R-1)(L-1) = (2-1)(2-1) = 1$

$C \times L$ interactions, $\quad (C-1)(L-1) = (3-1)(2-1) = 2$

$R \times C \times L$ interactions, $\quad (R-1)(C-1)(L-1) = (2-1)(3-1)(2-1) = 2$

and within cells, $\quad RCL(n-1) = 12 \times 5 = 60$

In this illustrative sample all three variables may be viewed as fixed, and s_w^2 becomes the appropriate error term for testing all effects. The following F ratios may be calculated:

$$F_r = \frac{s_r^2}{s_w^2} = \frac{7{,}667}{479} = 16.01 \qquad p < .01$$

$$F_c = \frac{s_c^2}{s_w^2} = \frac{11{,}815}{479} = 24.67 \qquad p < .01$$

$$F_l = \frac{s_l^2}{s_w^2} = \frac{9{,}730}{479} = 20.31 \qquad p < .01$$

$$F_{rc} = \frac{s_{rc}^2}{s_w^2} = \frac{68}{479} - .14 \qquad p > .05$$

$$F_{rl} = \frac{s_{rl}^2}{s_w^2} = \frac{9}{479} = .02 \qquad p > .05$$

$$F_{cl} = \frac{s_{cl}^2}{s_w^2} = \frac{376}{479} = .78 \qquad p > .05$$

$$F_{rcl} = \frac{s_{rcl}^2}{s_w^2} = \frac{112}{479} = .23 \qquad p > .05$$

In this illustrative example the difference in environments, strains, and early versus late blinded are highly significant. All interaction effects are not significant. In fact all interaction terms are somewhat less than would ordinarily be expected on a chance basis.

17.10 UNEQUAL NUMBERS IN THE SUBCLASSES

Difficulties associated with unequal numbers in the subclasses discussed in Section 16.11 apply also in the three-way classification cases. For small departures from equality the method of unweighted means as discussed in Section 16.12 may be used. The least-squares solution to the problem of unequal n's in the subclasses is described by Bancroft (1968). In general it is advisable to avoid unequal n's wherever possible.

BASIC TERMS AND CONCEPTS

Three-way classification

First-order interaction

Second-order interaction

EXERCISES

1 In a $5 \times 4 \times 3$ factorial experiment with 10 observations per cell, calculate the number of degrees of freedom associated with **(a)** the main effects, **(b)** interaction sums of squares, and **(c)** within-cells sums of squares.

2 Given the following cell means with five observations per cell:

L_1	C_1	C_2
R_1	5	15
R_2	10	5

L_2	C_1	C_2
R_1	10	20
R_2	15	10

Calculate the $R \times C$, $R \times L$, $C \times L$, and $R \times C \times L$ interactions.

3 In a three-way analysis of variance what is the correct error term for testing **(a)** the first-order interaction mean squares when all three variables are random, **(b)** the column sum of squares when rows are random and columns and layers are fixed, **(c)** the second-order interaction when all three variables are fixed.

4 The following are data for a $2 \times 2 \times 2$ factorial experiment with six observations in each of the eight experimental treatments.

	L_1				L_2			
	C_1		C_2		C_1		C_2	
R_1	5	8	8	19	7	18	25	35
	2	10	10	17	10	26	31	40
	4	3	23	16	25	31	16	18
R_2	22	16	4	8	31	22	35	18
	15	10	10	5	8	19	41	16
	8	26	11	12	15	27	32	20

Calculate (**a**) the sums of squares and (**b**) the mean squares for these data. On the assumption that the three variables are fixed, (**c**) test the significance of the main effects and interaction effects.

5 The following are data for a $3 \times 3 \times 2$ factorial experiment with nine observations in each of the 18 combinations of experimental treatments.

Layer 1

	C_1			C_2			C_3		
R_1	5	4	11	6	7	15	12	17	18
	1	2	1	2	1	5	13	16	4
	6	9	5	8	12	10	18	21	17
R_2	8	8	10	10	18	19	25	16	12
	4	2	10	15	17	5	17	18	13
	10	10	8	19	10	18	14	9	31
R_3	10	19	6	12	21	31	15	31	22
	15	16	18	11	19	32	10	19	11
	17	8	20	13	19	27	22	35	16

Layer 2

	C_1			C_2			C_3		
R_1	15	12	5	22	18	5	6	5	12
	18	16	18	17	17	4	2	4	2
	10	19	16	9	6	3	9	10	6
R_2	17	18	25	12	17	11	6	9	7
	20	22	16	20	16	12	3	5	2
	15	19	29	18	10	10	4	8	12
R_3	25	22	19	16	10	31	9	18	5
	16	31	12	12	19	30	10	15	12
	19	34	17	12	19	26	17	7	9

Calculate (**a**) the sums of squares and (**b**) the mean squares for these data. On the assumption that the three variables are fixed, (**c**) test the significance of the main effects and interaction effects.

ANSWERS TO EXERCISES

1

Source	df
R	4
C	3
L	2
$R \times C$	12
$R \times L$	8
$C \times L$	6
$R \times C \times L$	24
Within cells	540
Total	599

2

Source	SS
$R \times C$	562.50
$R \times L$	.00
$C \times L$	.00
$R \times C \times L$	.00

3 **a** $R \times C \times L$ interaction **b** $R \times C$ interaction
c Within cells

4

Source	SS	df	Mean square	F
Rows	12.00	1	12.00	.21
Columns	216.75	1	216.75	3.80
Layers	1,800.75	1	1,800.75	
$R \times C$	280.33	1	280.33	4.92†
$R \times L$	8.33	1	8.33	.15
$C \times L$	114.08	1	114.08	2.00
$R \times C \times L$	208.33	1	208.33	3.65
Within cells	2,281.30	40	57.03	

† Significant at the .05 level.

5

Source	SS	df	Mean square	F
Row	1,763.57	2	881.78	28.17†
Column	91.01	2	45.51	1.45
Layer	3.56	1	3.56	.11
$R \times C$	162.32	4	40.58	1.30
$R \times L$	36.78	2	18.39	.59
$C \times L$	2,507.70	2	1,253.85	40.06†
$R \times C \times L$	101.62	4	25.41	.81
Within cells	4,507.65	144	31.30	

† Significant at the .01 level.

18

ORTHOGONAL AND OTHER MULTIPLE-COMPARISON PROCEDURES

18.1 INTRODUCTION

On comparing k means, the analysis of variance may lead to a significant F test. This is an initial step in the analysis of the data, and further analysis may be appropriate. If the treatment, or independent variable, is a nominal variable, the investigator may wish to compare selected pairs of means, or to compare every mean with every other mean, there being $k(k-1)/2$ such comparisons, or to compare a particular subset of means with another subset. If the treatment variable is of the interval or ratio type, that is, quantitative, the investigator may wish to compare means two at a time or in subsets as in the nominal case, or she or he may wish to explore the nature of the functional relation between the treatment variable and the dependent variable. For example, does a linear or some type of curvilinear relation exist? Such a form of analysis is called trend analysis. Certain procedures used in comparing means two at a time, or in subsets, following an F test, and also the analysis of data for trend, belong essentially to the same general class and will be considered initially in this chapter. Several other common procedures used in making comparisons between means, which belong to a different class, will be considered later in the chapter.

18.2 A PRIORI VERSUS A POSTERIORI COMPARISONS

A distinction is commonly made between a priori comparisons and a posteriori comparisons. A priori comparisons are *planned* prior to the conduct of the experiment. They have direct relevance to the theory that gave rise to the conduct of the experiment. They are in effect the questions the investigator is hoping to answer by doing the experiment. A posteriori comparisons are *unplanned, post hoc,* or *postmortem* comparisons. These are comparisons the investigator decides to make after inspection of the data, and may be suggested by such inspection. From a logical viewpoint a priori comparisons may be applied whether or not the F test has led to the rejection of the null hypothesis. A posteriori comparisons are only applied following a significant F test. From a practical point of view the utility of the distinction between a priori and a posteriori comparisons may be questioned. Frequently in the analysis of experimental data the investigator may be rather unclear about whether certain comparisons were formulated on an a priori basis or not. In the experience of this writer the distinction between a priori versus a posteriori comparison, or planned versus unplanned comparisons, has a preciseness to it which is somewhat unrealistic in relation to the realities of much experimental work. At any rate the methods for making orthogonal comparisons described in the earlier part of this chapter logically apply to a priori or planned comparisons. The methods described later in the chapter apply to a posteriori or unplanned comparisons.

18.3 ORTHOGONAL COMPARISONS

Consider an experiment with four experimental groups and means $\bar{X}_1$, $\bar{X}_2$, $\bar{X}_3$, and $\bar{X}_4$. The investigator may wish, for example, to test the differences $\bar{X}_1 - \bar{X}_2$ or $\bar{X}_3 - \bar{X}_4$, or if the n's are equal, $\bar{X}_1 + \bar{X}_2 - \bar{X}_3 - \bar{X}_4$. These are called *comparisons* or *contrasts*. When more than one comparison is to be made, a distinction is drawn between orthogonal and nonorthogonal comparisons.

The word "orthogonal" means independent, and orthogonal comparisons are independent of each other. Nonorthogonal comparisons are not independent of each other. Such comparisons are correlated. In the analysis of variance for one-way classification, for example, the between-groups sum of squares has associated with it $k-1$ degrees of freedom. The essential notion in orthogonal comparisons is that the between-groups sum of squares can be partitioned into not more than $k-1$ independent parts. Each part has 1 degree of freedom associated with it. Each part or bit of the sum of squares contains information. We can, as it were, break down the between-groups sum of squares into as many independent bits of information as we have degrees of freedom.

A comparison or contrast is viewed as a weighted sum of means. Thus for a set of four means the difference $\bar{X}_1 - \bar{X}_2$ is the weighted sum of $\bar{X}_1$, $\bar{X}_2$, $\bar{X}_3$, and $\bar{X}_4$, the weights being $1, -1, 0$, and 0. Also the difference $\bar{X}_3 - \bar{X}_4$ is a weighted sum, the weights being $0, 0, 1$, and -1. Also a comparison of the kind $\bar{X}_1 + \bar{X}_2 - \bar{X}_3 - \bar{X}_4$ is a weighted sum, the weights being 1, 1, −1, −1. Let us designate these weights by the symbol c_{ij}, where the first subscript identifies the particular set of weights, which defines a comparison, and the second subscript identifies the particular weight in the set. A set of weights, c_{ij}, may be viewed as a dummy variable which is used in order to make certain forms of analysis possible. The weights c_{ij} are sometimes spoken of as coefficients. Note that $\Sigma c_{ij} = 0$.

The discussion to follow concerns itself with experimental designs where the groups are of equal size; that is, the n's are equal.

Two comparisons, or contrasts, for equal n's are said to be orthogonal when the sum of the cross products of the weights is also 0. Thus, when $\Sigma c_{ij}c_{kj} = 0$, two comparisons are orthogonal.

Consider the following sets of weights for four groups.

	Means			
	$\bar{X}_1$	$\bar{X}_2$	$\bar{X}_3$	$\bar{X}_4$
c_{1j}	1	−1	0	0
c_{2j}	0	0	1	−1
c_{3j}	1	0	0	−1

Here the comparison $\bar{X}_1 - \bar{X}_2$ and $\bar{X}_3 - \bar{X}_4$ are orthogonal because $\Sigma c_{1j}c_{2j} = (1)(0) + (-1)(0) + (0)(1) + (0)(-1) = 0$. The comparisons $\bar{X}_1 - \bar{X}_2$ and $\bar{X}_1 - \bar{X}_4$ are not orthogonal because $\Sigma c_{1j}c_{3j} = 1$ and not 0.

18.4 PARTITIONING THE SUM OF SQUARES

In Section 15.3 a deviation about a grand mean was viewed as comprising two parts as follows: $(X_{ij} - \bar{X}) = (X_{ij} - \bar{X}_j) + (\bar{X}_j - \bar{X})$. The total sum of squares was partioned into a within-group and a between-group sum of squares. In the present situation we wish to partition the between-group sum of squares still further into additive and independent parts.

We proceed by writing an equation as follows:

[18.1] $$\bar{X}_j' = a + b_1c_{1j} + b_2c_{2j} + \cdot\cdot\cdot + b_mc_{mj}$$

This is a multiple regression equation which predicts the group means. The value $\bar{X}_j'$ is a predicted group mean. It is the best estimate of the group mean which can be made using the regression equation. In this equation a is the point where the curve intercepts the X axis. The reader will recall that in Chapter 8 the general form of the regression equation used in predicting X from Y was $X' = a + bY$. In the multiple regression equation above, instead of a single independent variable Y, we have a set of variables $c_{1j}, c_{2j}, \ldots, c_{mj}$. These are orthogonal or independent sets of weights, however obtained. They constitute a set of uncorrelated variables. The b's in the above equation are regression weights determined by the method of least squares. Methods for calculating such weights are described in Chapter 26, which the reader may wish to refer to at this time.

Because a in equation [18.1] is equal to $\bar{X}$ for equal n's, we may write

$$(\bar{X}'_j - \bar{X}) = b_1c_{1j} + b_2c_{2j} + \cdots + b_mc_{mj} \tag{18.2}$$

We now write the identity

$$(X_{ij} - \bar{X}) = (X_{ij} - \bar{X}_j) + (\bar{X}_j - \bar{X}'_j) + (\bar{X}'_j - \bar{X}) \tag{18.3}$$

This identity states that the deviation of a particular score X_{ij} from the grand mean $\bar{X}$ may be viewed as comprising three parts: (1) a deviation of the score X_{ij} from the mean of the group $\bar{X}_j$ to which it belongs, (2) a deviation of $\bar{X}_j$ from the value predicted by the linear regression equation, and (3) a deviation of the predicted mean $\bar{X}'_j$ from the grand mean. We now substitute equation [18.2] in the above identity to obtain

$$(X_{ij} - \bar{X}) = (X_{ij} - \bar{X}_j) + (\bar{X}_j - \bar{X}'_j) + b_1c_{1j} + b_2c_{2j} + \cdots + b_mc_{mj} \tag{18.4}$$

This equation is squared, summed over the n cases in the jth group, the n's for the k groups being equal, and summed over the k groups. The cross-product terms vanish, and we obtain

$$\sum_{j=1}^{k}\sum_{i=1}^{n} (X_{ij} - \bar{X})^2 = \sum_{j=1}^{k}\sum_{i=1}^{n} (X_{ij} - \bar{X}_j)^2 + n\sum_{j=1}^{k} (\bar{X}_j - \bar{X}'_j)^2 + nb_1^2\sum_{j=1}^{k} c_{1j}^2 + nb_2^2\sum_{j=1}^{k} c_{2j}^2 + \cdots + nb_m^2\,\Sigma c_{mj}^2 \tag{18.5}$$

Thus the total sum of squares is partitioned into a within-group sum of squares, a sum of squares of deviations of group means from the value $\bar{X}'_j$ obtained from the regression equation, and a series of sums of squares each associated with a particular orthogonal comparison.

The reader should note that the between-groups sum of squares has associated with it $k - 1$ degrees of freedom. This means that if $k - 1$ orthogonal comparisons are made the deviation sum of squares $n\sum_{j=1}^{k} (\bar{X}_j - \bar{X}'_j)^2$ will be 0. With many sets of data, fewer than $k - 1$ comparisons are appropriate and a deviation sum of squares will result.

18.5 COMPUTATION FORMULAS FOR ORTHOGONAL COMPARISONS

The computation formulas for calculating the various regression components are readily shown to be

$$nb_1^2\sum_{j=1}^{k} c_{1j}^2 = \frac{\left(\sum_{j=1}^{k} c_{1j}T_j\right)^2}{n\sum_{j=1}^{k} c_{1j}^2} = \frac{n\left(\sum_{j=1}^{k} c_{1j}\bar{X}_j\right)^2}{\sum_{j=1}^{k} c_{1j}^2} \tag{18.6}$$

$$nb_2^2 \sum_{j=1}^{k} c_{2j} = \frac{\left(\sum_{j=1}^{k} c_{2j} T_j\right)^2}{n \sum_{j=1}^{k} c_{2j}^2} = \frac{n\left(\sum_{j=1}^{k} c_{2j} \bar{X}_j\right)^2}{\sum_{j=1}^{k} c_{2j}^2}$$

and so on.

If a single difference between two means only is under study, say, $\bar{X}_1 - \bar{X}_2$, the above formulas reduce to $n(\bar{X}_1 - \bar{X}_2)^2/2$, and the required F ratio for testing the significance of the difference is

[18.7] $$F = \frac{n(\bar{X}_1 - \bar{X}_2)^2}{2s_w^2}$$

with 1 degree of freedom associated with the numerator and $N - k$ associated with the within-groups variance estimate s_w^2.

18.6 ILLUSTRATIVE EXAMPLES OF ORTHOGONAL COMPARISONS

Table 18.1 shows an illustrative example of the computation required for orthogonal comparisons. In this example $n = 8$ and $k = 4$. The three comparisons made are $\bar{X}_1 - \bar{X}_2$, $\bar{X}_3 - \bar{X}_4$, and $\bar{X}_1 + \bar{X}_2 - \bar{X}_3 - \bar{X}_4$. These three comparisons are orthogonal. In this example the number of degrees of freedom associated with the between-group sum of squares is 3. The between-group sum of squares has been divided into three independent additive parts with 1 degree of freedom associated with each part. Table 18.2 shows the analysis-of-variance table. The first $\bar{X}_1 - \bar{X}_2$ is not significant. The third comparison $\bar{X}_1 + \bar{X}_2 - \bar{X}_3 - \bar{X}_4$ is highly significant; $p < .01$. Quite clearly in these data by far the greater part of the between-groups variation results because the means for groups I and II are substantially different from the means for groups III and IV.

18.7 COMPARISONS AS CORRELATIONS

Viewed as a correlational problem the question can be raised as to how well the orthogonal coefficients c_{ij} can predict the group means. The reader will recall that a correlation coefficient squared, r^2, is a proportion. It is the proportion of the variance of one variable that can be predicted from or attributed to another variable.

In making orthogonal comparisons, it can be readily shown that

[18.8] $$r^2 = \frac{\text{sum of squares for comparison}}{\text{between-group sum of squares}}$$

Here r is the correlation between the group means $\bar{X}_j$ and the coefficients c_{ij} for the ith comparison. The correlation coefficient squared, r^2, is the

Table 18.1

Illustrative Example of Orthogonal Comparisons

	Treatment				
	I	II	III	IV	
	5	4	5	14	$T = 294$
	7	12	8	7	$T^2/N = 2{,}701.13$
	6	8	17	6	$\sum T_j^2/n = 2{,}995.75$
	3	7	26	12	$\sum_{j=1}^{k}\sum_{i=1}^{n} X_{ij}^2 = 3{,}524$
	9	7	14	15	$\sum_{j=1}^{k} c_{1j}T_j = -14$
	7	6	8	19	$\sum_{j=1}^{k} c_{2j}T_j = 10$
	4	4	13	8	$\sum_{j=1}^{k} c_{3j}T_j = 94$
	2	9	11	11	
n	8	8	8	8	
T_j	43	57	102	92	$\sum_{j=1}^{k} c_{1j}^2 = 2$
$\bar{X}_j$	5.38	7.13	12.75	11.50	
c_{1j}	1	−1	0	0	$\sum_{j=1}^{k} c_{2j}^2 = 2$
c_{2j}	0	0	1	−1	
c_{3j}	1	1	−1	−1	$\sum_{j=1}^{k} c_{3j}^2 = 4$
$\sum_{i=1}^{n} X_{ij}^2$	269	455	1604	1196	

	Sum of squares	
Comparison 1	$(-14)^2/(8 \times 2)$	= 12.25
Comparison 2	$(10)^2/(8 \times 2)$	= 6.25
Comparison 3	$(94)^2/(8 \times 4)$	= 276.13
Within	3,524.00 − 2,995.75	= 528.25
Total	3,524.00 − 2,701.13	= 822.87

Table 18.2

Analysis of variance for orthogonal comparisons for the data of Table 18.1

Source of variation	Sum of squares	Degrees of freedom	Variance estimates
Comparison 1	12.25	1	12.25
Comparison 2	6.25	1	6.25
Comparison 3	276.13	1	276.13
Within	528.25	28	18.87
Total	822.87	31	

$F_1 = \frac{12.25}{18.87} = .65 \quad F_2 = \frac{6.25}{18.87} = .33 \quad F_3 = \frac{276.13}{18.87} = 14.63$

proportion of the variation in the group means that can be attributed to each of the separate comparisons. For the illustrative data of Table 18.2 the three values of r^2 and $r^2 \times 100$ are as follows:

	r^2	$r^2 \times 100$
Comparison 1	.04	4
Comparison 2	.02	2
Comparison 3	.94	94

Thus 94 percent of the variation between the group means can be attributed directly to comparison 3. The important feature of these data is that the first pair of means differs very substantially from the second pair. The question may be raised as to the relationship between the correlations of orthogonal coefficients with group means and the regression coefficient b_i in equation [18.4]. If this equation had been expressed in standard-score form, the values of b_i would in fact be the correlation coefficients. This, however, is not the case. Equation [18.4] is not in standard-score form, and no such simple relation exists. The relation between b_i and r_i is given by $b_i^2 = r^2\, \Sigma(\bar{X}_j - \bar{X})^2/\Sigma c_{ij}^2$, although this is of limited interest.

18.8 TREND ANALYSIS

In experiments where the treatment, or independent, variable is nominal, the analysis of the data cannot be extended beyond an F test applied to the group means and the comparison of means either two at a time or in subgroups. In some experiments the treatment variable is of the interval or ratio type. Examples are experiments on the behavioral effects of different dosages of a drug, different periods of practice in learning a task, or different numbers of reinforcements in conditioning or extinction. With such experiments investigators may extend their analysis to an examination of certain characteristics of the shape of the relation between the treatment variable and the experimental variable. Questions of the following type may be raised: Do the group means increase significantly in a linear fashion with increase in the treatment variable? Is a straight line a good fit to the group means, or do significant deviations from linearity exist? Do the group means increase and then decrease with increase in the treatment variable in an inverted U fashion? Trend analysis as described in this and the following sections provides answers to questions of this kind. It is an application of the analysis of variance.

In most applications of trend analysis the investigator is usually not interested in the development of a precise equation to describe the functional relation between the treatment variable and the experimental variable, although such an equation could readily be obtained. Usually concern is with questions of significance.

The procedures described in this and the following sections assume that the levels of the treatment variable are *equally spaced.* Equal spacing is not a necessary condition for the application of methods of trend analysis. It leads, however, to simplification in computation.

18.9 ORTHOGONAL POLYNOMIALS IN TREND ANALYSIS

The method of trend analysis described here represents a simple extension of the methods described above for making orthogonal comparisons or contrasts. Trend analysis is a particular case of the method of orthogonal comparisons. This particular case arises when specially chosen sets of weights or coefficients which are concerned with trend relations in the data are used in making the comparisons.

These weights are ordinarily called coefficients for orthogonal polynomials. To illustrate, consider the following coefficients for three groups or levels of the independent variable:

Linear	c_{1j}	−1	0	1
Quadratic	c_{2j}	1	−2	1

Here $\Sigma c_{1j} = \Sigma c_{2j} = 0$; also $\Sigma c_{1j}c_{2j} = 0$. The weights are orthogonal. The set c_{1j} is a set of linear coefficients. Its use will indicate whether the group means show a systematic tendency to increase or decrease in a linear fashion. The set c_{2j} is a set of quadratic coefficients. Its use will indicate the presence of a quadratic relation. Consider further the following set of coefficients for four groups, or levels, of the independent variable.

Linear	c_{1j}	−3	−1	1	3
Quadratic	c_{2j}	1	−1	−1	1
Cubic	c_{3j}	−1	3	−3	1

Here $\Sigma c_{1j} = \Sigma c_{2j} = \Sigma c_{3j} = 0$. Also $\Sigma c_{1j}c_{2j} = \Sigma c_{1j}c_{3j} = \Sigma c_{2j}c_{3j} = 0$. The three sets of coefficients are orthogonal. The set c_{1j} is the linear set, c_{2j} the quadratic, and c_{3j} the cubic. The number of times the signs change determines the degree of the polynomial. For the coefficients above, one sign change occurs in the linear set, two in the quadratic, and three in the cubic.

Consider further the set of weights for five groups or levels of the factor.

Linear	c_{1j}	−2	−1	0	1	2
Quadratic	c_{2j}	2	−1	−2	−1	2
Cubic	c_{3j}	−1	2	0	−2	1
Quartic	c_{4j}	−1	−4	6	−4	−1

Here an additional component is identified, namely, a quartic component. This component is concerned with whether the relation has the shape of a W or an inverted W. It may be informative to plot the above coefficients graphically against groups or levels as shown in Figure 18.1.

Essentially what trend analysis does from a correlational viewpoint is to correlate the group means with the sets of coefficients and test the significance of the correlations. If the correlation between the group means and a particular set of coefficients is high, then the group means have a shape, as it were, similar to that portrayed by the coefficients.

A description of the procedure for obtaining sets of orthogonal coefficients for use in trend analysis is beyond the scope of the present discussion. The procedure is described in Kendall (1952) and elsewhere. Table J of the Appendix shows a short table of coefficients for orthogonal polynomials. A more extended table will be found in Fisher and Yates (1963).

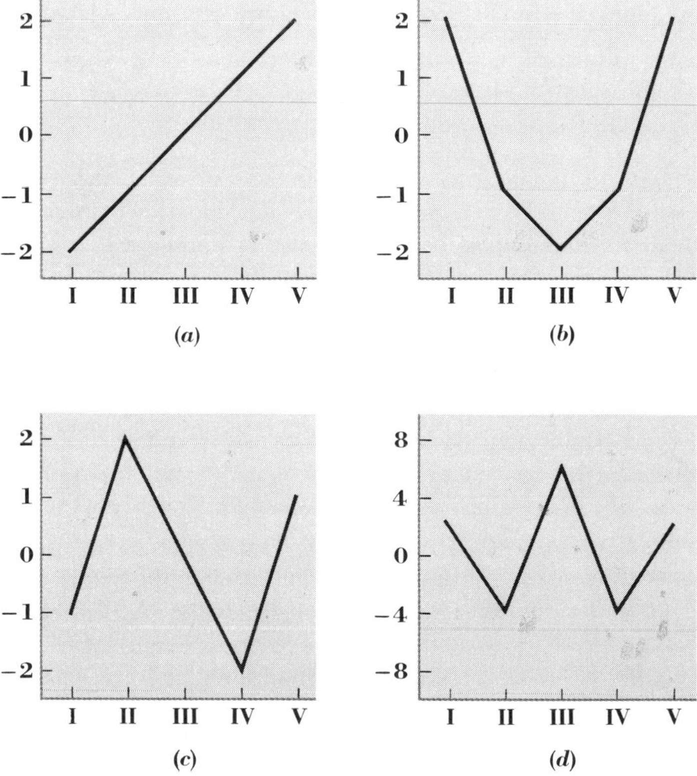

Figure 18.1 Coefficients of orthogonal polynomials for (*a*) linear, (*b*) quadratic, (*c*) cubic, and (*d*) quartic trend components plotted against levels.

18.10 ILLUSTRATIVE EXAMPLE OF TREND ANALYSIS USING ORTHOGONAL POLYNOMIALS

Table 18.3 shows an illustrative example of the computation required for a trend analysis using orthogonal polynomials. In this example $n = 8$, the groups being of equal size, and $k = 4$. Table 18.4 shows the corresponding analysis-of-variance table. In this example the linear component is significant with $p < .01$. Neither the quadratic nor the cubic regression component is significant. The reader will note that the deviation component is 0. This occurs because a cubic equation will always fit four points exactly, just as a linear equation will always fit two points exactly, and a quadratic equation three points. In this example with $k = 4$ the analysis has been carried as far as possible.

Table 18.3

Illustrative example of computation for trend analysis using orthogonal polynomials for the data of Table 18.1

	Treatment				
	I	II	III	IV	$\sum_{j=1}^{k} c_{1j}T_j = 192$
n	8	8	8	8	$\sum_{j=1}^{k} c_{2j}T_j = -24$
T_j	43	57	102	92	
$\bar{X}_j$	5.38	7.13	12.75	11.50	
c_{1j}	-3	-1	1	3	$\sum_{j=1}^{k} c_{3j}T_j = -86$
c_{2j}	1	-1	-1	1	
c_{3j}	-1	3	-3	1	
$\sum_{i=1}^{n} X_{ij}^2$	269	455	1,604	1,196	$\sum_{j=1}^{k} c_{1j}^2 = 20$
					$\sum_{j=1}^{k} c_{2j}^2 = 4$
					$\sum_{j=1}^{k} c_{3j}^2 = 20$

	Sum of squares	
Linear	$192^2/(8 \times 20)$	$= 230.40$
Quadratic	$(-24)^2/(8 \times 4)$	$= 18.00$
Cubic	$(-86)^2/(8 \times 20)$	$= 46.23$
Deviation	$822.87 - 528.24 - 230.40$ $- 18.00 - 46.23$	$= .00$
Within	$3{,}524.00 - 2{,}995.75$	$= 528.25$
Total	$3{,}524.00 - 2{,}701.13$	$= 822.87$

Table 18.4

Analysis of variance using trend analysis for data of Table 18.1

Source of variation	Sum of squares	Degrees of freedom	Variance estimate
Linear regression	230.40	1	$s_l^2 = 230.40$
Quadratic regression	18.00	1	$s_q^2 = 18.00$
Cubic regression	46.23	1	$s_c^2 = 46.23$
Deviation	.00	0	
Within	528.24	28	$s_w^2 = 18.87$
Total	822.87	31	

$F_1 = \frac{230.40}{18.87} = 12.21 \qquad F_2 = \frac{18.00}{18.87} = .95 \qquad F_3 = \frac{46.23}{18.87} = 2.45$

18.11 TREND ANALYSIS WITH UNEQUAL n's

If a single comparison only is under study, which in trend analysis may be linear, quadratic, cubic, or of a higher order, the n's may be equal or unequal. Questions of orthogonality arise of course only when two or more comparisons are made. For unequal n's two comparisons are orthogonal when $\Sigma c_{ij}c_{kj}/n_j = 0$. Ordinarily with unequal n's two comparisons will not be orthogonal except under rather special circumstances. For example, an experiment might be designed such that the n's are either equal or constitute a symmetrical set, such as 4, 10, 4 for $k = 3$, or 10, 20, 20, 10 for $k = 4$. For these values of n the linear and quadratic components are orthogonal and $\Sigma c_{1j}c_{2j}/n_j = 0$.

If a single comparison only is under consideration, say, for linear regression, the formula for calculating the required sum of squares for unequal n's is given by

[18.9] $$\sum_{j=1}^{k} n_j(\bar{X}'_j - \bar{X})^2 = \frac{\left(\sum_{j=1}^{k} c_{ij}T_j - T\sum_{j=1}^{k} n_j c_{ij}/N\right)^2}{\sum_{j=1}^{k} n_j c_{ij}^2 - \left(\sum_{j=1}^{k} n_j c_{ij}\right)^2 \Big/ N}$$

The deviation sum of squares may be obtained by subtracting the within-group and linear-regression sums of squares from the total sum of squares. For equal n's the above formula reduces to

$$\left(\sum_{j=1}^{k} c_{ij}T_j\right)^2 \Big/ n\,\Sigma c_{ij}^2$$

that is, it is the same as formula [18.6].

The application of the above formula as applied to linear trend is illustrated in Tables 18.5 and 18.6. The data are those used previously in Table 15.2.

Table 18.5

Computation for linear trend analysis with unequal n's using the data of Table 15.2

	Method				
	1	2	3	4	$N = 26$ $T = 146$ $T^2/N = 819.85$
$\bar{X}_j$	5.38	8.40	6.14	3.00	$\sum_{j=1}^{k} n_j c_{1j} = -4$
n_j	8	5	7	6	
T_j	43	42	43	18	$\sum_{j=1}^{k} c_{1j} T_j = -74$
c_{1j}	−3	−1	1	3	
$\sum_{i=1}^{n_j} X_{ij}^2$	269	364	287	68	$\sum_{j=1}^{k} n_j c_{1j}^2 = 138$
$\frac{T_j^2}{n_j}$	231.13	352.80	264.14	54.00	$\sum_{j=1}^{k} \sum_{i=1}^{n_j} X_{ij}^2 = 988$
					$\sum_{j=1}^{k} \frac{T_j^2}{n_j} = 902.07$

The total sum of squares is 168.15, as before, and the within-group sum of squares is 85.93. The linear-regression sum of squares is

$$\frac{\left(\sum_{j=1}^{k} c_{1j}T_j - T \sum_{j=1}^{k} n_j c_{1j}/N\right)^2}{\sum_{j=1}^{k} n_j c_{1j}^2 - \left(\sum_{j=1}^{k} n_j c_{1j}\right)^2 \Big/ N} = \frac{\{-74 - [146(-4)]/26\}^2}{138 - 16/26} = 19.34$$

The deviation sum of squares, obtained by subtraction, is 168.15 − 85.93 − 19.34 = 62.88. The F ratio for linear regression is significant at a little better than the .05 level. The F ratio for deviations from linear regression is significant at better than the .01 level. If the .05 level is accepted as a suitable level of significance, the conclusion is that the group means show a significant linear trend. Because of the large deviation component, however, the conclusion is that a straight regression line is not a

Table 18.6

Linear trend analysis for the data of Table 18.5

Source of variation	Sum of squares	Degrees of freedom	Variance estimate
Linear regression	19.34	1	$s_b^2 = 19.34$
Deviation	62.88	2	$s_d^2 = 31.44$
Within	85.93	22	$s_w^2 = 3.91$
Total	168.15	25	

$F_1 = \frac{19.34}{3.91} = 4.95 \qquad F_2 = \frac{31.44}{3.91} = 8.04$

good fit to the data. The relation between the treatment and the experimental variables may be represented more appropriately by a nonlinear regression line.

Quite clearly if the intention of the investigator is to make a number of comparisons, rather than a single comparison, the experiment should be designed such that the n's are equal. Otherwise the various components into which the sum of squares is partitioned will not ordinarily be orthogonal. In practical experimental work, situations arise where the n's are unequal. If the departures from equality are not gross, methods may be used identical to those described in Chapter 16 for making adjustments for unequal n's in the analysis of variance for two-way classification.

18.12 EXTENDED APPLICATIONS OF TREND ANALYSIS

The methods of trend analysis described in this chapter have application to experiments involving one-way classification. The methods may be applied to two-way, and higher-order, factorial experiments. With a two-way factorial experiment of the type described in Chapter 16, both row and column means may be analyzed for trend, using orthogonal polynomials as in the one-way classification case. The various regression and deviation components for rows and columns may be tested by using s_w^2 in the denominator of the F ratio. Also in factorial experiments the interaction between treatments may be analyzed for trend. In a two-way factorial experiment with R rows and C columns, the analysis of interaction for trend is a method for studying the response surface which the RC cell means define. Do all cell means fall in a plane? Do they form a response surface with quadratic characteristics? Questions of this general kind, and others, can be answered by the analysis of interaction for trend. Such methods are described in detail by Winer (1971).

18.13 LIMITATIONS OF TREND ANALYSIS

Trend analysis as described in this chapter has a number of limitations. First, the method requires that the levels of the treatment variable are equally spaced. The problem of unequal spacing can, however, be overcome by the selection of sets of coefficients that are themselves unequally spaced. Second, considerable caution should be exercised in drawing conclusions about the nature of the functional relation between the treatment variable and the dependent variable. Many experiments cover a limited range of levels of the treatment variable. The nature of the relation that exists within one range, or band, of the treatment variables may be quite different from that which exists within another range, or band. To illustrate, the shape of the relation in the population may be that of an inverted U. An experiment where the levels of the treatment variables are restricted to the lower ranges might indicate a positive linear relation, whereas an experiment restricted to the higher levels of the treatment variable might indicate the opposite. Third, the drawing of conclusions about the nature of a functional relation for a small or minimal number of levels is unwarranted. For example, to conclude that a linear relation exists when two levels only are available is obviously absurd. The investigator does not know what would have happened if other levels of treatment had been used beyond or between those studied in the experiment. If a particular functional relation is hypothesized, a sufficient number of levels should be used as to overdetermine the relation. Thus, if the investigator hypothesizes an inverted U relation, at least four or more levels of the treatment variable should be used, and not three.

18.14 OTHER MULTIPLE-COMPARISON PROCEDURES

The methods described above apply to a priori or planned orthogonal comparisons. Other methods for comparing means are frequently used which may be applied on an a posteriori basis, that is, after inspection of the data. These methods do not require the orthogonality of the comparisons made. Methods in common use, using the F test or a statistic known as the *studentized range,* have been developed by Scheffé (1953), Tukey (1949), Newman (1939), Keuls (1952), and Duncan (1955, 1957). Summaries and comparisons of these methods are given by Winer (1971) and Bancroft (1968). Different methods for making multiple comparisons of the type described here adopt different criteria for the rejection of the null hypothesis. The selection of an appropriate criterion in this context is a thorny problem, and no widely accepted method exists for deciding between some of these multiple-comparison procedures. The problem has been discussed by Tukey (1949), who drew a distinction between *per-comparison* error rate and *experiment-wise* error rate. The per-comparison error rate is the number of comparisons falsely declared significant divided by the total number of comparisons. The experiment-wise error rate is the number of experiments with at least one difference falsely declared significant divided by the total number of experiments. Two multiple-comparison methods are described below. One of these uses the F test and is due to Scheffé. The other uses the studentized range and is known as the Newman-Keuls method.

18.15 MULTIPLE COMPARISONS USING SCHEFFÉ'S METHOD

The method described here is due to Scheffé and uses the criterion that the probability of rejecting the null hypothesis when it is true, a Type 1 error, should not exceed .01 or .05, for example, for any of the comparisons made. This is a very rigorous criterion. To apply the method, follow these steps. *First,* calculate the following F ratio between pairs of means by using the within-group variance estimate s_w^2:

[18.10]
$$F = \frac{(\bar{X}_i - \bar{X}_j)^2}{s_w^2/n_i + s_w^2/n_j}$$

This is t^2 because $F = t^2$ for $k = 2$. *Second,* consult a table of F and obtain the value of F required for significance at the .05 or .01 level, or any desired level, for $df_1 = k - 1$ and $df_2 = N - k$. *Third,* calculate the quantity F', which is $k - 1$ times the F required for significance at the desired significance level; that is, $F' = (k - 1)F$. *Fourth,* compare the values of F and F'. For any difference to be significant at the required level, F must be greater than or equal to F'.

The above method may be illustrated with reference to the data of Table 15.2. The means for the four groups are $\bar{X}_1 = 5.38$, $\bar{X}_2 = 8.40$, $\bar{X}_3 = 6.14$, and $\bar{X}_4 = 3.00$; also, $n_1 = 8$, $n_2 = 5$, $n_3 = 7$, and $n_4 = 6$. The values of F are as follows:

Comparison	F
I, II	7.18
I, III	.55
I, IV	4.97
II, III	3.81
II, IV	20.34
III, IV	8.18

The values of F required for significance at the .05 and .01 levels, respectively, for $df_1 = 3$ and $df_2 = 22$ are 3.05 and 4.82. The values of F' required for significance at these levels are 9.15 and 14.46. The only comparison that achieves significance is that between groups II and IV. For this comparison the significance level is less than .01.

The Scheffé method may be used not only to compare means two at a time but to make any comparison at all. For example, given four groups, we may wish to compare the mean for the first two groups combined, $\bar{X}_{1+2} = (n_1\bar{X}_1 + n_2\bar{X}_2)/(n_1 + n_2)$, with the mean for the second two groups, $\bar{X}_{3+4} = (n_3\bar{X}_3 + n_4\bar{X}_4)/(n_3 + n_4)$. The F ratio in this case is

[18.11]
$$F = \frac{(\bar{X}_{1+2} - \bar{X}_{3+4})^2}{s_w^2/(n_1 + n_2) + s_w^2/(n_3 + n_4)}$$

The Scheffé method is more rigorous than other multiple comparison methods with regard to Type I error. It will lead to fewer significant differences. It is easy to apply. No special problems arise because of unequal n's. It uses the readily available F test. The criterion it employs in the evaluation of the null hypothesis is simple and readily understood.

It is not seriously affected by violations of the assumptions of normality and homogeneity of variance, unless these are gross. It can be used for making any comparison the investigator wishes to make.

Concern may attach to the fact that the Scheffé procedure is more rigorous than other procedures, and will lead to fewer significant results. Because this is so, the investigator may choose to employ a less rigorous significance level in using the Scheffé procedure; that is, the .10 level may be used instead of the .05 level. This is Scheffé's recommendation (1959). Readily available tables of F do not ordinarily contain critical values at the .10 level. Tables showing the .10 critical values are given in Fisher and Yates (1963) and also in Winer (1971).

18.16 MULTIPLE COMPARISONS USING THE STUDENTIZED RANGE

A number of methods for making multiple comparisons use a statistic known as the *studentized range*. Given k treatment means based on equal n's, the studentized range is simply the difference between the largest and the smallest treatment means divided by an estimate of the standard error associated with a single treatment mean; that is,

[18.12]
$$Q = \frac{\bar{X}_{\text{max}} - \bar{X}_{\text{min}}}{s_{\bar{x}}} = \frac{\bar{X}_{\text{max}} - \bar{X}_{\text{min}}}{\sqrt{s_w^2/n}}$$

For different values of k, and different degrees of freedom associated with s_w^2, the sampling distribution of Q is known. Table L shows the 95 and 99 percentile points for the distribution of Q for different k and degrees of freedom. These are the values required for the rejection of the null hypothesis at the .05 and .01 levels. The null hypothesis in this case is $H_0: \mu_1 = \mu_2 = \cdot\cdot\cdot = \mu_k$ or that the samples came from a single population with mean μ. The Q test could quite clearly be used as an alternate to the usual F test.

Multiple-comparison methods based on the studentized ranges compare the observed studentized ranges for pairs of means with determined criterion values of the studentized range. If the observed value of Q exceeds the criterion value of Q, the difference is said to be significant. Some of these methods are sequential. They involve ordering the means from low to high. The size of the difference required for significance changes with the separation of the means in the rank order. Other methods use a fixed interval: that is, a fixed criterion value of Q for all comparisons is used, regardless of the separation of the means in the rank order. All multiple-comparison methods using the studentized range apply only to equal n's. This detracts from their usefulness.

One commonly used multiple-comparison method using the studentized range is the Newman-Keuls method. This method uses the criterion that the probability of rejecting the null hypothesis when it is true should not exceed .01 or .05 for all ordered pairs, regardless of the number of steps they are apart. To apply this method the means are ranked from low to high. The studentized ranges are obtained for all $k(k-1)/2$ pairs of means, using $s_{\bar{x}} = \sqrt{s_w^2/n}$. Criterion values of Q at the .01 or .05 levels for $k = 2, 3, \ldots, k$, are obtained from Table L for the degrees of freedom associated with s_w^2. Denote these criterion values as $Q_2, Q_3, \ldots, Q_k$. Thus Q_2 is the criterion value for comparing two adjacent means, Q_3 for means separated by one intervening rank, Q_4 for means separated by two intervening ranks, and Q_k for means separated by $k-2$ intervening ranks. Comparisons of observed Q's with criterion values are made in a particular sequential manner. Values in the first row of the table of Q are compared with the criterion values, proceeding from right to left until a nonsignificant difference is found. No further comparisons are made in that row. This procedure is then applied to the second row, third row, and so on.

To illustrate, consider the following means for five groups of equal size, $n = 6$.

	I	II	III	IV	V
n	6	6	6	6	6
$\bar{X}_j$	12.56	9.54	3.00	6.45	8.32

The analysis-of-variance table for these data is

Source	Sum of squares	df	Variance estimate
Between	304.02	4	$76.01 = s_b^2$
Within	262.50	25	$10.50 = s_w^2$
Total	566.52	29	

The overall F here is $F = 76.01/10.50 = 7.24$, which is significant at better than the .01 level. The means are arranged in rank order from $\bar{X}_3 = 3.00$ to $\bar{X}_1 = 12.56$. The order of the means is $\bar{X}_3, \bar{X}_4, \bar{X}_5, \bar{X}_2, \bar{X}_1$. The difference between every mean and every other mean is calculated. These differences are divided by $s_{\bar{x}} = \sqrt{s_w^2/n} = \sqrt{10.50/6} = 1.324$ to obtain the studentized ranges Q, as follows:

Table of Q

	III	IV	V	II	I
III		2.61	4.02	4.92	7.22
IV			1.41	2.33	4.61
V				.92	3.20
II					2.28
I					

If the .05 significance level is adopted, criterion values of Q for a k of 2, 3, 4, and 5 and $df = 25$ are obtained from Table L. These are $Q_2 = 2.91$, $Q_3 = 3.52$, $Q_4 = 3.89$, and $Q_5 = 4.16$. First, we compare the top row of the above table of Q with the criterion values. The largest value 7.22 involves the comparison of III with I. The criterion here is $Q_5 = 4.16$. The observed value exceeds the criterion value and is declared significant. The next largest value in this row is 4.92, which is compared with $Q_4 = 3.89$ and is found to be significant. The next largest is 4.02, which is compared with $Q_3 = 3.52$ and is significant. The next value is 2.61, which when compared with 2.91 is not significant. We now proceed to the second row of the Q table, and compare 4.61 with $Q_4 = 3.89$, which is significant. The next value, 2.33, is not significant, and comparisons for that row are discontinued. Proceeding similarly, no other values in the table attain significance. Thus, in this example, the differences between III and I, between III and II, between III and IV, and between IV and I are significant at the .05 level. The reader should note that the above procedure requires that comparisons be discontinued for any row at the first nonsignificant Q value. This rule prevents the making of an inconsistent decision, that is, declaring that a larger difference is not significant whereas a smaller difference adjacent to it is significant. This circumstance could arise on occasion with the Newman-Keuls method if all Q values were compared directly with the criterion values.

The Duncan method, sometimes called the *Duncan new multiple-range test,* is a sequential test which is similar to the Newman-Keuls test. It differs, however, in the level of significance used. The Newman-Keuls test uses a significance level of .05 or .01 for $k = 2, 3, \ldots, k$. The Duncan method uses a significance level $1 - (1 - \alpha)^{k-1}$, where α is ordinarily .05 or .01. Thus for $\alpha = .01$ for two adjacent means, $k = 2$ and the significance level is $1 - .99 = .01$. For means separated by an intervening mean, $k = 3$ and the significance level is $1 - (.99)^2 = .02$. For means separated by two intervening means, $k = 4$ and the significance level is $1 - (.99)^3 = .03$. Thus the farther the means are apart in the rank ordering, the more lenient the standard of significance. The Duncan method requires a smaller studentized range for significance than does the Newman-Keuls method, except when $k = 2$. The method will yield more significant results than the Newman-Keuls method. The Duncan method requires the use of special tables.

Another multiple-comparison method is due to Tukey and has been called the *honestly significant difference method.* It uses a single criterion value of the studentized range, regardless of the separation of means in the rank order. It is a nonsequential method. The criterion value is Q_k, the value required for significance at the .05 or .01 levels for k and df, where k is the total number of means in the set and, as before, df is the number of degrees of freedom associated with s_w^2. This method will lead to fewer significant differences than either the Newman-Keuls or the Duncan method.

All multiple-comparison procedures that use the studentized range are limited in their application because they apply only to groups composed of equal n's. Bancroft (1968) has suggested an intuitive modification for dealing with unequal sample size. His proposal is to replace the n used in $s_{\bar{x}} = \sqrt{s_w^2/n}$ by $\bar{n}_h$, that is $s_{\bar{x}} = \sqrt{s_w^2/\bar{n}_h}$, where $\bar{n}_h$ is the harmonic mean of the number of observations in the groups, that is:

[18.13] $$\bar{n}_h = \frac{k}{(1/n_1) + (1/n_2) + \cdots + (1/n_k)}$$

Clearly this method should not be used where the n's differ appreciably from each other.

18.17 CONCLUDING OBSERVATIONS

The problem of choosing a particular a posteriori multiple-comparison procedure to apply to a particular set of experimental data has never been properly resolved. This choice is made by each investigator in the light of considerations which appear relevant to him or her. One consideration here is clearly the relative importance which attaches to Type I and Type II errors. The reader will recall that Type I error is the probability of asserting that a difference exists when no such difference exists. A Type II error is the probability of asserting that there is no difference when a difference does in fact exist. In terms of per-comparison Type I error, multiple-comparison procedures may be ordered from low to high as follows: Scheffé, Tukey, Newman-Keuls, and Duncan. The Scheffé method in any experiment will lead to the smallest number of significant differences whereas the Duncan method will lead to the largest number. In terms of Type II error, the order of the procedure is the reverse: Duncan, Newman-Keuls, Tukey, and Scheffé. Many investigators may wish to compromise between Type I and Type II errors and may choose to use the Scheffé procedure with a significance level greater than .05, say, .10, or the Newman-Keuls procedure.

BASIC TERMS AND CONCEPTS

Comparisons: multiple, a priori, a posteriori, orthogonal

Comparisons as correlations

Trend analysis

Orthogonal polynomials

Quadratic, cubic, and quartic relations

Scheffé method

Studentized range

Newman-Keuls method

Duncan multiple-range test

Tukey method

EXERCISES

1 Consider the following data:

	I	II	III	IV
n	15	15	15	15
T_j	87	160	188	66
$\bar{X}_j$	5.80	10.67	12.53	4.40
$\sum_{i=1}^{n} X_{ij}^2$	640	1,700	2,800	500

Apply a method of orthogonal comparisons to compare $\bar{X}_1$ with $\bar{X}_4$, $\bar{X}_2$ with $\bar{X}_3$, and $\bar{X}_1 + \bar{X}_4$ with $\bar{X}_2 + \bar{X}_3$.

2 Consider the following data:

	I	II	III
n	20	20	20
$\bar{X}_j$	7.40	10.50	12.65
$\sum_{i=1}^{n} X_{ij}^2$	2,156	2,150	2,760

Calculate (**a**) the within-group, linear-regression, and deviation sums of squares, (**b**) the three variance estimates, (**c**) F ratios for testing the significance of linear regression and the deviations from linear regression.

3 Obtain the slope of the linear regression line for the data of Exercise 1 above, assuming a unit difference between the levels of the treatment variable.

4 What are the coefficients for orthogonal polynomials c_{1j}, c_{2j}, and c_{3j} for $k = 6$?

5 Consider the following data:

	I	II	III	IV	V
n	10	10	10	10	10
$\bar{X}_j$	5.50	8.65	10.43	4.86	9.50
$\sum X_{ij}^2$	305	875	1,050	305	905

Calculate the linear-regression, quadratic-regression, cubic-regression, deviation, and within-groups variance estimates and apply the appropriate tests of significance.

6 Consider the following data:

	I	II	III	IV	V
n	5	5	5	5	5
$\bar{X}_j$	8.60	8.40	20.00	29.20	18.40

The within-group mean square $s_w^2 = 16.32$, and the overall F ratio is 23.81. Compare means two at a time by using Scheffe's procedure. Indicate which comparisons are significant at the .01 level.

7 Apply the Newman-Keuls procedure to the data of Exercise 6 and indicate which comparisons are significant at the .01 level.

8 Apply Scheffé's procedure to the data of Exercise 3, Chapter 15. Indicate which comparisons are significant at the .01 level.

ANSWERS TO EXERCISES

1

Source of variation	Sum of squares	Degrees of freedom	Variance estimate
Comparison 1	14.700	1	14.700
Comparison 2	26.133	1	26.133
Comparison 3	633.750	1	633.750
Within	782.067	56	13.965
Total	1,456.650	59	

$F_1 = \frac{14.700}{13.965} = 1.053$ $\qquad F_2 = \frac{26.133}{13.965} = 1.871$

$F_3 = \frac{633.750}{13.965} = 45.381\dagger$

† Highly significant.

2 **a** 565.35, 275.63, 3.00
b 9.92, 275.63, 3.00
c 27.79, .30

3 2.63

4 $k = 6$
C_{1j}: −5, −3, −1, 1, 3, 5
C_{2j}: 5, −1, −4, −4, −1, 5
C_{3j}: −5, 7, 4, −4, −7, 5

5

Linear:	17.72
Quadratic:	13.64
Cubic:	134.09
Deviation:	79.17
Within groups:	3.62

$F_1 = 4.90$ $(p < .05)$; $F_2 = 3.77$ $(p > .05)$; $F_3 = 37.04$ $(p < .01)$.

6

Comparison	F
I, II	.006
I, III	19.91
I, IV	65.01
I, V	14.71
II, III	20.61
II, IV	66.27
II, V	15.32
III, IV	12.97
III, V	.39
IV, V	17.87

$df_1 = 4$, $df_2 = 20$
$F' = 17.72$
Significant comparison $(p < .01)$
I, III; I, IV; II, III;
II, IV; IV, V

7

	II	I	V	III	IV
II		.111	5.535†	6.421†	11.513†
I			5.424†	6.310†	5.978†
V				.886	5.978†
III					5.092

$Q_2 = 4.02$ $Q_3 = 4.64$ $Q_4 = 5.02$ $Q_5 = 5.29$

† Significant at the .01 level.

8

Comparison	F
I and II	.26
I and III	.68
I and IV	2.06
II and III	2.14
II and IV	1.04
III and IV	5.90

$df_1 = 3$, $df_2 = 21$
$F'_{01} = 14.61$
No comparisons are significant at .01 level

19

REPEATED-MEASUREMENT AND OTHER EXPERIMENTAL DESIGNS

19.1 INTRODUCTION

In previous chapters the basic ideas involved in the design, analysis, and interpretation of one-way classification experiments and factorial experiments were discussed in considerable detail. In psychology and education considerable use is made, both through choice and necessity, of other experimental designs. The purpose of this chapter is to introduce the reader in an elementary way to a number of other experimental designs in common use. Some of these designs involve assumptions and present problems that are not involved in designs previously discussed. Some awareness and understanding of these assumptions and problems are essential to the proper use of these designs.

Research workers in psychology and education make frequent use of experimental designs in which measurements are repeated a number of times on the same subjects. These designs and the assumptions underlying their use are described in some detail in this chapter. Randomized block designs, designs with nested factors, and Latin square designs are also described.

The treatment of these designs in this chapter must of necessity be elementary. Some of these designs can be combined and extended in a variety of ways leading to experiments of much complexity. Because of the ready availability of computers for data analysis, the current trend in psychology and education is toward more complex designs. Questions can be raised regarding the merits of this trend toward complexity. A much more advanced treatment of the topics discussed in this chapter will be found in books by Winer (1971), Myers (1979), and other authors.

19.2 EXPERIMENTS WITH REPEATED MEASUREMENTS

Many experiments in psychology and education require the repeated measurement of the same subjects under a number of different conditions or treatments. Such experiments may be single-factor experiments in which each subject is tested or measured under a number of different experimental conditions. The simplest experiment of this kind would be one in which the same subjects are tested under two experimental conditions. Sometimes, when the same subjects are tested under a number of different treatments, the order of the presentation of treatments to subjects is randomized independently for each subject or a systematic plan for the ordering of the presentation of treatments to subjects is adopted. The purpose of either randomization or the use of a systematic plan for the ordering of treatments is to eliminate effects which might result from the order of the treatments. In some situations randomization is not appropriate because the different levels of the treatment variable have a natural order. This is the case where performance is measured at different time intervals, as, for example, in the study of changes in dark adaptation with time, or for different numbers of trials in a simple learning experiment.

Experiments of the type described above are called *one-factor experiments with repeated measurements.* In such experiments N subjects are measured under k conditions or treatments. The matrix of data thus obtained is a table of numbers containing N rows and k columns.

Repeated measurements may, however, be used in two-way classification or higher-order factorial experiments. For example, in a 2×2 factorial experiment four treatment combinations exist, four groups of experimental subjects are used, and each combination is applied to a different group of subjects. Experiments may be designed in which each of the N subjects receives all four treatment combinations. The matrix of data is a block of numbers containing two rows, two columns, and N layers, each layer corresponding to a subject. In general, for an $R \times C$ factorial experiment with repeated measurements the matrix of data is an $R \times C \times N$ block of numbers. The idea involved here can be extended to higher-order factorial experiments.

Two-way classification experiments may also be conducted with repeated measurements over one factor but not over the other factor. Consider an experiment involving two factors with three levels of one factor and two levels of the other. If this were an independent-groups factorial experiment, six treatment combinations and six groups of subjects would be used. The investigator may, however, decide to use two groups of subjects, with each of the two groups receiving only one level of one factor but all three levels of the other. Thus the experiment has repeated measurements over the factor with three levels, but not over the factor with two levels.

Experiments with repeated measurements have advantages and disadvantages. One advantage is that the measurements obtained under the different treatment conditions will in many experiments be highly correlated since they are made on the same subjects. The presence of these correlations will reduce the error term. Another advantage resides in the number of subjects. It may be more economical in terms of time and effort to test the same subjects under each treatment. A further point here is that the nature of certain experimental problems demands the use of repeated-measurement designs. One disadvantage of experiments with repeated measurements is that performance under prior treatments may affect performance under subsequent treatments due to either fatigue, practice, boredom, or some other circumstance. Effects resulting from such circumstances are sometimes called carry-over effects. An investigator may not be able to clearly decide whether the results observed under the different treatments are due to those treatments or are due to the carry-over effects. A further problem associated with repeated measurement designs is the assumption made in the analysis of data. This matter is discussed in some detail in Section 19.10.

19.3 ONE-FACTOR EXPERIMENTS WITH REPEATED MEASUREMENTS: COMPUTATION AND EXPECTATION OF MEAN SQUARES

As indicated above, the data resulting from a one-factor experiment with repeated measurements may be represented as a table of numbers in which rows represent experimental subjects and columns represent treatments; that is, the representation of the data is the same as that for the two-way classification with one observation per cell. The analysis of such data involves nothing new. The data are analyzed as in the two-way classification case with one observation per cell. The required computation formulas are given in Section 16.9. Three sums of squares result: sums of squares for subjects (rows), treatments (columns), and interaction.

For a one-factor experiment with repeated measurements, subjects constitute a random variable and treatments are usually viewed as fixed. The model is the mixed model for $n = 1$. The expectations of the mean squares are as follows:

Mean squares	Expectation of mean squares
Subjects, s_r^2	$\sigma_c^2 + C\sigma_a^2$
Treatments, s_c^2	$\sigma_e^2 + \sigma_{ab}^2 + R\sigma_b^2$
Interaction, s_{rc}^2	$\sigma_e^2 + \sigma_{ab}^2$

The proper error term for testing differences between treatments is s_{rc}^2; that is, $F_c = s_c^2/s_{rc}^2$. No unbiased test of individual differences between subjects is possible, unless it is assumed that the interaction term is 0. With nearly all sets of data this assumption is not warranted, because the performance of subjects under different pairs of treatments is correlated. Ordinarily in most experiments of this type individual differences between subjects are of limited interest anyway, because with most variables that are the object of study the investigator expects a priori substantial differences between subjects.

19.4 ILLUSTRATIVE EXAMPLE OF ONE-FACTOR EXPERIMENT WITH REPEATED MEASUREMENTS

Table 19.1 shows hypothetical data for a one-factor experiment with repeated measurements. Rows are individuals, and columns are treatments. The data are presumed to relate to a random sample of individuals tested under different treatment conditions. This is a mixed model. One basis of classification, the columns, is fixed. The other basis of classification, the rows, is random.

Table 19.1

Data for the analysis of variance with two-way classification: $n = 1$, scores for a sample of subjects tested under four different conditions

	Conditions					
Subject	A	B	C	D	$T_{r.}$	$\bar{X}_{r.}$
1	31	42	14	80	167	41.75
2	42	26	25	106	199	49.75
3	84	21	19	83	207	51.75
4	26	60	36	69	191	47.75
5	14	35	44	48	141	35.25
6	16	80	28	76	200	50.00
7	29	49	80	39	197	49.25
8	32	38	76	84	230	57.50
9	45	65	15	91	216	54.00
10	30	71	82	39	222	55.50
$T_{.c}$	349	487	419	715		$T = 1{,}970$
$\bar{X}_{.c}$	34.90	48.70	41.90	71.50		$\bar{X}_{..} = 49.25$

$\sum_{r=1}^{R} T_{r.}^2 = 394{,}350$ $\quad \sum_{c=1}^{C} T_{.c}^2 = 1{,}045{,}756$ $\quad \sum_{r=1}^{R} \sum_{c=1}^{C} X_{rc}^2 = 122{,}984$

Applying the appropriate computation formulas, the following sums of squares are obtained:

ROWS

$$\frac{1}{C}\sum_{r=1}^{R} T_{r.}^2 - \frac{T^2}{N} = \frac{394{,}350}{4} - \frac{(1{,}970)^2}{40} = 1{,}565.00$$

COLUMNS

$$\frac{1}{R}\sum_{c=1}^{C} T_{.c}^2 - \frac{T^2}{N} = \frac{1{,}045{,}756}{10} - \frac{(1{,}970)^2}{40} = 7{,}553.10$$

INTERACTION

$$\sum_{r=1}^{R}\sum_{c=1}^{C} X_{rc}^2 - \frac{1}{C}\sum_{r=1}^{R} T_{r.}^2 - \frac{1}{R}\sum_{c=1}^{C} T_{.c}^2 + \frac{T^2}{N}$$

$$= 122{,}984 - \frac{394{,}350}{4} - \frac{1{,}045{,}756}{10} + \frac{(1{,}970)^2}{40}$$

$$= 16{,}843.40$$

TOTAL

$$\sum_{r=1}^{R}\sum_{c=1}^{C} X_{rc}^2 - \frac{T^2}{N} = 122{,}984 - \frac{(1{,}970)^2}{40} = 25{,}961.50$$

Table 19.2 summarizes the analysis-of-variance data for this example. Because this is a mixed model with $n = 1$ and $F_r = s_r^2/s_{rc}^2 = .279$, no meaningful test of row effects is possible. The proper error term for column effects is s_{rc}^2. The F ratio for column effects is found to be 4.04. The F ratios required for significance with 3 and 27 degrees of freedom associated with the numerator and denominator, respectively, are 2.96 at the 5 percent and 4.60 at the 1 percent levels. Thus the column differences

Table 19.2

Analysis of variance for data of Table 19.1

Source of variation	Sum of squares	Degrees of freedom	Variance estimate
Rows	1,565.00	9	$173.89 = s_r^2$
Columns	7,553.10	3	$2{,}517.70 = s_c^2$
Interaction	16,843.40	27	$623.83 = s_{rc}^2$
Total	25,961.50		

$F_c = \frac{s_c^2}{s_{rc}^2} = 4.04 \qquad F_r = \frac{s_r^2}{s_{rc}^2} = .279$

are significant at the 5 percent level but fall short of significance at the 1 percent level.

19.5 TWO-FACTOR EXPERIMENTS WITH REPEATED MEASUREMENTS

In Sections 19.3 and 19.4 one-factor experiments with repeated measurements were considered. On occasion experiments are encountered that involve two factors with repeated measurements. Given R levels of one treatment and C levels of another, each subject may be tested under each of the RC treatments. If $R = 2$ and $C = 2$, the levels of R being R_1 and R_2 and of C being C_1 and C_2, there are four treatment combinations, R_1C_1, R_1C_2, R_2C_1, and R_2C_2. Each of N subjects might receive all the four treatments, the presentations being possibly, although not necessarily, arranged in random order for each subject.

Such data constitute an RCN block of numbers. Rows and columns are treatments, and layers are experimental subjects. These data are analyzed as in the triple-classification case with one observation in each cell. Use the computation formulas given in Section 17.8, writing $n = 1$. Seven sums of squares result: rows, columns, subjects, $R \times C$, $R \times S$, $C \times S$, and $R \times C \times S$. There is, of course, no within-cells sum of squares.

The model here is a mixed model with $n = 1$. Rows and columns will ordinarily be fixed variables. Layers, or subjects, is a random variable. For this model the expectation of the sums of squares and the degrees of freedom are as follows:

Mean square	Expectation of mean square	*df*
Rows, s_r^2	$\sigma_e^2 + C\sigma_{ac}^2 + NC\sigma_a^2$	$R - 1$
Columns, s_c^2	$\sigma_e^2 + R\sigma_{bc}^2 + NR\sigma_b^2$	$C - 1$
Subjects, s_s^2	$\sigma_e^2 + RC\sigma_c^2$	$N - 1$
$R \times C$, s_{rc}^2	$\sigma_e^2 + \sigma_{abc}^2 + N\sigma_{ab}^2$	$(R - 1)(C - 1)$
$R \times S$, s_{rs}^2	$\sigma_e^2 + C\sigma_{ac}^2$	$(R - 1)(N - 1)$
$C \times S$, s_{cs}^2	$\sigma_e^2 + R\sigma_{bc}^2$	$(C - 1)(N - 1)$
$R \times C \times S$, s_{rcs}^2	$\sigma_e^2 + \sigma_{abc}^2$	$(R - 1)(C - 1)(N - 1)$

Inspection of these expectations indicates that the appropriate error term for testing row effects is the $R \times S$ mean square, $F_r = s_r^2/s_{rs}^2$. The appropriate error term for testing column effects is the $C \times S$ mean square, $F_c = s_c^2/s_{cs}^2$. The appropriate error term for testing $R \times C$ interaction is the $R \times C \times S$ mean square, $F_{rc} = s_{rc}^2/s_{rcs}^2$. Unless the $R \times S$, $C \times S$, and $R \times C \times S$ interactions are assumed to be 0, which with most sets of data will not be the case, no unbiased test of differences between subjects, or $R \times S$ or $C \times S$ interactions, can be made. These are ordinarily not of interest.

19.6 ILLUSTRATIVE EXAMPLE OF TWO-FACTOR EXPERIMENT WITH REPEATED MEASUREMENTS

The following are fictitious illustrative data for a repeated-measurement experiment with six experimental subjects tested under $2 \times 3 = 6$ treatment combinations.

Subject 1

	C_1	C_2	C_3	
R_1	4	5	7	16
R_2	1	4	2	7
	5	9	9	

Subject 2

	C_1	C_2	C_3	
R_1	6	8	10	24
R_2	3	6	6	15
	9	14	16	

Subject 3

	C_1	C_2	C_3	
R_1	1	6	5	12
R_2	3	5	4	12
	4	11	9	

Subject 4

	C_1	C_2	C_3	
R_1	2	10	12	24
R_2	1	4	7	12
	3	14	19	

Subject 5

	C_1	C_2	C_3	
R_1	5	10	10	25
R_2	5	6	5	16
	10	16	15	

Subject 6

	C_1	C_2	C_3	
R_1	1	7	8	16
R_2	2	8	7	17
	3	15	15	

For computational purposes it is necessary to write down the totals for rows by columns summed over subjects, rows by subjects summed over columns, and subjects by columns summed over rows. Viewing the data as a cube of numbers, these are the numbers on the surface of the cube. The totals for rows by columns summed over subjects, $T_{rc.}$, are as follows:

	$T_{rc.}$ Columns			$T_{r..}$
Rows	19	46	52	117
	15	33	31	79
$T_{.c.}$	34	79	83	$196 = T_{...}$

The totals above are obtained by adding the cell totals over subjects. Thus $4 + 6 + 1 + 2 + 5 + 1 = 19$, and so on. Totals for rows $T_{r..}$, for columns $T_{.c.}$, and the grand total are shown. The totals for rows by subjects summed over columns are as follows:

$T_{r.s}$
Subjects

							$T_{r..}$
Rows	16	24	12	24	25	16	117
	7	15	12	12	16	17	79
$T_{.s.}$	23	39	24	36	41	33	$196 = T_{...}$

Here the number 16 in the top left cell is obtained by summing the cells for the first subject over columns; thus $4 + 5 + 7 = 16$. Likewise $1 + 4 + 2 = 7$, and so on. The totals for columns by subjects summed over rows, $T_{.cs}$, are as follows:

$T_{.cs}$
Subjects

							$T_{.c.}$
	5	9	4	3	10	3	34
Columns	9	14	11	14	16	15	79
	9	16	9	19	15	15	83
$T_{..s}$	23	39	24	36	41	33	$196 = T_{...}$

Here the values in the left-hand column in the above table are obtained by summing the cells for the first subject over rows; thus $4 + 1 = 5$, $5 + 4 = 9$, $7 + 2 = 9$, and so on.

Use is now made of the computational formulas given in Section 17.8. In the present example $n = 1$. Also a slight notational change has been made. In this example layers are subjects and the symbol S is used instead of L. In the formulas to follow S is the same as L in the formulas of Section 17.8. The factor S has N levels, where N is the number of subjects.

First we calculate *eight* quantities which are used in the computation formulas. For this illustrative example these are as follows:

$$\frac{1}{CN} \sum^{R} T_{r..}^2 = \frac{1}{3 \times 6} \times 19{,}930 = 1{,}107.22$$

$$\frac{1}{RN} \sum^{C} T_{.c.}^2 = \frac{1}{2 \times 6} \times 14{,}286 = 1{,}190.50$$

$$\frac{1}{RC} \sum^{N} T_{..s}^2 = \frac{1}{2 \times 3} \times 6{,}692 = 1{,}115.33$$

$$\frac{1}{N} \sum^{R} \sum^{C} T_{rc.}^2 = \frac{1}{6} \times 7{,}456 = 1{,}242.67$$

$$\frac{1}{C}\sum^{R}\sum^{N} T^2_{r.s} = \frac{1}{3} \times 3{,}540 = 1{,}180.00$$

$$\frac{1}{R}\sum^{C}\sum^{N} T^2_{.cs} = \frac{1}{2} \times 2{,}544 = 1{,}272.00$$

$$\sum^{R}\sum^{C}\sum^{N} X^2_{rcs} = 1{,}360.00$$

$$\frac{T^2}{RCN} = \frac{1}{36} \times 38{,}416 = 1{,}067.11$$

In the above computation the quantity $\sum^{R} T^2_{r..} = (117)^2 + (79)^2 = 19{,}930$; $\sum^{C} T^2_{.c.} = (34)^2 + (79)^2 + (83)^2 = 14{,}286$; and so on. Applying the computation formulas of Section 17.8, and remembering that $L = S$, the required sums of squares are as follows:

ROWS

$$\frac{1}{CN}\sum^{R} T^2_{r..} - \frac{T^2}{RCN} = 1{,}107.22 - 1{,}067.11 = 40.11$$

COLUMNS

$$\frac{1}{RN}\sum^{C} T^2_{.c.} - \frac{T^2}{RCN} = 1{,}190.50 - 1{,}067.11 = 123.39$$

SUBJECTS

$$\frac{1}{RC}\sum^{N} T^2_{..s} - \frac{T^2}{RCN} = 1{,}115.33 - 1{,}067.11 = 48.22$$

$R \times C$ INTERACTION

$$\frac{1}{N}\sum^{R}\sum^{C} T^2_{rc.} - \frac{1}{CN}\sum^{R} T^2_{r..} - \frac{1}{RN}\sum^{C} T^2_{.c.} + \frac{T^2}{RCN}$$

$$= 1{,}242.67 - 1{,}107.22 - 1{,}190.50 + 1{,}067.11 = 12.06$$

$R \times S$ INTERACTION

$$\frac{1}{C}\sum^{R}\sum^{N} T^2_{r.s} - \frac{1}{CN}\sum^{R} T^2_{r..} - \frac{1}{RC}\sum^{N} T^2_{..s} + \frac{T^2}{RCN}$$

$$= 1{,}180.00 - 1{,}107.22 - 1{,}115.33 + 1{,}067.11 = 24.56$$

$C \times S$ INTERACTION

$$\frac{1}{R}\sum^{C}\sum^{N} T^2_{.cs} - \frac{1}{RN}\sum^{C} T^2_{.c.} - \frac{1}{RC}\sum^{N} T^2_{..s} + \frac{T^2}{RCN}$$

$$= 1{,}272.00 - 1{,}190.50 - 1{,}115.33 + 1{,}067.11 = 33.28$$

$R \times C \times S$ INTERACTION

$$\sum^{R}\sum^{C}\sum^{N} X^2_{rcs} - \frac{1}{N}\sum^{R}\sum^{C} T^2_{rc.} - \frac{1}{C}\sum^{R}\sum^{N} T^2_{r.s} - \frac{1}{R}\sum^{C}\sum^{N} T^2_{.cs} + \frac{1}{CN}\sum^{R} T^2_{r..}$$

$$+ \frac{1}{RN}\sum^{C} T^2_{.c.} + \frac{1}{RC}\sum^{N} T^2_{..s} - \frac{T^2}{RCN} = 1{,}360.00 - 1{,}242.67$$

$$- 1{,}180.00 - 1{,}272.00 + 1{,}107.22 + 1{,}190.50$$

$$+ 1{,}115.33 - 1{,}067.11 = 11.27$$

TOTAL

$$\sum^{R}\sum^{C}\sum^{N} X^2_{rcs} - \frac{T^2}{RCN} = 1{,}360.00 - 1{,}067.11 = 292.89$$

The analysis-of-variance table for these data is given in Table 19.3. As indicated in Section 19.5 the appropriate error term for testing row effects is s_{rs}^2, for column effects s_{cs}^2, and for $R \times C$ interaction s^2_{rcs}. The F ratios are as follows:

$$F_r = \frac{s_r^2}{s_{rs}^2} = \frac{40.11}{4.91} = 8.17 \qquad p < .05$$

$$F_c = \frac{s_c^2}{s_{cs}^2} = \frac{61.70}{3.33} = 18.53 \qquad p < .01$$

$$F_{rc} = \frac{s_{rc}^2}{s^2_{rcs}} = \frac{6.03}{1.13} = 5.34 \qquad p < .05$$

In this illustrative example the column effects are significant at better than the .01 level, whereas row effects and $R \times C$ interaction fall between the .05 and .01 levels of significance.

Table 19.3

Analysis-of-variance table for illustrative examples of two-factor experiment with repeated measures

Source of variation	Sum of squares	Degrees of freedom	Variance estimate
Rows	40.11	1	$40.11 = s_r^2$
Columns	123.39	2	$61.70 = s_c^2$
Subjects	48.22	5	$9.64 = s_s^2$
$R \times C$	12.06	2	$6.03 = s_{rc}^2$
$R \times S$	24.56	5	$4.91 = s_{rs}^2$
$C \times S$	33.28	10	$3.33 = s_{cs}^2$
$R \times C \times S$	11.27	10	$1.13 = s^2_{rcs}$
Total	292.89	35	

19.7 TWO-FACTOR EXPERIMENTS WITH REPEATED MEASUREMENTS ON ONE FACTOR

A not uncommon type of experiment in psychological and educational research is a two-factor experiment with repeated measurements over one factor only. To illustrate, consider an experiment involving four different learning trials under two drug treatments. Two groups of n subjects each may be used. The first group may be tested on the four learning trials under the first drug treatment. The second group may be tested on the four learning trials under the second drug treatment. Designs of this type are sometimes called mixed designs, but this term should not be confused with mixed models, where the mixing is with respect to fixed and random factors rather than repeated- and nonrepeated-measurement factors.

The notation for such an experiment may be illustrated in the particular case where two experimental groups of three subjects are used with each subject measured under four experimental conditions. The data may be represented as follows:

		Subjects	C_1	C_2	C_3	C_4	Means
Group 1	R_1	1	X_{111}	X_{121}	X_{131}	X_{141}	$\bar{X}_{1.1}$
		2	X_{112}	X_{122}	X_{132}	X_{142}	$\bar{X}_{1.2}$
		3	X_{113}	X_{123}	X_{133}	X_{143}	$\bar{X}_{1.3}$
		Means	$\bar{X}_{11.}$	$\bar{X}_{12.}$	$\bar{X}_{13.}$	$\bar{X}_{14.}$	$\bar{X}_{1..}$
Group 2	R_2	4	X_{214}	X_{224}	X_{234}	X_{244}	$\bar{X}_{2.4}$
		5	X_{215}	X_{225}	X_{235}	X_{245}	$\bar{X}_{2.5}$
		6	X_{216}	X_{226}	X_{236}	X_{246}	$\bar{X}_{2.6}$
		Means	$\bar{X}_{21.}$	$\bar{X}_{22.}$	$\bar{X}_{23.}$	$\bar{X}_{24.}$	$\bar{X}_{2..}$
		Means	$\bar{X}_{.1.}$	$\bar{X}_{.2.}$	$\bar{X}_{.3.}$	$\bar{X}_{.4.}$	$\bar{X}_{...}$

Here triple subscripts are used. The first subscript identifies the row or group to which the subject belongs, the second subscript identifies the column or the level of the repeated measurement, the third subscript identifies the subject. For example, X_{214} is a measurement for the fourth subject in the second group at the first level of the repeated measurement.

For this type of experimental design the total sum of squares may be partitioned into two parts, a between-subjects and a within-subjects sum of squares. The between-subjects sum of squares can be further partitioned into two parts, a row sum of squares and a subjects-within-groups sum of squares. Denote this latter term by S/R. The within-subjects sum of squares can be further partitioned into three parts, a column sum of squares, a row-by-column interaction, and a third part which is a column-by-subject interaction pooled over groups or rows. Denote this latter term by $(S \times C)/R$. Thus, in effect, the total sum of squares is partitioned into five separate sums of squares. These sums of squares with the associated number of degrees of freedom and variance estimates are shown in Table 19.4.

Some comment on the sums of squares in Table 19.4 is appropriate. The meaning of the row, column, and interaction sums of squares is obvious. These are concerned with variability due to the main effects and the interaction between the main effects. The subjects-within-groups sum of squares, S/R, is simply the variability among subjects for the first group, added to the variability among subjects for the second group, and so on for all levels of R. It may be viewed as the variability among subjects with the variability due to row treatment effects, as it were, removed. The $(C \times S)/R$ term is a column-by-subject interaction for the first group, added to the column-by-subject interaction for the second group, and so on for all levels of R.

The numbers of degrees of freedom associated with row, column, and $R \times C$ interaction sums of squares are $R - 1$, $C - 1$, and $(R - 1)(C - 1)$, respectively. The S/R term has associated with it $n - 1$ degrees of freedom for each group, and for R groups the number of degrees of freedom is $R(n - 1)$. The $(C \times S)/R$ term involves the summing of the $C \times S$ interaction over R groups or levels. The number of degrees of freedom associated with each level is $(n - 1)(C - 1)$; consequently the total number of degrees of freedom associated with this term is $R(n - 1)(C - 1)$.

Table 19.4

Analysis of variance for two-factor experiments with repeated measurements on one factor

Source of variation	Sum of squares	*df*	Variance estimate
Between subjects	$C \sum^{R} \sum^{n} (\bar{X}_{r.s} - \bar{X}_{...})^2$	$Rn - 1$	s_b^2
Rows	$nC \sum^{R} (\bar{X}_{r..} - \bar{X}_{...})^2$	$R - 1$	s_r^2
S/R	$C \sum^{R} \sum^{n} (\bar{X}_{r.s} - \bar{X}_{r..})^2$	$R(n - 1)$	$s_{s/r}^2$
Within subjects	$\sum^{C} \sum^{R} \sum^{n} (X_{rci} - \bar{X}_{r.s})^2$	$Rn(C - 1)$	s_w^2
Columns	$nR \sum^{C} (\bar{X}_{.c.} - \bar{X}_{...})^2$	$C - 1$	s_c^2
$R \times C$	$n \sum^{R} \sum^{C} (\bar{X}_{rc.} - \bar{X}_{r..} - \bar{X}_{.c.} + \bar{X}_{...})^2$	$(R - 1)(C - 1)$	s_{rc}^2
$(C \times S)/R$	$\sum^{C} \sum^{R} \sum^{n} (X_{rci} - \bar{X}_{r.s} - \bar{X}_{rc.} + \bar{X}_{r..})^2$	$R(n - 1)(C - 1)$	$s_{cs/r}^2$
Total	$\sum^{R} \sum^{C} \sum^{n} (X_{rci} - \bar{X}_{...})^2$	$RCn - 1$	

The expectations of the variance estimates or mean squares for a two-factor experiment with repeated measurements on one factor, assuming the row and column treatment variables to be fixed, are as follows:

ROWS, s_r^2	$\sigma_e^2 + C\sigma_s^2 + nC\sigma_a^2$
S/R, $s_{s/r}^2$	$\sigma_e^2 + C\sigma_s^2$
COLUMNS, s_c^2	$\sigma_e^2 + \sigma_{bs}^2 + nR\sigma_b^2$
$R \times C$, s_{rc}^2	$\sigma_e^2 + \sigma_{bs}^2 + n\sigma_{ab}^2$
$(C \times S)/R$, $s_{cs/r}^2$	$\sigma_e^2 + \sigma_{bs}^2$

The quantities σ_a^2, σ_b^2, σ_{ab}^2, and σ_e^2 are variance components as described in Section 16.6. The quantity σ_s^2 is a variance component which is due to the variation of subjects within groups.

Examination of the above expectations indicates that the correct error term for testing row effects is the S/R variance estimate. The correct error term for testing column and $R \times C$ interaction effects is the $(C \times S)/R$ variance estimate.

19.8 COMPUTATION FORMULAS FOR TWO-FACTOR EXPERIMENTS WITH REPEATED MEASUREMENTS ON ONE FACTOR

Computation formulas may readily be obtained to compute the required sums of squares. As previously the sum of all nRC observations will be denoted by T. We denote the row and column totals by $T_{r..}$ and $T_{.c.}$, respectively. The total for the rth row and cth column summed over subjects is $T_{rc.}$. The total for any subject in any row summed over columns is $T_{r.s}$. The total for n subjects for any group summed over columns is $T_{rc.}$. Given this notation, the computation formulas are as follows:

BETWEEN SUBJECTS

[19.1] $$\frac{1}{C}\sum^{R}\sum^{n} T_{r.s}^2 - \frac{T^2}{nRC}$$

ROWS

[19.2] $$\frac{1}{nC}\sum^{R} T^2_{r..} - \frac{T^2}{nRC}$$

S/R

[19.3] $$\frac{1}{C}\sum^{R}\sum^{n} T^2_{r.s} - \frac{1}{nC}\sum^{R} T^2_{r..}$$

WITHIN SUBJECTS

[19.4] $$\sum^{R}\sum^{C}\sum^{n} X^2_{rci} - \frac{1}{C}\sum^{R}\sum^{n} T^2_{r.s}$$

COLUMNS

[19.5] $$\frac{1}{nR}\sum^{C} T^2_{.c.} - \frac{T^2}{nRC}$$

$R \times C$

[19.6] $$\frac{1}{n}\sum^{R}\sum^{C} T^2_{rc.} - \frac{1}{nC}\sum^{R} T^2_{r..} - \frac{1}{nR}\sum^{C} T^2_{.c.} + \frac{T^2}{nRC}$$

SC/R

[19.7] $$\sum^{R}\sum^{C}\sum^{n} X^2_{rci} - \frac{1}{C}\sum^{R}\sum^{n} T^2_{r.s} - \frac{1}{n}\sum^{R}\sum^{C} T^2_{rc.} + \frac{1}{nC}\sum^{R} T^2_{r..}$$

TOTAL

[19.8] $$\sum^{R}\sum^{C}\sum^{n} X^2_{rci} - \frac{T^2}{nRC}$$

In practical computation a number of these terms can be obtained by simple subtraction.

19.9 ILLUSTRATIVE EXAMPLE OF A TWO-FACTOR EXPERIMENT WITH REPEATED MEASUREMENTS ON ONE FACTOR

Table 19.5 shows illustrative data for a two-factor experiment with repeated measurements on one factor. Two groups of four subjects were used. Each subject was measured under five experimental conditions. The totals required for computational purposes are also shown in this table.

Table 19.5

Illustrative data for a two-factor experiment with repeated measurements on one factor

		Subjects	C_1	C_2	C_3	C_4	C_5	$T_{r.s}$
Group 1	R_1	1	2	7	6	7	9	31
		2	4	3	7	12	14	40
		3	7	6	4	12	10	39
		4	1	'3	3	6	6	19
		$T_{rc.}$	14	19	20	37	39	$T_{1..} = 129$
Group 2	R_2	1	4	4	7	9	1	25
		2	10	12	12	12	16	62
		3	8	7	8	12	10	45
		4	5	7	6	7	8	33
		$T_{rc.}$	27	30	33	40	35	$T_{2..} = 165$
		$T_{.c.}$	41	49	53	77	74	$T = 294$

From this table *six* quantities can be readily calculated as follows:

$$\frac{1}{C}\sum^{R}\sum^{n} T_{r.s}^2 = \tfrac{1}{5}[(31)^2 + (40)^2 + \cdots + (33)^2] = 2{,}405.20$$

$$\frac{1}{nC}\sum^{R} T_{r..}^2 = \frac{1}{4\times 5}[(129)^2 + (165)^2] = 2{,}193.30$$

$$\frac{1}{nR}\sum^{C} T_{.c.}^2 = \frac{1}{4\times 2}[(41)^2 + (49)^2 + \cdots + (74)^2] = 2{,}287.00$$

$$\frac{1}{n}\sum^{R}\sum^{C} T_{rc.}^2 = \tfrac{1}{4}[(14)^2 + (19)^2 + \cdots + (35)^2] = 2{,}347.50$$

$$\sum^{R}\sum^{C}\sum^{n} X_{rci}^2 = (2)^2 + (7)^2 + \cdots + (8)^2 = 2{,}664.00$$

$$\frac{T^2}{nRC} = \frac{(294)^2}{4\times 2\times 5} = 2{,}160.90$$

Applying the computation formulas given in Sec. 19.8, the required sums of squares are as follows:

BETWEEN SUBJECTS

$$\frac{1}{C}\sum^{R}\sum^{n} T_{r.s}^2 - \frac{T^2}{nRC} = 2{,}405.20 - 2{,}160.90 = 244.30$$

ROWS

$$\frac{1}{nC}\sum^{R} T_{r..}^2 - \frac{T^2}{nRC} = 2{,}193.30 - 2{,}160.90 = 32.40$$

S/R

$$\frac{1}{C}\sum^{R}\sum^{n} T^2_{r.s} - \frac{1}{nC}\sum^{R} T^2_{r..} = 2{,}405.20 - 2{,}193.30 = 211.90$$

WITHIN SUBJECTS

$$\sum^{R}\sum^{C}\sum^{n} X^2_{rci} - \frac{1}{C}\sum^{R}\sum^{n} T^2_{r.s} = 2{,}664.00 - 2{,}405.20 = 258.80$$

COLUMNS

$$\frac{1}{nR}\sum^{C} T^2_{.c.} - \frac{T^2}{nRC} = 2{,}287.00 - 2{,}160.90 = 126.10$$

$R \times C$

$$\frac{1}{n}\sum^{R}\sum^{C} T^2_{rc.} - \frac{1}{nC}\sum^{R} T^2_{r..} - \frac{1}{nR}\sum^{C} T^2_{.c.} + \frac{T^2}{nRC}$$
$$= 2{,}347.50 - 2{,}193.30 - 2{,}287.00 + 2{,}160.90 = 28.10$$

SC/R

$$\sum^{R}\sum^{C}\sum^{n} X^2_{rci} - \frac{1}{C}\sum^{R}\sum^{n} T^2_{r.s} - \frac{1}{n}\sum^{R}\sum^{C} T^2_{rc.} + \frac{1}{nC}\sum^{R} T^2_{r..}$$
$$= 2{,}664.00 - 2{,}405.20 - 2{,}347.50 + 2{,}193.30 = 104.60$$

TOTAL

$$\sum^{R}\sum^{C}\sum^{n} X^2_{rci} - \frac{T^2}{nRC} = 2{,}664.00 - 2{,}160.90 = 503.10$$

The analysis-of-variance table for these data is shown in Table 19.6. The degrees of freedom for rows are $R - 1 = 2 - 1 = 1$; for columns are, $C - 1 = 5 - 1 = 4$; and for $R \times C$ interaction are, $(R - 1)(C - 1) =$

Table 19.6

Analysis of variance for the data of Table 19.5

Source	Sum of squares	Degrees of freedom	Variance estimate
Between subjects	244.30		
Rows	32.40	1	$32.40 = s_r^2$
S/R	211.90	6	$35.32 = s^2_{s/r}$
Within subjects	258.80		
Columns	126.10	4	$31.53 = s_c^2$
$R \times C$	28.10	4	$7.03 = s_{rc}^2$
SC/R	104.60	24	$4.36 = s^2_{sc/r}$
Total	503.10	39	

$(2-1)(5-1) = 4$. The S/R term is the variation within subjects summed over groups, and the degrees of freedom are $R(n-1) = 2(4-1) = 6$. The SC/R term is the subject-by-column interactions summed over rows, and the degrees of freedom are $R(n-1)(C-1) = 2(4-1)(5-1) = 24$.

The appropriate error term for testing row effects is the S/R variance estimate. The appropriate error term for testing column effects and $R \times C$ interaction is the SC/R variance estimate. The F ratios are as follows:

$$F_r = \frac{s_r^2}{s_{s/r}^2} = \frac{32.40}{35.32} = .92 \qquad p > .05$$

$$F_c = \frac{s_c^2}{s_{sc/r}^2} = \frac{31.53}{4.36} = 7.23 \qquad p < .01$$

$$F_{rc} = \frac{s_{rc}^2}{s_{sc/r}^2} = \frac{28.10}{4.36} = 6.44 \qquad p < .01$$

In this illustrative example the row effects are not significant, whereas the column effects, row-by-column effects, and row-by-column interaction are significant at better than the .01 level.

19.10 ASSUMPTIONS UNDERLYING REPEATED MEASUREMENT DESIGNS

Questions may be raised regarding the assumptions involved in the valid use of the F test in repeated measurement designs. A discussion of these assumptions requires consideration of the matrix of covariances between treatments. To illustrate, in a one-factor repeated measurement design, measurements are made on N subjects under C treatments. The covariation between measurements made under different treatments may be studied. For example, if $C = 4$ the covariance table may be represented as follows:

	1	2	3	4
1	s_1^2	$r_{12}s_1s_2$	$r_{13}s_1s_3$	$r_{14}s_1s_4$
2	$r_{12}s_1s_2$	s_2^2	$r_{23}s_2s_3$	$r_{24}s_2s_4$
3	$r_{13}s_1s_3$	$r_{23}s_2s_3$	s_3^2	$r_{34}s_3s_4$
4	$r_{14}s_1s_4$	$r_{24}s_2s_4$	$r_{34}s_3s_4$	s_4^2

In this table variances, s_i^2, appear along the main diagonal and covariances, $r_{ij}s_is_j$, on either side of the main diagonal. If it is assumed that all the variances are equal, and also that all the covariances in the population sampled, then the population matrix may be written as follows:

	1	2	3	4
1	σ^2	$\rho\sigma^2$	$\rho\sigma^2$	$\rho\sigma^2$
2	$\rho\sigma^2$	σ^2	$\rho\sigma^2$	$\rho\sigma^2$
3	$\rho\sigma^2$	$\rho\sigma^2$	σ^2	$\rho\sigma^2$
4	$\rho\sigma^2$	$\rho\sigma^2$	$\rho\sigma^2$	σ^2

Here ρ is a population correlation coefficient, and $\rho\sigma^2$ a population covariance. When all variances and also all covariances are equal, that property is described as compound symmetry.

The statement has often been made that the valid use of the F test in repeated measurement designs required the assumption of compound symmetry. Huynh and Feldt (1970) have shown that under multivariate normality, this assumption is sufficient but not necessary. This means that the F test is valid under compound symmetry but may be valid under less stringent conditions as well. The sufficient and necessary condition is the equality of variances of differences between all treatment pairs. If $\sigma^2_{i-j} = \sigma_i^2 + \sigma_j^2 - 2\rho_{ij}\sigma_i\sigma_j$ is the variance of differences between treatment pairs i and j, this requirement is that all σ^2_{i-j} are equal. Methods exist for testing whether a sample covariance matrix departs significantly from compound symmetry, and also for testing the equality of the σ^2_{i-j}s.

Many sample covariance matrices will clearly not conform to the requirements of compound symmetry or the equality of variance. How does this affect the use of the F test used in evaluating the ratios of mean squares? Work by Box (1954) gives rise to practical procedures for dealing with this problem. Box has shown that the ratio of mean squares used in the simple repeated measurement design has an F distribution with $\epsilon(C - 1)$ and $\epsilon(C - 1)\ (R - 1)$ degrees of freedom. In the notation used here $R = N$ the number of subjects. The parameter ϵ, epsilon, is descriptive of the departure of the covariance matrix from compound symmetry. It varies from a maximum of $\epsilon = 1$, to $\epsilon = 1/(C - 1)$ as its lower bound. The parameter ϵ may be estimated from the sample covariance matrix. For estimates of $\epsilon < 1$, the F distribution may be validly used with a reduced number of degrees of freedom. For example, if $C = 5$, $R = N = 10$ and an estimated ϵ of .75, the Type I probability would be estimated from the F distribution using $.75\ (5 - 1) = 3$, and $.75(5 - 1)(10 - 1) = 27$ degrees of freedom, instead of the unreduced 4 and 36.

Without estimating ϵ, a rule-of-thumb procedure suggested by Geisser and Greenhouse (1958) may be applied, which is useful in practical situations. Since ϵ extends from 1 to $1/(C - 1)$, the number of degrees of freedom associated with F will extend from $(C - 1)$ and $(C - 1)(R - 1)$ when $\epsilon = 1$, to 1 and $(R - 1)$ when $\epsilon = 1/(C - 1)$. Proceed as follows: *First,* test for column effects using 1 degree of freedom associated with the numerator and $(R - 1)$ with the denominator. If the result is significant at the desired level, no further test is required because this is a conservative procedure and works against obtaining a significant difference. *Second,* if the result is not significant, test the F ratio using $(C - 1)$ and $(C - 1)(R - 1)$ degrees of freedom. If this is not significant, no further test is required because this is a liberal procedure and works in the direction of too many significant dfffer-ences. *Third,* if the conservative procedure indicates that the result is not significant and the liberal procedure indicates that it is significant, then ϵ must be estimated and the degrees of freedom adjusted accordingly.

The rule-of-thumb procedure described above may be extended to other repeated measurement designs. In a two-factor experiment with repeated measurements, proceed as follows. *First,* row and column effects are tested using an F ratio with 1 and $N-1$ degrees of freedom. If the result is significant, no further test is required. *Second,* if the result is not significant, proceed under the assumption that $\epsilon = 1$, using the unreduced *df.* If the result is not significant, no further test is required. *Third,* if these procedures do not permit a decision, an estimate of ϵ is required. Procedures of the above type may be extended to other repeated measurement designs.

Estimating ϵ from the covariance matrix is computationally quite simple. The procedure is described in Winer (1971) and Myers (1979). See also Huynh and Feldt (1976) for a discussion of bias in the estimation of ϵ and a proposed alternative method of estimation. When ϵ is estimated from the data, and reduced *df* used in the F test, the *df* may be fractional rather than whole numbers. Presumably the *df* may be rounded to the nearest whole numbers, or interpolation used. This may be untrustworthy if the *df* are small.

Procedures other than those described above may be used—see Huynh and Mandeville (1979). Further, an adjustment such as the Box test should be part of computer programs for repeated measurement designs.

19.11 RANDOMIZED BLOCK DESIGNS

Consider a one-way classification experiment of the type described in Chapter 15, where N subjects are assigned at random to k treatment groups. Assume that the groups are of equal size with n subjects in each of the k groups. Between-groups and within-groups sums of squares are obtained with $k-1$ and $N-k$ degrees of freedom, respectively. The precision of such experiments can be increased, sometimes substantially, by grouping the subjects into a number of blocks, using a variable that is known to be correlated with the dependent variable. For example, assume that four different methods of learning an artificial language are under investigation and 40 subjects are available. In the usual one-way classification experiment these 40 subjects would be allocated to the four methods at random resulting in 4 groups of 10 subjects each. Measures of scholastic achievement may, however, be available, and this variable may be thought to be correlated with the performance of subjects in learning the artificial language. Subjects may be divided into two groups, or blocks, of 20 subjects each. One group may be a high, the other a low, scholastic achievement group. The 20 subjects in each block may then be assigned at random to the four methods, resulting in eight groups of five subjects. Such an experiment is called a randomized block experiment. The essence of the idea of a randomized block experiment is that subjects can be grouped into blocks according to a known classification variable, which is correlated with the dependent variable. Subjects within blocks are then assigned at random to the k treatments. The analysis of data for such an experiment presents no problems. In the example above the analysis of data is the same as for a double-classification factorial experiment as described in Chapter 16. Four sums of squares are obtained. Given k treatments and B blocks the degrees of freedom associated with treatments, blocks, interaction, and within-cells sums of squares are $k-1$, $B-1$, $(k-1)(B-1)$, and $N-Bk$, respectively.

What is the purpose of a randomized block experiment? The primary purpose is to reduce the size of the error term used in the denominator of the F ratio, which for the fixed model is the within-cells mean square. The relative efficiency of the experiment is thereby increased in relation to the one-way classification experiment. If the blocking variable has a substantial correlation with the dependent variable, the sums of squares associated with blocks may prove to be of some appreciable size; also an interaction term of some magnitude may be found. The effect of this will be to reduce the size of the within-group sum of squares and the within-group mean square and, thereby, increase the likelihood of obtaining a significant difference for the main effect.

The reader should note that in the one-way classification experiment the number of degrees of freedom associated with the error term, the within-groups mean square, is $N-k$, whereas in the randomized block experiment the number of degrees of freedom associated with the error term is $N-Bk$. Thus in the randomized block experiment a loss in degrees of freedom associated with the error term occurs, which must be compensated for by the sum of squares associated with blocks and interaction. An informative discussion of this point will be found in Myers (1979). Myers' treatment of the subject shows that the relative efficiency of the randomized block experiment in relation to the usual one-way classification experiment will be greater than 1 whenever the F test of the combined block and interaction effects exceeds 1. The relative efficiency will increase as the sum of squares associated with blocks and interaction effects increases.

Because the degrees of freedom associated with the error term in a randomized block design are $N-Bk$, the power of the F test will decrease as the number of blocks increases. Also, as the number of blocks increase the within-cells sum of squares decreases. These are opposing effects which suggest that in a randomized block experiment some optimum number of blocks exists. This topic has been investigated by Feldt and Mahmoud (1958). The reader will find Myers' (1979) discussion of this topic helpful. The gist of the matter is that the optimum number of blocks is related to the correlation between the blocking variable and the dependent variable, sample size N, and the number of treatment levels k. The optimum number of blocks increases with increase in the correlation and sample size N and decreases with increase in the number of treatment levels. In the design of a randomized block experiment investigators should inform themselves of these matters and keep them in mind.

The blocking variable is usually a classification variable which is characteristic of the subjects and is in no way under the control of the investigator. Examples are sex, socioeconomic level, scholastic achievement, IQ, litter membership, strain, and so on. A blocking variable may be of the nominal, ordinal, or interval-ratio type. With ordinal or interval-ratio variables arbitrary groupings (such as high, medium, and low) are used as blocks.

In some experiments the blocking variable is of no interest to the investigator and is used purely for the purpose of error reduction. In other experiments the blocking variable may be of considerable intrinsic interest in itself and may be an integral part of the experiment. In such experiments error reduction may not be a matter of concern. Such experiments are in effect ordinary factorial experiments, where one or more of the variables are classification rather than treatment variables.

Randomized block experiments are on occasion conducted with one observation per cell. To illustrate, let the treatment variable be four different dosages of a drug intended to alleviate depression and let the blocking variable be a score on a depression scale administered prior to the administration of any drug. Let the dependent variable be a measure of motor performance, such as reaction time. If 20 subjects were available, these subjects could be divided into 5 blocks of 4 subjects each. The four subjects with the highest scores on the depression scale would constitute the first block, the next highest subjects the second block, and so on. Within each block subjects are assigned to treatments at random. The result is a 5×4 table of numbers with one observation per cell. Such data are analyzed using the two-way classification analysis with one observation per cell. The total sum of squares is partitioned into treatment, block, and interaction sums of squares. Care must be exercised in the choice of error term in applying the F ratio. Frequently, as in the illustrative example above, the blocking variable may be viewed as a random variable and the treatment variable as a fixed variable; that is, the model is mixed. The proper error term for testing treatment effects is the interaction mean square. No test of the effects due to blocks can be made. In general in randomized block designs investigators must concern themselves with whether the blocking variable may be viewed as fixed or random, and govern themselves accordingly in the choice of the appropriate error term.

19.12 EXPERIMENTS WITH NESTED FACTORS

Consider an experiment which is intended to investigate two different drugs, A and B, in the treatment of depressed patients. The dependent variable is a measure of improvement under the drug. Assume that the patients are under treatment by three different therapists. Such an experiment might be conducted as a 2×3 factorial experiment with six groups of n experimental subjects. All six treatment combinations are present in the experiment. In such a factorial experiment both factors are said to be crossed. In the example above each of the two drugs crosses, as it were, each of the three therapists, and each of the three therapists crosses the two drugs. This design enables an investigation of drug-therapist interaction.

Practical considerations may, however, prevent the experiment from being conducted in the manner described above. The experiment may be conducted in a somewhat different way. Instead of three therapists, six therapists may be used. Three therapists may treat subjects receiving drug A, each therapist treating n subjects, and three other therapists may treat subjects receiving drug B. Six groups of subjects are involved. Such an experiment may be represented as follows:

	Drug A	Drug B
Therapists	1 2 3	4 5 6
Groups	n n n	n n n

In this example therapists are said to be nested under drugs. Therapists constitute a nested factor. Experiments of this general type are spoken of as nested designs or hierarchial designs. In the simple example above the total sum of squares can be partitioned into a drug sum of squares, a therapist or nested-factor sum of squares, and a within-group sum of squares. No interaction can be obtained because therapists are not crossed with drugs but are nested under drugs. In such experiments the assumption is that interaction is either 0 or negligible.

Obviously designs of this type can become very complex. Consider an experiment with 2 drugs, 4 hospitals, and 12 therapists. Hospitals may be nested under drugs, and therapists under hospitals, in a hierarchical fashion as follows:

	Drug A		Drug B	
Hospitals	1	2	3	4
Therapists	1 2 3	4 5 6	7 8 9	10 11 12
Groups	n n n	n n n	n n n	n n n

Also, nested factors may occur within factorial designs. Consider an experiment with two drugs and two hospitals, with therapists completely nested under drugs and hospitals. Such an experiment would take the form:

		Drug A	Drug B
Hospital 1	Therapist Group	1 2 3 n n n	4 5 6 n n n
Hospital 2	Therapist Group	7 8 9 n n n	10 11 12 n n n

This is a 2×2 factorial experiment with therapists completely nested under the four treatment combinations.

Experiments may be designed where some factors are nested and others are crossed. Such experiments are called partially nested or partially hierarchical experiments. Also, experiments with nested factors may be designed with repeated measurements. Very complicated experiments involving nested factors may be conducted. A discussion of these complex designs will be found in more advanced texts such as Winer (1971).

In the sections to follow a simple example of a one-way classification experiment with one nested factor is given.

19.13 NOTATION AND COMPUTATION FORMULAS FOR EXPERIMENTS WITH NESTED FACTORS

The notation for an experiment with a nested factor is illustrated below. The main treatment effect is denoted by A and the nested factor by B. The symbol A will also be used to denote the number of levels of the A factor. Also the symbol B will be used to denote the number of levels of the B factor nested under each level of A. In the notational example below, two levels of A occur and three levels of B nested under each level of A. Thus six groups of n subjects are used. The notation is as follows:

	A_1			A_2		
B_{11}	B_{12}	B_{13}	B_{21}	B_{22}	B_{23}	
X_{111}	X_{121}	X_{131}	X_{211}	X_{221}	X_{231}	
X_{112}	X_{122}	X_{132}	X_{212}	X_{222}	X_{232}	
X_{113}	X_{123}	X_{133}	X_{213}	X_{223}	X_{233}	
...	...	...	...	...	...	
X_{11n}	X_{12n}	X_{13n}	X_{21n}	X_{22n}	X_{23n}	
$T_{11.}$	$T_{12.}$	$T_{13.}$	$T_{21.}$	$T_{22.}$	$T_{23.}$	
$\bar{X}_{11.}$	$\bar{X}_{12.}$	$\bar{X}_{13.}$	$\bar{X}_{21.}$	$\bar{X}_{22.}$	$\bar{X}_{23.}$	
$T_{1..}$ $\bar{X}_{1..}$			$T_{2..}$ $\bar{X}_{2..}$			$T_{...}\bar{X}_{...}$

In the above notation the first subscript identifies the level of A, the second subscript the level of B nested under A, and the third subscript identifies the member within the group. Thus the symbol X_{213} identifies the score for the third member of the first group nested under the second level of the A factor. Means and totals are also shown. The symbol $\bar{X}_{...}$ is as usual the grand mean, and T is the sum of all observations. If the symbol A is used to denote the number of levels of the A factor, and B the number of levels of the B factor nested under A, then the total number of groups is AB and the total number of observations is $ABn = N$.

In this simple type of nested design the total sum of squares is partitioned into three parts. The first is the sum of squares associated with the variation in means for the A factor. With A levels of the A factor this sum

Table 19.7

Analysis-of-variance table for one-way classification experiment with one nested factor

Source	Sum of squares	df	Variance estimate
A (main effect)	$nB \sum^{A} (\bar{X}_{a..} - \bar{X}_{...})^2$	$A - 1$	s_a^2
B (nested factor)	$n \sum^{A} \sum^{B} (\bar{X}_{ab.} - \bar{X}_{a..})^2$	$A(B - 1)$	s_b^2
Within cells	$\sum^{A} \sum^{B} \sum^{n} (X_{abi} - \bar{X}_{ab.})^2$	$AB(n - 1)$	s_w^2
Total	$\sum^{A} \sum^{B} \sum^{n} (X_{abi} - \bar{X}_{...})^2$	$N - 1$	

of squares has associated with it $A - 1$ degrees of freedom. The second sum of squares is that associated with the nested factor. This is the sum of squares associated with the variation in means of the nested factor at each level of the A factor. This sum of squares may be obtained separately for each level of the A factor and then pooled over all levels of the A factor. With B levels of the B factor, the number of degrees of freedom associated with this sum of squares at each level of A is $B - 1$, and for all levels of A the number of degrees of freedom is $A(B - 1)$. The third sum of squares is that for within groups and is simply the sum of squares within each group summed over all AB groups. This sum of squares has associated with it $AB(n - 1)$ degrees of freedom.

The analysis-of-variance table for a one-way classification experiment with one nested factor is shown in Table 19.7.

The required sums of squares may be calculated, using the following formulas:

A (MAIN EFFECT)

[19.9]
$$\frac{1}{nB} \sum^{A} T_{a..}^2 - \frac{T^2}{N}$$

B (NESTED FACTOR)

[19.10]
$$\frac{1}{n} \sum^{A} \sum^{B} T_{ab.}^2 - \frac{1}{nB} \sum^{A} T_{a..}^2$$

WITHIN CELLS

[19.11]
$$\sum^{A} \sum^{B} \sum^{n} X_{abi}^2 - \frac{1}{n} \sum^{A} \sum^{B} T_{ab.}^2$$

TOTAL

$$\sum^{A}\sum^{B}\sum^{n} X_{abi}^2 - \frac{T^2}{N} \qquad [19.12]$$

In this design if both the A factor and the nested factor B are fixcd the correct error term for testing both factors is the within-cell mean square. If A is fixed and B is random the correct error term for testing the A effect is the nested-factor mean square.

19.14 ILLUSTRATIVE EXAMPLE OF EXPERIMENT WITH NESTED FACTORS

Table 19.8 shows illustrative data, and Table 19.9 the results, for a one-way classification experiment with one nested factor. Here the A factor has three levels. Under each level of A two levels of the nested factor B occur. Thus $A = 3$, $B = 2$; also $n = 6$, and $ABn = N = 36$. The following quantities required in the computation formulas are calculated.

$$\frac{1}{nB}\sum^{A} T_{a..}^2 = \frac{1}{6 \times 2}[(111)^2 + (220)^2 + (374)^2] = 16{,}716.42$$

$$\frac{1}{n}\sum^{A}\sum^{B} T_{ab.}^2 = \tfrac{1}{6}[(49)^2 + (62)^2 + \cdots + (164)^2] = 16{,}918.83$$

$$\sum^{A}\sum^{B}\sum^{n} X_{abi}^2 = 19{,}181.00$$

$$\frac{T^2}{N} = \frac{(705)^2}{36} = 13{,}806.25$$

Table 19.8

Illustrative data for experiment with nested factor

	A_1		A_2		A_3	
	B_{11}	B_{12}	B_{21}	B_{22}	B_{31}	B_{32}
	7	5	21	24	31	34
	9	4	33	31	42	39
	6	16	14	14	56	26
	4	13	16	10	42	13
	11	14	10	19	18	24
	12	10	10	18	21	28
Totals	49	62	104	116	210	164
Means	8.17	10.33	17.33	19.33	35.00	27.33
Totals and means	111	9.25	220	18.33	374	31.17

$T = 705 \qquad \bar{X}_{...} = 19.58$

Table 19.9

Analysis of variance for data of Table 19.8

Source of variation	Sum of squares	Degrees of freedom	Variance estimate
A (main effect)	2,910.17	2	$1{,}455.09 = s_a^2$
B (nested factor)	202.41	3	$67.47 = s_b^2$
Within groups	2,262.17	30	$75.41 = s_w^2$
Total	5,374.75	35	

$F_a = \dfrac{s_a^2}{s_w^2} = 19.30 \qquad F_b = \dfrac{s_b^2}{s_w^2} = .89$

The required sums of squares are then as follows:

A (MAIN EFFECT)

$$\frac{1}{nB}\sum^{A} T_{a..}^2 - \frac{T^2}{N} = 16{,}716.42 - 13{,}806.25 = 2{,}910.17$$

B (NESTED FACTOR)

$$\frac{1}{n}\sum^{A}\sum^{B} T_{ab.}^2 - \frac{1}{nB}\sum^{A} T_{a..}^2 = 16{,}918.83 - 16{,}716.42 = 202.41$$

WITHIN GROUPS

$$\sum^{A}\sum^{B}\sum^{n} X_{abi}^2 - \frac{1}{n}\sum^{A}\sum^{B} T_{ab.}^2 = 19{,}181.00 - 16{,}918.83 = 2{,}262.17$$

TOTAL

$$\sum^{A}\sum^{B}\sum^{n} X_{abi}^2 - \frac{T^2}{N} = 19{,}181.00 - 13{,}806.25 = 5{,}374.75$$

If we assume that both the *A* factor and the nested factor *B* are fixed, the correct error term is the within-groups sum of squares. In this example the main effect is highly significant, whereas the nested factor is not only not significant but contributes very little to the total variation.

19.15 EXPERIMENTS USING LATIN SQUARE DESIGNS

Consider for illustrative purposes a simple repeated measurement experiment with three subjects and three treatments. The ordinal position in which the treatments are presented may be a matter of concern because of the possible presence of carry-over effects. Treatments may be assigned to subjects at random, or a systematic assignment of treatments to subjects may be used as follows:

	Ordinal position of treatment 1	2	3
S_1	A_1	A_2	A_3
S_2	A_2	A_3	A_1
S_3	A_3	A_1	A_2

Here A_1, A_2, and A_3 represent three treatment levels of a factor A. Note that each occurs once in each row and column. Note further that each level of A appears once in the first ordinal position, once in the second ordinal position, and once in the third ordinal position. The design is balanced. Such an arrangement is called a *Latin square*. Experiments that use such arrangements are said to have Latin square designs. In the simple design described above three factors are involved: subjects, ordinal position of treatments, and the factor A with three levels A_1, A_2, and A_3.

In general a Latin square may be viewed as an arrangement of k letters, or symbols, such that each letter, or symbol, occurs once in each row and column. For a 3×3 Latin square of the type described above, clearly other arrangements exist such that each letter, or symbol, appears once in each row and column. For example, rows and columns may be interchanged to obtain

A_2	A_1	A_3
A_1	A_3	A_2
A_3	A_2	A_1

For a 3×3 Latin square other arrangements can readily be identified. If the letters, or symbols, in the first row and first column of a Latin square are in their natural order, A_1, A_2, A_3, for example, the square is said to be in *standard form*. For a 3×3 Latin square, one square in standard form exists. The total number of possible 3×3 Latin squares is 12. For a 4×4 Latin square, four squares in standard form exist. The total number of possible 4×4 Latin squares is 576. As k increases the number of possible squares increases very rapidly. For example, the number of possible 6×6 Latin squares is 812,251,200. Tables of Latin squares are given in Fisher and Yates (1963) and elsewhere.

Given the large number of possible squares, how does the investigator decide which square to use in a particular experiment? This is done by selecting a Latin square of the required order at random from tables of Latin squares. Rows and columns are then randomized separately to obtain a randomization of the original square.

In psychological and educational research Latin square designs appear to serve three purposes. *First,* Latin square designs are used to control for possible bias resulting from variables, which may be thought to correlate with the dependent variable but are themselves not of primary concern in the investigation. This use is an alternative to randomization. *Second,* Latin square designs are used to simplify three-way and higher-order factorial experiments by using a reduced, and specifically selected, number of groups of subjects and treatment combinations. The purpose here is to achieve economy in cost and effort by using a fractional replication of the factorial experiment. *Third,* Latin squares are used in a variety of ways in experiments with repeated measurements to counterbalance the order of treatment effects. These three purposes of Latin square designs are discussed in more detail below.

The use of Latin square designs to control for bias introduced by variables which may be thought to correlate with the dependent variable may be illustrated as follows. Consider a drug experiment in which time of day at which subjects are tested is thought to affect the results. Four different times are used. Also some experimenter bias may be thought to be present among experimenters. Four different experimenters are used, E_1, E_2, E_3, and E_4, and four drug levels, A_1, A_2, A_3, and A_4. Such an experiment may involve the following design.

	Time of day			
	1	2	3	4
E_1	A_3	A_1	A_2	A_4
E_2	A_1	A_3	A_4	A_2
E_3	A_4	A_2	A_1	A_3
E_4	A_2	A_4	A_3	A_1

The experiment is balanced with respect to time of day and experimenters. Bias introduced by these variables is thought to be controlled. In this experiment 1 subject may be used for each cell, there being 16 subjects in all. In general, the number of subjects in such an arrangement is k^2. This experiment may, of course, be replicated, which means that the whole design is repeated over again as many times as desired. Thus instead of using one subject in each cell, any number of subjects, say, n, may be used. This is a replication n times, using the same Latin square.

Latin square designs may be used to simplify factorial experiments by reducing the number of groups of subjects and treatment combinations. Consider a $3 \times 3 \times 3$ factorial experiment. This experiment requires 27 groups of n subjects. Each group receives 1 of the 27 treatment combinations. Such an experiment may prove costly and difficult to complete. An alternative is to use a reduced number of groups of subjects and treatment combinations. The experiment is, thereby, an incomplete factorial experiment, incomplete in the sense that all treatment combinations are not represented. The question arises as to what particular subset of treatment combinations to use. One way of selecting a particular subset is to use a Latin square design. The experiment may take the following form:

	C_1	C_2	C_3
R_1	A_3	A_1	A_2
R_2	A_1	A_2	A_3
R_3	A_2	A_3	A_1

This is a three-factor experiment with three levels of each factor C_1, C_2, and C_3, R_1, R_2, and R_3, and A_1, A_2, and A_3. Here, however, 9 groups of subjects and treatment combinations are used instead of 27 in the complete $3 \times 3 \times 3$ factorial experiment. One group is selected from each row, column, and block of the complete factorial design. A certain balance is, thereby, achieved. The Latin square design is one-third of the complete factorial design.

The economies achieved in such a Latin square design are obtained at some cost in interpretation. The design assumes that the interactions between main effects are negligible. For example, the variance estimate for A can only be assumed to validly represent that main effect if the RC interaction is 0 or negligible. Otherwise the variation associated with the RC interaction is said to be confounded with the variation associated with A. In general Latin square designs of the type described should only be used when reasons exist to believe that the variation associated with the interactions between main effects is trivial or of no great magnitude.

Perhaps the most common use of the Latin square in experiments in psychology and education is to counterbalance the order of treatment effects in repeated measurement designs. Consider an experiment with four ordinal positions and four subjects. The following design may be used:

	Ordinal position of treatments 1	2	3	4
S_1	A_3	A_1	A_2	A_4
S_2	A_2	A_4	A_1	A_3
S_3	A_4	A_2	A_3	A_1
S_4	A_1	A_3	A_4	A_2

Four measurements are obtained for each subject in the ordinal positions indicated. Since four subjects may prove quite inadequate to test for the main effect A, four groups of n subjects may be used as follows:

	Ordinal position of treatments 1	2	3	4
Group 1	A_3	A_1	A_2	A_4
Group 2	A_2	A_4	A_1	A_3
Group 3	A_4	A_2	A_3	A_1
Group 4	A_1	A_3	A_4	A_2

In this design the n subjects in group 1 all receive the A treatment in the order A_3, A_1, A_2, and A_4; the n subjects in group 2 receive the A treatment in the order A_2, A_4, A_1, and A_3; and so on. This arrangement is a replication of the experiment n times using a single Latin square. Instead of replicating the experiment using a single Latin square, that is, the same Latin square for each replication, repeated-measurement experiments are sometimes conducted using different Latin squares that are independently randomized. To illustrate, suppose three replications are required. Twelve subjects are used. The first four subjects are assigned to treatments using a 4×4 Latin square. The second group of four subjects is assigned to treatments using a different independently randomized Latin square, and likewise for the third group of four subjects. Thus in this design three different Latin squares are used instead of a single Latin square. In experiments of the above type, separate groups of subjects are viewed as random samples drawn from a population, subjects being assigned to groups at random.

A variety of complex repeated measurement experiments involving Latin squares exists. In some experiments Latin square arrangements are incorporated as part of more complex designs. In using Latin square designs with repeated measurements the reader should remember that the basic assumption of homogeneity of covariance involved in all repeated-measurement designs applies. Also certain interaction terms are assumed to be negligible.

Latin square designs are illustrated below with two simple examples. One of these involves the use of a Latin square in an incomplete factorial experiment. The other involves repeated measurements.

19.16 ILLUSTRATIVE EXAMPLE OF A LATIN SQUARE DESIGN AS A FRACTIONAL REPLICATION OF A FACTORIAL EXPERIMENT

Consider a $3 \times 3 \times 3$ factorial experiment. Denote the factors by R, C, and A. Instead of using $3 \times 3 \times 3$ groups and treatment combinations, a Latin square design may be used to produce a fractional replication with nine groups and treatment combinations as follows:

	C_1	C_2	C_3
R_1	A_3	A_1	A_2
R_2	A_1	A_2	A_3
R_3	A_2	A_3	A_1

Table 19.10 shows illustrative data for such an experiment. Here $n = 4$. The total number of subjects in this experiment is $k^2n = N = 3 \times 3 \times 4 = 36$. Totals for each level of the R, C, and A factors are shown. These totals will be denoted by $T_{r...}$, $T_{.c..}$, and $T_{..a.}$, respectively. The grand total will be denoted simply by T. The individual cell totals are shown and will be denoted by $T_{rc..}$. A single observation is as usual represented by X_{rcai}. The following quantities are calculated:

$$\frac{1}{nk}\sum^{k} T_{r...}^2 = \frac{1}{4 \times 3}[(119)^2 + (198)^2 + (321)^2] = 13{,}033.83$$

$$\frac{1}{nk}\sum^{k} T_{.c..}^2 = \frac{1}{4 \times 3}[(148)^2 + (227)^2 + (263)^2] = 11{,}883.50$$

$$\frac{1}{nk}\sum^{k} T_{..a.}^2 = \frac{1}{4 \times 3}[(154)^2 + (223)^2 + (261)^2] = 11{,}797.17$$

$$\frac{1}{n}\sum^{k}\sum^{k} T_{rc..}^2 = \frac{1}{4}[(29)^2 + (26)^2 + \cdots + (99)^2] = 14{,}135.00$$

$$\frac{T^2}{N} = \frac{(638)^2}{36} = 11{,}306.78$$

$$\sum^{n}\sum^{k}\sum^{k} X_{rcai}^2 = [(2)^2 + (7)^2 + \cdots + (25)^2] = 17{,}590.00$$

Table 19.10

Data for a Latin square design as a fractional replicate of a factorial experiment

	C_1	C_2	C_3	$T_{r\ldots}$
R_1	A_3 2 7 8 12 (29)	A_1 1 5 4 16 (26)	A_2 8 18 12 26 (64)	119
R_2	A_1 1 6 8 14 (29)	A_2 4 8 26 31 (69)	A_3 16 11 41 32 (100)	198
R_3	A_2 10 24 20 36 (90)	A_3 7 28 46 51 (132)	A_1 10 36 28 25 (99)	321
$T_{.c..}$	148	227	263	638

$T_{..1.} = 26 + 29 + 99 = 154$
$T_{..2.} = 64 + 69 + 90 = 223$
$T_{..3.} = 29 + 100 + 132 = 261$

The required sums of squares are then obtained as follows:

ROWS

$$\frac{1}{nk} \sum^{k} T^2_{r\ldots} - \frac{T^2}{N} = 13{,}033.83 - 11{,}306.78$$
$$= 1{,}727.06$$

COLUMNS

$$\frac{1}{nk} \sum^{k} T^2_{.c..} - \frac{T^2}{N} = 11{,}883.50 - 11{,}306.78 - 576.72$$

A

$$\frac{1}{nk} \sum^{k} T^2_{..a.} - \frac{T^2}{N} = 11{,}797.17 - 11{,}306.78$$
$$= 490.39$$

RESIDUAL

$$\frac{1}{n}\sum^{k}\sum^{k} T^2_{rc..} - \frac{1}{nk}\sum^{k} T^2_{r...} - \frac{1}{nk}\sum^{k} T^2_{.c..} - \frac{1}{nk}\sum^{k} T^2_{..a.} + \frac{2T^2}{N} = 14{,}135.00$$

$$- 13{,}033.83 - 11{,}883.50 - 11{,}797.17 + 2 \times 11{,}306.78 = 34.06$$

WITHIN CELLS

$$\sum^{n}\sum^{k}\sum^{k} X^2_{rcai} - \frac{1}{n}\sum^{k}\sum^{k} T^2_{rc..} = 17{,}590.00 - 14{,}135.00 = 3{,}455.00$$

TOTAL

$$\sum^{n}\sum^{k}\sum^{k} X^2_{rcai} - \frac{T^2}{N} = 17{,}590.00 - 11{,}306.78 = 6{,}283.22$$

In this type of experiment the total sum of squares is partitioned into three sums of squares associated with the main effects, R, C, and A; a residual sum of squares, and a within-cells sum of squares. The residual sum of squares is that part of the variation associated with R, C, and A that is not accounted for by the sums of squares associated with R, C, and A. The statistical model underlying this design assumes zero interaction between R, C, and A. If interactions are present these interactions will be partially incorporated in the residual term and partially confounded with main effects. If the interactions are 0 the residual variance estimate should be about the same as the within-cells variance estimate. The degrees of freedom associated with R, C, and A are $R - 1$, $C - 1$, and $A - 1$, respectively. Since $R = C = A = k$, the number of degrees of freedom associated with the residual sum of squares is $(k - 1)(k - 2)$, and with the within-cells sum of squares is $k^2(n - 1)$. The analysis-of-variance table for the data of Table 19.10 is shown in Table 19.11.

Table 19.11

Analysis-of-variance table for the data of Table 19.10

Source of variation	Sum of squares	Degrees of freedom	Variance estimate
Rows	1,727.06	2	$863.53 = s_r^2$
Columns	576.72	2	$288.36 = s_c^2$
A	490.39	2	$245.20 = s_a^2$
Residual	34.06	2	$17.03 = s^2_{res}$
Within cells	3,455.00	27	$127.96 = s_w^2$
Total	6283.22	35	

$F_r = \frac{s_r^2}{s_w^2}$	$F_c = \frac{s_c^2}{s_w^2}$	$F_a = \frac{s_a^2}{s_w^2}$	$F_{\text{res}} = \frac{s^2_{\text{res}}}{s_w^2}$
$= 6.75$	$= 2.25$	$= 1.92$	$= .13$

In this design the correct error term for testing R, C, A, and residual effects is the within-cells variance estimate. If the residual sum of squares is significant, this clearly raises questions about the appropriateness of the model. In the present illustrative example the R effect is significant at better than the .01 level; also, C, A, and residual effects are not significant.

19.17 ILLUSTRATIVE EXAMPLE OF A LATIN SQUARE USED WITH REPEATED MEASUREMENTS

Consider a simple experiment involving repeated measurements in which four subjects are used. Each subject is tested four times on a factor A. A Latin square is used to assign treatments to subjects. Such data are illustrated in Table 19.12. In this Table, C refers to ordinal position in which levels of A are administered. Thus the first subject S_1 receives the four treatments on the A factor in the order A_2, A_4, A_1, A_3; the second subject S_2 in the order A_3, A_1, A_2, A_4; and so on. Table 19.12 shows row and column totals and totals for the A factor. The row (subject) and column (ordinal position) totals will be denoted by $T_{r..}$, and $T_{.c.}$, respectively, and A factor totals by $T_{..a}$. The total for all observations is T, and any particular observation is X_{rca}.

Table 19.12

Illustrative data for Latin square design with repeated measurements

	Ordinal position of treatments C_1	C_2	C_3	C_4	$T_{r..}$
S_1	A_2 10	A_4 21	A_1 5	A_3 14	50
S_2	A_3 12	A_1 7	A_2 11	A_4 19	49
S_3	A_1 6	A_3 16	A_4 24	A_2 12	58
S_4	A_4 22	A_2 8	A_3 17	A_1 9	56
$T_{.c.}$	50	52	57	54	213

$T_{..1} = 5 + 7 + 6 + 9 = 27$
$T_{..2} = 10 + 11 + 12 + 8 = 41$
$T_{..3} = 14 + 12 + 16 + 17 = 59$
$T_{..4} = 21 + 19 + 24 + 22 = 86$

The following quantities are calculated:

$$\frac{1}{k}\sum^{k} T_{r..}^2 = \tfrac{1}{4}[(50)^2 + (49)^2 + (58)^2 + (56)^2] = 2{,}850.25$$

$$\frac{1}{k}\sum^{k} T_{.c.}^2 = \tfrac{1}{4}[(50)^2 + (52)^2 + (57)^2 + (54)^2] = 2{,}842.25$$

$$\frac{1}{k}\sum^{k} T_{..a}^2 = \tfrac{1}{4}[(27)^2 + (41)^2 + (59)^2 + (86)^2] = 3{,}321.75$$

$$\frac{T^2}{N} = \frac{(213)^2}{16} = 2{,}835.56$$

$$\sum^{k}\sum^{k} X_{rca}^2 = 3{,}367.00$$

The required sums of squares are as follows:

ROWS (SUBJECTS)

$$\frac{1}{k}\sum^{k} T_{r..}^2 - \frac{T^2}{N} = 2{,}850.25 - 2{,}835.56 = 14.69$$

COLUMNS (ORDER)

$$\frac{1}{k}\sum^{k} T_{.c.}^2 - \frac{T^2}{N} = 2{,}842.25 - 2{,}835.56 = 6.69$$

A (MAIN EFFECT)

$$\frac{1}{k}\sum^{k} T_{..a}^2 - \frac{T^2}{N} = 3{,}321.75 - 2{,}835.56 - 486.19$$

RESIDUAL

$$\sum^{k}\sum^{k} X_{rca}^2 - \frac{1}{k}\sum^{k} T_{r..}^2 - \frac{1}{k}\sum^{k} T_{.c.}^2 - \frac{1}{k}\sum^{k} T_{..a}^2 + \frac{2T^2}{N}$$

$$= 3{,}367.00 - 2{,}850.25 - 2{,}842.25 - 3{,}321.75 + 2 \times 2{,}835.56$$
$$= 23.87$$

TOTAL

$$\sum^{k}\sum^{k} X_{rca}^2 - \frac{T^2}{N} = 3{,}367.00 - 2{,}835.56 = 531.44$$

The analysis-of-variance table for the data of Table 19.12 is shown in Table 19.13. The number of degrees of freedom associated with row (subjects), column (ordinal position), and the A factor is in each case $k - 1$, which in this example is $4 - 1 = 3$. The number of degrees of freedom associated with the residual sum of squares is $(k-1)(k-2)$, which here is $(4-1)(4-2) = 6$. The model underlying this application of the analysis of variance assumes zero interaction between the three factors involved in

Table 19.13

Analysis-of-variance table for the data of Table 19.12

Source of variation	Sum of squares	Degrees of freedom	Variance estimate
Rows (subjects)	14.69	3	$4.90 = s_r^2$
Columns (ordinal position)	6.69	3	$2.23 = s_c^2$
A (main effect)	486.19	3	$162.06 = s_a^2$
Residual	23.87	6	$3.98 = s_{res}^2$
Total	531.44	15	

$F_r = \frac{s_r^2}{s_{\text{res}}^2} = 1.23 \qquad F_a = \frac{s_a^2}{s_{\text{res}}^2} = 40.72 \qquad F_c = \frac{s_c^2}{s_{\text{res}}^2} = .56$

the experiment. If this is true the correct error term for testing all three effects is the residual variance estimate. Assuming this model in this example, subject and ordinal position effects are not significant; the A factor effect is highly significant.

According to Myers (1979) in this model if the interactions are not 0 the F test will be biased negatively and will lead to too few rejections of the null hypothesis. The presence of bias in testing differences between subjects is of little consequence because such differences are usually not of interest. If the interaction between C (ordinal position) and A (main effect) is not 0, this will bias the F ratios for C and A effects. Certainly in many experiments no prior reasons exist to suspect an interaction between C and A effects.

The above is a very simple example of a Latin square design with repeated measurements. A common extension of this design is to replicate the design by using groups of n subjects corresponding to rows instead of a single subject. Thus, referring to the illustrative example, instead of testing S_1 in the order A_2, A_4, A_1, and A_3, any number of subjects n may be tested in the order A_2, A_4, A_1, and A_3, and so on.

BASIC TERMS AND CONCEPTS

One-factor experiment with repeated measurements

Two-factor experiment with repeated measurements

Two-factor experiment with repeated measurements on one factor

Compound symmetry

Epsilon, ϵ

Randomized block design

Nested factors

Latin square design

EXERCISES

1 The following are measurements made on a sample of 12 subjects under 3 experimental conditions:

	Condition		
Subject	C_1	C_2	C_3
1	8	7	15
2	19	14	20
3	7	9	6
4	23	20	18
5	14	26	12
6	6	14	15
7	5	9	20
8	22	25	20
9	11	15	16
10	4	12	8
11	13	18	20
12	8	6	28
$T_{.c}$	140	175	198
$\bar{X}_{.c}$	11.67	14.58	16.50

Obtain the sums of squares and the variance estimates. Test the column means on the assumption that experimental condition is a fixed variable.

2 The following are data for a repeated-measurements experiment in which four experimental subjects are tested under four treatment combinations.

Subject 1

	C_1	C_2
R_1	5	10
R_2	17	19

Subject 2

	C_1	C_2
R_1	9	19
R_2	20	22

Subject 3

	C_1	C_2
R_1	8	22
R_2	23	28

Subject 4

	C_1	C_2
R_1	20	35
R_2	19	21

Apply an analysis of variance to these data, and test the significance of row and column effects.

3 The following are data for a two-factor experiment with repeated measurements on one factor. Three groups of two subjects are tested under four conditions.

	Subjects	C_1	C_2	C_3	C_4	$T_{r.s}$
	1	4	1	3	2	10
R_1	2	7	6	1	3	17
	$T_{rc.}$	11	7	4	5	$T_{1..} = 27$
	3	5	5	2	4	16
R_2	4	9	5	1	2	17
	$T_{rc.}$	14	10	3	6	$T_{2..} = 33$
	5	7	8	1	4	20
R_3	6	9	1	4	3	17
	$T_{rc.}$	16	9	5	7	$T_{3..} = 37$
	$T_{.c.}$	41	26	12	18	$T = 97$

Apply an analysis of variance to these data, and test the significance of row, column, and interaction effects.

4 The following are data for a 2×2 factorial experiment with repeated measurements on six subjects.

Subject 1

	C_1	C_2	
R_1	1	6	7
R_2	5	3	8
	6	9	

Subject 2

	C_1	C_2	
R_1	2	7	9
R_2	4	5	9
	6	12	

Subject 3

	C_1	C_2	
R_1	5	9	14
R_2	8	3	11
	13	12	

Subject 4

	C_1	C_2	
R_1	2	1	3
R_2	3	7	10
	5	8	

Subject 5

	C_1	C_2	
R_1	4	4	8
R_2	2	8	10
	6	12	

Subject 6

	C_1	C_2	
R_1	4	8	12
R_2	1	10	11
	5	18	

Apply an analysis of variance to these data.

5 The following are data for a two-factor experiment with repeated measurements on one factor:

	Subjects	C_1	C_2	C_3	C_4	Totals
R_1	1	12	10	9	4	35
	2	10	5	8	2	25
		22	15	17	6	60
R_2	3	16	14	10	1	41
	4	27	23	14	8	72
		43	37	24	9	113
R_3	5	14	12	9	6	41
	6	23	14	10	9	56
		37	26	19	15	97
	Totals	102	78	60	30	270

Complete the analysis of variance.

6 Complete the analysis of variance on the following data for an experiment with nested factors.

	A_1			A_2		
	B_{11}	B_{12}	B_{13}	B_{21}	B_{22}	B_{23}
	5	11	11	4	10	6
	6	14	12	1	12	11
	4	16	18	7	18	12
	3	22	25	8	15	20
	9	17	19	6	14	10
Totals	27	80	85	26	69	59
Means	5.40	16.00	17.00	5.20	13.80	11.80
Totals and means	192		12.80	154		20.53

7 Complete the analysis of variance on the following data for a Latin square design that is used as a fractional replicate of a factorial experiment.

	C_1	C_2	C_3	C_4	$T_{r...}$
R_1	A_1 12 16 25 32 (85)	A_4 10 16 24 20 (70)	A_3 9 12 15 10 (46)	A_2 4 7 10 12 (33)	234
R_2	A_4 12 15 20 28 (75)	A_2 18 10 6 15 (49)	A_1 24 8 16 5 (53)	A_3 10 12 18 15 (55)	232
R_3	A_2 10 6 12 10 (38)	A_3 9 9 8 7 (33)	A_4 6 5 8 2 (21)	A_1 4 5 7 1 (17)	109
R_4	A_3 4 6 7 2 (19)	A_1 4 7 7 1 (19)	A_2 3 4 5 2 (14)	A_4 4 6 3 2 (15)	67
$T_{.c..}$	217	171	134	120	642
$T_{..a.}$	174	134	153	181	

ANSWERS TO EXERCISES

1

Source	Sum of Squares	*df*	Variance estimate
Row	754.08	11	68.55
Column	142.17	2	71.09
Interaction	598.50	22	27.20
Total	1,494.75	35	

$F = 2.61$, $df_1 = 2$, $df_2 = 22$; $p > .05$

2

Source	Sum of squares	*df*	Mean square	*F*
Subjects	258.60	3	86.23	
Rows	105.06	1	105.06	1.42 (*SR*)
Columns	189.06	1	189.06	26.15† (*SC*)
$S \times R$	220.69	3	73.56	
$S \times C$	21.69	3	7.23	
$R \times C$	68.06	1	68.06	16.10† (*SRC*)
$S \times R \times C$	12.69	3	4.23	

Note: Error term shown in parentheses following *F* ratio. † Significant at the .05 level.

3

Source	Sum of squares	Degrees of freedom	Variance estimate	F
Between subjects	13.71			
Rows	6.34	2	3.17	1.29
S/R	7.37	3	2.46	
Within subjects	137.25			
Columns	78.79	3	26.26	4.37†
$R \times C$	4.33	6	.72	.12
SC/R	54.13	9	6.01	
Total	150.96	23		

$F_r = \frac{s_r^2}{s_{s/r}^2} = 1.29$ $F_c = \frac{s_c^2}{s_{sc/r}^2} = 4.37†$ $F_{rc} = \frac{s_{rc}^2}{s_{sc/r}^2} = .12$

† Significant at the .05 level

4

Source of variation	Sum of squares	Degrees of freedom	Variance estimate
Rows	1.50	1	1.50
Columns	38.00	1	38.00
Subjects	26.83	5	5.37
$R \times C$	.16	1	.16
$R \times S$	14.00	5	2.80
$C \times S$	27.00	5	5.40
$R \times C \times S$	57.84	5	11.57
Total	165.33		

$F_r = \frac{s_r^2}{s_{rs}^2} = .54$ $F_c = \frac{s_c^2}{s_{cs}^2} = 7.04†$ $F_{rc} = \frac{.16}{11.57} = .02$

† Significant at the .05 level.

5

Source	Sum of squares	Degrees of freedom	Variance estimate
Between subjects	345.50	5	
Rows	184.75	2	92.38
S/R	160.75	3	53.58
Within subjects	585.00	18	
Columns	460.50	3	153.50
$R \times C$	87.25	6	14.54
SC/R	37.25	9	4.14
Total	930.50	23	

$F_r = \frac{s_r^2}{s_{s/r}^2} = 1.72$ $F_c = \frac{s_c^2}{s_{sc/r}^2} = 37.07†$ $F_{rc} = \frac{s_{rc}^2}{s_{sc/r}^2} = 3.51‡$

† Significant at better than .01 level.
‡ Significant at .05 level.

6

Source of variation	Sum of squares	Degrees of freedom	Variance estimate
A (main effect)	48.14	1	48.14
B (nested factor)	615.73	4	153.93
Within group	389.60	24	16.23
Total	1,053.47	29	

$F_a = \frac{s_a^2}{s_w^2} = 2.97 \qquad F_b = \frac{153.93}{16.23} = 9.48†$

† Significant at better than the .01 level.

7

Source of variation	Sum of squares	Degrees of freedom	Variance estimate
Rows	1,369.31	3	456.44
Columns	352.81	3	117.60
A	85.06	3	28.35
Residual	151.75	6	25.29
Within cells	1,007.00	48	20.98
Total	2,965.93	63	

$F_r = \frac{s_r^2}{s_w^2} = 21.76 \qquad F_c = \frac{s_c^2}{s_w^2} = 5.61$

$F_a = \frac{s_a^2}{s_w'^2} = 1.35 \qquad F_{res} = \frac{s_{res}^2}{s_w^2} = 1.21$

20

ANALYSIS OF COVARIANCE

20.1 INTRODUCTION

One object of experimental design is to ensure that the results observed may be attributed within limits of error to the treatment variable and to no other causal circumstance. For example, the assignment of subjects to groups at random and the matching of subjects are experimental procedures whose purpose is to ensure freedom from bias. Situations arise, however, where one or more variables are uncontrolled because of practical limitations associated with the conduct of the experiment. A statistical rather than an experimental method may be used to "control" or "adjust for" the effects of one or more uncontrolled variables, and thereby permit a valid evaluation of the outcome of the experiment. The analysis of covariance is such a method.

To illustrate, an investigator may wish to compare three methods of teaching French. Each method is applied to a different group of subjects. Mean scores on a test of French achievement, following a period of instruction, are obtained for the three groups. Measures of intelligence may be available for all subjects. The three groups may differ on intelligence, and intelligence may be correlated with French achievement. Thus the investigator does not know the extent to which the differences in French achievement result from different methods of instruction or from differences in intelligence between the groups of subjects. The analysis of covariance may be used to compare the differences in French achievement between groups when the influence of intelligence has been statistically controlled or removed.

In psychology and education the analysis of covariance is primarily used as a procedure for the statistical control of an uncontrolled variable.

It may, however, serve other purposes. It may increase the precision of the experiment by reducing the error variance. It may assist understanding or raise questions regarding the nature of the treatment effect. If, for example, differences between treatment effects disappear or are negligible when the effects of the uncontrolled variable are removed, does this mean that differences in the dependent variable are a consequence of the uncontrolled variable? The method may also be used to test the homogeneity of a set of regression coefficients and to deal with problems associated with missing observations.

The analysis of covariance was developed by R. A. Fisher, and the first example of its use appeared in the literature in 1932. Many of its applications have been in agricultural experimentation. An examination of examples of its use in agriculture sometimes helps to clarify its application in the behavioral sciences.

In the analysis of covariance the uncontrolled variable is called the *covariate* or the *concomitant variable,* and is denoted by Y. The variable under study, the dependent variable, is denoted by X. The influence of the covariate is removed using a straightforward linear regression method. Residual sums of squares are used to provide variance estimates that are then employed in tests of significances. Adjusted means on X may be obtained. These adjusted means show what part of the variation in the means of X remains after the variation due to the covariate Y is removed. The analysis of covariance is in part a correlational procedure. It combines both regression and analysis of variance methods.

The statistical ideas and procedures used in the analysis of covariance are straightforward. The method, however, presents a variety of interpretative difficulties, which in some situations may either invalidate its use or raise serious questions about it. Undoubtedly many incorrect applications of the method have been made. It is not a foolproof substitute for the proper experimental control of uncontrolled variables. Despite the limitations of the method, in the practical conduct of research, situations arise where proper experimental control of one or more presumably relevant and important variables may not be possible. In such situations the investigator may simply discontinue the enquiry or may knowingly proceed and attempt to make the best possible interpretation of the results in the light of all available information. If this latter course is chosen, the analysis of covariance may prove helpful.

Initially in this presentation the analysis of covariance as a statistical procedure will be described. Interpretative problems associated with the use of the method will then be discussed.

20.2 NOTATION

An application of a simple analysis of covariance requires paired observations on k groups of experimental subjects. The number of pairs of observations in the k groups is denoted by $n_1, n_2, \ldots, n_k$. The paired observations are assumed to be paired samples drawn from k populations. The data may be represented as follows:

	Group 1		Group 2		Group k	
	Y_{11}	X_{11}	Y_{12}	X_{12}	Y_{1k}	X_{1k}
	Y_{21}	X_{21}	Y_{22}	X_{22}	Y_{2k}	X_{2k}
	Y_{31}	X_{31}	Y_{32}	X_{32}	Y_{3k}	X_{3k}
	. . .	. . .	. . .	. . .	. . .	. . .
	$Y_{n_1 1}$	$X_{n_1 1}$	$Y_{n_2 2}$	$X_{n_2 2}$	$Y_{n_k k}$	$X_{n_k k}$
Means	$\bar{Y}_1$	$\bar{X}_1$	$\bar{Y}_2$	$\bar{X}_2$	$\bar{Y}_k$	$\bar{X}_k$

In this notation X is the variable under study, the dependent variable, whereas Y is the uncontrolled variable, or covariate.

The group means on Y and X are represented by $\bar{Y}_1, \bar{Y}_2, \ldots, \bar{Y}_k$ and $\bar{X}_1, \bar{X}_2, \ldots, \bar{X}_k$, respectively. A dot notation $\bar{Y}_1, \bar{Y}_2$, and so on, where the dot represents the variable subscript, could have been used. In this chapter analysis of covariance is discussed for one-way classification only. The notation used is consistent with the notation used in the analysis of variance for one-way classification, discussed in Chapter 15. In a presentation of the analysis of covariance for two-way classification, a dot notation would prove helpful.

Some readers may feel that it would be more appropriate to use X to denote the covariate and Y the dependent variable. Here X is the dependent variable because this is consistent with the notation used elsewhere in this book. In its simplest form the analysis of covariance involves three variables, a covariate, a dependent variable, and an independent variable.

In the analysis of covariance, sums of products are considered. The sum of products for the observations in the jth group is denoted by

$$\sum_{i=1}^{nj} (X_{ij} - \bar{X}_j)(Y_{ij} - \bar{Y}_j)$$

The sum of products for all observations in the k groups, that is, the total sum of products, is

$$\sum_{j=1}^{k} \sum_{i=1}^{n_j} (X_{ij} - \bar{X})(Y_{ij} - \bar{Y})$$

20.3 PARTITIONING A SUM OF PRODUCTS

Before proceeding further with the discussion, it is useful to observe that with paired observations on k groups the total sum of products may be partitioned into within-groups and between-groups sums of products in a manner analogous to that in which a total sum of squares is partitioned into within-groups and between-groups sums of squares. As in the analysis of variance for one-way classification we may write

$$(X_{ij} - \bar{X}) = (X_{ij} - \bar{X}_j) + (\bar{X}_j - \bar{X})$$
$$(Y_{ij} - \bar{Y}) = (Y_{ij} - \bar{Y}_j) + (\bar{Y}_j - \bar{Y})$$

These are multiplied, summed over the n_j cases in the jth group, and summed over k groups. When this is done, two cross-product terms to the right vanish in the summation process, and we obtain

[20.1] $$\sum_{j=1}^{k}\sum_{i=1}^{n_j}(X_{ij}-\bar{X})(Y_{ij}-\bar{Y}) = \sum_{j=1}^{k}\sum_{i=1}^{n_j}(X_{ij}-\bar{X}_j)(Y_{ij}-\bar{Y}_j) + \sum_{j=1}^{k} n_j(\bar{X}_j-\bar{X})(\bar{Y}_j-\bar{Y})$$

The term to the left is a total sum of products of deviations about the grand means $\bar{X}$ and $\bar{Y}$. The first term to the right is the sum of products of deviations about the group means. It is the within-groups sum of products. The second term to the right is the sum of products of deviations between groups. It is the between-groups sum of products.

20.4 REGRESSION LINES

With data consisting of paired observations for k groups, a number of different regression lines may be identified. The slope of the regression line used in predicting X from a knowledge of Y, as given in Chapter 9, is

[20.2] $$b_{xy} = r\frac{s_x}{s_y} = \frac{\sum_{i=1}^{N}(X_{ij}-\bar{X})(Y_{ij}-\bar{Y})}{\sum_{i=1}^{N}(Y_{ij}-\bar{Y})^2}$$

This slope is obtained by dividing a sum of products by a sum of squares. If we divide the total within-groups and between-groups sums of products by the corresponding sums of squares, three regression coefficients are obtained. These are the slopes of three different regression lines.

The *first* is the total overall regression line for predicting X from a knowledge of Y based on all the observations put together. The slope of this line is

[20.3] $$b_t = \frac{\sum_{j=1}^{k}\sum_{i=1}^{n_j}(X_{ij}-\bar{X})(Y_{ij}-\bar{Y})}{\sum_{j=1}^{k}\sum_{i=1}^{n_j}(Y_{ij}-\bar{Y})^2}$$

A *second* regression line is the overall within-groups regression line. In predicting X from a knowledge of Y we may consider each of the k groups separately. Each group has its own within-group regression line with slope b_j. Information from these k separate regression lines may be pooled to obtain an overall within-groups regression line whose slope is given by

[20.4]
$$b_w = \frac{\sum_{j=1}^{k} \sum_{i=1}^{n_j} (X_{ij} - \bar{X}_j)(Y_{ij} - \bar{Y}_j)}{\sum_{j=1}^{k} \sum_{i=1}^{n_j} (Y_{ij} - \bar{Y}_j)^2}$$

The numerator of this equation is the within-groups sum of products, and the denominator is the within-groups sum of squares for Y. It should be noted here that the slopes of the individual group regression lines, b_1, $b_2, \ldots, b_k$, are estimates of population parameters $\beta_1, \beta_2, \ldots, \beta_k$. The pooling process used to obtain b_w involves an assumption of homogeneity of slope, that is, that $\beta_1 = \beta_2 = \ldots = \beta_k$. This is a basic assumption in the analysis of covariance.

A *third* regression line may be considered with slope b_b obtained by dividing the between-groups sum of products by the between-groups sum of squares for Y. The slope of this line is

[20.5]
$$b_b = \frac{\sum_{j=1}^{k} n_j(\bar{X}_j - \bar{X})(\bar{Y}_j - \bar{Y})}{\sum_{j=1}^{k} n_j(\bar{Y}_j - \bar{Y})^2}$$

Of particular interest in the analysis of covariance is the within-groups regression equation. This equation has the form

[20.6]
$$X''_{ij} = b_w(Y_{ij} - \bar{Y}_j) + \bar{X}_j$$

where b_w is the within-groups regression slope. As will be shown, the analysis of covariance makes use of the sum of squares of residuals of X about this regression line.

20.5 ADJUSTING THE SUM OF SQUARES OF X

Given measurements on Y and X for k groups, the total sum of squares for both Y and X may be partitioned into within-groups and between-groups sums of squares, using the analysis of variance methods described in Chapter 15. Also the total sum of products may be partitioned into within-groups and between-groups sums of products, as described in Section 20.3. How may the sum of squares on X be adjusted to allow for, or to remove, the influence of the variation of the uncontrolled variable Y?

Let us first consider, in general, the problem of calculating a sum of squares of residuals of X about the linear regression line used in predicting X from a knowledge of Y. The equation for this linear regression line may be written $X'_i = b_{xy}(Y_i - \bar{Y}) + \bar{X}$, where X'_i is a predicted value, and b_{xy} is the slope of the line. The sum of squares of residuals about this line is $\sum_{i=1}^{N} (X_i - X'_i)^2$. By substituting $b_{xy}\ (Y_i - \bar{Y}) + \bar{X}$ for X'_i in this sum of squares and using simple algebra, it is readily shown that the sum of squares of residuals is

$$[20.7] \qquad \sum_{i=1}^{N} (X_i - X_i')^2 = \sum_{i=1}^{N} [(X_i - \bar{X}) - b_{xy}(Y_i - \bar{Y})]^2$$

$$= \sum_{i=1}^{N} (X_i - \bar{X})^2 - \frac{\left[\sum_{i=1}^{N} (X_i - \bar{X})(Y_i - \bar{Y})\right]^2}{\sum_{i=1}^{N} (Y - \bar{Y})^2}$$

This latter expression says that in order to obtain a sum of squares of residuals on X, we subtract from the sum of squares of X a quantity which is equal to the square of the sum of products of X and Y divided by the sum of squares of Y. This quantity will be either 0 or positive. Thus in the nonzero case this procedure will always reduce the size of the sum of squares. A residual sum of squares obtained in this way is sometimes called a reduced sum of squares.

In the analysis of covariance we proceed by calculating an adjusted *total* sum of squares on X, using a regression line with slope b_t based on all the observations for the k groups put together. This adjusted total sum of squares is given by

$$[20.8] \qquad \sum_{j=1}^{k} \sum_{i=1}^{n_j} (X_{ij} - X_{ij}')^2 = \sum_{j=1}^{k} \sum_{i=1}^{n} [(X_{ij} - \bar{X}) - b_t(Y_{ij} - \bar{Y})]^2$$

$$= \sum_{j=1}^{k} \sum_{i=1}^{n_j} (X_{ij} - \bar{X})^2 - \frac{\left[\sum_{j=1}^{k} \sum_{i=1}^{n_j} (X_{ij} - \bar{X})(Y_{ij} - \bar{Y})\right]^2}{\sum_{j=1}^{k} \sum_{i=1}^{n_j} (Y_{ij} - \bar{Y})^2}$$

The next step is to calculate an adjusted *within-groups* sum of squares, using the within-group regression line with slope b_w. This sum of squares is given by

$$[20.9] \qquad \sum_{j=1}^{k} \sum_{i=1}^{n_j} (X_{ij} - X_{ij}'')^2 = \sum_{j=1}^{k} \sum_{i=1}^{n_j} [(X_{ij} - \bar{X}_j) - b_w(Y_{ij} - \bar{Y}_j)]^2$$

$$= \sum_{j=1}^{k} \sum_{i=1}^{n_j} (X_{ij} - \bar{X}_j)^2 - \frac{\left[\sum_{j=1}^{k} \sum_{i=1}^{n_j} (X_{ij} - \bar{X}_j)(Y_{ij} - \bar{Y}_j)\right]^2}{\sum_{j=1}^{k} \sum_{i=1}^{n_j} (Y_{ij} - \bar{Y}_j)^2}$$

Thus to calculate an adjusted within-groups sum of squares on X, we subtract from the within-groups sum of squares on X a quantity which is equal to the within-groups sum of products, squared, divided by the within-groups sum of squares on Y. The adjusted sum of squares for *between groups* is now obtained by subtracting the adjusted within-groups sum of squares on X from the adjusted total sum of squares.

20.6 DEGREES OF FREEDOM AND VARIANCE ESTIMATES

The numbers of degrees of freedom associated with the unadjusted and adjusted sums of squares on X are as follows:

	Unadjusted X	Adjusted X
Between	$k-1$	$k-1$
Within	$N-k$	$N-k-1$
Total	$N-1$	$N-2$

The number of degrees of freedom associated with the adjusted total sum of squares on X is $N-2$. This sum of squares consists of squared residuals about a linear regression line, and as such has $N-2$, and not $N-1$, degrees of freedom. Thus 1 degree of freedom is lost in the adjustment process. The number of degrees of freedom associated with the adjusted within-groups sum of squares is $N-k-1$. The number of degrees of freedom associated with the adjusted between-groups sum of squares is $k-1$, and is unchanged because the between-groups regression line did not enter into the calculation of the between-groups sum of squares, this sum of squares being obtained by subtraction.

The adjusted sums of squares on X are now divided by their associated degrees of freedom to obtain within-groups and between-groups variance estimates s_w^2 and s_b^2. The interpretation of these variance estimates is the same as in the analysis of variance, except that the null hypothesis under test relates to adjusted treatment means, that is, means that are free of the linear effect of the covariate.

To test the significance of the difference between the adjusted means of X, an F ratio s_b^2/s_w^2 is obtained. This ratio is interpreted with $df=k-1$ associated with the numerator and $df=N-k-1$ associated with the denominator.

20.7 COMPUTATION FORMULAS

Computation formulas for sums of squares for X and Y are given in Section 15.6. To simplify the notation for the computation formulas for sums of products, denote the sums of all the observations in the jth group for X and Y by T_{x_j} and T_{y_j}, respectively. Thus

$$\sum_{i=1}^{n_j} X_{ij} = T_{x_j} \qquad \sum_{i=1}^{n_j} Y_{ij} = T_{y_j}$$

Denote the sums of X and Y for all the observations together in the k groups by T_x and T_y. Thus

$$\sum_{j=1}^{k}\sum_{i=1}^{n_j} X_{ij} = T_x \qquad \sum_{j=1}^{k}\sum_{i=1}^{n_j} Y_{ij} = T_y$$

The sum of products for the jth group may be represented by

$$\sum_{i=1}^{n_j} X_{ij}Y_{ij} = T_{x_j y_j}$$

and the sum of products for all observations in the k groups by

$$\sum_{j=1}^{k}\sum_{i=1}^{n_j} X_{ij}Y_{ij} = T_{xy}$$

The computation formula for the *total* sum of products is

[20.10] $$\sum_{i=1}^{k}\sum_{i=1}^{n_j} (X_{ij}-\bar{X})(Y_{ij}-\bar{Y}) = T_{xy} - \frac{T_xT_y}{N}$$

The *within-groups* sums of products may be obtained by

[20.11] $$\sum_{j=1}^{k}\sum_{i=1}^{n_j} (X_{ij}-\bar{X}_j)(Y_{ij}-\bar{Y}_j) = T_{xy} - \sum_{j=1}^{k}\frac{T_{x_j}T_{y_j}}{n_j}$$

The *between-groups* sums of products is

[20.12] $$\sum_{j=1}^{k} n_j(\bar{X}_j-\bar{X})(\bar{Y}_j-\bar{Y}) = \sum_{j=1}^{k}\frac{T_{x_j}T_{y_j}}{n_j} - \frac{T_xT_y}{N}$$

The above formulas are applicable to groups of unequal or equal size. In the particular case where $n_1 = n_2 = \cdots = n$ we may, of course, write the term

[20.13] $$\sum_{j=1}^{k}\frac{T_{x_j}T_{y_j}}{n_j} = \frac{\sum_{j=1}^{k} T_{x_j}T_{y_j}}{n}$$

20.8 SUMMARY

In summary, to test the significance of the difference between k adjusted means on X, using the analysis of covariance, the following steps are involved:

1. Partition the total sum of squares on both Y and X into two components, a within-groups and a between-groups sum of squares, using the usual analysis-of-variance formulas.
2. Partition the total sum of products into two components, a within-groups and a between-groups sums of products.
3. Calculate an adjusted total sum of squares on X to remove the linear effects of the covariate Y.
4. Calculate an adjusted within-groups sum of squares on X, using the within-groups regression of X on Y.
5. Calculate an adjusted between-groups sum of squares by subtraction;

that is, subtract the adjusted within-groups sum of squares from the adjusted total sum of squares.

6 Obtain the variance estimates s_w^2 and s_b^2 by dividing the adjusted within-groups sum of squares on X by $df = N - k - 1$, and the between-groups by $df = k - 1$.

7 Test the significance of the adjusted means on X by referring $F = s_b^2/s_w^2$ to a table of F.

Table 20.1

Computation for the analysis of covariance: paired observations on Y and X for three groups

	Group I		Group II		Group III		
	Y	X	Y	X	Y	X	$N = 24$
	5	5	6	5	4	7	$T_y = 291$
	11	6	6	7	5	8	$\bar{Y} = 12.13$
	12	9	7	12	7	3	$T_y^2/N = 3{,}528.38$
	26	12	12	10	9	4	
	28	15	14	8	10	10	$\sum_{j=1}^{k}\sum_{i=1}^{n_j} Y_{ij}^2 = 4{,}777$
	24	16	9	4	6	5	
	11	18	8	15			$\sum_{j=1}^{k} \frac{T_{y_j}^2}{n_j} = 4{,}062.27$
	27	20	12	9			
	12	4					$T_x = 224$
	20	12					$T_x^2/N = 2{,}090.67$
n_j	10		8		6		$\sum_{j=1}^{k}\sum_{i=1}^{n_j} X_{ij}^2 = 2{,}618$
$T_{y_j} T_{x_j}$	176	117	74	70	41	37	
$\bar{Y}\ \bar{X}$	17.60	11.70	9.25	8.75	6.83	6.17	$\sum_{j=1}^{k} \frac{T_{x_j}^2}{n_j} = 2{,}209.57$
$\sum_{i=1}^{n_j} Y_{ij}^2 \ \sum_{i=1}^{n_j} X_{ij}^2$	3,720	1,651	750	704	307	263	$T_{xy} = 3{,}248$
$T_{x_j y_j}$	2,341		652		255		$\sum_{j=1}^{k} \frac{T_{x_j} T_{y_j}}{n_j} = 2{,}959.53$
$\frac{T_{y_j}^2}{n_j}\ \frac{T_{x_j}^2}{n_j}$	3,097.60 1,368.90		684.50 612.50		280.17 228.17		$\frac{T_x T_y}{N} = 2{,}716.00$

Sum of squares

	Y	X
Between	4,062.27 − 3,528.38 = 533.89	2,209.57 − 2,090.67 = 118.90
Within	4,777.00 − 4,062.27 = 714.73	2,618.00 − 2,209.57 = 408.43
Total	4,777.00 − 3,528.38 = 1,248.62	2,618.00 − 2,090.67 = 527.33

Sum of products

Between	2,959.53 − 2,716.00 = 243.53
Within	3,248.00 − 2,959.53 = 288.47
Total	3,248.00 − 2,716.00 = 532.00

20.9 ILLUSTRATIVE EXAMPLE: ANALYSIS OF COVARIANCE

Table 20.1 shows an artificial example illustrating the analysis of covariance. Information is available on two variables for $k = 3$ with unequal n's. The treatment means are $\bar{X}_1 = 11.70$, $\bar{X}_2 = 8.75$, and $\bar{X}_3 = 6.17$. Although these means differ appreciably, the means for the covariate Y also differ, these being $\bar{Y}_1 = 17.60$, $\bar{Y}_2 = 9.25$, and $\bar{Y}_3 = 6.83$. If a correlation of some magnitude exists between Y and X, we might anticipate that a substantial part of the variation in the X means will result from the differences in the Y means. All terms necessary for the direct calculation of sums of squares on X and Y and sums of products are given in Table 20.1.

Table 20.2 summarizes the analysis of covariance. The adjusted total sum of squares for X is $527.33 - 532.00^2/1{,}248.62 = 300.66$. The adjusted within-groups sum of squares is $408.43 - 288.47^2/714.73 = 292.00$. The adjusted between-groups sum of squares is $300.66 - 292.00 = 8.66$. The variance estimates are $s_b{}^2 = 4.33$ and $s_w{}^2 = 14.60$, and $F = 4.33/14.60 = .30$. This ratio is not significant, and, indeed, it falls substantially short of the value of F of unity expected under the null hypothesis. Quite clearly almost all the variation in the X means can be attributed to the influence of the uncontrolled variable Y.

It is of interest here to calculate directly the adjusted means on X. These adjusted values are given by

$$\bar{X}_j'' = b_w(\bar{Y} - \bar{Y}_j) + \bar{X}_j$$

In this sample $b_w = 288.47/714.73 = .404$. The three adjusted means are

$$\bar{X}_1'' = .404(12.13 - 17.60) + 11.70 = 9.49$$
$$\bar{X}_2'' = .404(12.13 - 9.25) + 8.75 = 9.91$$
$$\bar{X}_3'' = .404(12.13 - 6.83) + 6.17 = 8.31$$

Table 20.2

Analysis of covariance for data of Table 20.1

	Source of variation		
	Between	Within	Total
Sum of squares: Y	533.89	714.73	1,248.62
Sum of squares: X	118.90	408.43	527.33
Sum of products	243.52	288.47	532.00
Degrees of freedom	2	21	23
Adjusted sum of squares: X	8.66	292.00	300.66
Degrees of freedom for adjusted sum of squares	2	20	22
Variance estimates	$s_b{}^2 = 4.33$	$s_w{}^2 = 14.60$	

$F = 4.33/14.60 = .30$ $p > .05$

These adjusted means vary very little one from another, a fact which is clearly reflected in the small F ratio. We may safely conclude that the differences between the unadjusted means for X are due largely to the effects of Y.

20.10 HOMOGENEITY OF REGRESSION COEFFICIENTS

For certain purposes it may be a matter of interest to test the hypothesis that the slopes of the regression lines within the k groups are the same; that is, $H_0 : \beta_1 = \beta_2 = \cdot\ \cdot\ \cdot = \beta_k = \beta_w$. This hypothesis is assumed to be true in any application of the analysis of covariance.

To test this hypothesis, we require that the sum of squares for Y and X and the sum of products for each of the k treatments be given separately. These values are given by

$$\sum_{i=1}^{n_j} (X_{ij} - \bar{X}_j)^2 = \sum_{i=1}^{n_j} X_{ij}^2 - \frac{T_{x_j}^{\ 2}}{n_j}$$

$$\sum_{i=1}^{n_j} (Y_{ij} - \bar{Y}_j)^2 = \sum_{i=1}^{n_j} Y_{ij}^2 - \frac{T_{y_j}^{\ 2}}{n_j}$$

$$\sum_{i=1}^{n_j} (Y_{ij} - \bar{Y}_j)(X_{ij} - \bar{X}_j) = \sum_{i=1}^{n} X_{ij}Y_{ij} - \frac{T_{x_j}T_{y_j}}{n_j}$$

The next step is to calculate separately for each group an adjusted sum of squares on X in the manner previously described. This is done by subtracting from the sum of squares for X a quantity equal to the square of the sum of products divided by the sum of squares for Y. These adjusted sums of squares are summed over the k groups to obtain

$$\sum_{j=1}^{k} \left\{ \sum_{i=1}^{n_j} (X_{ij} - X_j)^2 - \frac{\left[\sum_{i=1}^{n_j} (Y_{ij} - \bar{Y}_j)(X_{ij} - \bar{X}_j) \right]^2}{\sum_{i=1}^{n_j} (Y_{ij} - \bar{Y}_j)^2} \right\} = A \qquad [20.14]$$

For convenience we denote this term by A. This term is a sum of squares of residuals about the k individual regression lines. It has associated with it $\sum^{k} (n_j - 2) = \sum^{k} n_j - 2k = N - 2k$ degrees of freedom. The number of degrees of freedom for each regression line is $n_j - 2$, and there are k such lines. We now obtain an adjusted within-groups sum of squares on X for all observations in the k groups in the manner described previously in this chapter. Denote this sum of squares by B. This sum of squares has $N - k - 1$ degrees of freedom.

To test for homogeneity of regression the following F ratio is calculated:

[20.15] $$F = \frac{(B - A)/(k - 1)}{A/(N - 2k)}$$

with $k - 1$ and $N - 2k$ degrees of freedom associated with the numerator and denominator, respectively.

The general rationale underlying the above F ratio is that when $H_0 : \beta_1 = \beta_2 = \cdot \cdot \cdot = \beta_k$ is true, the sum of squares of residuals about the individual group regression lines with slope b_j will be the same, within the limits of sampling error, as the sum of squares of residuals about a single within-groups regression line with slope b_w. Under this circumstance $B - A$ will depart from 0 only because of the presence of sampling error, and the expected value of the F ratio is unity. When $H_0 : \beta_1 = \beta_2 = \cdot \cdot \cdot = \beta_k$ is not true, B will tend to be greater than A and the expected value of F will be greater than unity.

Table 20.3 uses the data of Table 20.1 to illustrate a test of homogeneity of regression. Sums of squares on Y and X, sums of products, and adjusted sums of squares for X have been calculated for each of the three groups separately. The adjusted sums of squares are summed to obtain A. The Y, X, and sum-of-products columns are summed to obtain within-groups sums of squares and products. An adjusted within-groups sum of squares on X is calculated and denoted by B. An F ratio is calculated and is found to be .37, which is, of course, not significant.

In this example the regression slopes for the three groups are, respectively, $b_1 = .45$, $b_2 = .07$, and $b_3 = .08$. Also, $b_w = .404$. Given the small samples used in this illustrative example, the differences between such coefficients would not prove to be significant. In fact, the differences are less than chance expectation.

Table 20.3

Computation for test of homogeneity of regression coefficients using data of Table 20.1

Group	Sum of squares		Sum of products	Adjusted sum of squares for X
	Y	X		
1	622.40	282.10	281.80	154.51
2	63.50	91.50	4.50	91.19
3	26.83	34.83	2.17	34.65
Total	714.73	408.43	288.47	280.35 = A

$A = 280.35 \qquad B = 408.43 - 288.47^2/714.73 = 292.00$

$$F = \frac{(292.00 - 280.35)/(3 - 1)}{280.35/(24 - 6)} = \frac{5.83}{15.58} = .37$$

20.11 PROBLEMS OF INTERPRETATION

The statistical apparatus used in the analysis of covariance is simple. The appropriateness of the method and the interpretation of its results may present difficulties. The method makes the statistical assumption of homogeneity of regression coefficients. Apart from this, the following questions may be raised. *First,* is the covariate affected by treatments? *Second,* is the covariate itself a treatment variable? *Third,* does the calculation of adjusted means using linear extrapolation lead to meaningful and realistic results in the light of existing knowledge? *Fourth,* if intact groups are used, does this invalidate the experiment? *Fifth,* what is the nature of the causal relation between the covariate and the dependent variable?

The proper use of the method assumes that differences in adjusted means can be attributed directly within limits of error to treatment effects. In some experiments the covariate may be affected by treatments; consequently, differences among treatment means cannot unambiguously be related to treatments. If treatments affect the covariate values, the calculation of adjusted means will remove part of the treatment effects. If the covariate values are obtained prior to treatments or quite independently of them, then these covariate values will be unaffected by treatments.

Not uncommonly in the behavioral sciences the covariate and the dependent variable are both *intrinsic* attributes of the experimental unit, which is usually a human subject or an experimental animal. IQ, attitude, and achievement measures, for example, are intrinsic to the experimental subjects. The covariate may, however, be an *extrinsic* attribute. Such would be the case, for example, if the covariate were parental attitude, an index of socioeconomic status of the family, or number of siblings. Although bizarre exceptions come to mind, variation in extrinsic covariates cannot ordinarily be affected by treatments.

Consider the following example. The learning of a language, say French, is known to be affected by attitudes toward the French ethnolinguistic group. In an experiment on different methods of teaching French, attitude measures may be obtained prior to the initiation of instruction. These measures are the covariate, and are clearly unaffected by treatments. If attitude measures are obtained during or following the instruction, such measures may be affected by instruction and attitude change may be part of the treatment effect, since the process of learning French may result in attitude change toward the French ethno-linguistic group.

In some experiments the covariate may be an integral, though unplanned and uncontrolled, part of the treatment; that is, it may in effect be a treatment variable. Consider an experiment designed to investigate the effects of brain lesions at two separate brain loci on maze performance in experimental animals. Due to the operative procedures used, lesion size tends to be larger in one locus than the other. Do differences in maze performance result from locus of lesion or size of lesion? Let us suppose that mean maze performance scores are different for the two loci, but when lesion size is used as a covariate, no difference of consequence is found in the adjusted means. Can the assertion be made that the differences in means results from lesion size and has nothing to do with locus? The conclusion must be that the analysis of covariance provides no adequate basis for assertions regarding the relative causal effects of lesion size and locus. It is no substitute for a properly designed two-factor experiment in which both lesion size and locus are properly controlled by the investigator.

The analysis of covariance is applied on occasion in situations where the independent variable is not a set of experimental treatments, but a pre-existing natural grouping such as sexes, types of mental patient, or varieties of corn. In many instances these groupings are genetically determined. Applications of the analysis of covariance to this type of data are discussed by Lord (1967, 1969). He considers an experiment designed to study whether the diet provided in the residence halls of a university affects the weight of students, classified by sex. Weight at the beginning of the academic term is the covariate. Weight at the end of the term is the dependent variable. Sex is the independent variable. No differences whatever are found in before-and-after means for women or for men. One conclusion is that diet has no effect on weight. If, however, means are adjusted to make allowances for differences in initial weight, substantial differences are found between the sexes. A second conclusion is that men will show more weight gain than women when allowance is made for initial weight differences. Which conclusion is correct? The point here is that the two conclusions are answers to different questions. The first question is straightforward. Does the diet affect weight for men and women? The answer is no. The second question is hypothetical and may not be meaningful. If the mean weights for women and men were equal, would any weight change occur? This question is of dubious meaning. It raises a hypothetical speculation about a fictional population. The reason for this is that the adjusted means are obtained by linear extrapolation. In this example the extrapolation is to hypothetical samples, presumably drawn from hypothetical populations, in which the initial mean weights for women and men are the same. But no such populations may exist in nature. The extrapolation may be a statistical contrivance. No answers about populations as they actually exist may result from this procedure. The force of this argument becomes more compelling if, instead of women and men, we substitute mice and elephants. Speculation about mice that are about half the size of elephants may be an interesting intellectual exercise in science fiction. As a statement about a nonfictional world with normal mice and elephants, it has no relevance. Generally, in all applications of the analysis of covariance the meaningfulness of the extrapolation should be examined to ensure that the adjusted means could in fact occur. Note that in the sentence, "On the assumption that the covariate means are equal, then the dependent variable means are such and such . . . ," the problem is with the premise rather than the conclusion. Any assumption can of course be made, and the conclusion may be a perfectly logical deduction from it. The issue is whether or not the assumption is such that it leads to a conclusion that is not fictional. The point here has been stated succinctly by Scheffé (1959): ". . . such applications of the analysis of covariance require discretion, for the reason that the inferences may be correct answers to the wrong questions."

As mentioned above, in some applications of the analysis of covariance, intact groups are used. In educational research, for example, it is at times not possible to assign students to treatments at random. Intact classes of students must be used. Some authors argue that this practice is not warranted unless the class is used as the experimental unit. In the practical conduct of research this argument is perhaps overly rigorous. If the covariate is the major basis for assigning students to classes, then clearly the use of intact groups is inappropriate, because, in effect, the covariate is part of the treatment. If the covariate is not so used, and covariate differences result perhaps from a miscellany of idiosyncratic circumstances, then the use of intact groups may not necessarily invalidate the conclusions drawn. Of course, if covariate differences are large and exert a substantial influence on the means of the dependent variable, clearly the investigator must explore why this is so.

The use of intact groups is perhaps an extreme example of a common situation in experimental work. As discussed elsewhere in this book, the assumptions or conditions underlying the use of a statistical method are frequently not rigorously met in practice. The investigator may discard the data. Alternatively, using all information available, he or she may inquire how the conclusions drawn are affected by violation of the assumptions made or conditions required. This question may have a plausible answer. Answering such questions is one of the arts of the statistician.

As a purely statistical procedure the analysis of covariance makes no assumption about a causal relation between the covariate and the dependent variable. The covariate could be anything at all. Implicit in the meaningful use of the method, however, is the assumption that a *direct* causal relation exists between the covariate and the dependent variable. An example, sometimes quoted, is the relation between the prevalence of weeds and the yield of corn in an experiment involving different varieties. Prevalence of weeds and corn yield may be correlated, because both are correlated with a third variable, perhaps a soil fertility factor. Removing the weeds may in no way affect corn yield. Assuming a direct causal relation could lead to a costly, but completely irrelevant, program of weed control. Weeds as such may have little or nothing to do with corn yield. In any use of the analysis of covariance the investigator is well advised to reflect carefully on the nature of the causal relation between the covariate and the dependent variable.

In general the use of the analysis of covariance requires caution. It cannot be blindly applied. Certain statistical assumptions must be satisfied. Also a number of interpretative problems arise. In some instances these problems may be difficult to answer.

For further discussion on problems of interpretation in the analysis of covariance see Lord (1967, 1969), Scheffé (1959), Smith (1957), and Weisberg (1979).

20.12 BLOCKING AS AN ALTERNATIVE TO COVARIANCE ANALYSIS

An alternative to the analysis of covariance is a randomized blocks deisign as described in Section 19.11. This design has definite advantages over the analysis of covariance, provided, of course, that the experiment can be planned and conducted in this way in the first place.

One major advantage of a randomized blocks design is that it permits the study of interaction. The analysis of covariance assumes homogeneity of regression, usually linear regression. This assumption is equivalent to the assumption of zero interaction. Clearly if an interaction is thought to exist, a randomized blocks design is preferable.

A randomized blocks design controls the so-called "uncontrolled" variable, the blocking variable, by actually making the treatment groups equivalent on that variable. The design also makes the blocking variable and the independent variable independent of each other. In covariance analyses, if the analysis is to serve its purpose, the treatment groups will differ on the covariate; also the covariate and the independent variable will not be independent. The design controls the "uncontrolled" variable, the covariate, by assuming homogeneity of regression (linear in most cases) and by the statistical procedure of extrapolation. The analysis of covariance can be extended to deal with nonlinear regression involving quadratic or cubic equations, but the assumption of homogeneity of regression coefficients is still made.

With some sets of data, blocking the covariate after the data are collected may be a possibility, although remote. Attempts at after-the-fact blocking will usually lead to a variety of problems, such as unequal numbers in the cells, missing cell values, and the like.

In practice, one reason for using the analysis of covariance is that in a particular research context a randomized block experiment may be very difficult to conduct. Apart from problems of interpretation, the basic statistical assumption underlying the method is the homogeneity of regression assumption. This assumption should always be checked by applying the appropriate statistical test or by careful graphical inspection.

20.13 FURTHER USES OF THE ANALYSIS OF COVARIANCE

In this chapter we have considered situations involving only one covariate. The analysis of covariance may be used with more than one covariate. With such data the variables are combined using a multiple regression method. The description of the analysis of covariance in this chapter applies to one-way classification experiments. The method may be extended to higher order factorial experiments, to mixed designs with repeated measurements on one factor but not on another, and to other experimental situations as well. In most discussions of the analysis of covariance for two-way factorial experiments, equal numbers in the subclasses are assumed. The usual adjustments for unequal n's may be made. Descriptions of the procedures mentioned above are given in more advanced texts such as Winer (1971).

BASIC TERMS AND CONCEPTS

Statistical versus experimental control

Covariate

Sum of products

Partitioning the sum of products

Within-groups regression line

Between-groups regression line

Adjusted sum of squares

Homogeneity of regression

EXERCISES

1 The following are paired observations for three experimental groups:

	I		II		III	
	Y	X	Y	X	Y	X
	2	7	8	15	15	30
	5	6	12	24	16	35
	7	9	15	25	20	32
	9	15	18	19	24	38
	10	12	19	31	30	40
$\bar{Y}, \bar{X}$	6.60	9.80	14.40	22.80	21.00	35.00

In this example Y is the covariate or concomitant variable. Calculate the adjusted total, within-groups, and between-groups sums of squares on X, and test the significance of the differences between the adjusted means on X by using the appropriate F ratio.

2 For thc data of Exercise 1 above, calculate (**a**) the slope of the overall regression line b_t for predicting X from Y, (**b**) the slope of the overall within-groups regression line b_w, (**c**) the slope of the between-groups regression line b_b.

3 Calculate for Exercise 1 above the adjusted means on X.

4 Test the homogeneity of the slopes of the three within-groups regression lines in Exercise 1 above.

ANSWERS TO EXERCISES

1 Adjusted total sums of squares on $X = 319.13$
Adjusted within-groups sums of squares on $X = 128.73$
Adjusted between-groups sums of squares on $X = 190.41$
$F = 8.14$, $df_1 = 2$, $df_2 = 11$; $p < .01$

2 **a** 1.39 **b** .72 **c** 1.75

3 Group I, 15.14; group II, 22.51; group III, 29.95

4 $A = 125.54$, $B = 128.73$, $F = .15$; $p > .05$

NONPARAMETRIC STATISTICS

21

THE STATISTICS OF RANKS

21.1 INTRODUCTION

Ordinal, or rank-order, data may arise in a number of different ways. Quantitative measurements may be available, but ranks may be substituted to reduce arithmetical labor or to make some desired form of calculation possible. For example, measurements of height and weight may be obtained for a group of schoolchildren. A correlation between the paired measurements could readily be calculated. The investigator may, however, choose to substitute ranks for the measurements and to calculate a correlation between the paired ranks. In many situations where ranking methods are used, quantitative measurements are not available. The measuring operations used may be such that no comparative statements about the intervals between members are possible. For example, employees may be rank ordered by supervisors on job performance. Schoolchildren may be ranked by teachers on social adjustment. Different teas may be rank ordered by experienced judges on taste, or participants in a beauty contest may be rank ordered by judges on pulchritude. In such cases the data are comprised of sets of ordinal numbers, 1st, 2d, 3d, . . . , Nth. These are replaced by the cardinal numbers 1, 2, 3, . . . , N for purposes of calculation. The substitution of cardinal numbers for ordinal numbers always assumes equality of intervals. The difference between the first and second member is assumed equal to the difference between the second and third, and so on. This assumption underlies all coefficients of rank correlation. Because of difficulties associated with the measurement of psychological variables, statistical methods for handling rank-order data are of particular interest to psychologists.

Although rank correlation methods have been in use for many years, more recently extensive use has been made of ranks in dealing with many other statistical problems. Ranks are used, for example, in tests for comparing two correlated or independent samples, these being the analogs of t tests, and for a variety of other purposes. Such tests are described as *nonparametric* or *distribution-free* and are discussed in some detail in Chapter 22 to follow. Rank correlation methods are usually classed as nonparametric procedures.

21.2 INTEGERS

Ranks are represented by the integers 1, 2, 3, . . . , N. These are denoted by $X_1, X_2, X_3, \ldots, X_N$. The sum, and sum of squares, of the first N integers may be shown to be as follows:

[21.1] $$\sum_{i=1}^{N} X_i = \frac{N(N+1)}{2}$$

[21.2] $$\sum_{i=1}^{N} X_i^2 = \frac{N(N+1)(2N+1)}{6}$$

The mean of the first N integers is $\bar{X} = (N+1)/2$ and the variance, obtained by dividing the sum of squares about the mean by N, is $s^2 = (N^2 - 1)/12$. The mean is a simple function of the variance, $\bar{X} = 6s^2/(N-1)$. A knowledge of the above relations is sometimes useful.

21.3 MEASURES OF DISARRAY

Consider N individuals $A_1, A_2, A_3, \ldots, A_N$ ranked on two variables X and Y. The rankings on X may be denoted as $X_1, X_2, X_3, \ldots, X_N$ and those on Y as $Y_1, Y_2, Y_3, \ldots, Y_N$. Let the ranks for five individuals on X and Y be as follows:

X	1	2	3	4	5
Y	1	4	3	5	2

The X ranks are in their natural order. The Y ranks are not in their natural order. They exhibit a degree of disarray with respect to X. How may a measure of disarray in this situation be defined?

One commonly used measure of disarray is the sum of squares of differences between the paired ranks. This quantity may be designated by Σd^2. In the above example $\Sigma(X-Y)^2 = \Sigma d^2 = (0)^2 + (-2)^2 + (0) + (-1)^2 + (3)^2 = 14$. It is of interest to consider the minimum and maximum values of Σd^2. When members are ranked in the same order on both X and

Y, $\Sigma d^2 = 0$ and is a minimum. Thus if the ranks on X are 1, 2, 3, 4, 5 and on Y are 1, 2, 3, 4, 5, the differences are all 0. If the paired ranks are in an inverse order, the maximum degree of disarray, Σd^2 is a maximum. Thus if the ranks on X are 1, 2, 3, 4, 5 and on Y are 5, 4, 3, 2, 1, the differences d are $-4, -2, 0, 2$, and 4 and $\Sigma d^2 = 40$. No arrangement of Y with respect to X will produce a larger value of Σd^2. It may be shown that the maximum value of Σd^2 is given by

[21.3] $$\Sigma d^2_{\text{max}} = \frac{N(N^2 - 1)}{3}$$

If ranks on Y are arranged at random with respect to X, the expected value of Σd^2 is simply one-half Σd^2_{max} or

[21.4] $$E(\Sigma d^2) = \frac{N(N^2 - 1)}{6}$$

The statistic Σd^2 is only one of a number of measures of disarray which might be defined.

Another measure of disarray is the statistic S. Consider again the paired ranks on X 1, 2, 3, 4, 5 and on Y 1, 4, 3, 5, 2. The X ranks are in their natural order and the Y ranks exhibit a degree of disarray with respect to X. To calculate S we compare each rank on Y with every other rank, there being $N(N-1)/2$ such comparisons for N ranks. If a pair is ranked in its natural order, say, 1 and 4, a weight $+1$ is assigned. If a pair is ranked in an inverse order, say, 4 and 3, a weight -1 is assigned. The statistic S is the sum of such weights over $N(N-1)/2$ such comparisons. In the example above the weights are $+1, +1, +1, +1, -1, +1, -1, +1, -1, -1$, and $S = 2$. The maximum and minimum values of S may be considered. The maximum value occurs when both sets of ranks are in a natural order, all weights being $+1$, and is $N(N-1)/2$. The minimum value of S occurs when both sets of ranks are in an inverse order, all weights being -1, and is $-N(N-1)/2$. When ranks are arranged at random with respect to each other, the expected value of S is 0.

Both Σd^2 and S are used in the definition of measures of rank correlation and for other purposes as well.

21.4 SPEARMAN'S COEFFICIENT OF RANK CORRELATION ρ

The measure of disarray, Σd^2, is used in the definition of Spearman's coefficient of rank correlation. A coefficient of rank correlation is a statistic defined in such a way as to take a value of $+1$ when the paired ranks are in the same order, a value of -1 when the ranks are in an inverse order, and an expected value of 0 when the ranks are arranged at random with respect to each other. A definition of rank correlation which meets these requirements is

$$\rho = 1 - \frac{2\Sigma d^2}{\Sigma d^2_{\max}} \tag{21.5}$$

where ρ is the Greek letter rho. When the paired ranks are in the same order, $\Sigma d^2 = 0$ and $\rho = 1$. When the paired ranks are in an inverse order, $\Sigma d^2 = \Sigma d^2_{\max}$ and $\rho = -1$. In the case of independence, $2\Sigma d^2 = \Sigma d^2_{\max}$ and $\rho = 0$. By substituting $\Sigma d^2_{\max} = N(N^2 - 1)/3$ in the above formula, we get

$$\rho = 1 - \frac{6\Sigma d^2}{N(N^2 - 1)} \tag{21.6}$$

This is the usual way of writing Spearman's coefficient of rank correlation.

Pearson's product-moment correlation coefficient may be written in the form

$$r = \frac{\Sigma(X - \bar{X})(Y - \bar{Y})}{\sqrt{\Sigma(X - \bar{X})^2\,\Sigma(Y - \bar{Y})^2}} \tag{21.7}$$

Spearman's ρ is a particular case of the above formula. It is the particular case that arises when the variables are the first consecutive untied integers. If the above formula is applied directly to paired ranks, the result is identical with that obtained by applying the formula for ρ.

The calculation of ρ is illustrated in Table 21.1. The calculation is

Table 21.1

Calculation of Spearman's coefficient of rank correlation

	Rank		Difference	
Individual	X	Y	d	d^2
A_1	1	6	−5	25
A_2	2	3	−1	1
A_3	3	7	−4	16
A_4	4	2	2	4
A_5	5	1	4	16
A_6	6	8	−2	4
A_7	7	4	3	9
A_8	8	9	−1	1
A_9	9	5	4	16
A_{10}	10	10	0	0
Total			0	$\Sigma d^2 = 92$

$$\rho = 1 - \frac{6 \times 92}{10(100 - 1)} = .442$$

simple. We find the differences between the paired ranks, square them, sum to obtain Σd^2, and then apply the formula for ρ.

21.5 SPEARMAN'S ρ WITH TIED RANKS

In arranging the members of a group in order, a judge may be unable to discriminate between certain members. Where measurements are replaced by ranks, certain measurements may be equal. These circumstances give rise to tied ranks. If we attempt to replace the numbers 14, 19, 19, 22, 23, 23, 23, 25 by ranks, we observe immediately that 19 occurs twicc and 23 three times. Under these circumstances we assign to each member the average rank which the tied observations occupy. Thus 14 is ranked 1, the two 19s are ranked 2.5 and 2.5, the 22 is ranked 4, the three 23s are ranked 6, 6, and 6, and 25 is ranked 8. Having replaced the tied ranks by their average ranks, we proceed as before in the calculation of ρ. A calculation with tied ranks is illustrated in Table 21.2. If the ties are numerous, this type of adjustment for tied ranks may not prove altogether satisfactory.

The development of ρ from the ordinary product-moment r assumes that the ranks are the first N integers. Where tied ranks occur this is not so. Where a substantial number of tied ranks is found, the departure of the sum of squares of ranks from the sum of squares of the first N integers will

Table 21.2

Calculation of Spearman's coefficient of rank correlation with tied ranks

	Rank		Difference	
Individual	X	Y	d	d^2
A_1	1	8	−7	49.00
A_2	2.5	6.5	−4	16.00
A_3	2.5	4.5	−2	4.00
A_4	4.5	2	2.5	6.25
A_5	4.5	1	3.5	12.25
A_6	6	3	3	9.00
A_7	8	4.5	3.5	12.25
A_8	8	6.5	1.5	2.25
A_9	8	9	−1	1.00
A_{10}	10	10	0	.00
Total				$\Sigma d^2 = 112.00$

$$\rho = 1 - \frac{6 \times 112}{10(100 - 1)} = .321$$

be appreciable and the value of ρ will be thereby affected. While other procedures for correcting for ties may be used, one convenient approach is to calculate an ordinary product-moment correlation for the paired observations where average ranks have been substituted for ties.

21.6 TESTING THE SIGNIFICANCE OF SPEARMAN'S ρ

The problem of developing a test of significance for ρ is approached by considering the N factorial possible arrangements of Y with respect to a fixed ranking on X. Each arrangement is considered equally probable. For each arrangement either the quantity Σd^2 or ρ may be calculated. A frequency distribution can be made either of the Σd^2 or of the values of ρ. This is a null distribution and serves as a model for evaluating a particular observed value of Σd^2, or ρ, against the null hypothesis H_0 that no association between X and Y exists. If the particular observed arrangement of Y with respect to X, as reflected in a low Σd^2, or a high ρ, is improbable, say, $p \lessapprox .05$ or .01, then the null hypothesis is rejected. To illustrate, for $N = 2$, if X has the ranks 1, 2, only two arrangements of Y with respect to X are possible: 1, 2 and 2, 1. Only two values of ρ are possible, $+1$ and -1. For $N = 3$, if X has the ranks 1, 2, 3, there are six possible arrangements of Y and, as it turns out, four different values of ρ, -1, $-\frac{1}{2}$, $+\frac{1}{2}$, and $+1$. The sampling distribution of Σd^2 has been studied by Kendall (1970) and others. For $N = 7$ or 8 the distribution has a somewhat jagged or serrated appearance. The distribution is always symmetrical. As N increases in size, the distribution approaches the normal form.

Table G of the Appendix shows critical values of ρ for different values of N required for significance at various levels. Observe that for a small N, values of ρ of very substantial size must be obtained before we have adequate grounds for rejecting the hypothesis that no association exists between the rankings. For $N = 10$ we require a ρ equal to or greater than .564 before we can argue that a significant association exists in a positive direction at the 5 percent level.

With $N = 10$ or greater, we may test the significance of ρ by using a t given by

[21.8] $$t = \rho\sqrt{\frac{N-2}{1-\rho^2}}$$

This quantity has a t distribution with $N - 2$ degrees of freedom. For example, where $N = 10$ and $\rho = .564$, $t = 1.93$. For 8 degrees of freedom, the value of t at the .05 level is 2.31. For a two-tailed test we have insufficient grounds for arguing that the observed ρ is significantly different from 0. For a one-tailed test the observed ρ is significant at about 5 percent level.

21.7 KENDALL'S COEFFICIENT OF RANK CORRELATION

An alternative form of rank correlation, τ, or tau, has been developed by Kendall (1970). In the definition of Kendall's tau use is made of the measure of disarray, S. The maximum possible value of S is $N(N-1)/2$. Kendall's coefficient of rank correlation, τ, is defined as the obtained value of S divided by its maximum possible value; that is,

[21.9] $$\tau = \frac{S}{\frac{1}{2}N(N-1)}$$

The statistic τ has a value -1 when the paired ranks are in an inverse order, and a value of $+1$ when the paired ranks are in the same order. For X of 1, 2, 3, 4, 5 and Y of 1, 4, 3, 5, 2, $S = 2$, $N = 5$, and $\tau = \frac{2}{10} = .20$.

21.8 KENDALL'S τ WITH TIED RANKS

If ties occur, the convention is adopted, as in the calculation of Spearman's ρ, of replacing the tied values by the average rank. A comparison of two tied values on Y receives a weight of 0. If ties occur on X, a comparison of the corresponding paired Y values will also receive a weight of 0, regardless of whether the paired Y values are tied. Consider the following example with no ties on X and one tied pair on Y.

X	1	2	3	4	5	6
Y	2	3	4.5	4.5	1	6

On comparing each rank on Y with every other rank, and assigning a $+1$ for a pair in their natural order, a -1 for a pair in an inverse order, and a 0 for a tie, we obtain $+1$, $+1$, $+1$, -1, $+1$, $+1$, $+1$, -1, $+1$, 0, -1, $+1$, -1, $+1$, $+1$, and $S = 6$. Consider another example with tied values on both X and Y.

X	1.5	1.5	3	5	5	5
Y	2	3	4.5	4.5	1	6

Here the comparison on Y of 2 with 3 receives a weight of 0, because the order of the first two paired values on X is arbitrary. Similarly, comparisons on Y involving the last three values will receive weights of 0, because of the triplet of ties on X. In the above example the weights are 0, $+1$, $+1$, -1, $+1$, $+1$, $+1$, -1, $+1$, 0, -1, $+1$, 0, 0, 0, and $S = 4$.

To calculate tau with tied ranks, S is calculated in the manner described above, and the following formula applied:

[21.10] $$\tau = \frac{S}{\sqrt{[\frac{1}{2}N(N-1) - T_x][\frac{1}{2}N(N-1) - U_y]}}$$

In this formula $T_x = \frac{1}{2}\sum^{m} t(t-1)$ and $U_y = \frac{1}{2}\sum^{r} u(u-1)$. One ranking contains m sets of t ties; the other ranking, r sets of u ties. To illustrate, consider the example immediately above with ties on both X and Y. Here $T_x = \frac{1}{2}[2(2-1) + 3(3-1)] = 4$; also $U_y = \frac{1}{2}[2(2-1)] = 1$. In this example $N = 6$, and tau is as follows:

$$\tau = \frac{4}{\sqrt{[\frac{1}{2} \times 6(6-1) - 4][\frac{1}{2} \times 6(6-1) - 1]}} = .31$$

21.9 THE SIGNIFICANCE OF S AND τ

The sampling distribution of S is obtained by considering the N factorial arrangements of Y in relation to X. A value of S may be determined for each of the N arrangements. The distribution of these N factorial values is the sampling distribution of S. This distribution is symmetrical. Frequencies taper off systematically from the maximum value toward the tails. The distribution of S rapidly approaches the normal form. For $N \geq 10$ the normal approximation to the exact distribution is very close. The exact sampling distributions of S for $N = 4$ to $N = 10$ are given by Kendall (1970).

In testing the significance of the association between paired ranks, it is more convenient to apply a test directly to S rather than to τ. The variance of the sampling distribution of S without ties is given by

[21.11] $$\sigma_s^2 = \frac{N(N-1)(2N+5)}{18}$$

If the normal approximation to the exact sampling distribution of S is used, a correction for continuity should be applied. This is done by subtracting unity from the absolute value of S. To apply a significance test, we divide S, corrected for continuity, by the standard deviation of the sampling distribution to obtain the normal deviate z, as follows:

[21.12] $$z = \frac{|S| - 1}{\sqrt{N(N-1)(2N+5)/18}}$$

As usual, 1.96 and 2.58 are required for significance at the .05 and .01 levels, respectively, for a nondirectional test. To illustrate, consider the paired ranks:

X	1	2	3	4	5	6
Y	2	4	3	5	1	6

Here the weights are +1, +1, +1, −1, +1, −1, +1, −1, +1, +1, −1, +1, −1, +1, +1, and $S = 5$. Hence

$$z = \frac{5-1}{\sqrt{6(6-1)(2\times 6+5)/18}} = \frac{4}{5.323} = .751$$

Here the association between the paired ranks is clearly not significant.

Because ties are very common with problems involving ranks, it is useful to know the variance of the sampling distribution of S when ties occur. If ties occur in one set of ranks and not in the other, the variance of S becomes

[21.13] $$\sigma_s^2 = \tfrac{1}{18}\left[N(N-1)(2N+5) - \sum^{m} t(t-1)(2t+5)\right]$$

One set of ranks contains m sets of t ties. Note that the effect of ties is to reduce the variance σ_s^2. Examination of the above formula shows that the effect of ties is to reduce the variance by unity for each tied pair, by 3.67 for each triplet of ties, and by 8.67 for each quadruplet of ties. Thus we have available a very convenient correction procedure.

If ties occur in both sets of ranks, the variance of S is given by

[21.14] $$\sigma_s^2 = \tfrac{1}{18}\left[N(N-1)(2N+5) - \sum^{m} t(t-1)(2t+5) - \sum^{r} u(u-1)(2u+5)\right] + \frac{1}{9N(N-1)(N-2)}\left[\sum^{m} t(t-1)(t-2)\right]\left[\sum^{r} u(u-1)(u-2)\right] + \frac{1}{2N(N-1)}\left[\sum^{m} t(t-1)\right]\left[\sum^{r} u(u-1)\right]$$

In this formula one set of ranks contains m sets of t ties: the other set of ranks, r sets of u ties. This formula is the more general form for the variance of the distribution of S. As in the untied case, to apply a test of significance, we divide S, corrected for continuity, that is, $|S| - 1$, by the standard deviation of the sampling distribution σ_s to obtain the normal deviate z.

21.10 RANK CORRELATION WHEN ONE VARIABLE IS A DICHOTOMY

The statistic S, and the rank correlation coefficient tau, may be calculated when one set of ranks is a dichotomy. Consider an example given by Kendall (1970) for 15 boys and girls ranked on an examination. Here sex is a dichotomous variable. Let X be ranks on the examination and T, sex.

X	1	2	3	4	5	6	7	8	9	10	11	12	13	14	15
Y	B	B	G	B	G	G	B	B	B	G	B	G	B	G	G

Here the eight boys may be considered tied on Y; also the seven girls. Adopting the average-rank procedure, a rank of 4.5 may be assigned to each of the eight boys and a rank of 12 to each of the seven girls. The rank 4.5 is the average of the ranks 1 to 8, and the rank 12 is the average of the ranks 9 to 15. Thus we may write

X	1	2	3	4	5	6	7	8	9	10	11	12	13	14	15
Y	$4\frac{1}{2}$	$4\frac{1}{2}$	12	$4\frac{1}{2}$	12	12	$4\frac{1}{2}$	$4\frac{1}{2}$	$4\frac{1}{2}$	12	$4\frac{1}{2}$	12	$4\frac{1}{2}$	12	12

The statistic S calculated from these ranks is

$$S = 7 + 7 - 6 + 6 - 5 - 5 + 4 + 4 + 4 - 2 + 3 - 1 + 2 + 0 = 18$$

The coefficient tau is obtained by using formula [21.10] and is

$$\tau = \frac{18}{\sqrt{[\frac{1}{2} \times 15(15-1)][\frac{1}{2} \times 15(15-1) - 49]}} = .235$$

where $T_x = 0$ and $U_y = \frac{1}{2}(8 \times 7) + \frac{1}{2}(7 \times 6)$. In this example the rank 4.5 has been assigned to boys and the rank 12 to girls, resulting in a positive correlation. This, of course, is arbitrary. The ranks could have been reversed and a negative correlation obtained.

To test the significance of the association, the standard error σ_s is calculated and a normal deviate z obtained. In the present illustrative example X contains no ties, Y contains two groups of ties, and σ_s may be obtained using the square root of formula [21.13] as follows:

$$\sigma_s = \sqrt{\tfrac{1}{18}(15 \times 14 \times 35 - 8 \times 7 \times 21 - 7 \times 6 \times 19)} = 17.28$$

Subtracting unity from the absolute value of S as a continuity correction and dividing by σ_s, we obtain the normal deviate

$$z = \frac{|S| - 1}{\sigma_s} = \frac{17}{17.28} = .984$$

Clearly the association between ranks on the examination and sex is not significant.

When one variable is a dichotomy and no ties occur in the X ranks, S is reduced by unity as a continuity correction. When one variable is a dichotomy and the other contains m groupings of values of extent t_i, the absolute value of S is reduced by $(2N - t_1 - t_m)/2(m - 1)$, where $t_1 \geq 1$. Here t_1 and t_m are the number of tied values in the first and last groups. The reader should note also that when ties occur in the X variable, formula [21.14] and not formula [21.13] should be used in calculating σ_s.

21.11 COMPARISON OF ρ AND τ

The coefficients ρ and τ, although used for the same purpose, are on a somewhat different scale. If ρ and τ are calculated on the same data, the absolute value of ρ will usually be larger than the corresponding value of τ. Values of ρ and τ are highly correlated in samples from a bivariate normal population. When the correlation in the population is 0, the product-moment correlation between ρ and τ is .980 for $N = 5$ and approaches 1 as N approaches infinity. For practical purposes the correlation between the two statistics can be regarded as close to 1.

On considering the N factorial arrangements of one set of ranks in relation to another, the distributions of both ρ and τ are symmetrical and tend to the normal form for large N. The distribution of τ tends to approach the normal form more rapidly than that of ρ. The exact distributions are known for higher values of τ than of ρ. In general τ as a statistic is more amenable to mathematical manipulation than ρ. Problems resulting from tied values are more readily solved. Also, the measure of disarray S, used in the definition of τ, seems to have a degree of generality about it that does not characterize Σd^2. Thus S has a number of applications apart from its use in correlation.

21.12 THE COEFFICIENT OF CONCORDANCE W

For data comprising m sets of ranks, where $m > 2$, a descriptive measure of the agreement or concordance between the m sets is provided by Kendall's coefficient of concordance W. The data of Table 21.3 consist of six ranks assigned by four judges. These data were obtained in an investigation on interviewing technique. Four interviewers were required to interview six job applicants and rank order them on suitability for employment. If perfect agreement were observed between the four interviewers, one applicant would be assigned a 1 by all four. The sum of his or her ranks

Table 21.3
Ranks assigned to six job applicants by four interviewers

	Applicant					
Interviewer	a	b	c	d	e	f
A	6	4	1	2	3	5
B	5	3	1	2	4	6
C	6	4	2	1	3	5
D	3	1	4	5	2	6
R_j	20	12	8	10	12	22

would be 4. Another applicant would be assigned a 2 by all four interviewers. The sum of his or her ranks would be 8. The sum of ranks for the six applicants would be 4, 8, 12, 16, 20, and 24, not necessarily in that order. In general, when perfect agreement exists among ranks assigned by m judges, to N members, the rank sums are m, $2m$, $3m$, $4m$, . . . , Nm. The total sum of N ranks for m judges is $mN(N+1)/2$, and the mean rank sum is $m(N+1)/2$.

The degree of agreement between judges reflects itself in the variation in the rank sums. When all judges agree, this variation is a maximum. Disagreement between judges reflects itself in a reduction in the variation of rank sums. For maximum disagreement the rank sums will tend to be more or less equal. This circumstance provides the basis for the definition of a coefficient of concordance.

Let R_j represent the rank sum of the jth individual. The sum of squares of rank sums for N individuals is

[21.15] $$S = \Sigma\left(R_j - \frac{\Sigma R_j}{N}\right)^2$$

The maximum value of this sum of squares occurs when perfect agreement exists between judges and is equal to $m^2(N^3 - N)/12$. The coefficient of concordance W is defined as the ratio of S to the maximum possible value of S and is

[21.16] $$W = \frac{12S}{m^2(N^3 - N)}$$

When perfect agreement exists between judges, $W = 1$. When maximum disagreement exists, $W = 0$. That is, W does not take negative values. With more than two judges complete disagreement cannot occur. For example, if A and B are in complete disagreement and A and C are also in complete disagreement, then B and C must be in complete agreement.

In the example of Table 21.3 the rank totals are 20, 12, 8, 10, 12, and 22. The sum of ranks is 84. The mean rank total, the rank sum expected in the case of independence, is $\frac{84}{6} = 14$. The sum of squares of deviations about this mean is

$$S = (20-14)^2 + (12-14)^2 + (8-14)^2 + (10-14)^2 + (12-14)^2 + (22-14)^2 = 160$$

In our example $m = 4$ and $N = 6$ and the coefficient of concordance is

$$W = \frac{12 \times 160}{4^2(6^3 - 6)} = .571$$

The concordance among m sets of ranks may be described by calculating Spearman rank-order correlation coefficients between all possible pairs of ranks and finding the average value, denoted by ρ. This average is related to W. The relation is given by

[21.17] $$\bar{\rho} = \frac{mW - 1}{m - 1}$$

For the particular case where $m = 2$ the relation is $\rho = 2W - 1$. For $W = 0$, $\rho = -1$; for $W = .5$, $\rho = 0$; and for $W = 1$, $\rho = 1$.

21.13 THE COEFFICIENT OF CONCORDANCE WITH TIED RANKS

Where tied ranks occur, proceed as before and assign to each member the average rank which the tied observations occupy. If the ties are not numerous, we may compute W directly from the data without further adjustment. If the ties are numerous, a correction factor is calculated for each set of ranks. This correction factor is

[21.18] $$T = \frac{\Sigma(t^3 - t)}{12}$$

For example, if the ranks on X are 1, 2.5, 2.5, 4, 5, 6, 8, 8, 8, 10, we have two groups of ties, one of two ranks and one of three ranks. The correction factor for this set of ranks for X is

$$T = \frac{(2^3 - 2) + (3^3 - 3)}{12} = 2.5$$

A correction factor T is calculated for each of the m sets of ranks, and these are added together over the m sets to obtain ΣT. We then apply a formula for W in which this correction factor is incorporated. The formula is

[21.19] $$W = \frac{S}{\frac{1}{12}m^2(N^3 - N) - m\Sigma T}$$

The application of this correction tends to increase the size of W. The correction has a small effect unless ties are quite numerous.

21.14 THE SIGNIFICANCE OF THE COEFFICIENT OF CONCORDANCE W

For N of 7 or less, values of W required for significance at the 5 and 1 percent levels have been tabulated by Friedman (1940) and are reproduced in Kendall (1970) and Siegel (1956). A useful adaptation of these tables is given by Edward (1973). Critical values of W depend both on m, the number of sets of ranks, and on N, the number of ranks in each set. For N greater than 7, a χ^2 test may be applied. Calculate the quantity

$$\chi^2 = m(N-1)W \tag{21.20}$$

This has a chi-square distribution with $N-1$ degrees of freedom. For the data of Table 21.3, $S = 160$, $W = .571$, $m = 4$, and $N = 6$. Reference to Edwards' table provides critical values of .505 and .621 for significance at the 5 and 1 percent levels. If we apply the chi-square test to the same data, we obtain

$$\chi^2 = 4(6-1).571 = 11.42$$

For $df = 6 - 1 = 5$ the values of χ^2 required for significance are 11.07 and 15.09 at the 5 and 1 percent levels, and as before we are led to the conclusion of significant association at the 5 percent level. Of course, in this case the tabled values are to be preferred because N is less than 7. For N less than 7 the chi-square test will provide a very rough estimate of the required probabilities. Other procedures for testing the significance of W exist. For a more thorough discussion of this see Edwards (1973).

21.15 THE COEFFICIENT OF CONSISTENCE K

To obtain a ranking of objects on an attribute, the objects may be presented two at a time in all possible pairs and a judge required to make a choice on the presentation of each pair. Thus a choice is made between every object and every other object. This procedure is known as the *method of paired comparisons* and has been widely used in psychological work. The method is usually assumed to yield a more reliable ordering than that obtained by requiring a judge to order a whole group of objects directly. The number of possible pairs is the number of combinations of N things taken two at a time, or $N(N-1)/2$. As N increases, the number of comparisons increases very rapidly; consequently for large N the method is frequently impractical.

In the method of paired comparisons we may wish to ascertain the consistency of the choices made. Let A, B, and C be three objects. If A is preferred to B and B is preferred to C, consistency of judgment would require that A be preferred to C. If C is preferred to A, this latter choice is clearly inconsistent with the two previous choices. What meaning attaches to the presence of inconsistent choices? Let A, B, and C be red, blue, and yellow cards, each of a different saturation. A judge may prefer red to blue, blue to yellow, and then may indicate a preference of yellow to red. This inconsistent choice may result because the judge may be unable to discriminate and may indicate preferences in a more or less haphazard fashion. Many inconsistent choices in the method of paired comparisons result because the task requires a refinement of discrimination which is beyond the capacity of the judge. Inconsistent responses may also arise because the dimension of judgment has changed. The red card may be preferred to the blue and the blue to the yellow on the basis of hue. The yellow may be preferred to the red on the basis of saturation. A different dimension is used as a basis of choice and leads to the presence of an inconsistency. To illustrate further, an orange may be preferred to a peach because of its color, a peach may be preferred to a pear because of its flavor, a pear may be preferred to an orange because of its shape, and thus an inconsistency arises. Where inconsistencies are numerous, a question may attach to the meaning of the rank ordering of objects obtained. It is convenient to represent a choice A in preference to B by the notation $A \rightarrow B$ and a choice of B to A by $B \rightarrow A$. The sequence $A \rightarrow B \rightarrow C \rightarrow A$ is an inconsistent triplet, or triad, of choices. For any set of paired comparisons between N objects the number of inconsistent triads may be counted and used to define a coefficient of consistency of response.

Responses obtained by the method of paired comparisons may be represented in tabular fashion in the form of a response pattern as shown in Table 21.4. This table shows paired comparisons between nine objects, A, B, C, . . . , H, I. Object A is preferred to B, and a 1 is entered in the cell corresponding to row A and column B above the main diagonal. A complementary 0 is entered in column A and row B below the main diagonal. All other choices may be similarly represented. We note that where no response inconsistencies are present, all entries on one side of the main diagonal are 1s and all entries on the other side 0s. In this table the presence of some 0s above the main diagonal and the complementary 1s below it indicate the presence of inconsistencies. Let us now sum the rows of

Table 21.4

Response pattern for paired comparisons between nine objects and calculation of coefficient of consistence.

	A	B	C	D	E	F	G	H	I	Row sum R	$(R-\bar{R})^2$
A	·	1	0	0	1	1	1	0	1	5	1
B	0	·	1	1	1	1	0	1	1	6	4
C	1	0	·	0	0	1	1	1	1	5	1
D	1	0	1	·	1	1	1	1	1	7	9
E	0	0	1	0	·	1	1	1	0	4	0
F	0	0	0	0	0	·	1	1	1	3	1
G	0	1	0	0	0	0	·	1	1	3	1
H	1	0	0	0	0	0	0	·	0	1	9
I	0	0	0	0	1	0	0	1	·	2	4

$\bar{R} = 4 \qquad \Sigma(R-\bar{R})^2 = 30$

$$K = \frac{12\Sigma(R-\bar{R})^2}{N(N^2-1)} = \frac{12 \times 30}{9(9^2-1)} = .500$$

Table 21.4. If no inconsistencies were present, the row sums would be the numbers 8, 7, 6, 5, 4, 3, 2, 1, 0. Because of the presence of inconsistencies, the actual obtained numbers are 7, 6, 5, 5, 4, 3, 3, 2, 1, although not in that order. The effect of inconsistencies is to reduce the variability of the numbers obtained by adding up the rows of the response pattern. Denote a row sum by R. The mean of the row sums is $\bar{R} = \Sigma R/N$, which may be shown equal to $(N-1)/2$. The sum of squares of row sums is

[21.21] $$\Sigma(R-\bar{R})^2 = \Sigma R^2 - \frac{N(N-1)^2}{4}$$

It is appropriate to consider the maximum and minimum values of this sum of squares. The maximum value of $\Sigma(R-\bar{R})^2$ occurs when no inconsistencies are present in the response pattern and is equal to $N(N^2-1)/12$. The minimum value of $\Sigma(R-\bar{R})^2$ depends on whether N is odd or even. If N is *odd,* the minimum value of $\Sigma(R-\bar{R})^2$ is 0. If N is *even,* it may be shown that the minimum value of $\Sigma(R-\bar{R})^2$ is not 0, but is $N/4$. We then define a *coefficient of consistence* of response K as follows:

[21.22] $$K = \frac{\text{observed sum of squares} - \text{minimum sum of squares}}{\text{maximum sum of squares} - \text{minimum sum of squares}}$$

Simple substitution shows that if N is *odd*

[21.23] $$K = \frac{12\Sigma(R-\bar{R})^2}{N(N^2-1)}$$

and if N is *even*

[21.24] $$K = \frac{12\Sigma(R-\bar{R})^2 - 3N}{N(N^2-4)}$$

This is Kendall's coefficient of consistence. It has an expected value of 0 when responses are assigned at random, the case of maximal inconsistency, and 1 when no inconsistencies are present.

The calculation of K is illustrated in Table 21.4. In this example N is odd, $\Sigma(R-\bar{R})^2 = 30$, and $K = .500$.

How may the coefficient K be interpreted? The number of inconsistent triads of the kind $A \rightarrow B \rightarrow C \rightarrow A$ may be denoted by d, which is related to the coefficient K. It may be shown that when N is odd

[21.25] $$d = \frac{N(N^2-1)(1-K)}{24}$$

and when N is even

[21.26] $$d = \frac{N(N^2-4)(1-K)}{24}$$

In the example of Table 21.4, the number of inconsistencies d is found to be 15. The maximum possible number of inconsistencies is 30. Thus one-half the triadic relations are inconsistent, the other half consistent, and

$K = .50$. A K of .20 would mean that four-fifths of the relations were inconsistent and one-fifth consistent.

21.16 THE SIGNIFICANCE OF THE COEFFICIENT OF CONSISTENCE

The significance of the coefficient of consistence may be approached by considering the distribution of the number of triadic relations where choices are made at random. Kendall (1970) provides a table of probabilities that particular values of d will be attained or exceeded for $N = 2$ to 7. For $N > 7$, Kendall has shown that a χ^2 test may be used which provides approximate probabilities. The quantity

[21.27] $$\chi^2 = \frac{8}{N-4}\left(\frac{1}{4} C_3^N - d + \frac{1}{2}\right) + df$$

has an approximate χ^2 distribution with degrees of freedom given by

[21.28] $$df = \frac{N(N-1)(N-2)}{(N-4)^2}$$

The term C_3^N in the expression for χ^2 is the number of combinations of N things taken 3 at a time, or $N!/3!(N-3)!$. In using this test the required probability that a value of d equal to or greater than that obtained will result where choices are allotted at random is the *complement* of the probability of χ^2.

For the data of Table 21.4, $N = 9$ and $d = 15$. We have

$$df = \frac{9 \times 8 \times 7}{(9-4)^2} = 20.16$$

$$\chi^2 = \frac{8}{9-4}\left(\frac{1}{4} \times \frac{9!}{3!6!} - 15 + \frac{1}{2}\right) + 20.16 = 28.96$$

The probability associated with this value of χ^2 is about .90. This means that the significance level for d is about .10, the complement of .90. We conclude that the consistency represented in the data is not greater than we could reasonably expect on the assignment of choices at random. The coefficient of consistence $K = .50$ may be said to be not significantly different from 0.

BASIC TERMS AND CONCEPTS

Sum of the first N integers

Sum of squares of the first N integers

Measures of disarray, Σd^2 and S

Spearman's rank correlation

Tied ranks

Kendall's coefficient of rank correlation

Variance of sampling distribution of S

Relation of rho to tau

Coefficient of concordance

Coefficient of consistence

EXERCISES

1 Calculate Spearman's rank correlation coefficient for the following paired ranks:

X	1	2	3	4	5	6	7	8
Y	2	4	5	1	6	3	8	7

Does the coefficient obtained differ significantly from 0?

2 Convert the following measurements to ranks:

X	4	4	7	7	7	9	16	17	21	25
Y	8	16	8	8	16	20	12	15	25	20

Calculate Spearman's rank correlation coefficient. Does the coefficient obtained differ significantly from 0?

3 Are the following values of ρ significantly different from 0? **(a)** $\rho = .30$ for $N = 25$, **(b)** $\rho = .60$ for $N = 15$, **(c)** $\rho = .70$ for $N = 10$.

4 Calculate the statistics S for the following sets of paired ranks:

a

X	1	2	3	4	5	6
Y	6	4	2	1	3	5

b

X	1	2	3	4	5	6
Y	6	3.5	1.5	1.5	3.5	5

c

X	2	2	2	3	4	5
Y	6	3.5	1.5	1.5	3.5	5

d

X	2	2	2	5	5	5
Y	6	4	2	1	3	5

e

X	2	2	2	5	5	5
Y	6	3.5	1.5	1.5	3.5	5

f

X	2	2	2	5	5	5
Y	5	5	2	2	2	5

5 Calculate the sampling variances for the statistic S obtained for the sets of paired ranks in Exercise 4 above. In each case obtain the normal deviate z with a continuity correction.

6 Three judges rank-order a group of seven students on an examination as follows:

	Student						
Judge	*a*	*b*	*c*	*d*	*e*	*f*	*g*
A	1	2	3	4	5	6	7
B	2	3	4	5	1	7	6
C	5	4	1	2	3	6	7

Compute the Spearman rank coefficients between judges and the coefficient of concordance.

7 A supervisor ranks six employees A, B, C, D, E, and F on job performance, using the method of paired comparisons. The data are a follows: $A \rightarrow B$, $A \rightarrow C$, $A \rightarrow D$, $E \rightarrow A$, $F \rightarrow A$, $B \rightarrow C$, $D \rightarrow B$ $B \rightarrow E$, $B \rightarrow F$, $C \rightarrow D$, $C \rightarrow E$, $F \rightarrow C$, $D \rightarrow E$, $D \rightarrow F$, $E \rightarrow F$. Calculate the coefficient of consistence for these data. How may this coefficient be interpreted?

8 Calculate a coefficient of consistence for the table composed of the first five rows and columns of Table 21.4.

ANSWERS TO EXERCISES

1 $\rho = .643$, $p = .05$ for one-tailed test

2 $\rho = .594$, $p < .05$ for one-tailed test

3 **a** $p > .05$ for one-tailed test **b** $p < .05$ for one-tailed test **c** $p < .05$ for one-tailed test

4 **a** $S = -3$ **b** $S = -1$ **c** $S = 2$ **d** $S = -3$ **e** $S = -1$ **f** $S = -3$

5 **a** $\sigma_s^2 = 28.33$, $z = .376$ **b** $\sigma_s^2 = 26.33$, $z = .000$ **c** $\sigma_s^2 = 23.07$, $z = .208$
d $\sigma_s^2 = 21.00$, $z = .437$ **e** $\sigma_s^2 = 19.80$, $z = .000$ **f** $\sigma_s^2 = 16.20$, $z = .498$

6 $\rho_{ab} = .607$, $\rho_{ac} = .429$, $\rho_{bc} = .393$, $W = .651$

7 $K = .000$

8 $K = .400$

22

NONPARAMETRIC TESTS OF SIGNIFICANCE

22.1 INTRODUCTION

Many tests of significance involve assumptions about the nature of the distributions of the variables in the populations from which the samples are drawn. The t test and the analysis of variance, for example, assume normality of the parent distribution. In experimental work situations arise where either little is known about the population distribution of the dependent variable or this distribution is known to depart appreciably from the normal form. In such situations *nonparametric tests* may be appropriately used. Nonparametric tests make few assumptions about the properties of the parent distribution. Assumptions about the parent distribution are involved in nonparametric tests, but these are usually fewer in number, weaker, and easier to satisfy in data situations. Nonparametric tests are frequently spoken of as *distribution-free* tests. The implication is that they are free, or independent, of some characteristics of the population distribution.

The reader will recall the distinction between nominal, ordinal, interval, and ratio variables. Nonparametric methods are appropriate for nominal and ordinal data, parametric methods for interval and ratio data. In practice, nonparametric methods are frequently used with data of this latter type. The data are reduced to a form such that a nominal, or an ordinal, statistical procedure may be applied to them. An important class of nonparametric tests employs only the sign properties of the data. All observations above a fixed value, such as the median, may be assigned a plus, and all below, a minus. The original variable is replaced by, or transformed to, another variable which takes the sign values plus or minus.

Another class of nonparametric tests employs the rank properties of the data. The original observations are replaced by the numbers 1, 2, 3, . . . , N. Subsequent statistical manipulation and inferences are based on ranks.

In applying a conventional test of significance, such as a two-sample t test, estimates $\bar{X}_1$ and $\bar{X}_2$ of the population means μ_1 and μ_2 are used to test the null hypothesis $H_0: \mu_1 = \mu_2$ under the assumption that the distributions are normal and the variances are equal, $\sigma_1^2 = \sigma_2^2$. In many nonparametric procedures the null hypothesis under test is not formulated in terms of the parameters of parent populations, nor are estimates of population parameters calculated. A number of commonly used two-sample procedures test the null hypothesis in its most general form. This null hypothesis is that the samples come from populations with the same distribution. This is tested against the alternative that samples come from populations with different distributions. Of course, if we are willing to assume in a two-sample test that the distributions are normal and $\sigma_1^2 = \sigma_2^2$, then if the distributions are found to differ, the differences must be due to differences in location. Some nonparametric tests are tests of central position in a rank order and are in effect tests of medians. If the assumption is made that the parent populations are symmetrical, such tests become also tests of means.

The question arises as to the criteria to be applied in comparing alternate procedures for testing the same hypothesis. The procedures to be compared may be two nonparametric tests, or a nonparametric test and a test of the conventional type. Alternate tests of significance are usually compared in terms of efficiency. The reader will recall that in the discussion on estimation in Chapter 10 the concept of relative efficiency was introduced. This was defined as the ratio of two sampling variances and could be related to sample size. A median, for example, calculated on a sample of 100 cases has a sampling variance equal to that of a mean calculated on 64 cases. Thus alternate procedures for estimating population parameters have the same efficiency when their sampling variances are equal. In comparing alternate procedures for testing hypotheses a different approach is used. This approach uses the concept of the power of a test.

The power of a statistical test is the probability of rejecting the null hypothesis when that hypothesis is false. It is 1 minus the Type II error, or $1 - \beta$. The power of a statistical test depends on the level of significance, the alternate hypothesis H_1, and the sample size. Two tests A and B may be compared by considering the relative sample size required to make them equally powerful. The *relative efficiency* of the two tests is given by N_a/N_b, where N_b is the number of observations required to make test B as powerful as test A with N_a observations. Because this definition of relative efficiency depends, of course, also on the level of significance, the alternate hypothesis H_1, and the sample size, it is too complex for most practical purposes. This has led to the use of the *asymptotic relative ef-*

ficiency, which is the limiting value of the ratio N_a/N_b as N approaches infinity and H_1 approaches the null hypothesis H_0. Questions can be raised regarding the practical usefulness of the asymptotic relative efficiency, since in most cases interest does not reside in large sample size or in alternate hypotheses close to the null hypothesis. The asymptotic relative efficiency of many nonparametric tests is known relative to the most powerful test available, which for two-sample tests is frequently the t test. Comparisons with the t test, for example, are made for normal distributions where both a conventional and a nonparametric procedure may be applied.

For a comprehensive treatment of nonparametric tests the reader is referred to Siegel (1956) and to Bradley (1968). Both books contain useful tables.

22.2 A SIGN TEST FOR TWO INDEPENDENT SAMPLES

This test is known as the *median test.* It compares the medians of two independent samples. The null hypothesis is that no difference exists between the medians of the populations from which the samples are drawn. The corresponding parametric test is a t test for comparing the means of independent samples. The median test is based on the idea that in two samples drawn from the same population the expectation is that as many observations in each sample will fall above as below the joint median.

The data consist of two independent samples of N_1 and N_2 observations. To apply the median test the median of the combined $N_1 + N_2$ observations is calculated. In each sample, observations above the joint median are assigned a + and those at or below it a −. The number of + and − signs for each sample is ascertained. A χ^2 test is used to determine whether the observed frequencies of + and − signs depart significantly from expectation under the null hypothesis.

The following are observations for two independent samples:

Sample I	10	10	10	12	15	17	17	19	20	22	25	26
Sample II	6	7	8	8	12	16	19	19	22			

The median of the $N_1 + N_2$ observations is 16. Assigning a + to values above the median and a − to values at or below it, we obtain

Sample I	−	−	−	−	−	+	+	+	+	+	+	+
Sample II	−	−	−	−	−	−	+	+	+			

These data may be tabulated in the form of a 2 × 2 table as follows:

	+	−	
Sample I	7	5	12
Sample II	3	6	9
	10	11	

The value of χ^2 for this table is 1.29. The value of χ^2 required for significance at the 5 percent level is 3.84. Obviously, in this case we have no grounds for rejecting the null hypothesis that the samples came from populations with the same median. This is a two-tailed test.

The asymptotic relative efficiency of the sign test for independent samples when compared to the t test under the assumptions of normality and equal variance is $2/\pi = .637$.

22.3 A SIGN TEST FOR TWO CORRELATED SAMPLES

This test compares two correlated samples, and is applicable to data composed of N paired observations. The difference between each pair of observations is obtained. The null hypothesis is that the median difference between the pairs is 0. The test is based on the idea that under the null hypothesis the expectation is that half the differences between the paired observations will be positive and the other half negative. The symmetrical binomial $(1/2 + 1/2)^N$ is used to obtain the probabilities required for a one-tailed or a two-tailed test.

The following are paired observations, X and Y, for a sample of 10 individuals together with the sign of the difference between X and Y:

X	15	19	31	36	10	11	19	15	10	16
Y	19	30	26	8	10	6	17	13	22	8
Sign of $X - Y$	$-$	$-$	$+$	$+$	0	$+$	$+$	$+$	$-$	$+$

Under the null hypothesis the probability that X is greater than Y is equal to the probability that Y is greater than X, which in turn is equal to 1/2. The expected numbers of $+$ and $-$ signs are equal. In this example we have six plus signs, three minus signs, and one zero difference. The zero difference is discarded. From the binomial expansion $(1/2 + 1/2)^9$ we can ascertain the exact probability of obtaining six or more plus signs under the null hypothesis. This probability is .254. This is a one-tailed test. The probability of obtaining either six or more plus signs or six or more minus signs is .508. This is a two-tailed test. Clearly here we have no grounds for rejecting the null hypothesis.

Where N is not too small, the normal approximation to the binomial or χ^2 may be used, preferably with Yates' correction. In this case the expected values are $N/2$. In the above example the observed values are 6 and 3, the expected values are 4.5 and 4.5, the corrected observed values are 5.5 and 3.5, and $\chi^2 = .44$. The probability of obtaining a χ^2 equal to or greater than .44 under the null hypothesis is .507. Although N is small, this is in close agreement with the exact probability of .508 obtained from the binomial. The reader will recall that χ^2 provides the probability for a two-tailed test.

Instead of the χ^2 procedure described above, a computationally simpler method may be used. Obtain the difference between the number of + and − signs. Denote this difference by D. It may be shown that

$$z = \frac{|D| - 1}{\sqrt{N}}$$

approaches the normal form as N increases in size. This formula incorporates a continuity correction. Values of 1.96 and 2.58 are required for significance at the .05 and .01 levels of significance, respectively, for a nondirectional test. In the above example we have six plus signs and three minus signs. The difference $D = 6 - 3 = 3$, and

$$z = \frac{|3| - 1}{\sqrt{9}} = .67$$

Quite clearly this falls short of significance. The reader should recall that for $df = 1$ the quantity $z^2 = \chi^2$. In the above example, if the calculation is carried beyond two decimal places, $z^2 = .6667^2 = .4444$ and $\chi^2 = .4444$.

The asymptotic relative efficiency of this test when compared to the corresponding t test, and under the required assumptions, is $2/\pi = .637$.

22.4 A SIGN TEST FOR k INDEPENDENT SAMPLES

This is an obvious extension of the median test for two independent samples. The data comprise k samples of $n_1, n_2, \ldots, n_k$ observations. As before, the null hypothesis is that no difference exists in the medians of the populations from which the samples are drawn. The median of the combined $n_1 + n_2 + \cdots + n_k$ observations is calculated. For each sample, observations above the joint median are assigned a + and those either at or below the joint median a −. The data are arranged in a $2 \times k$ contingency table, and a χ^2 test applied.

The following are data for four samples.

Sample I	3	6	11	14	17	18	21	33
Sample II	3	3	4	5	5	8	9	14
Sample III	18	18	25	26	29	31		
Sample IV	14	18	20	22	22	25	27	35

The total number of observations is 30. The median is 18. Assigning a + to values above the median and a − to values at or below, we obtain

Sample I	−	−	−	−	−	−	+	+
Sample II	−	−	−	−	−	−	−	−
Sample III	−	−	+	+	+	+		
Sample IV	−	−	+	+	+	+	+	+

These data may be arranged in a 2 × 4 table as follows:

	+	−	
Sample I	2	6	8
Sample II	0	8	8
Sample III	4	2	6
Sample IV	6	2	8
	12	18	30

The value of χ^2 calculated on this table is 11.94. The number of degrees of freedom is $(4-1)(2-1)=3$. The value of χ^2 required for significance at the 1 percent level is 11.34. The observed value falls just above this.

22.5 A RANK TEST FOR TWO INDEPENDENT SAMPLES

The most commonly used rank test for comparing two independent samples is the Wilcoxon rank sum test. Tests which are equivalent to the rank sum test have been developed by Mann and Whitney, and others. The hypothesis under test here is that the two samples come from populations with the same distribution. If assumptions are made regarding the equivalence of shapes and variances of the two distributions, the Wilcoxon test becomes a test of central location.

In applying the Wilcoxon test the N_1 and N_2 observations are combined. The $N_1 + N_2$ observations are then arranged in order. A rank 1 is assigned to the smallest value, a rank 2 to the next smallest, and so on. The sum of ranks, R_1, is obtained for the *smaller* of the two samples, if the samples are unequal in size. If the samples are equal in size, either rank sum may be used. The sum of ranks R_1 is then evaluated in relation to its distribution, which is discussed below.

The model distribution against which particular values of R are evaluated is obtained by considering a finite population of integers comprised of $N_1 + N_2 = N$ members. These integers are 1, 2, 3, . . . , N. The problem of drawing samples of size N_1 from this population of $N_1 + N_2 = N$ members may be considered. The number of possible equiprobable samples is $C_{N_1}{}^{N}$. A value R_1 may be obtained for each sample and a frequency distribution made of the values of R_1. This distribution can be used to evaluate particular values of R_1. If a particular value of R_1 has a small probability in relation to this distribution, we reject the hypothesis that the two samples come from the same population. The reader should note that this distribution is closely related to the sampling distribution of means from a finite population discussed in Section 10.5. Here the sampling distribution of sums, and not means, is under consideration, and the population from which samples are drawn is one of integers extending from 1 to N. The mean of this distribution, $\bar{R}_1$, is N_1 times the mean of the $N_1 + N_2$ ranks and is

[22.1] $$\text{Mean} = \bar{R}_1 = \frac{N_1(N_1 + N_2 + 1)}{2}$$

The variance of the distribution of R_1 can be shown to be

[22.2] $$\text{Variance} = \sigma_{R_1}{}^2 = \frac{N_1 N_2 (N_1 + N_2 + 1)}{12}$$

The exact distributions of R_1 are known for N_1 and N_2 up to 25. The distribution approaches the normal form fairly rapidly. When both N_1 and N_2 are equal to or greater than 8 or 10, a large-sample procedure using the normal approximation and a continuity correction will lead to estimates of the required probabilities which do not differ much from those obtained from the exact distributions. The normal deviate z with a continuity correction is given by

[22.3] $$z = \frac{|R_1 - \bar{R}_1| - \frac{1}{2}}{\sqrt{\dfrac{N_1 N_2 (N_1 + N_2 + 1)}{12}}}$$

If this value is equal to or greater than 1.96 or 2.58, we reject the null hypothesis for a nondirectional test at the .05 or .01 level and accept the alternative hypothesis that the samples are from different populations. For a directional test the .05 and .01 levels are 1.64 and 2.33.

Consider the following observations:

Sample I	27	33	37	52	53	57	69	70	71	77		
Sample II	6	9	14	16	29	43	45	47	50	55	63	72

Assigning ranks, proceeding from the smallest to the largest values, we obtain

Sample I	5	7	8	13	14	16	18	19	20	22		
Sample II	1	2	3	4	6	9	10	11	12	15	17	21

The sum of ranks R_1 for sample I is 142. The mean of the distribution of R_1, that is, $\bar{R}_1$, is 115. The normal deviate is

$$z = \frac{|142 - 115| - \frac{1}{2}}{\sqrt{\dfrac{10 \times 12(10 + 12 + 1)}{12}}} = 1.75$$

Since this falls below 1.96, we have no grounds for rejecting the null hypothesis for a two-tailed test. The result is, however, significant at the 5 percent level for a one-tailed test.

When ties occur, the tied observations may be assigned the average of the ranks they would occupy if no ties had occurred. If ties are fairly numerous, a correction may be applied to the standard deviation in the denominator of the z ratio. Corrected for ties, that ratio becomes

[22.4]
$$z = \frac{|R_1 - \bar{R}_1| - \frac{1}{2}}{\sqrt{\frac{N_1 N_2}{N(N-1)}\left(\frac{N^3 - N}{12} - \Sigma T\right)}}$$

where $N = N_1 + N_2$ and $T = (t^3 - t)/12$, where t is the number of values tied at a particular rank. The summation of T extends over all groups of ties.

The above procedures use the normal approximation to the distribution of R_1. If an exact test is required, Table K may be used. This table shows the exact lower-tail critical values of R_1 for N_1 and N_2 up to 25 at probability levels equal to or less than .10, .05, .025, .01, .005, and .001. For example, the table shows a $p = .05$ for $R_1 = 19$ where $N_1 = 5$ and $N_2 = 5$. This means that the probability of obtaining a value equal to or less than R_1 in samples of this size is equal to or less than .05. This is a directional test which uses the lower tail of the distribution. Since the distribution is symmetrical, the corresponding upper-tail values are given by noting that a lower-tail value R_1 is $\bar{R}_1 - R_1$ points below the mean. The corresponding value above the mean is $\bar{R}_1 + (\bar{R}_1 - R_1) = 2\bar{R}_1 - R_1$. If R_1 is an upper-tail value, that is, if it is above the mean, Table K is entered with $2\bar{R}_1 - R_1$. The null hypothesis is rejected if it is smaller than the critical value. To assist calculation Table K shows values of $2\bar{R}_1$. In the illustrative example above, R_1 is 142 for $N_1 = 10$ and $N_2 = 12$. The mean value $\bar{R}_1$ is 115. The lower-tail value corresponding to R_1 is $2\bar{R}_1 - R_1 = 230 - 142 = 88$. Entering Table K, we note that p is slightly less than .05. Thus we may assert significance at this level for a directional test. For a nondirectional or two-tailed test, either R_1 or $2\bar{R}_1 - R_1$, whichever is appropriate, is referred to Table K and the probabilities in the table are doubled. The appropriate choice between R_1 and $2\bar{R}_1 - R_1$ is, of course, always the smaller value.

The above discussion sounds rather complex. In practice the procedure is simple. *First,* calculate R_1. Calculate $\bar{R}_1$ using formula [22.1]. *Second,* if R_1 is less than $\bar{R}_1$ refer R_1 to Table K to obtain the required probability. *Third,* if R_1 is greater than $\bar{R}_1$ calculate $2\bar{R}_1 - R_1$, and refer this quantity to Table K to obtain the required probability.

The Wilcoxon rank sum test is in effect the same as the Mann-Whitney U test. Mann and Whitney (1947) studied the distribution of a statistic U, which is related in a simple way to R_1.

[22.5]
$$U_1 = N_1 N_2 + \frac{N_1(N_1 + 1)}{2} - R_1$$

[22.6]
$$U_2 = N_1 N_2 + \frac{N_2(N_2 + 1)}{2} - R_2$$

U is the smaller of these two values. Tables showing critical values of U are reproduced in Siegel (1956).

The Wilcoxon rank sum test has an asymptotic relative efficiency when compared with the t test for independent samples of $3/\pi = .955$. This comparison assumes that the distributions are normal. If the distributions are rectangular, the asymptotic relative efficiency is 1.00. For certain other types of distributions the asymptotic relative efficiency is greater than 1.00. All the available evidence indicates that the Wilcoxon rank sum test is an excellent alternative to the t test.

22.6 A RANK TEST FOR TWO CORRELATED SAMPLES

The rank test described here is due to Wilcoxon and is usually called the *Wilcoxon matched-pairs signed-ranks test*. The data are a set of N paired observations on X and Y. The difference, d, between each pair is calculated. If the two observations in a pair are the same, then $d = 0$ and the pair is deleted from the analysis. Values of d may be either positive or negative. The d's are then ranked without regard to sign; that is, the absolute values $|X_i - Y_i|$ are ranked. A rank of 1 is assigned to the smallest d, of 2 to the next smallest, and so on. If two or more d's are tied, the practice usually adopted is to assign to the tied ranks the average of the ranks they would have been assigned if they had differed. The sign of the difference d is attached to each rank. If d is positive, the rank is positive; if d is negative, the rank is negative. Denote the sum of the positive ranks by W_+ and the sum of the negative ranks by W_-.

The model null distribution against which W_+ is evaluated is obtained in the following way. If the two samples X and Y are samples from the same population, then the probability that $X_i - Y_i$ is either plus or minus is 1/2. Also, the probability that the rank corresponding to $|X_i - Y_i|$ is either plus or minus is 1/2. All possible arrangements of plus or minus in relation to the N ranks may be considered. There are 2^N such arrangements. These are viewed as equiprobable. W_+ is calculated for each arrangement, and the frequency distribution of W_+ becomes the null distribution against which particular values of W_+ are evaluated. To illustrate, for $N = 3$ the number of arrangements of plus and minus is $2^3 = 8$. These are as follows:

1	2	3	W_+
+	+	+	6
−	+	+	5
+	−	+	4
+	+	−	3
−	−	+	3
−	+	−	2
+	−	−	1
−	−	−	0

Eight different values of W_+ occur, ranging from 0 to 6. In general W_+ will range from 0 to $N(N+1)/2$. It can be shown that the mean and variance of the distribution of W_+ are as follows:

[22.7] $$\text{Mean} = \bar{W}_+ = \frac{N(N+1)}{4}$$

[22.8] $$\text{Variance} = \sigma_{W_+}{}^2 = \frac{N(N+1)(2N+1)}{24}$$

The distribution of W_+ approaches the normal form fairly rapidly. Consequently for samples of reasonable size the normal approximation may be used. The normal deviate z is given by

[22.9] $$z = \frac{W_+ - \dfrac{N(N+1)}{4}}{\sqrt{\dfrac{N(N+1)(2N+1)}{24}}}$$

Values of 1.96 and 2.58 are, as usual, required for significance at the 5 percent and 1 percent levels for a nondirectional or two-tailed test.

For N up to 40, values of W_+ may be referred to Table I of the Appendix. This table provides the lower-tail cumulative probabilities for values of W_+ that lie close to the selected significance levels .05, .025, .01, and .005. To illustrate, for $N = 10$ and a significance level of .05, two values of W_+ are shown: 10 with an associated probability of .0420 and 11 with a probability of .0527. This means that the probability of obtaining a value of W_+ equal to or less than 10, when the null hypothesis is true, is .0420, and the probability of obtaining a value of W_+ equal to or less than 11 is .0527. This is a directional test. It uses the lower tail of the distribution. A directional test that uses the upper tail of the distribution may be required. Here we recall that the maximum value of W_+ is $N(N+1)/2$. Since the distribution is symmetrical, the corresponding upper-tail value is $[N(N+1)/2] - W_+$. This quantity is equal in absolute magnitude to the sum of the negative ranks $|W_-|$. To illustrate, suppose $N = 10$ and $W_+ = 50$. What is the probability of obtaining a value of W_+ equal to or greater than 50 when the null hypothesis is true? Here we calculate $[N(N+1)/2] - W_+ = 10 \times 11/2 - 50 = 5$. The value 5 is referred to Table I and a probability of .0098 is obtained. Thus under the null hypothesis the probability of obtaining a value of W_+ equal to or greater than 50 is in the neighborhood of .01. For a directional or two-tailed test we refer either W_+ or $[N(N+1)/2] - W_+$, whichever is the smaller, to the table and double the probabilities.

The discussion above is somewhat involved. In practice a simple procedure may be used. *First,* calculate W_+ and $|W_-|$. *Second,* take the smaller of these two quantities and refer this to Table I with the appropriate value of N.

The following are paired observations, X and Y, for a sample of 10 individuals:

X	15	19	31	36	10	11	19	15	10	16
Y	19	30	26	8	10	6	17	13	22	8
d	−4	−11	5	28	0	5	2	2	−12	8
Rank	−3	−7	4.5	9		4.5	1.5	1.5	−8	6

Values of d are calculated. One pair of observations is tied and is deleted from subsequent consideration. The d's are rank ordered by absolute magnitude. The lowest values are a pair of 2s. These are assigned rank values of 1.5. The sum W_+, the sum of positive ranks, is 27. The value $[N(N+1)/2] - W_+ = 9 \times 10/2 - 27 = 18$. Note that $|W_-|$, the absolute value of the negative ranks, is also 18. This value 18 is referred to Table I. No basis exists in this case for rejecting the null hypothesis for either a directional or a nondirectional test.

The asymptotic relative efficiency of the Wilcoxon signed-rank test relative to the t test is .955. This test can be viewed as a useful and satisfactory alternative to the t test.

22.7 A RANK TEST FOR k INDEPENDENT SAMPLES

A rank test for k independent samples is the Kruskal-Wallis (1952) one-way analysis of variance by ranks. This is a generalization of the Wilcoxon rank sum test to k groups. The null hypothesis is that the k independent samples of $n_1, n_2, \ldots, n_k$ members are from the same population. To apply the test all the observations for the k samples are ranked. The lowest value is assigned a rank of 1, the next lowest 2, and so on. The sum of the ranks, R_i, for each of the k samples is obtained. If all k samples are from the same population, the expectation is that the mean rank sums $\bar{R}_i$ will be equal for the k groups, and equal to the mean of the N ranks, which is $(N+1)/2$.

The null distribution here involves a consideration of the $N!$ arrangements of ranks in k groups. Each arrangement is regarded as equiprobable. For each arrangement the following statistic could be calculated:

[22.10] $$S = \sum^{k} n_i \left(\bar{R}_i - \frac{N+1}{2}\right)^2$$

The quantity in parentheses in the above expression is simply the squared difference between the means of the ith group and its expected value under the null hypothesis. Since the groups may be unequal in size, each squared difference is weighted according to group size, n_i, to obtain a final sum of squares, S. The statistic S could be calculated for each of the $N!$

arrangements of ranks and its frequency distribution examined. Some advantages attach to the study of a statistic H, which is closely related to S and is given by

[22.11] $$H = \frac{12S}{N(N+1)} = \frac{12}{N(N+1)} \sum^{k} n_i \left[\bar{R}_i - \frac{N(N+1)}{2}\right]^2$$

The distribution of H approximates the distribution of chi square with $k - 1$ degrees of freedom. For $k = 3$ and $n_i \lesseqgtr 5$, tables of critical values of H with exact probabilities have been prepared by Kruskal and Wallis. For larger values of k and n_i the chi-square approximation must be used. For computational purposes it is convenient to write H in the form

[22.12] $$H = \frac{12}{N(N+1)} \sum^{k} \left(\frac{R_i^2}{n_i}\right) - 3(N+1)$$

where R_i is the sum of ranks for the kth group.

When ties occur, the usual convention is adopted of assigning to the tied observations the average of the ranks they would otherwise occupy. The value of H is then divided by

$$1 - \frac{\Sigma T}{N^3 - N}$$

where $T = t^3 - t$, and t is the number of tied observations in a set. The quantity H corrected for ties is

[22.13] $$H = \frac{\dfrac{12}{N(N+1)} \displaystyle\sum_{i=1}^{k} \left(\frac{R_i^2}{n_i}\right) - 3(N+1)}{1 - \dfrac{\Sigma T}{N^3 - N}}$$

The correction for ties will increase the value of H.

The following are data for three samples:

Sample I	3	7	11	16	22	29	31	36	
Sample II	3	4	7	18	19	32			
Sample III	22	38	46	47	47	50	53	54	56

In this example $n_1 = 8$, $n_2 = 6$, $n_3 = 9$, and $N = 8 + 6 + 9 = 23$. All 23 observations are ranked to obtain

Sample I	1.5	4.5	6	7	10.5	12	13	15	
Sample II	1.5	3	4.5	8	9	14			
Sample III	10.5	16	17	18.5	18.5	20	21	22	23

The sums of ranks are calculated. These are $R_1 = 69.5$, $R_2 = 40$, and $R_3 = 166.5$. We note that we have four sets of ties of two observations each. $T = 2^3 - 2 = 6$, and for the four sets $\Sigma T = 24$. The value of H is then

$$H = \frac{\dfrac{12}{23(23+1)}\left(\dfrac{69.5^2}{8} + \dfrac{40^2}{6} + \dfrac{166.5^2}{9}\right) - 3(23+1)}{1 - \dfrac{24}{23^3 - 23}} = 13.88$$

In this example the effect of the correction for ties is negligible and may for all practical purposes be ignored. On reference to a table of χ^2 with $df = 2$, we note that an H of 13.88 is significant at better than the 1 percent level. We may then reject the hypothesis that the samples are from the same population.

The Kruskal-Wallis test has an asymptotic relative efficiency of .955 when compared with the F test resulting from the analysis of variance applied to independent samples. For $k = 2$ this test is equivalent to the Wilcoxon rank sum test.

22.8 A RANK TEST FOR *k* CORRELATED SAMPLES

A rank test for k correlated samples is the Friedman two-way analysis of variance by ranks (1937). The data are a set of k observations for a sample of N individuals. Such data arise in many experiments where subjects are tested under a number of different experimental conditions. The corresponding parametric test is an analysis of variance for two-way classification where observations are made on each of a group of individuals under more than two conditions. If there is reason to believe that the assumptions underlying the analysis of variance are not satisfied by the data, the Friedman rank method may be appropriate.

The data are arranged in a table containing N rows and k columns. The rows correspond to individuals, or groups, and the columns to experimental conditions. Table 22.1 shows such an arrangement of data for eight subjects tested under four experimental conditions. The observations in the rows are ordered as shown in Table 22.2. For example, the four observations in the top row are 4, 5, 9, and 3. These are replaced by the ranks 2, 3, 4, and 1. The ranks in each column are summed. If the samples are from the same population, the ranks in each column will be a random arrangement of the numbers 1, 2, 3, and 4. Under these circumstances the sums of ranks for columns will tend to be the same. If these sums differ significantly, the hypothesis that they are from the same population may be rejected.

The null distribution here involves a consideration of the $k!$ arrangements of ranks in any row. These are considered equiprobable. Given N rows, the number of possible equiprobable arrangements of ranks is $(k!)^N$.

Table 22.1
Material recalled after four time intervals for a group of eight subjects

	Time interval			
Subject	I	II	III	IV
1	4	5	9	3
2	8	9	14	7
3	7	13	14	6
4	16	12	14	10
5	2	4	7	6
6	1	4	5	3
7	2	6	7	9
8	5	7	8	9

For each of these arrangements a statistic S may be calculated, where

[22.14] $$S = \sum^{k} (R_i - \bar{R})^2$$

Here R_i = sum of ranks for the ith column
$\bar{R}$ = mean rank sum
S = sum of squares of rank sums about the mean rank sum

If the samples are for the same population, the expectation is that the R_i's are equal and the expected value of S is 0. At least in theory, for any N and k a frequency distribution may be made of the $(k!)^N$ values of S. This distribution may be used to evaluate particular values of S. If the probability associated with a particular value of S is small, the null hypothesis is rejected. For small values of k and N the exact distributions of S are

Table 22.2
Ranks assigned by rows for the data of Table 22.1

	Time interval			
Subject	I	II	III	IV
1	2	3	4	1
2	2	3	4	1
3	2	3	4	1
4	4	2	3	1
5	1	2	4	3
6	1	3	4	2
7	1	2	3	4
8	1	2	3	4
R_i	14	20	29	17

known. Bradley (1968) provides a table of exact critical values of S for $k = 3$ and N up to 15, and for $k = 4$ and N up to 8.

For values that lie outside the tabled values of S it is customary to use a statistic which is a function of S. This statistic is given by

[22.15] $$\chi_r^2 = \frac{12S}{Nk(k+1)}$$

This statistic has an approximate chi-square distribution with $k - 1$ degrees of freedom.

For computational purposes a more convenient way of writing χ_r^2 is

[22.16] $$\chi_r^2 = \frac{12}{Nk(k+1)} \sum^{k} R_i^2 - 3N(k+1)$$

For the data of Table 22.2 we have

$$\chi_r^2 = \frac{12}{8 \times 4(4+1)} (14^2 + 20^2 + 29^2 + 17^2) - 3 \times 8(4+1) = 9.45$$

This result for $df = 4 - 1 = 3$ falls between the .05 and .01 levels of significance. Actually it is a little above the 2 percent level. If this level of confidence is acceptable, we may conclude that the samples are not drawn from the same population. In this example $S = 126$. If this is referred to a table of exact critical values of S, as given in Bradley (1968), the associated probability is found to fall between .01 and .05, not far from the .02 level. The chi-square approximation is in close agreement with the more exact test.

The asymptotic relative efficiency of the Friedman test relative to the F test resulting from a two-way analysis of variance with one observation per cell is

$$\left(\frac{3}{\pi}\right)\left(\frac{k}{k+1}\right)$$

The efficiency of the test increases as k increases, and extends from .637 for $k = 2$ to a maximum of .955 for $k = \infty$. For $k = 2$ this test is the same as the sign test for correlated samples.

22.9 MONOTONIC TREND TEST FOR INDEPENDENT SAMPLES

In Chapter 19, methods of trend analysis for parametric data were described. Methods of trend analysis using ranks may be used. These are the nonparametric analogs of the tests described in Chapter 18. A test of monotonic trend is the nonparametric analog of a test of linear trend. This section describes a simple test of monotonic trend for independent samples, which employs the statistic S as used in the definition of Kendall's tau (see Chapter 21).

Comment on the concept of monotonicity is appropriate here. A function $Y = f(X)$ is said to be a *monotonic increasing function* if any increment in X is associated with an increment in Y. Also, if any increment in X is associated with a decrement in Y, the function is said to be a *monotonic decreasing function*. The magnitude of the increment or decrement in X associated with the increment or decrement in Y is irrelevant to the concept of a monotonic function. Monotonicity is an order concept.

A common type of experiment is one in which k treatments are applied to k independent groups of $n_1, n_2, \ldots, n_k$ members, and a measurement obtained for each member. Such data are frequently analyzed using the analysis of variance for one-way classification. The nonparametric analog of this is the Kruskal-Wallis one-way analysis of variance by ranks, described in Section 22.7. If, however, the k treatments exhibit an order, the question of monotonic trend may be raised. In effect, this question is answered by testing for significance the correlation between the ranks corresponding to the $n_1 + n_2 + \cdots + n_k = N$ measurements and the ranks for the k treatments. The ranks for treatments consist of k sets of tied members with $n_1, n_2, \ldots, n_k$ members, respectively, in each of the k sets. Thus for $k = 3$, and $n_1 = n_2 = n_3 = 10$, the ranks for treatments consist of 3 sets of 10 tied values.

To illustrate, the following are measurements for three independent samples obtained under three ordered treatments.

Sample I	3	7	11	16	22	29	31	36	
Sample II	3	4	7	18	19	32			
Sample III	22	38	46	47	47	50	53	54	56

These values are ranked as in the Kruskal-Wallis one-way analysis of variance by ranks. The values 1, 2, and 3 are assigned to indicate membership in the three samples. Thus the data are represented by a set of paired ranks as follows:

Sample I	X	1	1	1	1	1	1	1	1	
	Y	1.5	4.5	6	7	10.5	12	13	15	
Sample II	X	2	2	2	2	2	2			
	Y	1.5	3	4.5	8	9	14			
Sample III	X	3	3	3	3	3	3	3	3	3
	Y	10.5	16	17	18.5	18.5	20	21	22	23

The average rank procedure for tied ranks could, of course, be applied to the X variable, in which case the values 4.5, 11.5, and 19 would replace the values 1, 2, and 3. No advantage attaches to this. In practical computation the X variable need not be recorded at all. The X variable is shown here merely to illustrate the fact that the problem is a simple correlational one. A value of S as used in the definition of Kendall's tau, and described in Chapter 21 of this book, is calculated. The value of S in this example is as follows:

$$S = 14 + 10 + 9 + 9 + 4 + 3 + 3 + 1 + 9 + 9 + 9 + 9 + 9 + 7 = 105$$

The value 14 is obtained by comparing the initial Y value of 1.5 for sample I with all Y values for samples II and III. The Y value of 1.5 for sample I need not be compared with other Y values for that sample because all values in sample I are tied on X.

The sampling variance of S is obtained from formula [21.14]. In this example $N = 23$. The X variable contains three groups of ties—one group of eight ties, one of six ties, and one of nine ties. The Y variable contains four groups of tied pairs. The variance of S from formula [21.14] is 1,245.25, and the standard error is 35.28. Subtracting unity as a continuity correction and assuming normality of the sampling distribution, we obtain the normal deviate $z = 104/35.28 = 2.95$, which warrants rejection of the null hypothesis.

The steps involved in this procedure may be stated in summary as follows:

1 Rank all the observations from 1 to N on the Y variable, as in the Kruskal-Wallis one-way analysis of variance by ranks, substituting average ranks for tied values. Use the ranks for the order of treatments as the X variable, which will consist of as many sets of ties as there are treatments.

2 Calculate S in the usual way.

3 Calculate the sampling variance of S by applying formula [21.14].

4 Subtract unity from S as a continuity correction, and divide this by its standard error to obtain the normal deviate z.

For a more detailed discussion of nonparametric trend tests, see Ferguson (1965).

BASIC TERMS AND CONCEPTS

Nonparametric test

Relative efficiency

Sign test

Median test

Wilcoxon rank sum test

Mann-Whitney U test

Matched-pairs signed rank test

One-way analysis of variance by ranks

Two-way analysis of variance by ranks

Monotonic trend

EXERCISES

1 The following are data for two groups of experimental animals:

Group I	104	109	127	143	187	204	209	266	277
Group II	62	82	89	90	101	106	109	109	205

Apply a sign test without a continuity correction to test the hypothesis that the two samples come from populations with the same median.

2 The following are data for a sample of nine animals tested under control and experimental conditions:

Control	21	24	26	32	55	82	46	55	88
Experimental	18	9	23	26	82	199	42	30	62

Test the significance of the difference between the two medians using a sign test.

3 The following data are for four groups of experimental animals:

Group I	5	7	16	14	19
Group II	8	15	18	20	24
Group III	17	21	22	25	29
Group IV	23	27	28	31	32

Apply a sign test to test the hypothesis that the four samples come from populations with the same median.

4 Apply the Wilcoxon rank sum test to the data of Exercise 1 above.

5 Apply the Wilcoxon matched-pairs signed-ranks test to the data of Exercise 2 above.

6 Apply a Kruskal-Wallis one-way analysis of variance by ranks to the data of Exercise 3 above.

7 Apply a Friedman two-way analysis of variance by ranks to the following data:

	Treatment			
Subject	I	II	III	IV
1	5	9	4	1
2	6	8	7	3
3	9	10	8	7
4	5	10	4	2
5	8	6	4	1
6	10	8	7	5
7	14	12	13	10

8 Apply a monotonic trend test to the data of Exercise 3 above.

ANSWERS TO EXERCISES

1 $\chi^2 = 8.10, p < .01$

2 $p = .18$

3 $\chi^2 = 8.00, p < .01$

4 $z = 2.65, p < .01$

5 $W_+ = 28, p > .05$

6 $H = 12.87, p < .01$

7 $\chi_r^2 = 15.86, p < .01$

8 $z = 3.87, p < .01$

PSYCHOLOGICAL TEST AND MULTIVARIATE STATISTICS

23

TEST CONSTRUCTION STATISTICS

23.1 INTRODUCTION

Many commonly used psychological tests of ability, achievement, and aptitude are constructed of items which permit two categories of response only, either a *pass* or a *fail*. With such items a weight of 1 is commonly assigned for a pass and a weight of 0 for a failure. Such items are ordinarily spoken of as *dichotomous* items. A person's score, X, on a test of n items is simply the number of items done correctly, although, as will be subsequently shown, certain corrections or transformations may be applied to this score. A substantial body of theory, and statistical technique, has been developed which is concerned with the construction, standardization, accuracy of measurement, and validation of such tests. This body of theory and technique is commonly spoken of as *psychological test*, or *mental test*, theory. For a comprehensive discussion of this subject the reader is referred to Magnusson (1967), Lord and Novick (1968), and Allen and Yen (1979). A few elementary aspects only of the subject are discussed here.

23.2 THE MEAN AND VARIANCE OF TEST ITEMS

Consider a test of n items administered to a sample of N individuals. The number of items passed by each individual may be obtained. These are the test scores for the N individuals denoted by $X_1, X_2, \ldots, X_N$. In addition to this the number of individuals passing each of the n items may be obtained. These numbers, which may be denoted by $P_1, P_2, \ldots, P_n$, are ordinarily divided by N, the number of individuals, to obtain the proportions $p_1, p_2, \ldots, p_n$. With tests of ability these proportions are spoken of as the *difficulty values* of the items and are presumed to describe the difficulty of the item. Clearly, if few people in the sample under consideration pass the item, say, $p_i = .05$ or $p_j = .10$, the item is a difficult one. If a high proportion in the sample of people pass the item, say, $p_i = .80$ or $p_j = .90$, the item may be viewed as an easy item. Note that the proportion is inversely related to difficulty: the higher the proportion, the easier the item. Note further that the proportion p_i is obtained by adding together all the scores of 1 and 0 for a particular item and dividing by the number of individuals in the sample. The quantity p_i is, therefore, the *arithmetic mean* of the individual item.

Each of N individuals obtains a score of 1 or 0 on a particular item, say, item i, whose mean is p_i. The variance of the individual item may be considered, where in this case the variance is defined as $s^2 = \Sigma(X - \bar{X})^2/N$. If we substitute 1s or 0s for the X's, and p_i for $\bar{X}$, it can be shown that the individual-item variance is given by $s_i^2 = p_i q_i$, where $q_i = 1 - p_i$. The item standard deviation is given by $s_i = \sqrt{p_i q_i}$.

The difficulty value of an item, the item mean, is obviously not independent of the item variance. The variance is a maximum when $p_i = .50$, and $s_i^2 = .50 \times .50 = .25$ and departs from this maximum as p_i departs from .50. The item variance approaches 0 as p_i approaches 1.00 or 0.

23.3 CORRELATION BETWEEN TWO TEST ITEMS: THE PHI COEFFICIENT

In psychological test work the correlation between two test items is the usual product-moment correlation between two variables, where the variables are restricted to the integers 1 or 0. This statistic is the phi coefficient and is applicable to 2 × 2 tables only. It is related to χ^2. Although it may be used to describe the relation between any two dichotomous variables, its most common application is in describing the relation between two test items.

One formula for calculating the phi coefficient, or ϕ, is

[23.1] $$\phi = \frac{BC - AD}{\sqrt{(A + B)(C + D)(A + C)(B + D)}}$$

where A, B, C, and D are the four cell frequencies. The term in the denominator of the above expression is the square root of the product of the four marginal totals.

Table 23.1 shows a 2 × 2 table illustrating the relationship between two psychological test items. The value of ϕ based on this table is .376. The reader will note that in this example the two underlying variables may be regarded as continuous. The categories "pass" and "fail" may be con-

Table 23.1

Computation of phi coefficient of correlation between two test items

Frequency

		Item 2 Fail	Item 2 Pass	
Item 1	Pass	11 (*A*)	19 (*B*)	30
	Fail	15 (*C*)	5 (*D*)	20
		26	24	50

Proportion

		Item 2 Fail	Item 2 Pass	
Item 1	Pass	.22 (*a*)	.38 (*b*)	.60 (p_1)
	Fail	.30 (*c*)	.10 (*d*)	.40 (q_1)
		.52 (q_2)	.48 (p_2)	

$$\phi = \frac{19 \times 15 - 11 \times 5}{\sqrt{30 \times 20 \times 24 \times 26}} = .376$$

sidered a dichotomy of an underlying continuous ability variable. Individuals above a certain threshold value on the ability variable pass the item; those below it fail the item. The phi coefficient may be appropriately used where the variables are not continuous.

The phi coefficient is related to χ^2 calculated on a 2 × 2 table by the expression

$$\phi = \sqrt{\frac{\chi^2}{N}}$$

[23.2] or

$$\chi^2 = N\phi^2$$

Any formula for calculating χ^2 for a 2 × 2 table may with minor modification be used for calculating ϕ.

Alternative formulas for computing ϕ may be stated. Let us represent the proportion passing item i by p_i and those failing by q_i, where $p_i = 1 - q_i$. Similarly, the proportion passing item j is p_j and the proportion failing q_j. The proportion passing both items i and j is represented by p_{ij}. The ϕ coefficient of correlation between two test items may then be written as

[23.3]

$$\phi = \frac{p_{ij} - p_i p_j}{\sqrt{p_i p_j q_i q_j}}$$

For the example of Table 23.1, $p_{ij} = .38$ and the phi coefficient is

$$\phi = \frac{.38 - .60 \times .48}{\sqrt{.60 \times .52 \times .40 \times .48}} = .376$$

which checks with the result previously obtained. When one of the variables is evenly divided, $p_i = q_i = .50$, the formula for ϕ simplifies to

[23.4]

$$\phi = \frac{2p_{ij} - p_j}{\sqrt{p_j q_j}}$$

When both variables are evenly divided and $p_i = q_i = p_j = q_j = .50$, the formula becomes

[23.5] $$\phi = 4p_{ij} - 1$$

The phi coefficient has been widely used in statistical work associated with psychological tests. Usually when investigators speak of the correlation between dichotomously scored test items, the reference is to the phi coefficient.

The phi coefficient is a particular case of the product-moment correlation coefficient. If we assign integers, say, 1 and 0, to represent the two categories of each variable and calculate the product-moment correlation coefficient in the usual way, the result will be identical with ϕ.

The phi coefficient has a minimum value of -1 in the case of perfect negative and a maximum value of $+1$ in the case of perfect positive association. These limits, however, can be attained only when the two variables are evenly divided; that is, $p_i = q_i = p_j = q_j = .50$. When the variables are the same shape, $p_i = p_j$ and $q_i = q_j$, but are asymmetrical, $p_i \neq q_i$ and $p_j \neq q_j$, one or the other of the limits -1 or $+1$ may be attained but not both. The maximum and minimum values of phi are clearly influenced by the marginal totals. Consider the following 2×2 tables:

1

	−	+	
+	0	50	50
−	50	0	50
	50	50	

2

	−	+	
+	50	0	50
−	0	50	50
	50	50	

3

	−	+	
+	20	60	80
−	20	0	20
	40	60	

4

	−	+	
+	40	40	80
−	0	20	20
	40	60	

In Tables 1 and 2 both variables are evenly divided and coefficients of $+1$ and -1 are possible. Table 3 represents the maximum positive association possible, given the restriction of the marginal totals. The phi coefficient is .613. Table 4 shows the most extreme negative association possible with the same marginal totals. The phi coefficient is $-.403$. For this particular set of marginal totals phi can extend from a minimum of $-.403$ to a maximum of .613.

While the influence of the marginal totals on the range of values of phi may in some of its applications prove to be a disadvantage, this effect is in no way inconsistent with correlation theory. If a correlation coefficient is viewed as a measure of the efficacy of prediction, then perfect prediction in both a positive and a negative direction is possible only when the two distributions have the same shape and are symmetrical. If one variable is normally distributed and the other is rectangular, perfect prediction of the one from the other is not possible and the correlation coefficient reflects this fact. Perfect prediction in one direction requires symmetry also. The phi coefficient, although affected by the marginal totals, is a measure of the efficacy of prediction. From this viewpoint it quite rightly reflects the loss in degree of prediction resulting from the lack of concordance of the two marginal distributions.

Because $\chi^2 = N\phi^2$, we can readily test the significance of ϕ by referring $N\phi^2$ to a chi-square table with 1 degree of freedom. When $df = 1$, $\sqrt{\chi^2}$ is a normal deviate and we may refer $\phi\sqrt{N}$ to tables of the normal curve. In sampling from a population where no association exists, the distribution of ϕ should be approximately normal with a standard error of $1/\sqrt{N}$. Of course, all considerations pertaining to small frequencies (Section 13.9) apply here. Clearly, N should not be too small.

23.4 RESPONSE PATTERNS

The responses of N subjects on a test of n items may be represented in the form of a table containing n rows and N columns. The elements in this table are 1s and 0s. Such an arrangement of the data is spoken of as a *response* or *answer pattern*. Table 23.2 shows a hypothetical response pattern for a test of 5 items administered to a sample of 10 individuals.

Difficulty values are shown in the column to the right of Table 23.2 and scores are shown in the bottom row. The mean score $\bar{X}$ is 2.80. Note that $\bar{X} = \Sigma X/N = \Sigma p_i = 2.80$. The mean score on the test is the sum of the difficulty values. The variance of scores on the test, where the variance is defined as $s_x{}^2 = \Sigma(X - \bar{X})^2/N$, is 2.16.

Table 23.2

Response pattern for a test of 5 items administered to a sample of 10 subjects

		Individuals										
		1	2	3	4	5	6	7	8	9	10	p_i
Items	1	1	1	1	·	1	1	·	1	1	1	.80
	2	1	1	1	1	·	1	1	·	·	·	.60
	3	1	1	·	1	1	1	1	·	·	·	.60
	4	1	1	1	1	·	·	·	·	·	·	.40
	5	1	1	1	·	1	·	·	·	·	·	.40
	X_i	5	5	4	3	3	3	2	1	1	1	$\bar{X} = 2.80$

23.5 RELATION BETWEEN THE VARIANCE OF TEST SCORES AND THE PROPERTIES OF ITEMS

In test construction a knowledge of the relations between the variance of test scores and the characteristics of test items is of interest. To ascertain the nature of these relations each item on a psychological test is viewed as being interrelated with every other item, the nature of these interrelations being most usefully represented by the interitem covariances. The interitem covariance between two items i and j is given by $r_{ij}s_is_j = p_{ij} - p_ip_j$. The table, or matrix, of item variances and interitem covariances may be represented as follows:

$$\begin{bmatrix} s_1^2 & r_{12}s_1s_2 & \cdots & r_{1n}s_1s_n \\ r_{12}s_1s_2 & s_2^2 & \cdots & r_{2n}s_2s_n \\ \cdots & \cdots & \cdots & \cdots \\ r_{1n}s_1s_n & r_{2n}s_2s_n & \cdots & s_n^2 \end{bmatrix}$$

Such a table of variances and covariances is called a *covariance matrix*. This matrix is symmetrical with item variances along the diagonal and interitem covariances along both sides of the diagonal.

The reader may recall from Section 8.6 that the variance of the sum of a set of variables is the sum of all the elements in the covariance matrix. In the present context this means that the sum of all the elements in the interitem covariance matrix is the variance s_x^2 of scores on the test as a whole. This relation may be written as

$$s_x^2 = \sum_i^n s_i^2 + 2 \sum_{i,j}^{n(n-1)/2} r_{ij}s_is_j \qquad [23.6]$$

This formula describes the relation between the variance of the test and the properties of the individual test items which comprise the test. It expresses the test variance as a function of the item variances and the interitem covariances. For the illustrative response pattern in Table 23.2 for a test of 5 items administered to a sample of 10 subjects the interitem covariance matrix is

.16	−.08	−.08	−.02	.08
−.08	.24	.14	.16	.06
−.08	.14	.24	.06	.06
−.02	.16	.06	.24	.14
.08	.06	.06	.14	.24

The elements in the diagonal are the item variances, $s_i^2 = p_iq_i$. The elements on either side of the diagonal are the item covariances, $r_{ij}s_is_j = p_{ij} - p_ip_j$. The sum of all the elements in this matrix is 2.16, which is the variance of test scores, s_x^2.

The relation shown in formula [23.6] between test variance and the variances and covariances of the items which comprise the test has important implications for test construction. The purpose of any test is the differentiation of individuals or the description of individual differences. In the construction of a test an important consideration is to ensure that the test as a whole discriminates, or shows variation, between those individuals to whom it can be appropriately administered. Clearly a test on which every individual made a zero score, a perfect score, or the same score, and for which $s_x^2 = 0$, would serve no useful purpose. Test makers, therefore, in the development of tests attempt to ensure that the test scores have a fairly large variance in relation to the number of items the test contains. Formula [23.6] shows that the larger the variance of a particular item, the greater the contribution tends to be of that item to the variance of the test as a whole. This has led some test makers to suggest that only items whose difficulty values were not distantly removed from .50, say, between .30 and .70, be included in a test. Formula [23.6] shows also that the greater the interitem covariance between two items, the greater the contribution of those two items to the test variance. This has led test makers in their test construction procedures to include those items which have the higher interitem covariances and correlations.

23.6 INTERNAL CONSISTENCY

Inspection of the response pattern in Table 23.2 shows that individual 4 failed item 1, which had a difficulty value of .80. On the other hand the same individual passed items 2, 3, and 4 with difficulty values of .60, .60, and .40, respectively. His performance on item 1 is clearly inconsistent with his performance on items 2, 3, and 4. Likewise individual 5 failed item 2 with a difficulty value of .60 and passed item 5, a more difficult item, with a difficulty value of .40. This again is an inconsistency of response. If an individual who obtains a score X_i on a test obtains this score by passing the X_i easier items on the test and failing the $n - X_i$ more difficult items, his performance contains no inconsistencies. If all N individuals

Table 23.3

Internally consistent response pattern for a test of 5 items administered to a sample of 10 subjects.

		Individuals										
		1	2	3	4	5	6	7	8	9	10	p_i
Items	1	1	1	1	1	1	1	1	1	·	·	.80
	2	1	1	1	1	1	1	·	·	·	·	.60
	3	1	1	1	1	1	1	·	·	·	·	.60
	4	1	1	1	1	·	·	·	·	·	·	.40
	5	1	1	1	1	·	·	·	·	·	·	.40
	X_i	5	5	5	5	3	3	1	1	·	·	$\bar{X} = 2.80$

taking a test obtain their score X_i in this way, and no inconsistencies are present in the response pattern, that response pattern may be spoken of as an *internally consistent* pattern. With real data, response patterns that exhibit no inconsistencies do not ordinarily occur. Response patterns for different tests exhibit varying numbers of inconsistencies and varying degrees of internal consistency.

Table 23.3 shows an example of an internally consistent response pattern. The difficulty values of the items are the same as those in Table 23.2. Table 23.3 shows what the response pattern for the data of Table 23.2 would have been had no inconsistencies been present. Although the item difficulties are the same for Table 23.3 as for Table 23.2, the test scores are different. The test scores for internally consistent pattern of Table 23.3 are more variable. The variance of scores for Table 23.3 is $s_x^2 = 3.90$, as compared with a variance $s_x^2 = 2.16$ of Table 23.2. The effect of inconsistencies in the response pattern is to reduce the variance of test scores.

What would the variance of test scores be if all responses were assigned at random, given the restriction of the difficulty values? Under these conditions we would expect, since the responses are random, that all the interitem correlations and covariances would be 0. The expected value of the variance under these conditions is $s_x^2 = \Sigma s_i^2 = \Sigma p_i q_i$. Thus the test variance is the sum of the elements in the diagonal of the covariance matrix. All elements on either side of the diagonal are 0. For the data of Table 23.2 the expected variance, if all responses are assigned at random, is $s_x^2 = 1.12$. For any given set of difficulty values the test variance may vary from a minimum of $s_x^2 = \Sigma s_i^2$ to a maximum value which is the value of the variance which would result if the response pattern contained no inconsistencies.

23.7 THE ITEM SELECTION PROBLEM

Test makers in the construction of psychological tests have devised methods for pretesting test items. Not infrequently a preliminary form is prepared which contains a much larger number of items than will be included in the final test. If a test of 50 items is required, the test maker may begin with an initial collection of 100 items. These are administered on a trial basis to a sample of the population on which the test will ultimately be used. How does the test maker decide which are the better 50 items of the 100 items available? What considerations lead to the selection of one item and the rejection of another?

In some cases external criteria are available. Such criteria may be measures of job performance, average grades following one year at the university, or other indices of performance. Each test item may be correlated with the criterion variable, and the correlation coefficients used in the selection of items. Presumably, if the ultimate objective is to construct a test to predict the criterion variable, those items would be included in the final test that have a high correlation with that criterion. Items with a low correlation with the criterion would be rejected. At times criterion groups are available. The criterion may be membership in a top, middle, or low group on job performance; a group of psychotic patients and a group of normal subjects; groups of individuals who complete a four-year university course and those who do not. Here again correlation coefficients, or other statistics, may be used to describe how the test items discriminate between groups.

In many situations no external criterion against which the test items can be validated is available. Under these circumstances a criterion of internal consistency is commonly used; that is, the test maker by one means or another selects a subset of items that exhibits fewer inconsistencies in its response pattern than many of the other subsets that might have been selected. The problem may be stated in this way. For any group of n items a subset k may be selected in C_k^n possible ways. By what method may a set of k items be selected from the C_k^n possible sets such that the internal consistency for the set of k items selected is greater than the internal consistency for any of the $C_k^n - 1$ remaining sets? Many methods of item selection in common use, which are applied in the absence of an external criterion, provide *approximate* solutions to this problem. The subset of items selected is ordinarily not the most internally consistent subset, but is one of the more internally consistent subsets. Although an exact solution to the item selection problem as stated above is possible, its application would be very laborious except for quite small sets of items.

Although a great many methods of item selection exist, the most commonly used involves calculating the correlations between the test items and scores on the whole test. Since the items are dichotomously scored and scores on the whole test may be viewed for all practical purposes as continuous, a form of correlation is required which describes the relation between a dichotomous, or two-categoried, variable and a continuous variable. The forms of correlation used for this purpose are the point biserial correlation and the biserial correlation.

23.8 POINT BISERIAL CORRELATION

Point biserial correlation provides a measure of the relation between a continuous variable, such as scores on a test, and a two-categoried, or dichotomous, variable, such as "pass" or "fail" on a psychological test item. The data when arranged in the form of a frequency distribution compose a table comprised of R rows and 2 columns. Although commonly used as a measure of the correlation between test scores and test items, point biserial correlation may be used in other situations as well. For example, the continuous variable may be scores on a psychological test, and the dichotomous variable may be male or female, or high school graduates and university graduates, or a group of normal persons and a group of neurotics.

Point biserial correlation is a product-moment correlation. If we assign a 1 to individuals in one category and a 0 to individuals in the other and calculate the product-moment correlation, the result is a point biserial coefficient. Weights other than 1 and 0 may be assigned to the categories. The coefficient is not dependent on the weights assigned.

The formula for point biserial r is

[23.7] $$r_{pbi} = \frac{\bar{X}_p - \bar{X}_q}{s_x} \sqrt{pq}$$

In this formula s_x is the standard deviation of scores on the continuous variable, defined as $\Sigma(X - \bar{X})^2/N$. If the continuous variable is a test, s_x is the standard deviation of test scores. The quantities p and q are the proportions of individuals in the two categories of the dichotomous variable. If the dichotomous variable is a test item, p is the proportion of individuals who pass the item and q is the proportion who fail. $\bar{X}_p$ and $\bar{X}_q$ are the mean scores on the continuous variable of individuals within the two categories. Again, if the continuous variable is a set of test scores, $\bar{X}_p$ is the mean score of those who pass the item and $\bar{X}_q$ is the mean score of those who fail.

The calculation of point biserial correlation from ungrouped data is illustrated in Table 23.4. This table shows hypothetical scores on a test, and on a test item, for a group of 14 individuals. The mean score, $\bar{X}_p$, on the test for the six individuals who pass the item is 38.17, and the mean $\bar{X}_q$ for the eight individuals who fail the item is 23.88. The standard deviation of test scores is 18.77. The proportion of individuals who pass and fail the item, p and q, are, respectively, .43 and .57. The point biserial correlation is .38. In this example the point biserial correlation coefficient is a measure of the capacity of the test to discriminate between two groups. This statistic can always be interpreted as a measure of the degree to which the continuous variable differentiates, or discriminates, between the two categories of the dichotomous variable. The reader should note in Table 23.4 that if the eight individuals who fail the item were the eight individuals making the eight lowest scores on the test, and the six individuals who pass the item were the six individuals making the six highest scores on the test, then the point biserial correlation would assume a maximum value for the data. Also, if "pass" and "fail" were arranged at random in relation to test scores, the point biserial correlation would have an expected value of zero.

An alternative method of calculating point biserial correlation is the formula

[23.8] $$r_{pbi} = \frac{\bar{X}_p - \bar{X}}{s_x} \sqrt{\frac{p}{q}}$$

where $\bar{X}$ is the mean of all scores on the continuous variable, and s_x, $\bar{X}_p$, p, and q are as defined in formula [23.7].

Point biserial correlation is not independent of the proportions in the two categories. When $p = q = .50$, its maximum and minimum values will differ from those which would be obtained when, say, $p = .20$ and $q = .80$. The maximum value of r_{pbi} never reaches +1; the minimum value never

Table 23.4
Calculation of point biserial correlation from ungrouped data

Individual	Inventory score	Item score
1	6	0
2	8	1
3	8	0
4	11	0
5	16	1
6	25	0
7	27	0
8	31	0
9	31	1
10	39	0
11	44	0
12	50	1
13	56	1
14	68	1

Mean score for those who pass: $\bar{X}_p = 38.17$
Mean score for those who fail: $\bar{X}_q = 23.88$
$s_x = 18.77 \qquad p = 6/14 = .43 \qquad q = 8/14 = .57$

$$r_{pbi} = \frac{38.17 - 23.88}{18.19}\sqrt{.43 \times .57} = .38$$

reaches -1. In predicting a two-categoried variable from a continuous variable, perfect prediction is possible and occurs when the two frequency distributions do not overlap. Perfect prediction of a continuous variable from a two-categoried variable is obviously impossible. Some error in prediction must always occur in predicting a variable which may take a wide range of values from a variable which may take two values only. The point biserial correlation coefficient reflects this fact. It is worth noting here that the regression line obtained by calculating the means of the two columns is of necessity linear, there being only two points. The regression line obtained by calculating the means of the rows cannot be linear except under certain special circumstances.

To test the significance of r_{pbi} from 0 the situation may be treated as one requiring a comparison of the two means $\bar{X}_p$ and $\bar{X}_q$. The appropriate value of t may be written

[23.9] $$t = r_{pbi}\sqrt{\frac{N-2}{1-r_{pbi}^2}}$$

The number of degrees is $N-2$. This is a two-tailed test. For large N the quantity $1/\sqrt{N}$ may be used as the standard error of r_{pbi} in testing the significance of the difference from 0.

Biserial correlation, as distinct from point biserial correlation, provides a measure of the relation between a continuous variable and a dichotomous variable on the assumption that the variable underlying the dichotomy is continuous and normal. This statistic now appears to be infrequently used. A discussion of it may be found in previous editions of this book.

23.9 CONTRIBUTION OF ITEMS TO THE TEST VARIANCE

The proportion of individuals passing an item, p_i, has been discussed as a measure of item difficulty. Point biserial and biserial correlation have been discussed as measures of the discriminatory capacity of test items. Difficulty and discriminatory capacity are dual criteria in item selection. Because we have two criteria, a problem arises. Is item i with $p_i = .50$ and $r_{pbi} = .40$ a better item than item j with $p_j = .40$ and $r_{pbi} = .50$? The relative importance of item difficulty and item discriminatory capacity in the selection of items has been viewed as a problem by some test makers.

This problem has a simple solution which involves combining difficulty and discriminatory capacity into a single index. This index is the contribution of the individual item to the variance of test scores. The presumption here is that in the construction of tests we wish, as it were, to acquire or capture variance. An item that contributes more to the total variance is presumed to be a better item than one that contributes less.

In Section 23.5 the total test variance was shown to be the sum of all the elements in the interitem covariance matrix. The sum of a column, or row, in the covariance matrix may be viewed as the contribution of the individual item to the total variance. This is one way only of defining the contribution of the item to the total variance. For the illustrative numerical example of a covariance matrix in Section 23.5 the column sums for the five items are as follows:

Item	1	2	3	4	5
Sum	.06	.52	.42	.58	.58

The total variance is $s_x^2 = 2.16$. Item 1 is viewed as contributing .06 to this; item 2, .52; item 3, .42; and so on.

It can be readily shown that the contribution of item i to the total test variance is the item-test covariance $r_{ix}s_is_x$, where r_{ix} is the point biserial correlation coefficient. The total variance is the sum of the covariances for the n items; that is,

$$s_x^2 = \sum_{i=1}^{n} r_{ix}s_is_x \tag{23.10}$$

If we multiply the formula for the point biserial correlation coefficient by s_is_x, we obtain the item-test covariance, which is

$$r_{ix}s_is_x = (\bar{X}_p - \bar{X}_q)pq \tag{23.11}$$

Thus the contribution of the individual item to the total variance is, very simply, the difference between the means of those who pass and those who fail the item multiplied by pq.

In the covariance term $r_{ix}s_is_x$, s_x is the same for all items. This circumstance has led to the use of $r_{ix}s_i$, instead of $r_{ix}s_is_x$, as an index of item selection. The quantity $r_{ix}s_i$ has been referred to by some writers as the *reliability index*. If we divide formula [23.10] by s_x, we note that

[23.12] $$s_x = \sum_{i=1}^{n} r_{ix}s_i$$

Whereas the sum of the n terms $\sum^{n} r_{ix}s_is_x$ is the test variance, the sum of the n terms $\sum^{n} r_{ix}s_i$ is the test standard deviation.

The values of $r_{ix}s_i$ are slightly more difficult to compute than those of $r_{ix}s_is_x$, since $r_{ix}s_i = (\bar{X}_p - \bar{X}_q)pq/s_x$. The statistic $r_{ix}s_is_x$, or its simple modification $r_{ix}s_i$, is probably the most useful single index for the selection of items in common use.

BASIC TERMS AND CONCEPTS

Dichotomous variable

Difficulty value

Mean of a test item

Variance of a test item

Phi coefficient

Response pattern

Item covariance

Internal consistency

Point biserial correlation

Reliability index

EXERCISES

1 The following is the response pattern for a test of 5 items administered to a sample of 12 subjects.

Items	Subjects 1	2	3	4	5	6	7	8	9	10	11	12	P_i
1	1	1	·	1	·	1	1	1	1	1	·	1	9
2	1	1	1	·	1	·	·	1	1	1	1	·	8
3	1	·	1	1	1	1	1	·	·	·	·	·	6
4	1	1	1	1	·	·	·	·	·	·	·	·	4
5	1	1	·	·	1	·	·	·	·	·	·	·	3
X_i	5	4	3	3	3	2	2	2	2	2	1	1	$\bar{X} = 2.50$

Test scores X_i are shown along the bottom of the table. The number of individuals passing each item, P_i, are shown to the right. Calculate (**a**) the item means and (**b**) the item variances.

2 Calculate (**a**) the interitem covariances and (**b**) the intercorrelations for the data of Exercise 1.

3 If the response pattern of Exercise 1 had contained no inconsistencies, what would the variance of scores have been?

4 If the responses in the rows of the response pattern of Exercise 1 had been assigned at random, what would be the expected variance of scores?

5 Calculate the point biserial correlation between the individual items and the total test scores for the data of Exercise 1.

6 For the data of Exercise 1 calculate the contribution of each item to the total test variance.

7 For the data of Exercise 1 calculate the contribution of each item to the total test standard deviation.

ANSWERS TO EXERCISES

1 **a** .750, .667, .500, .333, .250
b .188, .222, .250, .222, .188

2 **a**

	1	2	3	4	5
1					
2	−.083				
3	−.042	−.083			
4	.000	.030	.083		
5	−.021	.083	.042	.084	

b

	1	2	3	4	5
1					
2	.405				
3	−.194	−.350			
4	.000	.136	.350		
5	−.112	.405	.194	.410	

3 3.583

4 1.070

5 .087, .316, .447, .790, .776

6 .042, .169, .250, .419, .376

7 .038, .151, .224, .375, .336

24

ERRORS OF MEASUREMENT

24.1 THE NATURE OF ERROR

The measurements obtained in the conduct of experiments are subject to error of greater or lesser degree. In measuring the activity of a rat, the intelligence of a child, or the response latency of an experimental subject, we may assume that the individual measurements are subject to some error. In general, the concept of error always implies a true, fixed, standard, or parametric value which we wish to estimate and from which an observed measurement may differ by some amount. The difference between a true value and an observed value is an error. If we represent a particular observation by X_i, the true value which it purports to estimate by T_i, and an error by e_i, we may write

[24.1] $$e_i = X_i - T_i$$

where e_i may take either positive or negative values.

A distinction may be made between *systematic* and *random* error. Observations which consistently overestimate or underestimate the true value are subject to systematic error. A stopwatch which underestimates time intervals will yield observations with systematic errors. A random error exhibits no systematic tendency to be either positive or negative and is assumed to average to 0 over a large number of subjects or trials. Random errors are also assumed to be uncorrelated both with true scores and with each other. The discussion in this chapter is concerned exclusively with random errors.

Any definition of error as the difference between an observed and true value is meaningless unless a precise definition is attached to the concept of true value. In theory a true value is sometimes conceptualized as the mean of an indefinitely large number of measurements of an attribute made under conditions such that the true value remains constant, and the procedures used in making the measurements do not change from trial to trial in any known systematic fashion. In mathematical language the true value may be defined as

$$T_i = \lim_{K \to \infty} \frac{\sum_{j=1}^{k} X_j}{K}$$

where X_j refers to the jth measurement. Thus the true value is the limit approached by the arithmetic mean as the number of repeated observations K is increased indefinitely. This concept of true value is appropriate for the measurements of physical quantities. For example, a yardstick may be used to measure the length of a desk. The measurement procedure may be repeated many times, and the variation in the observations attributed to error. It may be assumed that a considerable number of repeated observations may be made under fairly constant conditions, neither the desk nor the yardstick changing in any systematic way. By increasing the number of observations and taking their mean, the error in estimating the true value may be reduced. Theoretically, this error may be made as small as we like by increasing the number of observations. As the number of observations becomes indefinitely large, the mean approaches the true value as a limit.

Questions may be raised about the appropriateness of this concept of true value in the measurement of psychological quantities. Clearly, in the measurement of human behavior the making of a large number of repeated observations is usually not possible. The attribute being measured may fluctuate or change markedly with time, or the process of repeated measurement may modify the attribute under study. For example, in measuring the intelligence of a child, it is obviously out of the question to administer the same intelligence test 100 times to obtain an estimate of error. Quite apart from the labor involved in such estimation, the results obtained would be invalidated by practice, fatigue, and other effects. This circumstance has given rise in psychological work to a variety of procedures for estimating error other than by a series of repeated measurements. Despite the operational impracticality in psychology of estimating error by making a large number of repeated measurements, the concept of true score as the mean of an indefinitely large number of such measurements is still an important concept in the study of errors of measurement. Here we note that the role of true score is analogous to that of population parameter in sampling statistics. The difference between the sample statistic and the population parameter is a sampling error. By increasing sample size the magnitude of sampling error is reduced. For an infinite population an unbiased sample statistic will approach the population parameter as a limit as the sample becomes indefinitely large. A sampling error is an error associated with a statistic based on a sample of observations. An error of measurement is usually construed to be an error associated with a particular observation which is an estimate of a true value. In most instances both population parameters and true values cannot be known but can only be estimated from fallible data. This circumstance does not detract from the meaningfulness of these concepts, nor does it prevent the making of meaningful statements about the magnitude of error.

24.2 EFFECT OF MEASUREMENT ERROR ON THE MEAN AND VARIANCE

Consider a population of measurements. Each measurement is subject to error and may be written as $X_i = T_i + e_i$, where X_i is the observed and T_i the true measurement. By summation over all members of the population we obtain $\Sigma X_i = \Sigma T_i + \Sigma e_i$. If we assume that measurement error is random, and as often positive as negative, we may write $\Sigma e_i = 0$. Consequently, the sum of measurements subject to error is equal to the sum of true measurements. It follows also that the means of the observed and true values are equal, both being equal to the population mean μ. We conclude that measurement error exerts no systematic effect on the arithmetic mean. A mean based on a sample of N measurements will exhibit no systematic tendency to be either greater than or less than the mean of true measurements. The expectations of the mean of observed and true scores are equal to the population mean μ; that is,

[24.2] $$E(\bar{X}) = E(\bar{T}) = \mu$$

Measurement error exerts an effect on the sampling variance of the arithmetic mean. This point is discussed in Section 24.6.

Measurement error exerts a systematic effect on the variance. We may write $(X_i - \mu) = (T_i - \mu) + e_i$. If we square this identity, sum over all members of the population, and divide by N_p, where N_p is the number of members in the population, we obtain

$$\frac{\Sigma(X_i - \mu)^2}{N_p} = \frac{\Sigma(T_i - \mu)^2}{N_p} + \frac{\Sigma e_i^2}{N_p} + \frac{2\Sigma(T_i - \mu)e_i}{N_p}$$

On the assumption that measurement errors are random and uncorrelated with true scores, the third term to the right is equal to 0, and we may write

[24.3] $$\sigma_x^2 = \sigma_T^2 + \sigma_e^2$$

Thus the variance of observed scores is equal to the variance of true scores plus the variance of the errors of measurement. For a fixed σ_T^2, the more inaccurate the measurements the greater the value of σ_e^2 and the greater the variance σ_x^2.

24.3 THE RELIABILITY COEFFICIENT

Consider a situation where each member of a population has been measured on two separate occasions. Two observations are available for each member. Both are presumed to be measures of the same attribute, and both are subject to error. We may write

$$X_{i1} = T_i + e_{i1}$$
$$X_{i2} = T_i + e_{i2}$$

In deviation form these become

$$(X_{i1} - \mu) = (T_i - \mu) + e_{i1}$$
$$(X_{i2} - \mu) = (T_i - \mu) + e_{i2}$$

By multiplying these two equations, summing over a population of N_p members, and dividing by $N_p\sigma_1\sigma_2$, we obtain

$$\rho_{xx} = \frac{\Sigma(X_{i1} - \mu)(X_{i2} - \mu)}{N_p\sigma_1\sigma_2}$$
$$= \frac{\Sigma(T_i - \mu)^2 + \Sigma e_{i1}e_{i2} + \Sigma e_{i1}(T_i - \mu) + \Sigma e_{i2}(T_i - \mu)}{N_p\sigma_1\sigma_2}$$

On the assumption that errors are random and uncorrelated with each other and with true scores, the three terms on the right in the numerator are equal to 0. Because the paired observations are measures of the same attribute, $\sigma_1 = \sigma_2$. Also $\Sigma(T_i - \mu)^2 = N_p\sigma_T^2$. Hence, writing $\sigma_1 = \sigma_2 = \sigma_x$,

[24.4] $$\rho_{xx} = \frac{\sigma_T^2}{\sigma_x^2}$$

where ρ_{xx} is the reliability coefficient. The reliability coefficient is a simple proportion. It is the proportion of obtained variance that is true variance. If $\sigma_x^2 = 400$ and $\sigma_T^2 = 360$, the reliability coefficient $\rho_{xx} = .90$. This means that 90 percent of the variation in the measurements is attributable to variation in true score, the remaining 10 percent being attributable to error. Where sample estimates are used we may write

[24.5] $$r_{xx} = \frac{s_T^2}{s_x^2}$$

where r_{xx} is the sample estimate of the reliability coefficient.

24.4 METHODS FOR ESTIMATING RELIABILITY

Above, the reliability coefficient has been discussed without reference to methods for obtaining such coefficients in practice. A number of different practical methods for estimating reliability are used. These methods are as follows:

1 **Test-retest method** The same measuring instrument is applied on two occasions to the same sample of individuals. When the instrument is a psychological test, the test is administered twice to a sample of individuals and the scores correlated.

2 **Parallel-forms method** Parallel or equivalent forms of a test may be administered to the same group of subjects, and the paired observations correlated. Criteria of parallelism are required.

3 **Split-half method** This method is appropriate where the testing procedure may in some fashion be divided into two halves and two scores obtained. These may be correlated. With psychological tests a common procedure is to obtain scores on the odd and even items.

4 **Internal-consistency methods** These are used with psychological tests comprised of a series of items, usually dichotomously scored, a 1 being assigned for a pass and a 0 for a failure. These methods require a knowledge of certain test-item statistics.

The interpretation of a reliability coefficient depends on the method used to obtain it. When the same test is administered twice to the same group with a time interval separating the two administrations, some variation, fluctuation, or change in the ability or function measured may occur. The departure of r_{xx} from unity may be construed to result in part from error and in part from changes in the ability or function measured. With many psychological tests the value of r_{xx} will show a systematic decrease with increase in the time interval separating the two administrations. When the time interval is short, memory effects may operate. The subject may recall many of his previous responses and proceed to reproduce them. A spuriously high correspondence between measurements obtained at the two testings may thereby result. Regardless of the time interval separating the two testings, varying environmental conditions such as noise, temperature, and other factors may affect the result obtained. Likewise, varying physiological factors, fatigue and the like, may exert an influence.

In estimating reliability by the administration of parallel or equivalent forms of a test, criteria of parallelism are required. Test content, type of item, instructions for administering, and the like, should be similar for the different forms. Also the parallel forms should have approximately equal means and standard deviations. In addition, the intercorrelations should be equal. Thus with three parallel tests the intercorrelations should be such that $r_{12} = r_{13} = r_{23}$. A discussion of criteria for parallel tests is given by Gulliksen (1950). Situations arise where a large pool or population of test items is available. Samples of items may be drawn at random. Each sample of items is a randomly parallel form. This approach to the development of parallel tests has been studied at length by Lord (1955a, 1955b).

In many situations a single administration only of a test may be possible. The test is divided into two halves. A not uncommon procedure is to divide a test into odd and even items. Scores are obtained on the two halves, and these are correlated. The result is a reliability coefficient for a half test. Given a reliability coefficient for a half test, the reliability coefficient for a whole test may be estimated using the Spearman-Brown formula. This formula is

[24.6] $$r_{xx} = \frac{2r_{hh}}{1 + r_{hh}}$$

where r_{hh} is the reliability of a half test. If, for example, $r_{hh} = .80$, then $r_{xx} = .89$. The Spearman-Brown formula provides an estimate of the reliability of the whole test. It estimates what the reliability would be if each test half were made twice as long.

The split-half method should not be used with highly speeded test material. Obviously, if a test comprises easy items, and a subject is required to complete as many items as possible within a limited time interval, and all or nearly all items are correct, the scores on the two halves would be about the same and the correlation would be close to $+1.00$.

A method of obtaining reliability coefficients using test-item statistics has been developed by Kuder and Richardson (1937). Many psychological tests are constructed of dichotomously scored items. An individual either passes or fails the item. A 1 is assigned for a pass, and 0 for a failure. The score is the number of items done correctly. The proportion of individuals passing item i is denoted by the symbol p_i, and the proportion failing, by q_i, where $q_i = 1 - p_i$. An estimate of reliability is given by

[24.7] $$r_{xx} = \frac{n}{n-1} \frac{s_x^2 - \sum_{i=1}^{n} p_i q_i}{s_x^2}$$

where n = number of items

s_x^2 = variance of scores on test defined as $\Sigma(X - \bar{X})^2/N$

$p_i q_i$ = product of proportion of passes and fails for item i

$\sum_{i=1}^{n} p_i q_i$ = sum of these products for n items

This formula is frequently referred to as Kuder-Richardson formula 20. The coefficient r_{xx} computed by this formula will take values ranging from 0 to unity. If the responses of individuals to the test items are assigned at random, the expectation of s_x^2 is equal to $\sum_{i=1}^{n} p_i q_i$ and the expectation of r_{xx} is 0. If all items are perfectly correlated, a situation which can only arise when all have the same difficulty, $r_{xx} = 1$. The correlation between items is the phi coefficient.

If all assumptions implicit in the split-half method of estimating reliability coefficients are satisfied, the split-half and Kuder-Richardson formula 20 will yield identical results. Because these assumptions are rarely, if ever, satisfied in practice, differences in the coefficients obtained will result. One difficulty with the split-half method is that a test may be split in a great many ways, yielding many different values of r_{xx}. It may be shown that if a test is split in all possible ways, the average of all the split-half reliability coefficients with the Spearman-Brown correction is the Kuder-Richardson formula 20. This coefficient has a simple unique value for any particular test.

The Kuder-Richardson formula 20 is a measure of the internal consistency, or homogeneity, or scalability, of the test material. In this context these three terms may be considered synonymous. If the items on a test have high intercorrelations with each other and are measures of much the same attribute, then the reliability coefficient will be high. If the intercorrelations are low, either because the items measure different attributes or because of the presence of error, then the reliability coefficient will be low.

The Kuder-Richardson formula 20 may be applied to tests comprising items which elicit more than two categories of response. Personality and interest inventories and attitude scales frequently permit three or more response categories. For a dichotomously scored item we note that p_iq_i is the item variance s_i^2 and $\sum_{i=1}^{n} p_iq_i = \sum_{i=1}^{n} s_i^2$, the sum of the item variances. For an item with more than two response categories, where each category has been assigned a weight, the individual item variances may be calculated and their sum may be substituted in Kuder-Richardson formula 20 for $\sum_{i=1}^{n} p_iq_i$. Consider a test comprising statements which elicit the possible responses "agree," "undecided," "disagree." Let p_1, p_2, and p_3 be the proportion of individuals responding in the three categories. If weights 3, 2, 1 or $+1$, 0, -1, or any other system of weights, are assigned to the categories, the item variance may be calculated. These may be summed, and the sum substituted for $\sum_{i=1}^{n} p_iq_i$. The quantity s_x^2 is, of course, the variance of scores obtained by summing items with the assigned weights.

On the assumption that all test items are of equal difficulty, a simplified form of the Kuder-Richardson formula may be obtained for use with dichotomously scored test items. This formula may be written as

[24.8] $$r_{xx} = \frac{n}{n-1}\left[1 - \frac{\bar{X}(n-\bar{X})}{ns_x^2}\right]$$

where $\bar{X}$ = mean test score
s_x^2 = variance

This formula is referred to as Kuder-Richardson formula 21. The formula may be derived using the assumptions implicit in the concept of randomly parallel tests (Section 24.9).

24.5 EFFECT OF TEST LENGTH ON THE RELIABILITY COEFFICIENT

In discussing split-half reliability, a formula was given for estimating the reliability of a whole test from the reliability of a half test. This formula is a particular case of a more general Spearman-Brown formula for estimating increased reliability with increased test length. The more general formula is

[24.9] $$r_{kk} = \frac{kr_{xx}}{1 + (k - 1)r_{xx}}$$

where r_{xx} = an estimate of reliability of a test of unit length
r_{kk} = reliability of test made k times as long

If $r_{xx} = .60$ and the test is made four times as long, the reliability coefficient r_{kk} for the lengthened test is estimated as .86. From a theoretical point of view a test may be made as reliable as we like by increasing its length. Practical considerations, of course, restrict test length.

Because reliability is a function of test length, reliability coefficients calculated on tests of different lengths are, for certain purposes, not directly comparable. If, for example, we wish to compare the reliability of different types of test material, we presumably would require measures which were independent of the differing lengths of the tests. One procedure here is to use the Spearman-Brown formula and calculate reliability coefficients for a standard test of 100 items. If a test has 40 items, then a value of $k = 100/40 = 2.50$ would be used in estimating the reliability of the standard test. If another test has 150 items, then $k = 100/150 = .67$, and so on. Thus a comparison of the reliabilities of different tests may be made which is independent of differing test lengths.

24.6 EFFECT OF MEASUREMENT ERROR ON THE SAMPLING VARIANCE OF THE MEAN

Because measurement error affects the variance of a set of measurements it will also affect the sampling variance of the mean. The sampling variance of the arithmetic mean may be written as

[24.10] $$\sigma_{\bar{x}}^2 = \frac{\sigma_x^2}{N} = \frac{\sigma_T^2}{N} + \frac{\sigma_e^2}{N}$$

The component σ_T^2/N is the sampling variance of the means of samples of true measurements, and σ_e^2/N is the component of the sampling variance attributable to measurement error. While measurement error exerts no systematic effect on the sample mean as an estimate of μ, such error increases the variation in sample means with repeated sampling. The increase in sampling variance over that with no measurement error present is σ_e^2/N.

The ratio of the sampling variance of the mean of true scores to the sampling variance of the mean of obtained scores is the reliability coefficient. Thus

[24.11] $$\rho_{xx} = \frac{\sigma_T^2}{\sigma_x^2} = \frac{\sigma_T^2/N}{\sigma_x^2/N} = \frac{\sigma_{\bar{T}}^2}{\sigma_{\bar{x}}^2}$$

This means that the reliability coefficient may be interpreted as descriptive of the loss in efficiency of estimation resulting from measurement error. To illustrate, a mean calculated on a sample of 100 cases, where $\rho_{xx} = .80$, has a sampling variance equal to that of a mean calculated on a sample of 80 cases where $\rho_{xx} = 1.00$. The loss in efficiency of estimation resulting from measurement error amounts to 20 cases in 100.

24.7 EFFECT OF ERRORS OF MEASUREMENT ON THE CORRELATION COEFFICIENT

Errors of measurement tend to reduce the size of the correlation coefficient. The correlation between true scores will tend to be greater than the correlation between obtained scores. If ρ_{xy} is the correlation between X and Y in the population, the relation between the correlation of true and obtained scores is given by

[24.12] $$\rho_{T_x T_y} = \frac{\rho_{xy}}{\sqrt{\rho_{xx}\rho_{yy}}}$$

where $\rho_{T_x T_y}$ = correlation between true scores
ρ_{xx} = reliability of X
ρ_{yy} = reliability of Y

This formula is known as the *correction for attenuation.* Errors tend to attenuate the correlation coefficient between obtained scores from the correlation between true scores.

The corresponding sample form of the correction for attenuation is

[24.13] $$r_{T_x T_y} = \frac{r_{xy}}{\sqrt{r_{xx}r_{yy}}}$$

To illustrate, let $r_{xy} = .60$, $r_{xx} = .80$, and $r_{yy} = .90$. The correlation between true scores on X and Y, estimated by the above formula, is .707. The correlation may be viewed as attenuated from .707 to .60 because of errors of measurement. The squares of these coefficients yield a better appreciation of the loss in predictive capacity due to errors of measurement. The squares of .707 and .60 are .50 and .36. We conclude that the presence of errors of measurement results in 14 percent loss in predictive capacity. If the correlation between two variables is low, the correlation will not be markedly increased by improvements in reliability. If the correlation is high, improving reliability may result in substantial gains in the prediction of one variable from another.

Because the correlation between true scores can never exceed unity, the maximum correlation between two variables arises where $r_{T_x T_y} = 1$. Under this circumstance $r_{xy} = \sqrt{r_{xx} r_{yy}}$. This is an estimate of the maximum correlation between X and Y. If $r_{xx} = .80$ and $r_{yy} = .90$, the maximum possible correlation between X and Y is estimated as $\sqrt{.80 \times .90} = .85$.

24.8 RELIABILITY OF DIFFERENCE SCORES

Situations arise where the difference between two sets of measurements is defined as a score. The two measurements may be initial, or prestimulus, values and values obtained in the presence of a stimulus factor. If differences are obtained between standard scores on X on Y, that is, between z_x and z_y, the reliability of the differences may be estimated by

[24.14] $$r_{dd} = \frac{r_{xx} + r_{yy} - 2r_{xy}}{2 - 2r_{xy}}$$

where r_{xx} and r_{yy} = reliability coefficients for X and Y
r_{dd} = reliability of difference $z_x - z_y$

For fixed values of r_{xx} and r_{yy} the reliability of the difference will decrease with increase in r_{xy} from 0. If $r_{xx} = .90$ and $r_{yy} = .80$, for $r_{xy} = .80$ the reliability of differences $r_{dd} = .25$. For $r_{xy} = 0$, $r_{dd} = .85$. As r_{xy} departs in a positive direction from 0, the error variance accounts for an increasing proportion of the total variance of differences, with a resulting decrease in reliability. The point to note here is that differences scores may be grossly unreliable and should be used only after careful scrutiny of the data. When the correlation between the two variables is reasonably high, it is probable that with many sets of data most of the variance of differences is error variance.

24.9 THE STANDARD ERROR OF MEASUREMENT

Because $\rho_{xx} = \sigma_T^2/\sigma_x^2$ and $\sigma_x^2 = \sigma_T^2 + \sigma_e^2$, we may write

[24.15] $$\rho_{xx} = 1 - \frac{\sigma_e^2}{\sigma_x^2}$$

[24.16] and $$\sigma_e = \sigma_x \sqrt{1 - \rho_{xx}}$$

This latter formula is the standard error of measurement. Where s_x and r_{xx} are used as estimates of σ_x and ρ_{xx}, we obtain

[24.17] $$s_e = s_x \sqrt{1 - r_{xx}}$$

as the corresponding sample estimate. If it may be assumed that errors of measurement are independent of the magnitude of test score, then s_e may be used as the standard error associated with a single score and interpreted in the same way as the standard error of any statistic. On the assumption of a normal-curve approximation, the 95 and 99 percent confidence intervals of an individual's score X_i are estimated by $X_i \pm 1.96s_e$ and $X_i \pm 2.58s_e$, respectively. With most psychological tests, however, errors of measurement are not independent of the magnitude of test score. The standard error is higher in the middle-score range and diminishes in size as the score departs from the average. Because of this the use of s_e to estimate confidence intervals for particular scores may yield misleading results. The variance s_e^2 is a sort of average value, and s_e when applied to particular scores has meaning only in relation to scores near the average.

The problem of the standard error of measurement associated with psychological test scores has been investigated by Lord (1955a, 1955b). Lord defines the standard error of measurement as the standard deviation of scores an individual might be expected to obtain on a large number of randomly parallel test forms. The assumption is that the ability of the individual remains unchanged and is not affected by practice, fatigue, and the like. Randomly parallel forms are viewed as composed of items drawn at random from a large pool or population of items. The items are scored 1 for a pass and 0 for a failure, a score on a test being the sum of item scores. The proportion of items in the population which individual i can do correctly is θ_i. The true score of individual i for a test of n items is $T_i = n\theta_i$. The number of items done correctly by individual i for a random sample of n items is X_i. The standard deviation of the sampling distribution of the X_i's is the standard error. This is obtained from the standard deviation of the binomial and is given by

$$\begin{aligned} \sigma_e(X_i) &= \sqrt{n\theta_i(1-\theta_i)} \\ &= \sqrt{\frac{1}{n} T_i(n - T_i)} \end{aligned} \tag{24.18}$$

An individual's score X_i may be used as an estimate of T_i. Introducing the factor $n/(n-1)$ to obtain an unbiased estimate yields

$$s_e(X_i) = \sqrt{\frac{X_i(n - X_i)}{n - 1}} \tag{24.19}$$

This formula may be used for estimating the standard error of a test score X_i. Where $n = 100$ and $X_i = 50$, $s_e(X_i) - 5.02$. Where $X_i = 80$ and $n = 100$, $s_e(X_i) = 4.02$. The standard error diminishes in size as the more extreme values are approached.

Lord (1955a) has shown that if s_e^2 is taken as the average of $s_e^2(X_i)$ and substituted in $r_{tt} = 1 - s_e^2/s_t^2$, unbiased variance estimates being used throughout, Kuder-Richardson formula 21, described in Section 24.4, is obtained.

In most practical situations where parallel tests are used, the tests are not randomly parallel in the strictest sense. The items are matched to some extent. The standard error for such tests will be less than that estimated by $s_e(X_i)$. Thus $s_e(X_i)$ in most situations will tend to be a moderate overestimate. It is of interest to note that $s_e(X_i)$ is independent of the characteristics of the items which a test comprises, provided, of course, that these are scored 1 for a pass and 0 for a failure.

24.10 CONCLUDING OBSERVATIONS

The theory and method associated with the study of measurement error in psychology have been developed in relation to psychological testing. Much of this theory and method is generally applicable to measurements of all kinds. Little attention has been directed to the study of measurement error by experimental psychologists. It is probable that in much work in the field of human and animal learning, fairly gross error attaches to many of the measurements made. Reliability coefficients less than .50 are not uncommon, and coefficients of 0 are perhaps not isolated curiosities. The errors which attach to measurements in the field of animal experimentation are known quite often to be substantial. Low reliability does not necessarily invalidate a technique as a device for drawing valid inferences. Low reliability may be compensated for by increase in sample size. An unreliable technique used with a small sample is, however, capable of detecting gross differences only, and the probability of not rejecting the null hypothesis when it is false may be high. When significant results are reported with an unreliable technique on a small sample, the treatment applied is usually exerting a gross effect.

A common type of experimental design requires the making of measurements on an experimental group in the presence of a treatment and on a control group in the absence of the treatment. Although substantive evidence is lacking, it is probable that in many experiments the measurements are less reliable under the experimental than under the control conditions, one of the effects of the treatment being to increase measurement error. It seems probable that this effect is more likely to occur when the treatment is in the nature of a gross assault on the normal functioning of the organism, as is the case with certain drugs, stress agents, and operative procedures. Experimental situations may be found where the treatment may increase rather than decrease the reliability of the measurements. This author can recall one experiment where the important effect of the treatment was to stabilize and make more reliable the responses of the experimental animals.

The discussion of measurement error given in this chapter is of necessity brief and incomplete. The most comprehensive, and theoretically most advanced, discussion available on measurement error as applied to psychological tests is found in Lord and Novick (1968). A straightforward and more elementary discussion of this topic is given by Magnusson (1967). See also Allen and Yen (1979).

BASIC TERMS AND CONCEPTS

Error of measurement

True score

Reliability coefficient

Test-retest method

Split-half method

Spearman-Brown formula

Kuder-Richardson formula

Test length and reliability

Correction for attenuation

Difference scores

Standard error of measurement

EXERCISES

1 For $r_{xx} = .90$ and $s_x = 15$, estimate the variance of true scores and the error variance. What percentage of the obtained variance is due to error?

2 The following are correlations between half tests: .30, .50, .72, .80, .96. Find reliability coefficients for the whole tests.

3 The following are difficulty values for a test of 20 items:

(1) .97	(6) .53	(11) .50	(16) .04
(2) .95	(7) .75	(12) .55	(17) .35
(3) .76	(8) .40	(13) .42	(18) .27
(4) .80	(9) .82	(14) .30	(19) .15
(5) .60	(10) .20	(15) .15	(20) .09

The standard deviation of test scores is 6.5. Calculate reliability coefficients using both Kuder-Richardson formulas 20 and 21. Explain the difference between the two coefficients.

4 For a particular test, $r_{xx} = .50$. What is the effect on the reliability coefficient of making the test five times as long?

5 The sampling variance of an arithmetic mean of a test is 6.2 where $r_{xx} = .80$. What part of the sampling variance is due to sampling error, and what part to measurement error? If the test were made three times as long, what proportion of the sampling variance of the mean of the lengthened test would be due to measurement error?

6 Estimate the correlation between true scores on X and Y where $r_{xy} = .60$, $r_{xx} = .80$, and $r_{yy} = .90$. What is the maximum possible correlation between X and Y?

7 For the data of Exercise 6 above, calculate the reliability of difference scores in standard-score form between X and Y.

8 Estimate the standard error associated with the individual scores 7, 26, and 44 for a test of 50 items.

ANSWERS TO EXERCISES

1 $s_T^2 = 202.50$, $s_e^2 = 22.50$; 10 percent

2 .462, .667, .837, .889, .980

3 .968, .928

4 $r_{kk} = .833$

5 $s_T^2/N = 4.96$, $s_e^2/N = 1.24$. The proportion due to error will be .077.

6 .707, .849

7 .625

8 $s_e(7) = 2.75$, $s_e(26) = 3.57$, $s_e(44) = 2.32$

25

SCORE TRANSFORMATIONS: NORMS

25.1 INTRODUCTION

Many varieties of transformations are used in the interpretation and analysis of statistical data. A transformation is any systematic alteration in a set of observations whereby certain characteristics of the set are changed and other characteristics remain unchanged. The representation of a set of observations X as deviations from the mean $X - \bar{X} = x$ is a simple transformation. The mean of the transformed values is 0. All other characteristics of the transformed values are the same as those of the original values. The variability, skewness, and kurtosis remain unchanged. The ordinal properties of the data are preserved. The rank ordering of the observations is the same as before. The transformation of a variable X to standard-score form $(X - \bar{X})/s = z$ results in a change both in mean and standard deviation. The mean of the transformed values is 0, and the standard deviation is unity. Skewness, kurtosis, and rank order are unchanged.

Certain commonly used transformations change the shape of the frequency distribution of the variable. The variable may, for example, be transformed to the normal form. This may involve not only a change in mean and standard deviation but also a change in skewness and kurtosis. The original observations may be negatively skewed and leptokurtic. The transformed values may be normally distributed, or approximately so. This type of transformation does not change the rank order of the observations. The transformations most commonly used by psychologists that alter the shape of the frequency distribution are to the normal and rectangular forms. The conversion of a set of observations to percentile ranks is a transformation to a rectangular distribution.

The conversion of a set of frequencies $f_1, f_2, f_3, \ldots, f_k$ to proportions by dividing each frequency by N, or to percentages by dividing by N and multiplying by 100, is a simple transformation. The ordering of the transformed values is the same as the ordering of the original frequencies. If each frequency is divided by different values of N, say, $N_1, N_2, N_3, \ldots, N_k$, then the transformed values will quite probably have an order different from the original values. The conversion of a mental age to an IQ by dividing by chronological age and multiplying by 100 is a transformation which changes the ordinal properties of the data. In converting mental ages to IQs, not only is the order changed, but also the mean, standard deviation, skewness, and kurtosis. The transformed values have a mean of 100 in the standardization group and are approximately normally distributed with a known standard deviation.

Raw scores on psychological tests are usually highly arbitrary. The values of the mean, standard deviation, and possible range of scores reside in large measure in the predilection of the test constructor. Unless the mean, standard deviation, and something about the shape of the score distribution are known, no proper interpretation can be attached to the original, or raw, scores. Such scores are frequently transformed to normal distributions with an agreed mean and standard deviation. For example, a psychological test when administered to a representative sample of individuals from the population for which the test is intended may have a mean of 37 and a standard deviation of 9.6 and be positively skewed. Scores may be transformed to a normally distributed variable with a mean of 100 and a standard deviation of 16. Scores thus transformed to the normal form immediately take on meaning. If an individual has a score of 116, we know that he or she is one standard deviation unit above the average. Because the scores are normally distributed we know that his or her performance is better than that of about 84 percent of the population and below the performance of about 16 percent of the population. The procedure for developing such a transformation is known as *standardization*. A psychological test is said to be standardized when transformed scores are available, based on a reference group of acceptable size. The transformed scores themselves are called *norms*. An individual's score takes on meaning in relation to a standard, or normative, group. Tests are frequently standardized to permit age allowances. This means in effect that separate norms have been prepared for each age group. The average child in each age group may have a mean transformed score of, say, 100. The standard deviation of scores for each age group may be 16. Thus a younger child may make a lower raw score than an older child but have a considerably higher transformed score. Intelligence quotients are transformed scores which make adjustments for the differing chronological ages of children taking the test. Intelligence quotients are presumed to be independent of chronological age within an accepted age range. Most published tests are accompanied by manuals containing conversion tables which permit the transformation of raw scores to standardized scores. Both normal and rectangular transformations are used in test standardization.

25.2 TRANSFORMATIONS TO STANDARD MEASURE

A standard score is a deviation from the mean divided by the standard deviation; thus $z = (X - \bar{X})/s$. The mean is the origin, and the standard deviation is the unit of measurement. Thus a particular value is z standard deviation units above or below the mean. The mean of z scores is 0, and the standard deviation is unity. The skewness and kurtosis of the distribution are unchanged. The distribution of z scores has the same shape as the distribution of X. Standard scores on two or more variables are directly comparable only in the sense that they have the same mean and standard deviation.

A standard-score transformation does not change the proportionality of scale intervals. If X_1, X_2, and X_3 are three measurements in raw-score form and z_1, z_2, and z_3 are the same three measurements in standard-score form, then

$$\frac{X_1 - X_2}{X_2 - X_3} = \frac{z_1 - z_2}{z_2 - z_3}$$

This means that the *relative* distances between the variate values remain unchanged under a standard-score transformation. Let X_1, X_2, and X_3 be 20, 30, and 50. If $\bar{X} = 40$ and $s = 15$, then z_1, z_2, and z_3 become -1.33, $-.67$, and $.67$. We note that

$$\frac{20 - 30}{30 - 50} = \frac{-1.33 + .67}{-.67 - .67} = .50$$

Standard scores involve the use of decimals and plus and minus signs. This is sometimes inconvenient. Also the range of values will seldom exceed the limits -3 and $+3$. It is not uncommon to select an arbitrary origin and standard deviation to ensure that all, or nearly all, the measures have a plus sign and that decimals are eliminated. For this purpose a mean of 50 and a standard deviation of 15 are sometimes used. If z' denotes this type of score, then

$$\begin{aligned} z' &= 50 + 15\,\frac{X - \bar{X}}{s} \\ &= 50 + 15z \end{aligned}$$

To change the standard deviation we multiply every standard score by 15. To change the origin we merely add 50. Values of z' are rounded to the nearest integer. In comparing performance on a series of tests, standard-score values z' are more convenient than z. Of course, any other mean and standard deviation could be selected.

25.3 PERCENTILE POINTS AND PERCENTILE RANKS

In the standardization of psychological tests transformations to percentile ranks have frequent application. Such transformations are rectangular. Each percentile rank has the same frequency of occurrence. The frequency distribution is flat.

A clear distinction must be made between *percentile points* and *percentile ranks.* If k percent of the members of a sample have scores less than a particular value, that value is the kth percentile point. It is a value of the variable below which k percent of individuals lie. On an examination, if 85 percent of individuals score less than 60, then 60 is the 85th percentile point. If a frequency distribution is represented graphically and ordinates raised at all percentile points, the total area under the frequency distribution is divided into 100 equal parts.

Percentile points may be represented by the symbols $P_0, P_1, P_2, \ldots, P_{100}$. The points P_0 and P_{100} are limits which include all members of the sample. A percentile rank, as distinct from a percentile point, is a value on the transformed scale corresponding to the percentile point. If 60 is a score below which 85 percent of individuals fall, then 85 is the corresponding percentile rank. As in all transformations, values on the original scale correspond to certain values on the transformed scale. In the present context the values on the original scale are percentile points, the corresponding values on the transformed scale are percentile ranks.

The reader will recall that the median is a value of the variable above and below which 50 percent of cases lie. The median is the 50th percentile point, P_{50}. The upper quartile is a value of the variable above which 25 percent of cases and below which 75 percent of cases lie; conversely for the lower quartile. The upper quartile is the 75th percentile point, or P_{75}, and the lower quartile is the 25th percentile point, or P_{25}. Decile points are sometimes used. These, as the name implies, involve a dividing into tenths. A *decile point* is a value of the variable below which a certain percentage of individuals fall, the percentage being taken in units of 10. Decile ranks are transformed values corresponding to the decile points and taking the integer values 1 to 10. The median is the 5th decile. An ordinate at the median divides the area under the frequency distribution into 2 equal parts; ordinates at the upper quartile, the median, and the lower quartile divide the area into 4 equal parts; ordinates at the decile points divide the area into 10 equal parts; ordinates at the percentile points divide the area into 100 equal parts.

For small N the computation of percentile points and percentile ranks is not a very meaningful procedure. Given the scores 8, 17, 23, 42, 61, and 63, obviously little meaning could possibly attach to P_{30} or P_{80}. The conversion of these scores to percentile ranks would be a somewhat spurious procedure, with no advantage over ordinary ranks.

25.4 COMPUTATION OF PERCENTILE POINTS AND RANKS—UNGROUPED DATA

To illustrate the computation of percentile points and ranks for ungrouped data, consider the psychological-test scores tabulated in Table 25.1. We adopt the convention that any score value X has exact limits given by $X - .5$ and $X + .5$. The variable is presumed to be continuous. Thus the score 116 has exact limits 115.5 and 116.5. This convention is the same as that used in determining the exact limits of class intervals. Let us now calculate P_{40}, the 40th percentile point, the point below which 40 percent of individuals lie. $N = 60$, and 40 percent of this is 24. The 24th individual has a score of 110, the exact upper limit is 110.5, and this is taken as the point on the test scale below which 24 individuals lie. Thus $P_{40} = 110.5$. Note in this case that the 25th individual has a score of 111. The exact lower limit of this score is also 110.5. Consider now the calculation of P_{20}. We require a point on the test scale below which 12 and above which 48 individuals lie. The score of the 12th individual is 98 with an upper limit of 98.5. We note also that the score of the 13th individual is 100 with a

Table 25.1

Psychological-test scores for a group of 60 children arranged in order

Individual	Score	Individual	Score	Individual	Score
1	83	21	110	41	123
2	88	22	110	42	124
3	88	23	110	43	124
4	91	24	110	44	125
5	91	25	111	45	125
6	93	26	112	46	125
7	93	27	114	47	126
8	93	28	115	48	126
9	97	29	116	49	127
10	98	30	116	50	128
11	98	31	116	51	130
12	98	32	117	52	130
13	100	33	118	53	131
14	101	34	119	54	132
15	103	35	120	55	135
16	107	36	121	56	135
17	107	37	122	57	136
18	108	38	123	58	136
19	109	39	123	59	136
20	110	40	123	60	139

lower limit of 99.5. Presumably the value P_{20} falls somewhere between 98.5 and 99.5. It is indeterminate. As an arbitrary working procedure the percentile P_{20} may be taken halfway between these two values. Thus $P_{20} = 99.0$. To illustrate the handling of ties in the computation of percentile points let us calculate P_{10}. A score is required below which 6 and above which 54 individuals fall. We note that individuals 6, 7, and 8 have the same score, 93. Thus three individuals have scores within the exact limits 92.5 and 93.5. Since we require a point below which 6 individuals fall, we interpolate one-third of the way into this interval. One-third of this interval is .33, and $P_{10} = 92.50 + .33 = 92.83$. With the above data P_0 may be taken as the lower exact limit of the lowest score, or 82.5. Similarly, P_{100} may be taken as the upper exact limit of the highest score, or 139.5.

The calculation of percentile ranks as distinct from percentile points is the reverse of the above process. Above we calculated scores corresponding to particular ranks. We may now attend to the calculation of ranks corresponding to particular scores. To illustrate, consider individual 32 in Table 25.1. This individual is 32d from the bottom. His test score is 117. The number of individuals scoring below 117 is 31. The percentage below is $\frac{31}{60} \times 100 = 51.67$. The number scoring above 117 is 28. The percentage is $\frac{28}{60} \times 100 = 46.67$. These two percentages do not add to 100. Individual 32 occupies $\frac{1}{60} \times 100 = 1.67$ percent of the total scale. His percentile rank falls between 51.67 and $51.67 + 1.67 = 53.33$. We may take the midpoint of this interval as the required percentile rank. Thus the percentile rank corresponding to score 117 is

$$51.67 + \frac{1.67}{2} = 52.50$$

This method assumes that any rank R covers the interval $R - .5$ and $R + .5$.

Consider the question of ties. We note that five individuals score 110. The number of individuals scoring below 110 is 19, or $\frac{19}{60} \times 100 = 31.67$ percent of the total. The number scoring above 110 is 36, or $\frac{36}{60} \times 100 = 60.00$ percent. The number occupying the score position 110 is 5, or $\frac{5}{50} \times 100 = 8.33$ percent. The required percentile rank may be taken as the midpoint of the interval 31.67, and $31.67 + 8.33 = 40.00$. Thus the percentile rank of the score 110 is $31.67 + \frac{8.33}{2} = 35.83$.

Percentile ranks may be obtained by using simple formula

$$\text{PR} = 100\,\frac{R - .5}{N} \qquad [25.1]$$

where R = rank of individual, counting from the bottom
N = total number of cases

Where ties occur, R is taken as the average rank which the tied observations occupy. The average rank of the five individuals who score 110 is 22, and the corresponding percentile rank is, as before,

$$100 \frac{22 - .5}{60} = 35.83$$

Percentile ranks are ordinarily rounded to the nearest whole number. Thus the rank 35.83 becomes 36.

25.5 CALCULATION OF PERCENTILE POINTS AND RANKS—GROUPED DATA

The calculation of percentile points and ranks for grouped data will be discussed with reference to the data of Table 25.2. Cumulative frequencies are recorded in column 3, and cumulative percentages in column 4.

Let us calculate P_{25}. $N = 200$, and 25 percent of N is 50. We observe that the 50th case falls within the interval 65 to 69. The exact limits of this interval are 64.5 and 69.5. We must now interpolate within the interval to locate a point below which 50 cases fall. We note that 36 cases fall below and 26 cases within the interval containing P_{25}. To arrive at the 50th case we require 14 of the cases within the interval. Thus we take $\frac{14}{26}$ of the interval 64.5 to 69.5. This is $\frac{14}{26} \times 5 = 2.69$. We add this to the lower limit of the interval to obtain P_{25}, which is $64.5 + 2.69 = 67.19$.

The following formula may be used to calculate percentile points:

[25.2] $$P_i = L + \frac{pN - F}{f_i} h$$

Table 25.2

Cumulative frequencies and percentages of test scores

1 Class interval	2 Frequency	3 Cumulative frequency	4 Cumulative percentage
95–99	1	200	100.0
90–94	6	199	99.5
85–89	8	193	96.5
80–84	33	185	92.5
75–79	40	152	76.0
70–74	50	112	56.0
65–69	26	62	31.0
60–64	14	36	18.0
55–59	10	22	11.0
50–54	6	12	6.0
45–49	4	6	3.0
40–44	2	2	1.0
Total	200		

where $P_i = k$th percentile point
$p =$ proportion corresponding to ith percentile point; thus if $i = 62$, $p = .62$
$L =$ exact lower limit of interval containing P_i
$F =$ sum of all frequencies below L
$f_i =$ frequency of interval containing P_i
$h =$ class interval

For P_{25} in Table 25.2 we have $L = 64.5$, $p_i = .25$, $F = 36$, $f_i = 26$, and $h = 5$. Thus

$$P_{25} = 64.5 + \frac{.25 \times 200 - 36}{26} 5 = 67.19$$

This result is identical with that obtained previously for P_{25}. The reader will observe that for P_{50} this formula is the same as that given previously for calculating the median from grouped data.

The calculation of percentile ranks is the reverse of the above procedure. The cumulative percentages shown in column 4 of Table 25.2 are the percentile ranks corresponding to the exact top limits of the intervals. Thus 56.0 is the percentile rank corresponding to the percentile point 74.5, the exact top limit of the interval 70 to 74. Likewise 11.0 is the percentile rank corresponding to the percentile point 59.5, the exact top limit of the interval 55 to 59. The percentile rank of any score may be obtained by interpolation. What is the percentile rank corresponding to the score 81? The score 81 falls within an interval with exact limits 79.5 and 84.5. It is 1.5 score units above the bottom of this interval. The lower limit has a percentile rank of 76.0 and the upper limit 92.5. Thus we have two points on the score scale corresponding to two points on the percentile-rank scale. Five units on the score scale are equal to $92.5 - 76.0 = 16.5$ units on the percentile-rank scale, and 1.5 units on the score scale are equal to $(92.5 - 76.0)\,1.5/5 = 4.95$ units on the rank scale. We now take $76.0 + 4.95 = 80.95$ as the percentile rank of the score 81. Rounding this to the nearest integer we obtain a rank of 81. It is pure coincidence that in this case the percentile rank is numerically equal to the score.

The steps involved in finding percentile ranks from grouped data may be summarized as follows:

1 Find the exact lower limit of the interval containing the score X whose percentile rank is required.

2 Find the difference between X and the lower limit of the interval containing it.

3 Divide this by the class interval and multiply by the percentage within the interval.

4 Add this to the percentile rank corresponding to the bottom of the interval.

Usually, where percentile ranks are calculated we are interested in preparing a table for converting any score value to percentile ranks. Thus for every possible score, we require the corresponding percentile rank. This may be done by systematically computing all percentile-rank values in the manner described above. A somewhat easier procedure is to make a graphic plotting on suitable graph paper of cumulative percentages against the corresponding upper limits of the class intervals. Score values are plotted on the horizontal axis, and cumulative percentages on the vertical axis. The points may be joined by straight lines. Percentile ranks corresponding to scores may then be read directly from the graph. If the points are joined by straight lines, these rank values will be the same, within limits of error, as those obtained by linear interpolation directly on the numerical values. If the sample is small, the points when plotted may show considerable irregularity and it may be advisable to fit a smoothed curve to the data. The fitting of a smoothed curve by freehand methods is accurate enough for most practical purposes. A procedure related to the method described above is to calculate certain selected percentile points and then interpolate either numerically or graphically between these points. The percentile points $P_{10}, P_{20}, P_{30}, \ldots, P_{90}$ may be calculated. To achieve greater accuracy at the tails of the distribution it may be desirable to calculate P_1 and P_5; also P_{95} and P_{99}.

25.6 NORMAL TRANSFORMATIONS

The transformation of a variable to the normal form is a frequent procedure in test standardization and correlational analysis. Not uncommonly, test norms are normal transformations of the original raw scores with arbitrarily selected means and standard deviations. A type of normal transformation used by educators is a T score. These T scores are normally distributed, usually with a mean of 50 and a standard deviation of 10. A normal transformation with a mean of 100 and a standard deviation of 15, or thereabouts, resembles an IQ scale.

Transforming a set of scores to the normal form is a relatively simple procedure. Every percentile rank corresponds to a point on the base line of the unit normal curve measured from a mean of 0 in standard deviation units. A percentile rank of 50 corresponds to the zero point. A rank of 60 is .25 standard deviation units above the mean. A rank of 70 is .52 standard deviation units above the mean. Table 25.3 shows points on the base line of the unit normal curve corresponding to selected percentile ranks. These and other points are readily obtained from any table of areas under the normal curve (Table A of the Appendix).

In summary, the steps used in transforming a variable to the normal form are as follows. Percentile ranks corresponding to certain points on the score scale may be calculated. A table of areas under the normal curve is used to find the points on the base line of the unit normal curve

Table 25.3

Points on the base line of the unit normal curve corresponding to selected percentile ranks

Percentile rank	Standard deviation
99	+2.33
95	+1.65
90	+1.28
80	+ .84
70	+ .52
60	+ .25
50	.00
40	− .25
30	− .52
20	− .84
10	−1.28
5	−1.65
1	−2.33

corresponding to these percentile ranks. These points correspond to the percentile points on the original score scale. Thus a correspondence is established between a set of points on the original score scale and points on a normal distribution of zero mean and unit standard deviation. Percentile ranks are stepping-stones in establishing this correspondence. The normal standard scores are multiplied by a constant to obtain any desired standard deviation of the transformed values. A constant is usually added to produce a change in means, thus eliminating negative signs. A transformed value corresponding to any score value on the original scale may be obtained by interpolation.

Some freedom of choice is possible in the selection of a set of points on the score scale with associated percentile ranks. *First,* we may use the exact top limits of the intervals and obtain the corresponding percentile ranks from the cumulative-percentage frequencies. *Second,* we may take the midpoints of the class intervals and obtain percentile ranks corresponding to these. *Third,* we may use a selected set of percentile points with associated percentile ranks. Thus P_{10}, P_{20}, P_{30}, . . . , P_{90} may be used. P_1, P_5, and P_{95}, P_{99} may be added at the tails as a refinement. *Fourth,* we may select certain equally spaced points on the normal standard-score scale and ascertain their percentile ranks and the corresponding percentile-point scores. These equally spaced points may, for example, be −2.5, −2.0, −1.5, . . . , +1.5, +2.0, +2.5. The difference between the four alternatives outlined above is a matter of units. The first uses units of class interval of the original variable, a unit extending from the top of one interval to the top of the next. The second also uses units of class interval of the original variable, a unit extending from the midpoint of one interval to the midpoint of the next. The third, excluding the tails, uses equal units on the percentile-rank scale. The fourth alternative uses equal units on the normal standard-score scale. While minor advantages may be claimed for one procedure in preference to another, the differences, where N is fairly large, are trivial. Any one of the four procedures is satisfactory enough for most practical purposes.

To illustrate the transformation of a set of scores to the normal form we shall use the second alternative and take the midpoints of the class intervals with their corresponding percentile ranks. Table 25.4 shows a frequency distribution of test scores. Column 2 shows the exact midpoints of the class intervals. Column 3 shows the frequencies. Column 4 shows the cumulative frequencies to the midpoints. These are the cumulative frequencies to the bottom of the interval plus half the frequencies within the interval. The number of cases below the interval 60 to 64 is 22. The number within the interval is 14. Half this number is 7. The cumulative frequency to the midpoint is $22 + 7 = 29$. Column 5 shows the cumulative percentage frequencies to the midpoints. These cumulative percentage frequencies are percentile ranks corresponding to the midpoints of the intervals. The numbers in column 6 are points on the base line of the unit normal curve in standard deviation units from a zero mean. The percentage of the area of the unit normal curve falling below a standard score of 2.81 is 99.75, the percentage below a standard score of 2.06 is 98.00, and so on. These values are normalized standard scores corresponding to the

Table 25.4

Illustration of the transformation of scores to a normal distribution—data of Table 25.2

Class interval 1	Midpoint 2	Frequency 3	Cumulative frequency to midpoint 4	Cumulative percentage to midpoint 5	Normal standard deviation unit z 6	T score: $z \times 10$ 7	T score: $z \times 10 + 50$ 8
95–99	97	1	199.5	99.75	2.81	28.1	78.1
90–94	92	6	196.0	98.00	2.06	20.6	70.6
85–89	87	8	189.0	89.50	1.25	12.5	62.5
80–84	82	33	168.5	84.25	1.00	10.0	60.0
75–79	77	40	132.0	66.00	.41	4.1	54.1
70–74	72	50	87.0	43.50	−.16	−1.6	48.4
65–69	67	26	49.0	24.50	−.69	−6.9	43.1
60–64	62	14	29.0	14.50	−1.06	−10.6	39.4
55–59	57	10	17.0	8.50	−1.37	−13.7	36.3
50–54	52	6	9.0	4.50	−1.70	−17.0	33.0
45–49	47	4	4.0	2.00	−2.05	−20.5	29.5
40–44	42	2	1.0	.50	−2.58	−25.8	24.2
Total		200					

midpoints of the original score intervals. Thus we have a set of values on the original scale paired with a set of values on a normal transformed scale. Transformed values corresponding to any score on the original scale may be obtained by either arithmetical or graphic interpolation.

Table 25.4 shows a *T*-score transformation. In column 7 the standard scores of column 6 are multiplied by 10, thus yielding transformed scores with a standard deviation of 10. In column 8 a constant value 50 is added to the values of column 7, thus changing the origin from 0 to 50 and eliminating negative values. If we had multiplied by 15 and added 100, the transformed values would have a standard deviation of 15 and a mean of 100. Any other standard deviation and mean could be used.

25.7 THE STANINE SCALE

During World War II the United States Army Air Force Aviation Psychology Program used a stanine scale. Scores on psychological tests were converted to stanines. A stanine scale is an approximately normal transformation. A coarse grouping is used, only nine score categories being allowed. The transformed values are assigned the integers 1 to 9. The mean of a stanine scale is 5, and the standard deviation is 1.96. The percentages of cases in the stanine-score categories from 1 to 9 are 4, 7, 12, 17, 20, 17, 12, 7, and 4. Thus 4 percent have a stanine score 1, 7 percent a score 2, 12 percent a score 3, and so on. If a set of scores is ordered from the lowest to the highest, the lowest 4 percent assigned a score 1, the next lowest a score 2, the next lowest a score 3, and the process continued until the top 4 percent receives a score of 9, the transformed scores are roughly normal and form a stanine scale. Stanine scores correspond to equal intervals in standard deviation units on the base line of the unit normal curve. A stanine of 5 covers the interval from -25 to $+.25$ in standard deviation units. Roughly 20 percent of the area of the unit normal curve falls within this interval. A stanine of 6 covers the interval $+.25$ to $+.75$ in standard deviation units. Roughly 17 percent of the area of the unit normal curve falls within this interval. The interval used is one-half a standard deviation unit; a stanine of 9 includes all cases above $+2.25$, and a stanine of 1 all cases below -2.25 standard deviation units. Test scores can rapidly be converted to stanines. A stanine transformation is a simple method of converting scores to an approximate normal form. The grouping, although coarse, is sufficiently refined for many practical purposes.

BASIC TERMS AND CONCEPTS

Transformations

Standardization

Norms

Standard measure

Percentile points

Percentile ranks

Normal transformations

Stanines

EXERCISES

1 State the difference between percentile points and percentile ranks.

2 For the data of Table 25.1, compute (**a**) percentile points P_{25}, P_{50}, and P_{75} and (**b**) percentile ranks for scores 103, 123, and 136.

3 For the data of Table 25.2, compute (**a**) percentile points P_{10}, P_{30}, and P_{80} and (**b**) percentile ranks for the scores 59, 74, and 82.

4 Develop a T-score transformation for the data of Table 3.1

5 Develop a stanine transformation for the data of Table 3.1.

ANSWERS TO EXERCISES

1 The kth percentile point is a value such that k percent of the members of a sample have less than that value. A percentile rank is a value on a transformed rectangular scale which extends from 0 to 100.

2 **a** $p_{25} = 105.00$ **b** PR (103) = 24.17
$p_{50} = 116.17$ PR (123) = 65.00
$p_{75} = 125.17$ PR (136) = 95.83

3 **a** $p_{10} = 56.32$ **b** PR (59) = 10.50
$p_{20} = 65.71$ PR (74) = 53.50
$p_{80} = 80.71$ PR (82) = 84.25

4

Class interval	Midpoint	Frequency	T score
45–49	47	1	74.8
40–44	42	2	69.4
35–39	37	3	65.6
30–34	32	6	61.8
25–29	27	8	58.1
20–24	22	17	53.2
15–19	17	26	45.9
10–14	12	11	37.1
5–9	7	2	27.8
0–4	2	0	
Total		76	

5

Stanine	Exact limits of interval
9	39.43–
8	32.53–39.43
7	26.08–32.53
6	21.44–26.08
5	17.85–21.44
4	15.36–17.85
3	12.39–15.36
2	9.97–12.39
1	– 9.97

26

SELECTED MULTIVARIATE METHODS

26.1 INTRODUCTION

The term *multivariate statistics* conventionally refers to a broad class of correlational statistical methods used in the analysis of data comprising more than two variables, sometimes many. Many of these methods require elaborate arithmetical computation. The availability of computers has greatly increased their use.

In psychology, education, and possibly all the social and biological sciences, the phenomena under study are often very complex. They can be described, predicted, and understood, if at all, only by the study of many variables. An example is the field of human intelligence. The term *intelligence* is at times used to denote a number of variables that are distinguishable, although not necessarily independent or uncorrelated. These may include verbal, numerical, reasoning, spatial, perceptual, and other abilities. These variables in turn may be thought to be dependent on both environmental and genetic influences. An inquiry into some aspect of human intelligence frequently involves complex networks of interrelated variables. The investigator must proceed with caution. Progrcss requires an understanding of multivariate methods.

One major purpose of multivariate statistics is simplification. A study may involve a hundred variables and thousands of measurements. Such data require methods of analysis that describe the data in forms sufficiently simple that the information they contain can be grasped by the investigator.

Many multivariate methods rest on complex mathematical foundations and require much computation. Despite this, the nature of these methods, what they accomplish, and their interpretations are simple and can be readily understood by anyone with a knowledge of elementary statistics.

The discussion of multivariate statistics presented here avoids introducing matrix algebra, which is commonly used by writers on this topic. The present writer has found a knowledge of the *correlation of sums* useful in treating and conceptualizing multivariate problems. A brief digression on the correlation of sums is presented here.

26.2 THE CORRELATION OF SUMS

In Chapter 8 on correlation, the variance of the sum of two variables, X and Y, was considered. A simple extension of the argument provides a way of obtaining and conceptualizing the variance of sums of any number of variables. If the variances and covariances are written in the form of a table, with variances s_i^2 along the diagonal and covariances $r_{ij}s_is_j$ on both sides of the diagonal, the variance of sums is obtained simply by adding together all the elements in the table.

Consider now the correlation of sums with, for example, three variables, X_1, X_2, and X_3. How may we obtain, for example, the correlation of X_1 with the sum of X_2 and X_3? Here again the variances and covariances are written in a covariance table as follows:

	1	2	3
1	s_1^2	$r_{12}s_1s_2$	$r_{13}s_1s_3$
2	$r_{12}s_1s_2$	s_2^2	$r_{23}s_2s_3$
3	$r_{13}s_1s_3$	$r_{23}s_2s_3$	s_3^2

This table is partitioned into quadrants. Denote the sums of all terms in these quadrants by

A	C
C	B

The correlation between X_1 and $X_2 + X_3$ is then given by

$$\frac{C}{\sqrt{AB}}$$

Correlations obtained by the above method, using a covariance table, are correlations between *raw scores*, that is, between a raw score on X_1 and the sum of raw scores on X_2 and X_3.

Let us now write a table with 1s down the diagonal and correlations on both sides of the diagonal. Partition the table as before:

	1	2	3
1	1	r_{12}	r_{13}
2	r_{12}	1	r_{23}
3	r_{13}	r_{23}	1

The formula $C/\sqrt{AB}$ now provides the correlation between a standard score z_1 and the sum of standard scores z_2 and z_3.

In multivariate statistics the concept of a *weight* is commonly used. Different weights may be assigned to scores on different variables and a weighted sum of scores obtained. This weighted sum may then be correlated with another variable or with the weighted sum of another set of variables. The correlations used are product-moment correlations. In the above discussion on the correlation between X_1 and $X_2 + X_3$, these denoting raw scores, the variables are weighted in a manner proportional to their standard deviations. In effect the standard deviations act as weights. In standard-score form, the standard deviations are unity. The variables are equally weighted. Any other system of weights could be used. For example, weights can be obtained that will maximize the correlation between variable 1 and the weighted sum of variables 2 and 3.

Let us illustrate the correlation of sums with a numerical example. Consider three variables with standard deviations $s_1 = 2$, $s_2 = 3$, and $s_3 = 4$, and correlations $r_{12} = .5$, $r_{13} = .7$, and $r_{23} = .3$. The standard deviations may be written as weights along the top and to the side of the correlation table as follows:

	2	3	4
2	1.0	.5	.7
3	.5	1.0	.3
4	.7	.3	1.0

If we cross-multiply the rows and columns by the standard deviations, or weights, we obtain the following covariance table for the three variables:

4.0	3.0	5.6
3.0	9.0	3.6
5.6	3.6	16.0

The correlation between X_1 and $X_2 + X_3$ is then given by $C/\sqrt{AB}$ and is $8.6/\sqrt{4 \times 32.3} = .757$. If standard scores are used and the variables are assigned equal weight, the correlation between z_1 and $z_2 + z_3$ is $1.2/\sqrt{1 \times 2.6} = .744$.

Other weights may be applied. Weights may be calculated that make the correlation between z_1 and the weighted sum of z_2 and z_3 as large as possible. Multiple regression, discussed later in this chapter, accomplishes this result.

Although the weights used in the above examples are positive numbers, note that negative weights also may be used. For example, with three variables in standard-score form, z_1, z_2, and z_3, the third variable may be assigned a weight of -1 rather than $+1$. The correlation obtained by using $C/\sqrt{AB}$ is then the correlation of z_1 with $z_2 - z_3$.

The method discussed above, illustrated with three variables, may be extended to any number of variables. Consider a set of k variables divided into two subsets of m and $k-m$ variables. The correlation table may then be written in the usual way and partitioned into quadrants with m variables in one set and $k-m$ in the other. If the variables are weighted according to their standard deviations, the correlation of the sum of raw scores on the set of m variables with the set of $k-m$ variables is obtained by the formula $C/\sqrt{AB}$. With unit weights the correlation obtained is between the sum of two sets of standard scores. Other weights may be used. With some sets of data, although not all, the set of m variables, where $m > 1$, may be designated as dependent variables and the remaining $k-m$ as independent variables. Whether these designations are appropriate depends on the nature of the data.

Apart from any computational use, the correlation of sums as represented by the expression $C/\sqrt{AB}$ is a useful conceptual device. It may be used to obtain, or simply write down, many commonly used formulas, which otherwise can only be derived by tedious algebra.

26.3 PARTIAL CORRELATION

Let us assume that a test of intelligence and a test of psychomotor ability have been administered to a group of children showing considerable variation in age. Both intelligence and psychomotor ability increase with age. On the average 10-year-old children are more intelligent than 6-year-old children. They also have more highly developed psychomotor abilities. Scores on the two tests will correlate with each other because both are correlated with age. Partial correlation may be used with such data to obtain a measure of correlation with the effect of age eliminated or removed.

What is meant by *eliminating,* or *removing,* the effect of a third variable? These terms in the present context have a precise statistical meaning. Let X_1, X_2, and X_3 be three variables. All or part of the correlation between X_1 and X_2 may result because both are correlated with X_3. The reader will recall from previous discussion on correlation that a score on X_1 may be divided into two parts. One part is a score predicted from X_3. The other part is the residual, or error of estimate, in predicting X_1 from X_3. These two parts are independent, or uncorrelated. Similarly, a score on X_2 may be divided into two parts, a part predictable from X_3 and a residual, or error of estimate, in predicting X_2 from X_3. The correlation between the two sets of residuals, or errors of estimate, in predicting X_1 from X_3 and X_2 from X_3 is the partial correlation coefficient. It is the part of the correlation which remains when the effect of the third variable is eliminated, or removed.

The formula for calculating the partial correlation coefficient to eliminate a third variable is

[26.1] $$r_{12.3} = \frac{r_{12} - r_{13}r_{23}}{\sqrt{(1 - r_{13}^2)(1 - r_{23}^2)}}$$

The notation $r_{12.3}$ means the correlation between residuals when X_3 has been removed from both X_1 and X_2. This is sometimes called a first-order partial correlation coefficient.

Let X_1 and X_2 be scores on an intelligence and a psychomotor test for a group of school children. Let X_3 be age. Let the correlation between the three variables be as follows: $r_{12} = .55$, $r_{13} = .60$, and $r_{23} = .50$. The partial correlation coefficient is

$$r_{12.3} = \frac{.55 - .60 \times .50}{\sqrt{(1 - .60^2)(1 - .50^2)}} = .36$$

Using a variance interpretation, the proportion overlap between X_1 and X_2 is $r_{12}^2 = .55^2 = .303$. The proportion overlap with X_3 eliminated is $r_{12.3}^2 = .36^2 = .130$. The proportion overlap which results from the effects of age is $.303 - .130 = .173$. It would also be appropriate to state that the percentage of the total association present resulting from the effect of age is $\frac{.173}{.303} \times 100 =$ 57 percent. The remaining percent of the association results from other factors.

Partial correlation may be used to remove the effect of more than one variable. The partial correlation between X_1 and X_2 with the effects of both X_3 and X_4 removed is

[26.2] $$r_{12.34} = \frac{r_{12.4} - r_{13.4}r_{23.4}}{\sqrt{(1 - r_{13.4}^2)(1 - r_{23.4}^2)}}$$

This is a *second-order* partial correlation coefficient. Because of difficulties of interpretation, partial correlation coefficients involving the elimination of more than one variable are infrequently calculated.

A t test may be used to test whether a partial correlation coefficient is significantly different from 0. The required t is

[26.3] $$t = \frac{r_{12.3}}{\sqrt{(1 - r_{12.3}^2)/(N - 3)}}$$

This may be referred to a table of t with $N - 3$ degrees of freedom.

26.4 PART CORRELATION

With three variables X_1, X_2, and X_3, partial correlation is the correlation between residuals on X_1 and X_2 with X_3 removed from both variables by linear regression. A situation may be considered where X_3 is removed from one variable only, say X_2, and X_1 is then correlated with the residual of X_2. This type of correlation is called *part correlation* or *semipartial correlation.* The part correlation between X_1 and the residuals on X_2 with X_3 removed is given by

$$r_{1(2.3)} = \frac{r_{12} - r_{13}\, r_{23}}{\sqrt{1 - r_{23}{}^2}} \qquad [26.4]$$

If $r_{12} = .55$, $r_{13} = .60$, and $r_{23} = .50$, the part correlation is .289.

An application of part correlation arises where measurements have been obtained prior to an experimental treatment and subsequent measurements on the same subjects under the experimental treatment. It may be informative to calculate the correlation between some other variable and the measurements obtained under treatment with the effects of the initial values removed by linear regression. Part correlation is appropriate here.

26.5 MULTIPLE REGRESSION AND CORRELATION

The correlation coefficient may be used to predict or estimate a score on an unknown variable from knowledge of a score on a known variable. The regression equation in standard-score form is $z_1' = r_{12}z_2$, where z_1' is a predicted or estimated standard score. In this situation we have one dependent and one independent variable. If $z_2 = 1.2$ and $r_{12} = .80$, the best estimate of an individual's standard score on variable 1 is $z_1' = .80 \times 1.2 = .96$. The estimate is that the individual is .96 standard deviation units above the average.

We may consider a situation where we have one dependent and two independent variables. The dependent variable may be a measure of scholastic success. The independent variables may be two psychological tests used at university entrance. The dependent variable is spoken of as the *criterion.* The two independent variables are *predictors.* How may scores on the two predictors be combined to predict scholastic success? The correlation between the three variables may be arranged in a small table. Let these correlations be as follows:

	1	2	3
1	1.0	.8	.3
2	.8	1.0	.5
3	.3	.5	1.0

Variable 1 is the criterion, and variables 2 and 3 are the predictors. Note that 1.0s have been entered along the main diagonal. In estimating standard scores on 1 from standard scores on 2 and 3 separately, the two regression equations are $z_1' = .8z_2$ and $z_1' = .3z_3$. Variable 2 is a much better predictor than variable 3. Presumably, by employing a knowledge of both 2 and 3, a better estimate of the criterion may be obtained.

How may a set of weights be obtained which will maximize the correlation between the criterion and the sum of scores on the dependent variables? Let us represent weights by the symbols β_2 and β_3. An estimated standard score on 1 is then given by $z_1' = \beta_2 z_2 + \beta_3 z_3$. We wish to calculate weights β_2 and β_3 such that the correlation between z_1 and z_1' is a maximum. Mathematically, the problem reduces to the calculation of weights which will minimize the average sum of squares of differences between the criterion score z_1 and the estimated criterion score z_1'. We require values of β_2 and β_3 such that

$$\frac{1}{N} \sum (z_1 - z_1')^2 = \text{a minimum}$$

The values of β_2 and β_3 are multiple regression weights for standard scores. They are sometimes called *beta* coefficients.

With three variables the values of β_2 and β_3 are given by

[26.5] $$\beta_2 = \frac{r_{12} - r_{13}r_{23}}{1 - r_{23}^2}$$

[26.6] $$\beta_3 = \frac{r_{13} - r_{12}r_{23}}{1 - r_{23}^2}$$

In the above example

$$\beta_2 = \frac{.8 - .3 \times .5}{1 - .5^2} = .867$$

$$\beta_3 = \frac{.3 - .8 \times .5}{1 - .5^2} = -.133$$

Let us write these weights above and to the side of the correlation table and multiply the rows and columns as follows:

		.867	−.133
	1.0	.8	.3
.867	.8	1.0	.5
−.133	.3	.5	1.0

1.000	.694	−.040
.694	.752	−.058
−.040	−.058	.018

The sums of the elements in the four quadrants are

1.000	.654
.654	.654

The correlation between the criterion and the weighted sum is

$$\frac{C}{\sqrt{AB}} = \frac{.654}{\sqrt{.654}} = \sqrt{.654} = .809$$

This is a multiple correlation coefficient and may be denoted by R. No other system of weights will yield a higher correlation between the criterion and the weighted sum of predictors.

Note that the sum of elements in the top right quadrant of the weighted correlation table is equal to the sum in the lower right, or $C = B$. This circumstance will occur if the weights used are multiple regression weights. It provides a check on the calculation. We note also that $R^2 = C$ and $R = \sqrt{C}$. Thus the multiple correlation coefficient may be obtained by the formula

[26.7] $$R = \sqrt{\beta_2 r_{12} + \beta_3 r_{13}}$$

This is the commonly used formula for calculating a multiple correlation coefficient.

In our example the multiple correlation is .809. The correlation of variable 2 with the criterion is .8. The addition of the third variable increases prediction very slightly. In a practical situation the third variable could safely be discarded as contributing a negligible amount to the efficacy of prediction.

26.6 THE REGRESSION EQUATION FOR RAW SCORES

The equation $z_1' = \beta_2 z_2 + \beta_3 z_3$ is a regression equation in standard-score form. It will yield the best possible linear prediction of a standard score on 1 from standard scores on 2 and 3. In practice, we usually require a regression equation for predicting a raw score on 1 from a raw score on 2 and 3. Let X_1' be a predicted raw score on 1, and X_2 and X_3 the obtained raw scores on 2 and 3. The estimated standard score z_1' and the observed standard scores z_2 and z_3 may be written as

$$z_1' = \frac{X_1' - \bar{X}_1}{s_1}$$

$$z_2 = \frac{X_2 - \bar{X}_2}{s_2}$$

$$z_3 = \frac{X_3 - \bar{X}_3}{s_3}$$

By substituting these values in the regression equation in standard-score form, we obtain

$$\frac{X_1' - \bar{X}_1}{s_1} = \beta_2 \frac{X_2 - \bar{X}_2}{s_2} + \beta_3 \frac{X_3 - \bar{X}_3}{s_3}$$

Rearranging terms and writing the expression explicit for X_1' yields

$$X_1' = \beta_2 \frac{s_1}{s_2} X_2 + \beta_3 \frac{s_1}{s_3} X_3 + \left(\bar{X}_1 - \beta_2 \frac{s_1}{s_2} \bar{X}_2 - \beta_3 \frac{s_1}{s_3} \bar{X}_3\right) \qquad [26.8]$$

This is a regression equation in raw-score form. It may be used to predict a raw score on 1 from a raw score on 2 and 3. The values $\beta_2 s_1/s_2$ and $\beta_3 s_1/s_3$ act as weights. The quantity to the right in parentheses is a constant.

In the example of the previous section $\beta_2 = .867$ and $\beta_3 = -.133$. Let us assume that $s_1 = 5$, $s_2 = 10$, $s_3 = 20$; also $\bar{X}_1 = 20$, $\bar{X}_2 = 40$, and $\bar{X}_3 = 60$. The regression equation in raw-score form is written as

$$\begin{aligned} X_1' &= (.867)\tfrac{5}{10}X_2 + (-.133)\tfrac{5}{20}X_3 + [20 - (.867)\tfrac{5}{10}(40) - (-.133)\tfrac{5}{20}(60)] \\ &= .434X_2 - .033X_3 + 4.62 \end{aligned}$$

26.7 THE GEOMETRY OF MULTIPLE REGRESSION

Given two variables X_1 and X_2, each pair of observations may be plotted as a point on a plane. If interest resides in predicting one variable from a knowledge of another, a straight regression line may be fitted to the points and this line used for prediction purposes.

Given three variables X_1, X_2, and X_3, each triplet of observations may be plotted as a point in a space of three dimensions as shown in Figure 26.1. Instead of two axes at right angles to each other, we now have three.

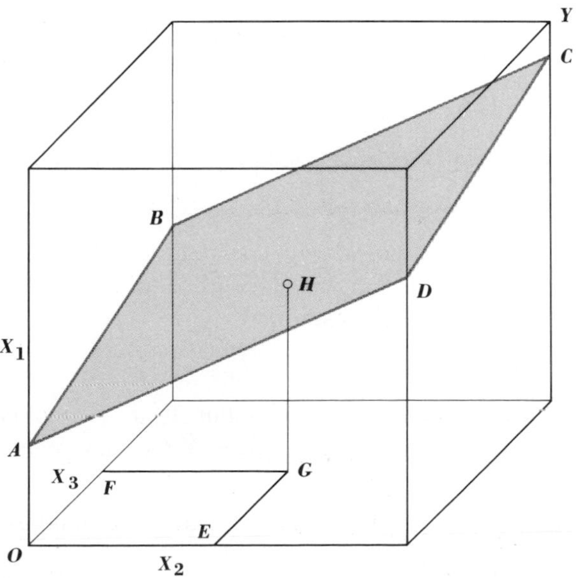

Figure 26.1 Geometrical representation of multiple regression. $ABCD$ is a multiple regression plane.

All triplets of observations may be plotted as points. If the correlations between the three variables are all positive, the assembly of points will show some tendency to cluster along the diagonal of the space of three dimensions extending from the origin O to Y. A plane may be fitted to the assembly of points. With two variables a regression line is fitted to points in a two-dimensional space. With three variables a regression plane is fitted to points in a three-dimensional space. In Figure 26.1 this plane is represented by $ABCD$. With two variables the regression equation is the equation for a straight line and is of the type $X_1' = b_2X_2 + a$, where b_2 is the slope of the line and a is the point where the line intercepts the X_1 axis. With three variables the regression equation is the equation for a plane and is of the type $X_1' = b_2X_2 + b_3X_3 + a$. Here b_2 is the slope of the line AD in Figure 26.1 and b_3 is the slope of the line AB. The constant a is the point where the plane intercepts the X_1 axis. In Figure 26.1 it is the distance AO.

Consider now a particular individual. Represent her score on X_2 by OE and on X_3 by OF. We locate the point G in the plane of X_2 and X_3 and proceed upward until we reach the point H in the regression plane $ABCD$. The distance GH is the best estimate of the individual's score on X_1 given her scores on X_2 and X_3. It is the best estimate in the sense that the regression plane is so located as to minimize the sums of squares of deviations from it parallel to the X_1 axis.

The reader will observe that the three-variable case is a simple extension of the two-variable case. A plane is used instead of a straight line. With four or more variables the idea is essentially the same. With four variables, in effect, we plot points in a space of four dimensions and fit a three-dimensional hyperplane to these points. By increasing the number of variables we may complicate the arithmetic. We do not complicate the idea.

26.8 MORE THAN THREE VARIABLES

In the discussion above we have considered the multiple regression case with three variables only, one criterion and two predictors. With k variables the multiple regression equation in standard-score form is

[26.9] $$z_1' = \beta_2 z_2 + \beta_3 z_3 + \cdots + \beta_k z_k$$

The raw-score form of this equation may be obtained, as previously, by substituting for the values of z_i the values $(X_i - \bar{X}_i)/s_i$ and rearranging terms. We thereby obtain

[26.10] $$X_1' = \beta_2 \frac{s_1}{s_2} X_2 + \beta_3 \frac{s_1}{s_3} X_3 + \cdots + \beta_k \frac{s_1}{s_k} X_k + A$$

where A is given by

[26.11] $$A = \bar{X}_1 - \beta_2 \frac{s_1}{s_2} \bar{X}_2 - \beta_3 \frac{s_1}{s_3} \bar{X}_3 - \cdot \cdot \cdot - \beta_k \frac{s_1}{s_k} \bar{X}_k$$

The multiple correlation coefficient is given by

[26.12] $$R = \sqrt{\beta_2 r_{12} + \beta_3 r_{13} + \cdot \cdot \cdot + \beta_k r_{1k}}$$

Thus to calculate this coefficient we multiply each correlation of a predictor with the criterion by its corresponding regression coefficient, sum these products, and take the square root.

A number of computational procedures exist for calculating the required regression weights with more than three variables. One such method is described in previous editions of this book. Such calculations are now done by computer.

26.9 THE SIGNIFICANCE OF A MULTIPLE CORRELATION COEFFICIENT

An F ratio may be used to test whether an observed multiple correlation coefficient is significantly different from 0. The required value of F is given by the formula

[26.13] $$F = \frac{R^2/g}{(1 - R^2)/(N - g - 1)}$$

where R = multiple correlation coefficient
N = number of observations
g = number of independent variables or predictors

The table of F is entered with $df_1 = g$ and $df_2 = N - g - 1$.

26.10 SHRINKAGE IN MULTIPLE CORRELATION

The multiple correlation coefficient is a measure of the efficacy of prediction for a particular sample. It is not, however, an unbiased estimate of the population correlation coefficient. Also, if the multiple regression weights calculated on one sample are applied to a second sample, the correlation between the weighted predictors and the criterion in the second sample will be less than the multiple correlation originally calculated on the first sample. The bias in the multiple correlation results because the process of determining regression weights, which minimize the average squared error, takes advantage, as it were, of the idiosyncracies of the sample. The extent of the bias depends on the population values of the multiple correlation coefficient, the sample size, and the number of predictor variables. The phenomenon under discussion here is commonly referred to as *shrinkage*. One estimate of the population value of the squared multiple correlation coefficient in the population is given by the formula

[26.14] $$R'^2 = 1 - \frac{N-1}{N-g-1}(1-R^2)$$

where R^2 = squared multiple correlation in the sample
N = sample size
g = number of predictors

The extent of the bias in R^2, as an estimate of the corresponding population value, decreases with increase in sample size and increases with increase in the number of predictors. For $R = .707$, $R^2 = .500$, $N = 26$, and $g = 10$, $R'^2 = .167$ and $R' = .409$. In this example the shrinkage in the multiple correlation coefficient is substantial. Herzberg (1969) in a computer simulation study has investigated formula [26.14] for estimating the squared multiple correlation. His conclusion is that this formula provides quite a reasonable estimate of the squared population multiple correlation.

On occasion multiple regression weights calculated on one sample are applied to a second sample. This procedure is known as *cross-validation*. The correlation between the criterion and the weighted predictors is known as the *cross-validity* and is denoted by r_c. The quantity r_c is generally an underestimate of the population multiple correlation. The bias tends to decrease with sample size and increase with the number of predictors.

26.11 STEPWISE MULTIPLE CORRELATION

Frequently in practical work the greater part of the prediction achieved can be attributed to a small number of variables, and the inclusion of additional variables contributes small and diminishing amounts to prediction. A procedure used here is called *stepwise multiple correlation*. In applying this procedure the best predictor is paired with every other predictor in turn and a multiple correlation calculated for each pair of predictors. If g is the number of predictors, this procedure requires the calculation of $g - 1$ multiple correlation coefficients. When the best pair of predictors has been selected, this pair may be combined with each of the remaining $g - 2$ variables, and a further $g - 2$ multiple correlation coefficients calculated. Thus the best triplet of predictors may be selected. This procedure may be continued until the inclusion of additional variables adds nothing of much significance to the multiple correlation coefficient. In using stepwise multiple correlation, tests of significance may be applied to determine whether the addition of one or more variables adds significantly to the multiple correlation. Care must be exercised in the application of such tests. Some of the commonly used tests lead to a biased result—see Wilkinson (1979). The above procedures require a substantial amount of computation which can only be readily done by computer. Note should be made of the fact that stepwise multiple correlation can be done in a descending fashion. Instead of finding the best two predictors and ascertaining in an ascending way what contributions are made to prediction when additional variables are added, we may begin with all g predictors and ascertain progressively what degree of prediction is lost when additional variables are dropped.

26.12 THE DISCRIMINANT FUNCTION WITH TWO CRITERION GROUPS

Not uncommonly the criterion, or dependent, variable consists of two groups of subjects, such as students who pass or fail a university program or hospital patients diagnosed as paranoid and nonparanoid schizophrenics. Assume that data are available on g independent variables. How may weights be assigned to the predictor, or independent, variables in order that the weighted sum of scores will provide the best possible differentiation or discrimination between the two criterion groups? The solution to this problem is a particular case of multiple regression. Membership in the two criterion groups, denoted A and B, are coded 1 and 0. Thus 1 denotes membership in group A, 0 membership in group B. The correlations between the dichotomous criterion and the independent variables are calculated using point biserial correlation. Multiple regression weights and a multiple regression equation are then obtained in the usual way. The multiple regression equation thus obtained is called a discriminant function. The predicted scores, or weighted scores, for the members of both group A and group B constitute a continuous variable.

How may a cutting score, a score that maximally discriminates between the two groups, be obtained? This may be done by simple inspection, although more complex methods may be used. The predicted scores for groups A and B are overlapping frequency distributions, if some degree of discrimination is present. The cutting score, the score that minimizes the number of incorrect assignments to groups, is about at the point where the two distribution curves intercept each other.

26.13 CANONICAL CORRELATION

Multiple correlation is used to predict one criterion, or dependent, variable from $k - 1$ independent variables or predictors. Multiple correlation may be generalized to situations with more than one, say m, dependent variables and $k - m$ independent variables. Weights may be obtained that maximize the correlation between the sum of weighted scores for the set of m variables with the sum of weighted scores for the set of $k - m$ variables. This form of correlation is known as *canonical correlation*. Multiple correlation is a particular case of canonical correlation. It is the particular case where $k = 1$. A canonical correlation is an ordinary product-moment correlation between two sums of weighted scores.

The calculation of a canonical correlation is a complex task, even in the simple case where the two sets consist of two variables only, that is, $m = 2$ and $k - m = 2$. Such correlations can ordinarily in practice only be obtained by computer. Canonical correlation was developed by Hotelling (1935), but was infrequently used because of difficulties of computation. One of its earlier applications was to obtain weights that would maximize the reliability of a test battery. Unlike multiple correlation where only one correlation coefficient exists, more than one canonical correlation may be obtained. For $k = 2$ and $m = 2$, two canonical correlations exist. With many sets of data, only the first canonical correlation will prove of interest.

In the previous section the discriminant function with two criterion groups was discussed. Canonical correlation may be used to obtain weights that discriminate between more than two groups. With two criterion groups, a simple code, 1 and 0, was used to denote group membership. With three criterion groups, two coded variables may be used to denote membership in the three groups. The appropriate method of coding is described in Section 27.2 of this book. The required weights are then obtained by canonical regression analyses. With two criterion groups, the discriminant function is a particular case of multiple regression. With three or more groups, the discriminant function is a particular case of canonical regression.

BASIC TERMS AND CONCEPTS

Multivariate statistics

Correlation of sums

Partial correlation

Second-order partial correlation

Part correlation

Multiple regression equation

Multiple correlation

Shrinkage in multiple correlation

Cross validation

Stepwise multiple correlation

Discriminant function

Canonical correlation

EXERCISES

1 Given the correlations $r_{12} = .70$, $r_{13} = .50$, and $r_{23} = .60$, compute $r_{12.3}$. What percentage of the association between variables 1 and 2 results because of the effect of variable 3?

2 The mean and standard deviation of a criterion variable are $\bar{X}_1 = 24.56$ and $s_1 = 4.52$. The means and standard deviations for two predictor variables are $\bar{X}_2 = 36.48$, $\bar{X}_3 = 16.95$, and $s_2 = 5.49$, $s_3 = 3.66$. The correlations are $r_{12} = .70$, $r_{13} = .65$, and $r_{23} = .33$. Compute (**a**) the correlation between standard scores on the criterion and the sum of standard scores on the two predictors, (**b**) the correlation between raw scores on the criterion and the sum of raw scores on the two predictors, (**c**) the multiple regression equation in standard-score form, (**d**) the multiple regression equation in raw-score form, (**e**) the multiple correlation coefficient.

3 The following are intercorrelations between first-year university averages and five university entrance examinations. Means and standard deviations are also given:

	X_1	X_2	X_3	X_4	X_5	X_6	$\bar{X}_i$	s_i
X_1	1.00	.62	.55	.43	.38	.09	72.61	6.56
X_2	.62	1.00	.72	.55	.43	.49	62.50	4.28
X_3	.55	.72	1.00	.55	.36	.47	58.65	5.33
X_4	.43	.55	.55	1.00	.65	.50	65.80	5.77
X_5	.38	.43	.36	.65	1.00	.20	69.75	3.91
X_6	.09	.49	.47	.50	.20	1.00	71.80	4.45

The multiple regression weights for predicting X_1 from the remaining variables are $\beta_2 = .439$, $\beta_3 = .268$, $\beta_4 = -.071$, $\beta_5 = .400$, $\beta_6 = -.296$. Obtain (**a**) the correlation between a standard score on the criterion variable and the sum of standard scores on the predictors, (**b**) the multiple regression equation in raw score form, (**c**) the multiple correlation coefficient.

ANSWERS TO EXERCISES

1 31.4 percent

2 **a** .828 **b** .826 **c** $z_1' = .545z_2 + 470z_3$
d $X_1' = .499X_2 + .580X_3 - 1.65$ **e** .830

3 **a** .537
b $X_1' = 673X_2 + .330X_3 - .081X_4 + 672X_5 - .436X_6 - 61.59$
c .771

27

MULTIPLE REGRESSION AND THE ANALYSIS OF VARIANCE

27.1 INTRODUCTION

An experiment explores the nature of the relations between variables. Consider a simple experiment involving two groups, an experimental and a control group. Here two means, $\bar{X}_1$ and $\bar{X}_2$, are usually calculated and a t test applied to test the significance of the difference between them. The null hypothesis is $H_0 : \mu_1 = \mu_2$. Two variables are involved, a dependent variable and an independent variable comprising two categories, the presence and the absence of a treatment. Instead of proceeding in the manner described above, weights of 1 and 0 may be used to represent the two categories of thc independent variable. These weights are arbitrary. Other weights may be used. A point biserial correlation coefficient, which is a simple product-moment correlation, may then be calculated. This correlation describes the strength of the relation between the two variables. A procedure may then be applied to test the significance of this correlation from 0, the null hypothesis being $H_0 : \rho = 0$. These two approaches lead to identical results. In this situation testing $H_0 : \mu_1 - \mu_2$ is thc same thing as testing $H_0 : \rho = 0$. Instead of a t test, an analysis of variance may be used to test the difference between the two means. As shown in Scction 15.9 the t test and the analysis of variance are equivalent procedures. It is of interest to note that in the analysis-of-variance approach the ratio of the within-group sum of squares to the total sum of squares, which is the correlation ratio $\eta^2_{y.x}$, is the same as the point biserial correlation squared; that is, for this situation $r^2_{pbi} = \eta^2_{y.x}$. The basic point illustrated here is that a problem involving means can be converted to a problem of correlation; that is, the situation can be conceptualized in correlational terms.

Can a correlational approach, instead of the analysis of variance, be used to analyze the data for more complex experiments? Within recent years this question has aroused some interest. Correlational methods have been devised which are the equivalent of conventional analysis-of-variance procedures. These methods use multiple correlation and regression and are viewed as providing a more general analytic system than the analysis of variance. The topic is discussed in detail in an article by Cohen (1968). A recent book on multiple regression as a general analytic system is that by Kerlinger and Pedhazur (1973).

The question can be raised as to why the relation between the analysis of variance and multiple regression has not been fully explored until this late date. The reason is historical. Multiple regression in its various ramifications has been viewed as the method commonly used by psychologists and educators in the psychometric tradition with an interest in measurement, prediction, and applied problems in general. Concern has been with the study of the relations between variables which exist in nature. The analysis of variance has been viewed as a method in the experimental tradition, where concern has been with the manipulation of variables by the experimenter. Another factor in the recent interest in multiple regression methods resides in the availability of computers which make possible forms of analysis which hitherto were not possible.

The purpose of this chapter is to introduce the student to a few introductory ideas involved in the correlational analysis of data commonly analyzed by conventional analysis-of-variance techniques. Quite apart from practical applications, a study of the correlational approach leads to insights which may broaden the understanding of the student in data analysis.

27.2 CODING A NOMINAL VARIABLE

Mention was made above of the relation between a t test, or the corresponding analysis of variance, where the null hypothesis is $H_0 : \mu_1 = \mu_2$, and the point biserial correlation, where the null hypothesis is $H_0 : \rho = 0$. In calculating the correlation, arbitrary weights of 1 and 0 are assigned to represent membership in the two categories. The assignment of weights 1 and 0 is a simple form of coding. If the two groups are denoted by A_1 and A_2, the weight 1 means membership in A_1, and 0 means nonmembership in A_1 or membership in A_2. If, however, a nominal variable is composed of 3, 4, or k categories, how may weights be assigned to represent category membership? How may such a variable be coded? Three methods for coding a nominal variable are discussed below. These are called *dummy* coding, *contrast* coding, and *orthogonal* coding.

Dummy coding involves the assignment of the weights 1 and 0 to represent membership in the categories of the nominal variable. If the variable has three categories, say, A_1, A_2 and A_3, 1s are used to represent membership in category A_1 and 0s in categories A_2 and A_3. Such a code generates a dummy variable which defines being a member of A_1, and not being a member of A_1. A second dummy variable is generated by assigning 1s to category A_2 and 0s to categories A_1 and A_3. This variable identifies membership in A_2, and nonmembership in A_2. Such a code may be represented as follows:

	X_1	X_2
A_1	1	0
A_2	0	1
A_3	0	0

All the information contained in the three-category nominal variable is represented, as it were, by the two dummy variables X_1 and X_2. Note that a third dummy variable is not required. Membership in A_3 is identified by not being a member of either A_1 or A_2. Dummy coding for a variable with four categories involves the assignment of weights as follows:

	X_1	X_2	X_3
A_1	1	0	0
A_2	0	1	0
A_3	0	0	1
A_4	0	0	0

In general, if the nominal variable is composed of k categories or groups, the number of dummy variables required to completely represent the information in the nominal variable will be $k - 1$. For equal n's the correlation between any two dummy variables is the negative of the reciprocal of their number; that is, $r_{ij} = -1/(k-1)$.

Contrast coding is similar to the procedure described in Section 18.3 where weights are assigned to portray contrasts or comparisons between means. Contrast coding implies that the investigator is interested in some particular set of contrasts or comparisons. The weights 1, 0, and -1 are used to create a set of $k - 1$ variables which represent the nominal variable. For example, given a four-category nominal variable, the weights 1, 0, and -1 may be used as follows:

	X_1	X_2	X_3
A_1	1	0	1
A_2	−1	1	1
A_3	0	−1	−1
A_4	0	0	−1

The variable X_1 implies that the investigator is interested in a contrast or comparison of A_1 with A_2. The variable X_2 implies a contrast or comparison of A_2 with A_3, whereas X_3 implies a contrast or comparison of A_1 and A_2 with A_3 and A_4. Note that the variables generated by contrast coding are not necessarily independent of each other but may be correlated as in the above example.

In orthogonal coding numbers are assigned to represent the k categories of the nominal variable in such a way that the resulting $k - 1$ variables are orthogonal or independent of each other. The term orthogonal is used here with precisely the same meaning as in Section 18.3. Any numerals whatsoever may be used provided the resulting variables are orthogonal. An example of orthogonal coding using the weights 1, 0, and −1 for four categories is as follows:

	X_1	X_2	X_3
A_1	1	0	1
A_2	−1	0	1
A_3	0	1	−1
A_4	0	−1	−1

If k is fairly large a problem arises in determining a coding system which generates a set of $k - 1$ orthogonal variables. This problem is discussed in some detail in Section 27.5. Note that the coefficients of orthogonal polynomials, used in the analysis of data for trend, constitute one class of sets of orthogonal variables. These coefficients are ordinarily used, however, when the independent variable is not nominal but is of the interval-ratio type. Orthogonal coding has an elegance which does not attach to dummy or contrast coding. Also it has some powerful algebraic advantages, which the reader will subsequently see.

27.3 CODING AND MULTIPLE REGRESSION

Consider data ordinarily analyzed by a one-way analysis-of-variance procedure. The independent variable, which is here assumed to be nominal, is coded using either dummy, contrast, or orthogonal coding, resulting in a set of $k-1$ coded variables. These variables may be correlated with the dependent variable Y and with each other. A multiple correlation squared, R^2, between the dependent variable and the $k-1$ coded variables may be computed. The value R^2 is the proportion of the variation in the dependent variable which can be attributed to the independent variable. A test of significance may be applied to R^2 using formula [26.13]. The procedure results in an F ratio identical to that obtained by an analysis of variance for one-way classification. It is an alternate road to the same result. When used in this context, R^2 is a measure of the strength of association between the dependent and independent variables and is identical with the correlation ratio $\eta^2_{y.x}$. Thus a test of the null hypothesis for means $H_0: \mu_1 = \mu_2 = \cdot \cdot \cdot = \mu_k$ can be translated into a null hypothesis stated in correlational terms as $H_0: R^2_{\text{pop}} = 0$.

The use of an orthogonal coding system simplifies both computation and interpretation. In multiple regression, when the independent variables are orthogonal, or uncorrelated with each other, the multiple regression weights β are simply the correlations of the independent variables with the dependent variable when the data are in standard-score form. This can be ascertained by simple inspection of the formulas used for calculating values of β. Also, R^2 is the sum of the squares of the correlations of the independent variable with the dependent variable. If Y denotes the dependent variable, and the independent variable is coded as $k-1$ orthogonal variables, then

$$R^2 = r_{y1}{}^2 + r_{y2}{}^2 + \cdot \cdot \cdot + r^2_{y(k-1)}$$

Thus R^2 may be viewed as partitioned into $k-1$ additive parts. The part $r_{y1}{}^2$ is the proportion of the total variation in Y attributable to the first coded variable, $r_{y2}{}^2$ the proportion attributable to the second coded variable, and so on. In using a multiple regression approach in a two-way or higher-order factorial experiment, some of the $r_{yi}{}^2$ will represent main effects, and others interaction effects. These separate parts may be tested for significance by using an F ratio. This leads to a result identical to that reached by the analysis of variance.

27.4 ILLUSTRATIVE EXAMPLE: ONE-WAY CLASSIFICATION USING DUMMY CODING

Table 27.1 shows a simple fictitious analysis of variance for one-way classification. Three experimental groups are used, denoted by A_1, A_2, and A_3. The number of subjects in each group is 5. The dependent variable is

Table 27.1
Analysis of variance: One-way classification

	A_1	A_2	A_3
	18	11	4
	19	16	8
	15	13	10
	23	12	9
	11	10	6
	86	62	37
$\bar{Y}_i$	17.200	12.400	7.400

Source of variation	Sum of squares	Degrees of freedom	Variance estimate
Between	240.133	2	$120.067 = s_b^2$
Within	125.200	12	$10.433 = s_w^2$
Total	365.333	14	

$F = \frac{120.067}{10.433} = 11.508 \qquad \eta^2_{y \cdot x} = .657$

denoted by Y. The ratio $F = 11.508$ which is significant at better than the .01 level. The correlation ratio is $\eta^2_{y.x} = .657$, which means that a proportion .657 of the variation in Y can be attributed to the independent variable.

Table 27.2 shows the dummy coding for the data of Table 27.1. The two coded variables are denoted by X_1 and X_2. The three correlations r_{y1}, r_{y2}, and r_{12} are shown, as are the standard deviations. The regression weights are calculated by using formulas [26.5] and [26.6], making the appropriate adjustments in the subscripts, and are $\beta_1 = .9361$ and $\beta_2 = .4776$. Using formula [26.7], the multiple correlation squared is $R^2 = .6573$. The F ratio obtained from formula [26.13] is $F = 11.5080$ with 2 degrees of freedom associated with the numerator and 12 degrees of freedom associated with the denominator.

In this example the multiple regression equation for predicting Y from X_1 and X_2 can readily be shown to be

$$Y' = 9.7978X_1 + 4.9984X_2 + 7.4013$$

If $X_1 = 1$ and $X_2 = 0$ are substituted in this equation the mean on Y for the A_1 group is obtained. This is $\bar{Y}_1 = 17.200$. If $X_1 = 0$ and $X_2 = 1$ are substituted, the mean on Y for the A_2 group is obtained. This is $\bar{Y}_2 = 12.400$. If $X_1 = 0$ and $X_2 = 0$ are substituted the mean on Y for the A_3 group is obtained. This is $\bar{Y}_3 = 7.400$.

Table 27.2

Dummy coding for the data of Table 27.1

	Y	X_1	X_2
	18	1	0
	19	1	0
A_1	15	1	0
	23	1	0
	11	1	0
	11	0	1
	16	0	1
A_2	13	0	1
	12	0	1
	10	0	1
	4	0	0
	8	0	0
A_3	10	0	0
	9	0	0
	6	0	0

$r_{y1} = .6972$ $\quad r_{y2} = .0095$ $\quad r_{12} = -.5000$

$s_y = 5.1083$ $\quad s_1 = .4879$ $\quad s_2 = .4879$

$\beta_1 = .9361$ $\quad \beta_2 = .4776$ $\quad R^2 = .6573$

$$F = \frac{R^2/g}{(1-R^2)/(N-g-1)} = \frac{.6573/2}{(1-.6573)/(15-2-1)} = 11.5080$$

For this illustrative example the analysis of variance and the multiple regression approach lead to an identical F ratio. The two procedures are in effect the same.

27.5 ILLUSTRATIVE EXAMPLE: FACTORIAL EXPERIMENT USING ORTHOGONAL CODING

Table 27.3 shows fictitious data for a 3×2 factorial experiment with two observations in each cell. The analysis-of-variance table is shown; also the F ratio for testing row, column, and interaction effects. The proportion of the total variation in Y accounted for by row, column, and interaction effects may be obtained by dividing the sums of squares for rows, columns, and interaction, respectively, by the total sum of squares. The proportion for rows is .1667, columns .7696, and interaction .0021. Thus 16.67 percent of the variation may be attributed to the row effect, 76.96 percent to the column effect, and .21 percent to the interaction.

Table 27.4 shows orthogonal coding for the data of Table 27.3. The

Table 27.3

Analysis of variance: Two-way factorial experiment

	C_1	C_2	C_3
R_1	4 8	10 16	25 32
R_2	12 18	20 24	35 36

Source of variation	Sum of squares	Degrees of freedom	Variance estimates
Rows	208.333	1	208.333
Columns	962.000	2	481.000
Interaction	2.667	2	1.333
Within cells	77.000	6	12.833
Total	1,250.000	11	

$F_r = \frac{208.333}{12.833} = 16.234 \qquad F_c = \frac{481.000}{12.833} = 37.481$

$F_{rc} = \frac{1.333}{12.833} = .104$

Table 27.4

Orthogonal coding for the data of Table 27.3

	Y	X_1	X_2	X_3	X_4	X_5
R_1C_1	4	1	1	1	1	1
	8	1	1	1	1	1
R_1C_2	10	1	−1	1	−1	1
	16	1	−1	1	−1	1
R_1C_3	25	1	0	−2	0	−2
	32	1	0	−2	0	−2
R_2C_1	12	−1	1	1	−1	−1
	18	−1	1	1	−1	−1
R_2C_2	20	−1	−1	1	1	−1
	24	−1	−1	1	1	−1
R_2C_3	35	−1	0	−2	0	2
	36	−1	0	−2	0	2
r_{yi}		−.4082	−.2800	−.8314	.0000	−.0462
r_{yi}^2		.1667	.0784	.6912	.0000	.0021

$\Sigma(Y - \bar{Y})^2 = 1{,}250 \qquad R^2 = \Sigma r_{yi}^2 = .9384$

data comprise six groups with two subjects, and two measurements, in each group. The row variable has two categories only, R_1 and R_2. The first coded variable is obtained simply by contrasting R_1 with R_2. Thus 1 is used to denote membership in R_1 and -1 to denote membership in R_2. Only one coded variable is required to account for the variation in Y due to the row variable. The column variable has three categories C_1, C_2, and C_3, and two coded variables are required to account for the variation in Y due to this variable. The first of these, X_2, is obtained by contrasting C_1 with C_2; that is, 1 is used to denote membership in C_1 and -1 in C_2. The second, X_3, is obtained by contrasting C_1 and C_2 with C_3. Thus 1 denotes membership in C_1 or C_2, and -2 denotes membership in C_3. In this example the interaction sum of squares has 2 degrees of freedom, and two coded variables X_4 and X_5 are required to account for interaction. These are obtained by multiplication. The coded variable X_4 is obtained by multiplying X_1 and X_2, and X_5 is obtained by multiplying X_1 and X_3. The result is a set of five coded variables each corresponding to 1 degree of freedom. In general, in a two-way factorial experiment the number of coded variables will equal the total number of degrees of freedom associated with row, column, and interaction effects. Note that this set of five coded variables is orthogonal. All sums and sums of products are 0.

Table 27.4 shows also the correlations of the coded variables with the dependent variable Y, the correlations squared, and the multiple correlation squared. Because the coded variables X_i are orthogonal the multiple correlation squared is simply the sum of the squared correlations between the coded variables and Y; that is,

$$R^2 = \Sigma r_{yi}^2 = .1667 + .0784 + .6912 + .0000 + .0021 = .9384$$

R^2 is the proportion of the total variation in Y accounted for by combined row, column, and interaction effects. An F ratio may be calculated by using formula [26.13] as follows:

$$F = \frac{R^2/g}{(1 - R^2)/(N - g - 1)} = \frac{.9384/5}{(1 - .9384)/(12 - 5 - 1)} = 18.2804$$

Here g is the number of independent coded variables. This F ratio has 5 degrees of freedom associated with the numerator and 6 with the denominator. It is highly significant. This F ratio is of limited interest. Ordinarily interest resides in the significance of row, column, and interaction effects, separately, rather than combined. Row, column, and interaction effects may be tested separately by using a simple modification of the above procedure. The coded variable X_1 is concerned with the row effect, X_2 and X_3 with the column effect, and X_4 and X_5 with the interaction effect. The proportion of the variation in Y accounted for by the row effect is .1667. The proportion accounted for by the column effect is $.0784 + .6912 = .7696$. For interaction the proportion is $.0000 + .0021 = .0021$.

To calculate F ratios associated with the row effect, for example, we replace R^2/g in the numerator of the F ratio by the proportion of the variance accounted for by the row effect divided by the associated number of degrees of freedom. The denominator remains as before. Proceed similarly for column and interaction effects. The row, column, and interaction F ratios are then as follows:

$$F_r = \frac{.1667/1}{(1-.9384)/(12-5-1)} = 16.233$$

$$F_c = \frac{.7696/2}{(1-.9384)/(12-5-1)} = 37.481$$

$$F_{rc} = \frac{.0021/2}{(1-.9384)/(12-5-1)} = .103$$

An alternate procedure for obtaining these F ratios is to multiply the sum of squares for Y by the proportions associated with row, column, and interaction effects. The sum of squares associated with these effects is thereby obtained. Thus the sum of squares for rows is $.1667 \times 1250 = 208.333$. The sums of squares for columns and interaction may be similarly obtained. Given these sums of squares, row, column, and interaction effects may be tested for significance by using the usual analysis-of-variance procedure.

Table 27.5 shows in summary form the multiple regression analysis for the data of Table 27.3. Apart from minor differences due to the rounding of decimals, the F ratios obtained by using a multiple regression method are identical with those obtained by using the analysis of variance. The two methods are equivalent and lead to the same result.

Table 27.5

Multiple regression analysis for the data of Table 27.3

Source of variation	Proportion of variation	df	$\frac{\text{Proportion}}{df}$
Regression R^2	.9384	5	
Rows	.1667	1	.1667
Columns	.7696	2	.3848
Interaction	.0021	2	.0011
Within cells	.0616	6	.0103
Total	1.0000	11	

$F_r = \frac{.1667}{.0103} = 16.233$ $\quad F_c = \frac{.3848}{.0103} = 37.481$ $\quad F_{rc} = \frac{.0011}{.0103} = .103$

Note: The values in this table were calculated to six decimals and rounded to four: consequently slight discrepancies in decimals occur.

27.6 EXTENDED APPLICATIONS

This chapter provides a brief introduction to the multiple regression analysis of data usually analyzed by the analysis of variance. The two approaches are equivalent. The multiple regression approach can be extended far beyond the simple applications discussed in this chapter. It can be applied to multiple comparison, trend analysis, the analysis of covariance, problems of unequal n's, and other situations. The multiple regression approach appears at times to simplify problems, leading to a more straightforward interpretation of the data. This statement applies, for example, to covariance analysis. The method is useful also when the independent variables are not nominal but are of the interval-ratio type. Modern computers handle very readily the data for regression analysis. For an extended discussion of multiple regression analysis, the reader is referred to Kerlinger and Pedhazur (1973).

BASIC TERMS AND CONCEPTS

Dummy coding

Contrast coding

Orthogonal coding

Partitioning R^2

EXERCISES

1 The following are data for a simple experiment with three groups of subjects.

	A_1	A_2	A_3
	2	4	19
	6	8	20
	7	6	15
T	15	18	54
$\bar{Y}_i$	3.00	6.00	18.00

Use a multiple regression approach with dummy coding to test the null hypothesis for means.

2 Write down the set of orthogonal coded variables for a $2 \times 2 \times 2$ factorial experiment.

3 The following are data for a $2 \times 2 \times 2$ factorial experiment with two observations in each cell.

	L_1 C_1	L_1 C_2			L_2 C_1	L_2 C_2	
R_1	2 5	8 12	27	R_1	3 8	10 15	36
R_2	6 7	14 20	47	R_2	9 12	20 23	64
	20	54	74		32	68	100

Use a multiple regression approach to test row, column, and interaction effects.

ANSWERS TO EXERCISES

1 $R^2 = .897$, $F = 26.167$; $p < .01$

2

	X_1	X_2	X_3	X_4	X_5	X_6	X_7
$R_1C_1L_1$	1	1	1	1	1	1	1
$R_1C_2L_1$	1	−1	1	−1	1	−1	−1
$R_2C_1L_1$	−1	1	1	−1	−1	1	−1
$R_2C_2L_1$	−1	−1	1	1	−1	−1	1
$R_1C_1L_2$	1	1	−1	1	−1	−1	−1
$R_1C_2L_2$	1	−1	−1	−1	−1	1	1
$R_2C_1L_2$	−1	1	−1	−1	1	−1	1
$R_2C_2L_2$	−1	−1	−1	1	1	1	−1

3

Source of variation	Proportion of variance	df	$\frac{\text{Proportion}}{df}$
Regression R^2	.8875		
R	.2492	1	.2492
C	.5301	1	.5301
L	.0731	1	.0731
$R \times C$	.0277	1	.0277
$R \times L$	.0069	1	.0069
$C \times L$	.0004	1	.0004
$R \times C \times L$	.0000	1	.0000
Within cells	.1125	8	.0141
Total	1.0000		

$F_r = 17.72$† $F_c = 37.69$† $F_1 = 5.20$ $F_{rc} = 1.97$
$F_{r1} = .49$ $F_{c1} = .03$ $F_{rc1} = .00$

† Significant at the .01 level.
All values originally calculated to six decimals.

28

FACTOR ANALYSIS

28.1 INTRODUCTION

Factor analysis is a multivariate statistical method which is used in the analysis of tables, or matrices, of correlation coefficients. These coefficients are usually product-moment correlation coefficients, although other measures or indices of association may be used. In many of the applications of factor analysis the variables are scores on educational or psychological tests. The method had its origins in education and psychology. Factor analytic methods are, however, quite general and can be applied to correlations between variables of any type, e.g., economic, anthropological, physiological, meteorological, or physical. Direct inspection of any large matrix of correlation coefficients shows immediately that no simple intuitive interpretation of the pattern of relations among the variables is possible. A method of analysis is required which will assist the investigator in reaching a meaningful interpretation of the ways in which the variables are related. Factor analysis is a method for doing this.

In multiple regression, concern is with prediction. A distinction is made between a dependent variable and a set of independent variables. Factor analysis is usually applied to data where a distinction between dependent and independent variables is not meaningful. Concern is with the description and interpretation of interdependencies within a single set of variables.

Factor analysis achieves its purpose in two ways. First, it reduces the original set of variables to a smaller number of variables called *factors*. Second, factors acquire meaning because of structural properties that may exist within the set of relationships. The process of *reducing the number of variables*, and the *concept of structure*, are basic to an understanding of factor analysis. Their meaning is elaborated later in this chapter.

Factor analysis as a method originated with C. Spearman. In a paper on the theory of intelligence published in 1904, Spearman argued that "all branches of intellectual activity have in common one fundamental function (or group of functions) whereas the remaining or specific elements of the activity seem in every case to be wholly different from that in all others." Spearman analyzed tables of intercorrelations between psychological tests and purportedly was able to show that the intercorrelations could be accounted for in terms of one general factor common to all tests and factors which were specific, or unique, to each test. This was known as the *theory of two factors*. The general factor was referred to as g. It is of interest to note that Spearman was primarily concerned with psychological theory and viewed the factor analyses of tables of intercorrelations as providing evidence to support his theory. Spearman's theory had neurophysiological aspects. He linked his general factor with energy, which was thought to serve the whole cortex or nervous system, and linked his specific factors with particular groups of neurons, which served particular kinds of operations.

Following Spearman, much work on factor analysis was done by many workers, including Godfrey H. Thomson, L. L. Thurstone, H. Hotelling, J. P. Guilford, and R. B. Cattell. More recent contributors include D. N. Lawley, J. B. Carroll, L. Guttman, J. L. Horn, K. G. Joreskog, L. R. Tucker, H. F. Kaiser, and others. At an early stage Spearman's theory of two factors was the object of much criticism and was quickly discarded, although the concept of a general factor common to many types of cognitive activity still persists and will probably continue to do so. L. L. Thurstone in effect generalized the factor analytic method and invented multiple-factor analysis, which could be applied to data involving any number of factors.

L. L. Thurstone and other early factor analysts were involved in prodigious arithmetical calculation. This calculation is now quickly done on any computer. The use of factor analysis is on the increase.

The purpose of the present chapter is to explain in a simple way the basic ideas involved in factor analysis, so that a student may understand and use the method. This is no easy pedagogical task. At present the term *factor analysis* does not refer to a single method. A variety of related methods are subsumed under the term. This chapter describes the ideas and methods most commonly used. The student who wishes to pursue the study of factor analysis in greater depth is referred to the comprehensive treatment of the subject by Harman (1967) and the more recent works by Gorsuch (1974) or Harris (1975). The present discussion follows Harman's statistical notation. See also Mulaik (1972).

28.2 REDUCING THE NUMBER OF VARIABLES

In the discussion on subjective probability (Chapter 6) it was noted that the estimation and use of subjective probabilities, including a priori probabilities, occur with great frequency. This is an important activity of mind. Likewise, reducing a large, perhaps complex set of variables to a smaller set or to a single variable is a common place subjective process. The smaller set is usually viewed as more basic. Causal and explanatory properties may be attributed to it. The assertion that a student did well on all his or her university courses because of high intelligence and hard work is an example of such a subjective reduction. Performance on 30 or 40 university courses is related, presumably in a causal sense, to two more basic variables, intelligence and work. Another example arises when individuals' attitudes toward a wide range of political and social events are attributed subjectively to an assessment of those individuals' location on a single liberal-conservative or left-right variable. The field of cultural stereotypes provides a further example. A great variety of qualities may be attributed to an individual because of a single variable such as racial origin, religion, or skin color—a subjective general factor phenomenon. In summary, the subjective reduction of large sets of variables describing complex situations to a smaller set which possesses explanatory properties is a common phenomenon, the purpose of which is to create understanding, or its illusion, through simplification.

Factor analysis is the operational correlate of the subjective reductionism described above in much the same sense that a relative frequency is the operational correlate of a subjective probability. The investigator selects a complex process such as human intelligence, academic achievement, learning, perception, or hyperactivity in children. He defines variables, perhaps a large number, which are thought to describe or relate to aspects of the process. He obtains measurements on and intercorrelates these variables. He performs a factor analysis which reduces the number of variables to a smaller number, called factors. Meaning is attached to the factors. Properties, explaining the original process, may be assigned to these factors.

The method used by factor analysts to reduce the original set of variables to a smaller set is complex. Some insights into how this is done may be obtained by considering two examples. The first uses partial correlation; the second is geometrical.

Consider a set of five variables with intercorrelations as follows:

	1	2	3	4	5
1					
2	.50				
3	.60	.30			
4	.70	.35	.42		
5	.80	.40	.56	.48	

Partial correlation may be used to remove the effect of variable 1 on the correlations between variables 2, 3, 4, and 5. For example,

$$r_{23.1} = \frac{r_{23} - r_{12}\, r_{13}}{\sqrt{(1 - r_{12}^2)(1 - r_{13}^2)}} = \frac{.30 - .50 \times .60}{\sqrt{(1 - .25)(1 - .36)}} = 0$$

Because this partial correlation is 0, variable 1 may be thought to account for the correlations between 2 and 3. Similarly, removing the effect of variable 1 makes all the other partial correlations 0. This means that the relations between the five variables may be accounted for, attributed to, or explained by a single variable. Although we have measurements on five variables, one variable only is needed to explain the relations between them. If, for example, variable 1 were chronological age, and 2, 3, 4, and 5 were measures of motor performance for a group of school children, then one variable only, namely age, accounts for the correlations. Thus five variables are reduced to a single variable.

Consider a geometrical illustration with paired observations as follows:

X	Y
1.5	3.0
2.0	4.0
2.5	5.0
3.0	6.0
3.5	7.0

These pairs may be represented as points on graph paper as shown in Figure 28.1*a*. Two numbers locate each point. The points fall along a straight line. Let us now rotate the frame of reference, that is, the axes, X and Y, to a new location as shown in Figure 28.1*b*. Denote the axes in the new location as X' and Y'. Note that all points fall along Y'. The coordinates of the points in relation to X' and Y' are

X'	Y'
.00	3.35
.00	4.47
.00	5.59
.00	6.71
.00	7.83

The location of each point may now be described by a single number measured along the Y' axis. In effect the two variables X and Y have been replaced by, or reduced to, a single variable Y'. Note that plotting the points in relation to X and Y uses a space of two dimensions. Representing the points as distances from an origin along the Y' axis uses a space of one dimension only.

The ideas present in the above discussion can be extended to any number of variables. For example, measurements on three variables may be plotted as points in a space of three dimensions. If it so happened that these points fell along a straight line passing through the origin (and in standard-

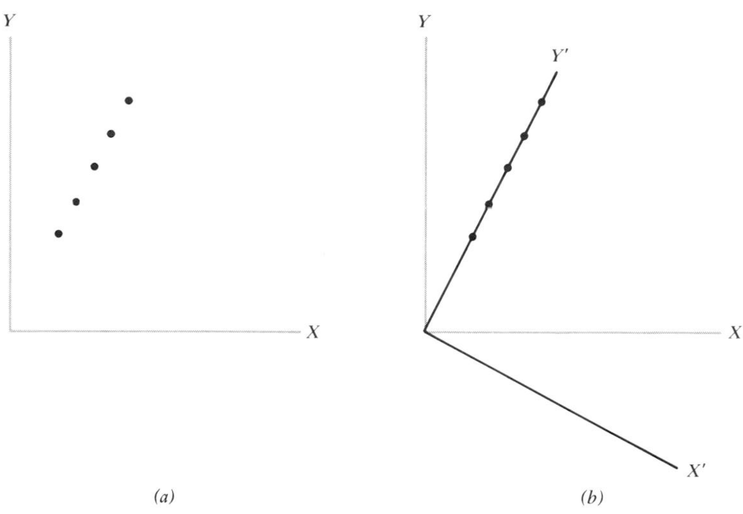

Figure 28.1 Rotation of axes to obtain a simplified description.

score form, lines representing points pass through the origin), then a system using three variables and a space of three dimensions is reduced to a system using one variable and a space of one dimension. If the points fell not on a straight line but on a plane of two dimensions passing through the origin, then two variables would locate each point, and the original three variables would reduce to two.

In general, measurements on n variables may be represented as points in a space of n dimensions. The arrangement of points may occupy, usually in an approximate way only, a space of m dimensions, where $m < n$. Thus a set of n variables is reduced to a smaller set, m. Note that the word approximate has been used. With real data the fit is not exact but approximate. Questions arise regarding the closeness of the approximations. Should n variables be replaced by m, or $m - 1$ or $m + 1$, or some other number?

28.3 THE CONCEPT OF STRUCTURE

Factor analysis is concerned with the discovery and description of structure within a set of relations. What does *structure* mean? The concept in statistics has the same meaning as in Gestalt psychology, where structure refers to the configurational properties of a percept or experience. These determine or compel the meaning of the percept or experience.

Consider a simple example from statistics. Measurements on two variables X and Y may be plotted as points on graph paper. The points may arrange themselves in a haphazard or random fashion. Such an arrangement is without structure. Alternatively, the points may tend to arrange themselves along a straight line. The points exhibit a linear structure. The concept of linearity emerges and is determined by the configurational properties of the points, that is, by the way in which the points arrange themselves with respect to each other. The linearity present in the arrangement of points is not dependent on the location of the X and Y axes. These serve as a frame of reference. The points may, of course, exhibit other structures—quadratic, cubic, or other more complex shapes. As the arrangement of points departs from a random state, structure emerges. Structure is a departure from randomness.

The reader will recall from Chapter 9 that the correlation between two variables may be represented geometrically by two vectors. A vector is a line with direction and length. If the vectors have unit length, the correlation coefficient is the cosine of the angle between two vectors. Thus a correlation of .707 is represented by two vectors at a 45° angle to each other, and a correlation of −.500 by a 120° angle, and so on. The correlations between three variables may be represented by three vectors. The idea may be extended to any number of vectors.

In the geometrical model used in factor analysis the lengths of the vectors have meaning. Denote the length of a vector by the symbol h. Variable j may be represented by a vector of length h_j, and variable k by a vector of length h_k. Denote the angle between the two vectors by ϕ_{jk}. The correlation coefficient between two variables is equal to the length of the two vectors multiplied by the angular separation between them; that is $r_{jk} = h_j h_k \cos \phi_{jk}$. For a simple proof of this result see Thurstone (1947). Obviously if the vectors are of unit length, $h_j = h_k = 1$, and the correlation is the cosine of the angle between the vectors.

Any table of intercorrelations between n variables may be conceptualized geometrically by a vector model. The vectors have varying lengths, $h_1, h_2, \ldots, h_n$, and varying angular separations, ϕ_{jk}. The table of correlations and the vector model are two ways—one numerical, the other geometrical—of representing the same set of relations. These are two ways of saying the same thing.

Just as a set of points in two dimensions may arrange themselves at random or exhibit a structure such as linearity, so a set of n vectors with $n(n-1)/2$ angular separations may be either random or exhibit structural properties. For example, subsets of vectors may arrange themselves in clearly identifiable clusters. Subsets may be found to fall in clearly identifiable planes. The configuration of vectors may be structured in many ways. Recall that the n vectors may occupy, at least approximately, a space of fewer than n dimensions, perhaps substantially fewer. Figure 28.2 illustrates a number of possible arrangements of vectors.

The purpose of factor analysis is to describe the configuration of vectors in a parsimonious way and also in such a way as to reveal its structural properties. The structural properties determine or compel the meaning of the factors.

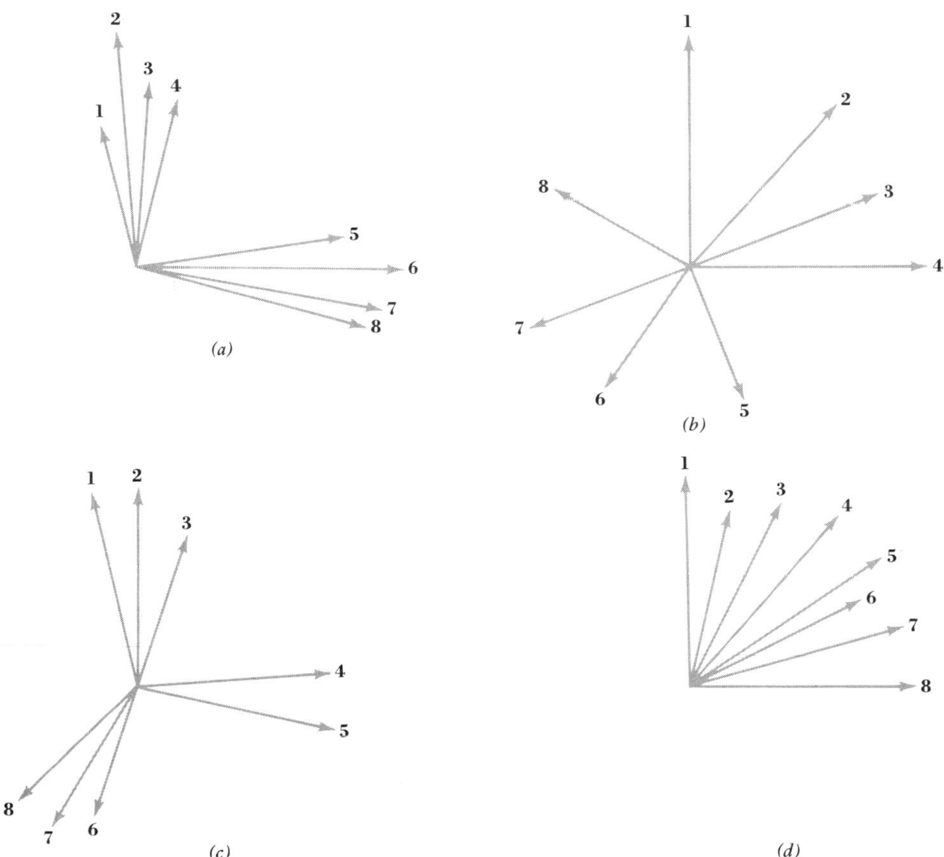

Figure 28.2 Different configurations of vectors.

28.4 BASIC EQUATIONS

The basic factor analytic model is that a score of individual i on variable j can be conceptualized as the weighted sum of scores on a smaller number of derived variables, called *factors*. This is a linear model, and may be expressed in standard-score form as follows:

$$z_{ji} = a_{j1}F_{1i} + a_{j2}F_{2i} + \cdots + a_{jm}F_{mi} + d_jU_{ji} \quad [28.1]$$

The quantity z_{ji} is the standard score of individual i on variable j. F_{1i} is the standard score of individual i on the first common factor; F_{2i} is his standard score on the second common factor; and F_{mi} is his standard score on the mth common factor. The quantity U_{ji} is the standard score of individual i on what is called a *unique* factor; that is, a factor that is involved in a single variable only, in this case the variable j. The coefficients $a_{j1}, a_{j2}, \ldots, a_{jm}$ are factor loadings. These are weights which attach to the common-factor scores. The coefficient d_j is the weight which attaches to the unique-factor score. Factor analysis is concerned primarily with the determination of the coefficients, or loadings, $a_{j1}, a_{j2}, \ldots, a_{jm}$. It is not usually concerned with estimating the factor scores F_{ji}, although methods for such estimation exist.

A factor must be viewed as a variable like any other, but with a simple difference. Most variables involve direct measurement. Factors are derived by a process of analysis from a set of variables obtained by direct measurement. Nonetheless an individual may be said to have a score on a factor in the same sense that the individual has a score on a test or examination.

The analogic relation between equation [28.1] and an ordinary multiple regression equation is of interest. A multiple regression equation in standard-score form may be written

[28.2] $$z_1' = \beta_2 z_2 + \beta_3 z_3 + \cdots + \beta_m z_m$$

where z_1' is a predicted standard score on the dependent variable; $z_2, z_3, \ldots, z_m$ are the standard scores on the independent variables; and $\beta_2, \beta_3, \ldots, \beta_m$ are weights so determined as to maximize the correlation between the dependent variable and the weighted sum of scores on the independent variables. What are the points of similarity and difference between a multiple regression equation in [28.2] and the factor analytic equation in [28.1]? First, the multiple regression equation states that a person's predicted score on a dependent variable may be conceptualized as the weighted sum of standard scores on a number of independent variables, or predictors. Similarly, the factor analytic equation states that a person's standard score on one variable may be conceptualized as the weighted sum of standard scores on a number of other variables, called *factors*. Second, both the regression coefficients, β_j, and the factor loadings, a_{ji}, are weights. In multiple regression, however, a distinction is made between a dependent variable and a set of independent variables. The values of β_j are determined in order to maximize the prediction of the former from the latter. In factor analysis no distinction is made between a dependent variable and a set of independent variables. This distinction is without meaning. The investigation begins with a set of interdependent variables which cannot be classified as dependent and independent. The values, a_{ji}, as will subsequently be seen, are determined by the structure, organization, or patterning that exists among the variables. Multiple regression is concerned with prediction, and equation [28.2] is a prediction equation. Factor analysis is concerned with the discovery and description of structure in complex patterns of relations. Third, it is worth nothing that in a multiple regression equation the independent variables $z_2, z_3, \ldots, z_m$ are usually not independent of each other, but are correlated. In the factor analytic equation, with some sets of data the factors are defined in such a way as to be independent or uncorrelated. This means that the factor scores $F_{1i}, F_{2i}, \ldots, F_{mi}$ are uncorrelated with each other. With other sets of data correlated factors may be used. The F_{ji} are not independent but are correlated. When the factors are uncorrelated, they are said to be *orthogonal*. A method that leads to a set of uncorrelated factors is spoken of as an *orthogonal solution*. When the factors are correlated, they are said to be *oblique*. A method that leads to a set of correlated factors is spoken of as an *oblique solution.*

Equation [28.1] expresses a standard score z_{ji} as a linear function of weighted factor scores. Alternatively, the model may be written in a form which expresses the variable z_j as a linear function of weighted factors, the factors being viewed as variables. Thus

[28.3] $$z_j = a_{j1}F_1 + a_{j2}F_2 + \cdots + a_{jm}F_m + d_jU_j$$

Here the $a_{j1}, a_{j2}, \ldots, a_{jm}$ are factor loadings and $F_1, F_2, \ldots, F_m$ are the factors as variables. For a set of n variables this model may be written as

[28.4] $$\begin{array}{l} z_1 = a_{11}F_1 + a_{12}F_2 + \cdots + a_{1m}F_m + d_1U_1 \\ z_2 = a_{21}F_1 + a_{22}F_2 + \cdots + a_{2m}F_m + d_2U_2 \\ \cdots\cdots\cdots\cdots\cdots\cdots\cdots\cdots\cdots\cdots \\ z_n = a_{n1}F_1 + a_{n2}F_2 + \cdots + a_{nm}F_m + d_nU_n \end{array}$$

Such an arrangement is called a *factor pattern.* Usually a factor pattern is written with the factor loadings recorded in columns with the number of common factors shown at the top of the column. To illustrate, the following is a factor pattern for three factors and five variables. The value h_j^2 is

	Factors			
Variables	I	II	III	h_j^2
1	a_{11}	a_{12}	a_{13}	h_1^2
2	a_{21}	a_{22}	a_{23}	h_2^2
3	a_{31}	a_{32}	a_{33}	h_3^2
4	a_{41}	a_{42}	a_{43}	h_4^2
5	a_{51}	a_{52}	a_{53}	h_5^2

known as the *communality,* and is the sum of the squares of the common-factor loadings. The communality is discussed in a section to follow.

The factor loadings are ordinary correlation coefficients. Thus a_{11} is the correlation of the first variable with factor I, a_{12} the correlation of the first variable with factor II, and so on.

28.5 COMPONENTS OF VARIANCE

Equation [28.1] is in standard-score form. Thus the scores z_{ji} and the factor scores F_{ji} have zero mean and unit variance. By squaring both sides of equation [28.1], summing over N cases, dividing by N, and assuming that factor scores are uncorrelated, it can be readily shown that

[28.5] $$s_j^2 = 1 = a_{j1}^2 + a_{j2}^2 + \cdots + a_{jm}^2 + d_j^2$$

This equation states that the total variance of a test j may be partitioned into additive parts. The component of variance due to the first factor is a_{j1}^2, that due to the second factor is a_{j2}^2, and so on. The a_{jk} are correlation coefficients between the variables and the factors. Using the variance interpretation of the correlation coefficient, the a_{jk}^2 may be interpreted as the proportion of the total variance that may be attributed to the different factors.

In the terminology of factor analysis different components of variance have been assigned different names. The *communality* of a variable, usually represented by the symbol h_j^2, is the sum of the squares of the common factor loadings, a common factor being one that is common to more than one variable in the set. Thus

[28.6] $$h_j^2 = a_{j1}^2 + \cdot \cdot \cdot + a_{jm}^2$$

The communality is that part of the variance that may be attributed to common factors. The part of the variance that is left over, and that cannot be attributed to common factors, is called the *uniqueness* and is represented by the symbol d_j^2. The uniqueness is sometimes partitioned into two components, *specificity*, b_j^2, and *error variance*, e_j^2. The specificity is that part of the total variance which is due to factors that are specific to the particular variable and not due to measurement error. Since all measurement involves error in some degree, in all applications of factor analysis part of the uniqueness will be due to measurement error. If the reliability coefficient for a particular variable is known, the error variance may be readily calculated from the formula $e_j^2 = 1 - r_{jj}$, where r_{jj} is a reliability coefficient for test j. The quantity e_j^2 is nothing other than the error variance associated with a single test score in standard-score form.

28.6 REPRODUCING THE CORRELATION COEFFICIENTS

A factor analysis begins with a table of correlation coefficients between variables. The analysis represents the variables as the weighted sum of factors. What relation exists between the factors and the original correlation coefficients? How may the correlation coefficient be accounted for, as it were, by the common factor loadings? Or, stated in different terms, how good a fit is the linear model to the observed correlations?

The correlation between two variables in standard-score form is given by

[28.7] $$r_{jk} = \frac{\sum^{N} z_{ji} z_{ki}}{N - 1}$$

We may write the following:

[28.8] $$\begin{aligned} z_{ji} &= a_{j1}F_{1i} + a_{j2}F_{2i} + \ldots + a_{jm}F_{mi} + d_jU_{ji} \\ z_{ki} &= a_{k1}F_{1i} + a_{k2}F_{2i} + \ldots + a_{km}F_{mi} + d_kU_{ki} \end{aligned}$$

These two equations, which are also in standard-score form, may be multiplied together, summed over N individuals, and divided by $N - 1$. The common factors are uncorrelated with each other. The unique factors are uncorrelated with each other and with the common factors. Given these simplifying assumptions, the following result is obtained:

[28.9] $$r'_{jk} = a_{j1}a_{k1} + a_{j2}a_{k2} + \cdots + a_{jm}a_{km}$$

This equation, which applies only to uncorrelated factors, shows how the correlation coefficient between two variables may be reproduced from the factor loadings. The quantity $a_{j1}a_{k1}$ is viewed as the contribution of the first factor to the correlation, $a_{j2}a_{k2}$ as the contribution of the second factor to the correlation, and so on. We have used r'_{jk} to refer to the reproduced correlation coefficient to distinguish it from r_{jk}, the observed correlation. The reproduced correlation coefficient between two variables is the sum of the products of the common factor loadings. With real data, because of sampling or other kinds of error, the correlation reproduced from the factor loadings will differ in some degree from the observed correlations. The differences between the observed and the reproduced correlations are known as *residual correlations*, $\bar{r}_{jk} = r_{jk} - r'_{jk}$. The magnitude of the residual correlations indicates the extent to which the factors account for the observed correlations, or how good a fit the linear model is to the observed correlations.

28.7 AN ILLUSTRATIVE EXAMPLE

An example illustrating points raised thus far may prove useful. Consider the following table of correlations between six variables:

	1	2	3	4	5	6
1						
2	.61					
3	.48	.45				
4	.09	.27	.36			
5	−.05	.14	.27	.45		
6	−.36	−.05	.15	.48	.61	

Before the method has been discussed, this table may be factor analyzed, and the following factor pattern obtained.

	Factors		
Variable	I	II	h_j^2
1	.90	−.20	.85
2	.70	.10	.50
3	.60	.30	.45
4	.30	.60	.45
5	.10	.70	.50
6	−.20	.90	.85

Two factors only are required to account for the correlations. The original six variables are reduced to two. The factor loadings in the factor pattern

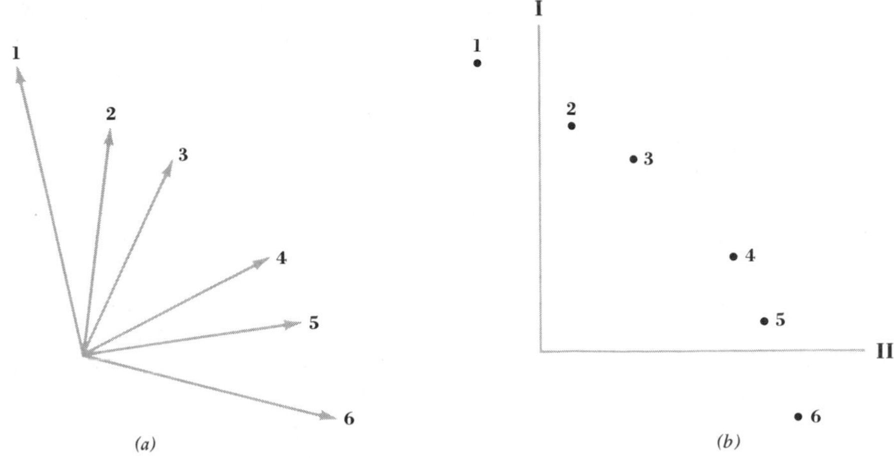

Figure 28.3 *(a)* Geometrical representation of intercorrelations, and *(b)* factor representation. In *(b)* the factor loadings are the projections of the terminal points of the vectors on axes I and II.

are correlation coefficients. Thus variable 1 has a correlation of .90 with factor I, and −.20 with factor II. According to the variance interpretation of the correlation coefficient, 81 percent of the variance of variable 1 is accounted for by factor I, and 4 percent by factor II. The communality, the sum of the squares of the common factor loadings, for variable 1 is $(.90)^2 + (-.20)^2 = .85$. This means that variable 1 has a proportion .85, or 85 percent, of its variance in common with the other variables in the set. The factor loadings reproduce the correlations exactly. Thus $r_{12} = a_{11}a_{21} + a_{12}a_{22} = .90 \times .70 + (-.20) \times .10 = .61$. All other correlations may be similarly obtained. The part of r_{12} accounted for by factor I is $.90 \times .70 = .63$, and by factor II $(-.20) \times .10 = -.02$. With fallible data the factor loadings would not reproduce the correlations exactly. Small residuals, some positive and some negative, would remain.

The table of intercorrelations, and the factors may be represented geometrically as shown in Figure 28.3.

This geometrical model is two-dimensional, there being two factors only. The vector lengths are the square roots of the communalities. The angular separation between vectors can be obtained using the equation $r_{ij} = h_i h_j \cos \phi_{ij}$. Thus $r_{12} = .61$, $h_1 = .9220$, $h_2 = .8367$ $\cos \phi_{ij} = .7908$, and ϕ is roughly 38°. Of course, arguments may be advanced for placing the reference axes in some other location or using oblique rather than orthogonal factors. Discussion relevant to these considerations is presented later in this chapter.

28.8 METHODS OF FACTORING

Geometrically the factorization of a table of correlations involves inserting a set of reference axes in the configuration of vectors and describing the terminal points of the vectors in relation to the reference axes. The projections of the terminal points of the vectors on the reference axes are the factor loadings. If the factors are uncorrelated, the reference axes are at right angles to each other or orthogonal. Frequently the reference axes are not at right angles to each other, but are oblique. The factors are correlated.

A variety of methods exist for inserting reference axes in a configuration of vectors and obtaining a factor matrix which reproduces approximately the correlation matrix. Quite clearly an indefinitely large number of solutions to this problem exist, since the reference axes could be located anywhere at all. This circumstance has led to a two-stage process in factor analysis. The first stage involves locating reference axes, or a frame of reference, using one of a number of methods that meet certain statistical criteria. A factor matrix is thereby obtained. Such solutions are referred to by Harman (1967) as *direct* solutions. The *number of factors* is determined at the direct-solution stage. Although such solutions may have interesting statistical properties, the factors obtained are ordinarily not viewed as amenable to interpretation. If the variables are psychological variables, no psychological meaning can in most instances be attached to them.

One direct method of factoring which in the past has been widely used is the centroid solution, developed by L. L. Thurstone. Many examples of factor analysis reported in the earlier literature use the centroid solution. The centroid solution involves placing the first reference axis through the centroid of the configuration of vectors; obtaining a table of residual correlations, which are subject to certain adjustments; placing the second factor through the centroid corresponding to the table of residual correlations; and continuing the process until the magnitude of the residuals can be considered inconsequential. The centroid solution is an approximation to the principal factor solution, or the method of principal components. This at the present time is the most widely used method for obtaining a direct solution. Another direct solution, developed by Lawley (1940), may be obtained using the method of maximum likelihood. See also Lawley and Maxwell (1963).

The second stage of factor analysis requires the rotation of the reference axes to a new position such that the factors can be interpreted. The factor pattern resulting from this rotation is referred to by Harman as a derived solution. A variety of methods for obtaining derived solutions exist. All such solutions involve locating the reference axes in a position that is determined by the structural properties of the configuration of vectors. In the earlier history of factor analysis the rotations were done by plotting factor loadings on graph paper, two factors at a time. The angles of rotation were estimated visually. Analytical methods of rotation were later developed. These methods define objective criteria for rotation which, when minimized or maximized, fix the location of the frame of reference.

Before proceeding with a discussion of the principal-factor solution and the most commonly used methods of rotations, a comment on the estimation of communalities is appropriate.

28.9 THE COMMUNALITIES

The *communality*, h^2, is the proportion of the variance of a variable that is common to other variables in the set. It is the sum of the squares of the common factor loadings. Initially these values are unknown and must be estimated from the data. Estimates of communalities are inserted as diagonal elements in the table of correlations before proceeding with the factor analytic solution. A variety of methods for estimating communalities have been used. One arbitrary method takes the largest correlation in each row or column of the correlation matrix as an estimate. Iterative methods may be used. Communality estimates may be inserted in the diagonal, then a factor analytic procedure applied, and the sum of squares of common factor loadings used as new estimates of the communalities. This procedure may be repeated until an acceptable degree of convergence is obtained. One of the more widely used estimates of the communality, which has much to commend it, is the squared multiple correlation, R_j^2, of every variable with the $n - 1$ remaining variables. The squared multiple correlations, $R_1^2, R_2^2, \ldots, R_n^2$, of each variable with the $n - 1$ remaining variables can be readily obtained by computer. R_j^2 is the proportion of the variance of one variable that can be accounted for by the appropriately weighted sum of the remaining variables. As such it appears to provide a plausible communality estimate. Its use is not without difficulty. The reader will recall that R is not an unbiased estimate of the population multiple correlation; the magnitude of the bias is related to sample size and the number of predictors. Another suggestion is to use the test-retest reliability, r_{jj}, if available, as the communality estimate. It may be shown that R_j is the lower bound, and r_{jj} the upper bound, of the communality; that is, $R_j \leq h_j^2 \leq r_{jj}$.

The estimation of communality is closely related to the number of factors. The communalities are usually conceptualized as those diagonal elements that will enable the off-diagnoal correlations to be accounted for in the most parsimonious way, that is, by the smallest number of factors. With fallible data, subject to sampling error, the meaning of the "smallest number of factors," is not a simple matter. An elaboration of it is beyond the scope of the present elementary discussion. For a detailed description of the topic the reader may consult Gorsuch (1974), Harris (1975), and Mulaik (1972).

28.10 PRINCIPAL FACTOR SOLUTION

The most commonly used *direct* method of factoring in use at the present time is the principal factor solution. This method was developed by Hotelling in 1933. Because of the laborious nature of the arithmetical calculations involved, it was not widely used until electronic computers became available.

The geometrical model for the principal factor solution is of interest. Measurements on two variables for a sample of N members can be plotted as points in relation to two orthogonal reference axes. The result is the usual scatter diagram. If a correlation exists between the two variables, the arrangement of points will be elliposidal in form. Given more than two variables, say n, the points may be plotted with reference to n reference axes. The result is an elliposidal swarm of points. The principal factors correspond to the principal axes of this ellipsoid. Questions may be raised regarding the dimensionality of the ellipsoid. If 1's are inserted in the diagonals of the correlation matrix prior to factoring, the ellipsoid will have n principal axes, and n components will result. If estimates of the communalities are placed in the diagonals, m factors will result, where m is ordinarily less than n.

The identification of the principal axes of an ellipsoidal swarm of points can be shown to reduce algebraically to the identification of a set of factors in decreasing order of their contribution to the total communality, assuming that estimates of the communalities are used in the diagonal of the correlation matrix.

The brief discussion here attempts to present the algebraic rationale of a solution whereby the factors are obtained successively. In the computational procedures which are ordinarily employed in practice, the factors are obtained simultaneously. The solution amounts to the identification of a factor F_1, which maximizes the quantity

[28.10] $$V_1 = a_{11}^2 + a_{21}^2 + \cdots + a_{n1}^2$$

where V_1 is the contribution of F_1 to the total communality. The maximization of V_1 is subject to the condition that the sum of the products of the factor loadings will reproduce the correlation coefficients. Following the calculation of F_1 a matrix of first-factor residuals is obtained. The contribution of the first factor to r_{12} is $a_{11}a_{21}$, and the residual correlation is $r_{12} - a_{11}a_{21}$. In general, the first-factor residual correlations are given by $r_{jk} - a_{j1}a_{k1}$. A table of residual correlations results. This is a table of correlations with the influence of the first factor removed and with residual communalities in the diagnonal. The second factor, F_2, is obtained by maximizing

[28.11] $$V_2 = a_{12}^2 + a_{22}^2 + \cdots + a_{n2}^2$$

where V_2 is the contribution of the second factor to the communality. Again a table of residuals is obtained which shows those parts of the correlation that remain after the second factor has been removed. This procedure is continued until the part of the total communality that remains can be viewed as trivial, whatever this might mean. No rigorous method exists for determining the number of factors. A rule commonly employed in the principal factor solution is that the contribution of the factor to the total communality should not be less than 1. This rule is, of course, arbitrary.

28.11 ILLUSTRATIVE EXAMPLE OF PRINCIPAL-FACTOR SOLUTION

The principal-factor solution will be illustrated using data originally gathered by Holzinger and Swineford (1939). These data are reproduced from Harman (1967).

Twenty-four psychological tests were administered to a group of 145 children in grades seven and eight in a suburb of Chicago. The 24 psychological tests, together with means, standard deviations, and reliability coefficients, are shown in Table 28.1. The intercorrelations between these 24 psychological tests are shown in Table 28.2. Note that correlation coefficients below the main diagonal are shown. The table is, of course, symmetric about the main diagonal.

The data of Table 28.2 were analyzed using the principal-factor solution. In applying this method, squared multiple correlations (SMCs) have been

Table 28.1

Basic statistics for twenty-four psychological tests*

Test X_j	Mean $\bar{X}_j$	Standard deviation s_j	Reliability coefficient r_{ii}
1. Visual Perception	29.60	6.90	.756
2. Cubes	24.84	4.50	.568
3. Paper Form Board	15.65	3.07	.544
4. Flags	36.31	8.38	.922
5. General Information	44.92	11.75	.808
6. Paragraph Comprehension	9.95	3.36	.651
7. Sentence Completion	18.79	4.63	.754
8. Word Classification	28.18	5.34	.680
9. Word Meaning	17.24	7.89	.870
10. Addition	90.16	23.60	.952
11. Code	68.41	16.84	.712
12. Counting Dots	109.83	21.04	.937
13. Straight-Curved Capitals	191.81	37.03	.889
14. Word Recognition	176.14	10.72	.648
15. Number Recognition	89.45	7.57	.507
16. Figure Recognition	103.43	6.74	.600
17. Object-Number	7.15	4.57	.725
18. Number-Figure	9.44	4.49	.610
19. Figure-Word	15.24	3.58	.569
20. Deduction	30.38	19.76	.649
21. Numerical Puzzles	14.46	4.82	.784
22. Problem Reasoning	27.73	9.77	.787
23. Series Completion	18.82	9.35	.931
24. Arithmetic Problems	25.83	4.70	.836

* Reproduced from Harry H. Harman's *Modern factor analysis*, 2d ed., University of Chicago Press, Chicago, 1967, with permission of the author and publisher.

used as estimates of the communalities. These are the squared multiple correlations of each variable with the $n-1$ remaining variables. Table 28.3

Table 28.2

Intercorrelations of twenty-four psychological tests for 145 children*

Test: j	1	2	3	4	5	6	7	8	9	10	11	12	13	14	15	16	17	18	19	20	21	22	23	24
1	—	—	—	—	—	—	—	—	—	—	—	—	—	—	—	—	—	—	—	—	—	—	—	—
2	.318	—	—	—	—	—	—	—	—	—	—	—	—	—	—	—	—	—	—	—	—	—	—	—
3	.403	.317	—	—	—	—	—	—	—	—	—	—	—	—	—	—	—	—	—	—	—	—	—	—
4	.468	.230	.305	—	—	—	—	—	—	—	—	—	—	—	—	—	—	—	—	—	—	—	—	—
5	.321	.285	.247	.227	—	—	—	—	—	—	—	—	—	—	—	—	—	—	—	—	—	—	—	—
6	.335	.234	.268	.327	.622	—	—	—	—	—	—	—	—	—	—	—	—	—	—	—	—	—	—	—
7	.304	.157	.223	.335	.656	.722	—	—	—	—	—	—	—	—	—	—	—	—	—	—	—	—	—	—
8	.332	.157	.382	.391	.578	.527	.619	—	—	—	—	—	—	—	—	—	—	—	—	—	—	—	—	—
9	.326	.195	.184	.325	.723	.714	.685	.532	—	—	—	—	—	—	—	—	—	—	—	—	—	—	—	—
10	.116	.057	−.075	.099	.311	.203	.246	.285	.170	—	—	—	—	—	—	—	—	—	—	—	—	—	—	—
11	.308	.150	.091	.110	.344	.353	.232	.300	.280	.484	—	—	—	—	—	—	—	—	—	—	—	—	—	—
12	.314	.145	.140	.160	.215	.095	.181	.271	.113	.585	.428	—	—	—	—	—	—	—	—	—	—	—	—	—
13	.489	.239	.321	.327	.344	.309	.345	.395	.280	.408	.535	.512	—	—	—	—	—	—	—	—	—	—	—	—
14	.125	.103	.177	.066	.280	.292	.236	.252	.260	.172	.350	.131	.195	—	—	—	—	—	—	—	—	—	—	—
15	.238	.131	.065	.127	.229	.251	.172	.175	.248	.154	.240	.173	.139	.370	—	—	—	—	—	—	—	—	—	—
16	.414	.272	.263	.322	.187	.291	.180	.296	.242	.124	.314	.119	.281	.412	.325	—	—	—	—	—	—	—	—	—
17	.176	.005	.177	.187	.208	.273	.228	.255	.274	.289	.362	.278	.194	.341	.345	.324	—	—	—	—	—	—	—	—
18	.368	.255	.211	.251	.263	.167	.159	.250	.208	.317	.350	.349	.323	.201	.334	.344	.448	—	—	—	—	—	—	—
19	.270	.112	.312	.137	.190	.251	.226	.274	.274	.190	.290	.110	.263	.206	.192	.258	.324	.358	—	—	—	—	—	—
20	.365	.292	.297	.339	.398	.435	.451	.427	.446	.173	.202	.246	.241	.302	.272	.388	.262	.301	.167	—	—	—	—	—
21	.369	.306	.165	.349	.318	.263	.314	.362	.266	.405	.399	.355	.425	.183	.232	.348	.173	.357	.331	.413	—	—	—	—
22	.413	.232	.250	.380	.441	.386	.396	.357	.483	.160	.304	.193	.279	.243	.246	.283	.273	.317	.342	.463	.374	—	—	—
23	.474	.348	.383	335	.435	.431	.405	.501	.504	.262	.251	.350	.382	.242	.256	.360	.287	.272	.303	.509	.451	.503	—	—
24	.282	.211	.203	248	.420	.433	.437	.388	.424	.531	.412	.414	.358	.304	.165	.262	.326	.405	.374	.366	.448	.375	.434	—

* Reproduced, with minor modification, from Harry H. Harman's *Modern factor analysis*, 2d ed., University of Chicago Press, Chicago, 1967, with permission of the author and publisher.

Table 28.3

Principal-factor solution for twenty-four psychological tests (communality estimates: SMCs)*

	Common factors					Communality	
Test	P_1	P_2	P_3	P_4	P_5	Original	Calculated
1	.595	−.039	.369	−.184	−.073	.511	.531
2	.374	.026	.270	−.147	.121	.300	.250
3	.433	.115	.396	−.112	−.276	.440	.446
4	.501	.108	.290	−.178	.044	.409	.380
5	.701	.312	−.273	−.050	.003	.673	.666
6	.683	.404	−.213	.067	−.102	.677	.690
7	.676	.412	−.284	−.082	−.046	.684	.716
8	.680	.206	−.088	−.115	−.118	.564	.540
9	.690	.446	−.212	.076	.036	.713	.727
10	.456	−.469	−.446	−.128	.105	.579	.654
11	.589	−.372	−.198	.076	−.185	.541	.565
12	.448	−.491	−.154	−.263	.044	.537	.537
13	.590	−.268	.018	−.300	−.255	.539	.575
14	.435	−.063	−.012	.418	−.057	.358	.371
15	.390	−.102	.055	.362	.101	.293	.307
16	.512	−.098	.325	.259	.006	.429	.444
17	.471	−.212	−.036	.388	−.087	.412	.426
18	.521	−.331	.118	.145	.028	.443	.417
19	.450	−.115	.110	.167	−.178	.367	.287
20	.623	.135	.142	.049	.252	.464	.492
21	.596	−.220	.076	−.140	.204	.473	.471
22	.600	.103	.138	.053	.142	.449	.413
23	.685	.063	.160	−.096	.154	.561	.532
24	.635	−.169	−.192	−.009	.079	.527	.475
V_p	7.665	1.672	1.208	.920	.447	11.943	11.912
$100V_p/11.943$	64.2	14.0	10.1	7.7	3.7	100.	99.7

* Reproduced from Harry H. Harman's *Modern factor analysis,* 2d ed., University of Chicago Press, Chicago, 1967, with permission of the author and publisher.

shows the principal-factor solution obtained for these data. Five factors have been extracted. The body of the table shows the factor loadings for the 24 tests on the five factors. The rows at the bottom of the table show the contribution of each factor to the total communality. The contribution of each factor is simply the sum of the squares of the factor loadings. The percent of the total communality contributed by each factor is also shown. To the right of Table 28.3 the original SMC communalities are shown together with the communalities obtained following the calculation by summing the squares of the factor loadings for the individual rows.

The reader should note that the contribution of the factors decreases from the first to the fifth factor. The first factor accounts for 64.2 percent of the total communality, the second for 14.0 percent, and the fifth for only 3.7 percent. Note also that all loadings for the first factor are positive, whereas subsequent factors contain both positive and negative loadings.

The data of Table 28.3 are not amenable to any psychological interpretation. Before any meaning can be assigned to factors, the factors obtained by the principal-factor solution must be subject to further analysis.

28.12 ANALYTICAL METHODS OF ROTATION

The factors obtained by the principal factor or other direct solutions are not ordinarily viewed as amenable to interpretation. A rotation of the reference axes is made to a new position and derived factors obtained. Meaning is then assigned to these factors. Before the development of computers the rotation of axes was done using laborious graphic methods. The points represented the terminal points of the vectors. If a group of tests, or vectors, clustered together, or seemed to belong together, a reference axes or factor was placed through the cluster. The object of the enterprise was to obtain as many high positive and near zero loadings as possible. By this means it was thought that the rotated factors would be descriptive of the structural properties of the vector configuration. Criteria guiding the rotational process were proposed. One set of criteria, that of *simple structure,* was proposed by Thurstone (1947). These criteria specified required numbers of zero and nonzero loadings in the rows and columns of the factor pattern after rotation.

In the early 1950s it became apparent that a more objective and analytical method for rotation was required. Methods of rotation were developed by Carroll (1953), Ferguson (1954), Kaiser (1958), and others. These methods involved essentially the same idea, with modification. This idea was stated by Ferguson (1954) as follows:

> Consider a point p, plotted in relation to two orthogonal reference axes. These axes may be rotated into an indefinitely large number of positions, yielding, thereby, an indefinitely large number of descriptions of the location of the point. Which position provides the most parsimonious description? Intuitively it appears that the most parsimonious description will result when one or other of the axes passes through the point, the point being, thereby, described as a distance measured from an origin. Likewise as one or other of the reference axes is rotated in the direction of the point the product of the two coordinates grows smaller. This product is a minimum and equal to zero when one of the axes passes through the point, and is a maximum when a line from the origin and passing through the point is at a 45° angle to both axes. These circumstances suggest that some function of the product of the two coordinates might be used as a measure of the amount of parsimony associated with the description of the point. Note, further, that this product is an area, and is a measure of the departure of the point from the reference frame. Consider now any set of points with both positive and negative coordinates, the usual factor case. Here the sum of products is inappropriate because a zero sum can result when the sums of negative and positive products are equal. Here following the usual statistical convention we may use the sum of squares of products of coordinates, or some related quantity, as a measure of parsimony.

The problem was to define an appropriate descriptive measure of the departure of the axes of reference from the vector configuration. The axes could then be rotated to maximize or minimize this measure. The problem is directly analogous to fitting a straight line to a set of points. This may be

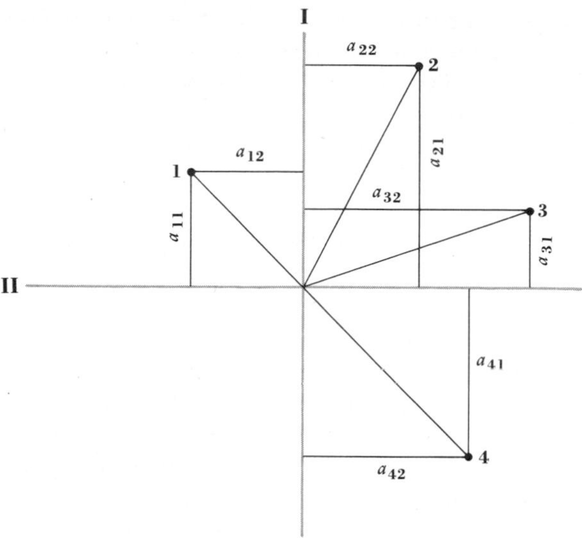

Figure 28.4 Departure of points from reference frame. The sum of squares of the areas $a_{11}a_{12}$, $a_{21}a_{22}$, $a_{31}a_{32}$, and $a_{41}a_{42}$ is a measure of the departure of points from the reference frame.

done visually, or the line may be located objectively using a sum of squares criterion. In the rotational problem the purpose is to fit a set of reference axes, usually in multidimensional space, to the vector configuration. This may be done by inspection, or analytically by defining an appropriate criterion which is then manipulated algebraically.

The criterion originally proposed may be illustrated with reference to two factors and four tests as shown in Figure 28.4. Here a measure of the departure of the terminal points of the vectors is the sum of the squares of the areas.

$$(a_{11}a_{12})^2 + (a_{21}a_{22})^2 + (a_{31}a_{32})^2 + (a_{41}a_{42})^2$$

For m factors and n variables we consider the $m(m-1)/2$ sum of products of pairs of coordinates, which may be written as

[28.12] $$\sum_{p<q=1}^{m} \sum_{j=1}^{n} (a_{jp}a_{jq})^2$$

The a_{jp} and a_{jq} are unrotated factor loadings. If b_{jp} and b_{jq} are now taken to denote factor loadings following rotation, the problem of rotation becomes one of locating the reference axes in such a position that the quantity

[28.13] $$\sum_{p<q=1}^{m} \sum_{j=1}^{n} (b_{jp}b_{jq})^2$$

is a *minimum*. By some simple algebra it can be shown that *minimizing* this quantity is the same thing as maximizing

[28.14] $$Q = \sum_{p=1}^{m} \sum_{j=1}^{n} b_{jp}{}^4$$

Thus the problem becomes one of locating the reference axes in a position such that the sum of all factor loadings in the factor matrix raised to the fourth power is a maximum. The statistic Q is a measure of parsimony. It is a measure of the departure of the terminal points of the vectors from the reference frame. It is a measure in the least square sense of the goodness of fit of the reference frame to the configuration. One difficulty with the Q criterion relates to the weighting of variables. This method weights each variable in a manner proportional to the square of its communality. This means that a variable with a communality of .80 will exert four times the weight in determining the final solution that a test with a communality of .40 will exert. A possible solution here is to use "normalized factor" loadings, that is, factor loadings adjusted in such a way that the sum of their squares for each variable, the communality, is equal to 1. In the geometrical model this amounts to extending all vectors to unit length. The rotation of normalized factor loadings assigns equal weight to each of the variables.

All analytical methods of rotation in common use involve, with modification, the essential ideas described above. Probably the method most widely accepted at present is a modification proposed by Kaiser (1958), who was led to consider the simplicity of a factor matrix. He defined the simplicity of any factor as the variance of its squared factor loadings and the simplicity of the factor matrix as the sum of these variances over all factors. This quantity, which is known as the *raw varimax criterion*, is as follows:

[28.15]
$$s^2 = \frac{1}{n} \sum_{p=1}^{m} \sum_{q=1}^{n} b_{jp}^4 - \frac{1}{n^2} \sum_{p=1}^{m} \left(\sum_{j=1}^{n} b_{jp}^2 \right)^2$$

This criterion weights each variable in a manner proportional to the square of the communality. Since empirically this weighting system did not yield a satisfactory result, Kaiser proposed a "normal" varimax criterion which may be written as follows:

[28.16]
$$V = n \sum_{p=1}^{m} \sum_{j=1}^{n} (b_{jp}/h_j)^4 - \sum_{p=1}^{m} \left(\sum_{j=1}^{n} b_{jp}^2/h_j^2 \right)^2$$

Here all factors are normalized prior to rotation. All communalities are 1, all vectors are of unit length, and all variables receive equal weight. The normal varimax criterion is commonly used in factor analytic studies at the present time.

28.13 ILLUSTRATIVE EXAMPLE OF THE NORMAL VARIMAX METHOD OF ROTATION

The normal varimax method of rotation will be illustrated with reference to the principal-factor solution for 24 psychological tests as shown in Table 28.3. The first four factors of Table 28.3 were rotated in order to maximize the normal varimax criterion of equation [28.16]. The fifth factor of Table

28.3 was excluded from this analysis because of its small contribution to the total communality.

Table 28.4 shows the rotated factors. All loadings greater than .400 are shown in boldface type. The contributions of each factor to the total communality are shown at the bottom of the table. The communalities calculated for the four factors, instead of the original five, are shown to the right of the table.

The rotated factors of Table 28.4 are amenable to psychological interpretation. Inspection of the tests used in this study suggests that the first factor is a verbal factor. Tests 5, 6, 7, 8, and 9 have high loadings on this factor. These tests are very largely tests of verbal ability. Inspection of the tests suggests that the second factor is a speed factor. The tests with the highest loadings on this factor are tests 10, 11, 12, and 13. The third factor is a deduction factor, and the tests loading highest in this factor are tests 1, 3, 4, and 23. The fourth factor has its highest loadings in tests 14, 15, 16, 17, and 18, and may be identified as a memory factor.

Table 28.4

Normal varimax solution for twenty-four psychological tests

Test	Verbal M_1	Speed M_2	Deduction M_3	Memory M_4	Calculated communality
1	.152	.204	**.658**	.165	.526
2	.111	.089	**.458**	.069	.235
3	.144	−.013	**.579**	.118	.370
4	.231	.090	**.557**	.080	.378
5	**.745**	.220	.199	.149	.666
6	**.764**	.078	.199	.225	.680
7	**.803**	.155	.196	.084	.714
8	**.577**	.233	.345	.137	.526
9	**.795**	.048	.203	.224	.726
10	.172	**.762**	−.090	.157	.643
11	.191	**.572**	.102	.396	.530
12	.018	**.704**	.178	.083	.535
13	.168	**.552**	**.413**	.078	.510
14	.207	.081	.060	**.561**	.368
15	.125	.082	.105	**.513**	.297
16	.073	.059	**.414**	**.514**	.444
17	.142	.222	.061	**.588**	.419
18	.026	.356	.285	**.455**	.416
19	.130	.168	.250	.385	.256
20	.384	.100	**.426**	.300	.429
21	.171	**.437**	**.405**	.212	.429
22	.351	.113	**.407**	.301	.392
23	.368	.225	**.521**	.225	.508
24	.351	**.479**	.176	.292	.469
V_p	3.613	2.616	2.934	2.300	11.462
$100V_p/11.462$	31.522	22.823	25.598	20.066	100.

28.14 CORRELATED FACTORS

Many factor analyses use orthogonal or uncorrelated factors. If factors are clearly correlated, the use of an orthogonal frame of reference will provide a distorted, or at least imprecise, description of the configuration of vectors. Such would be the case, for example, if the vectors were arranged in two clusters with, say, roughly a 45° angle separating them. Under such circumstances, oblique axes will provide a more parsimonious description of the vector configuration than orthogonal axes, or so it would appear.

Given a set of correlated factors, the table of correlations between factors may be factor analysed. This is called second-order factor analysis. An example is found in the work of R. B. Cattell. His distinction between fixed and crystallized intelligence is based on second-order factor analysis. Fixed and crystallized intelligence appear as two important second-order factors.

BASIC TERMS AND CONCEPTS

Factor analysis

General factor

Specific factor

Theory of two factors

Reducing the number of variables

Structure

Vector

Vector model

Configuration of vectors

Factor

Factor loading

Orthogonal factors

Oblique factors

Factor pattern

Communality

Specificity

Error variance

Reproducing the correlations

Residual correlations

Reference axes

Rotation of reference axes

Estimating the communalities

Centroid solution

Principal factor solution

Analytical methods of rotation

Varimax method of rotation

Correlation between factors

Second-order factor analysis

EXERCISES

1 If there is one general factor only, the general factor loading for variable i may be calculated by the formula:

$$g_i = \sqrt{\frac{r_{ij}r_{ik}}{r_{jk}}}$$

Calculate the **(a)** general and **(b)** specific factor loadings from the following table of intercorrelations.

	1	2	3	4	5
1					
2	.72				
3	.56	.63			
4	.16	.18	.14		
5	.18	.18	.14	.04	

2 Two variables, z_1 and z_2, may be expressed as linear functions of weighted factors as follows:

$$z_1 = .40F_1 + .60F_2 - .05F_3 + .10F_4 + .78U_1$$
$$z_2 = .05F_1 + .70F_2 - .10F_3 + .60F_4 \cdots + .37U_2$$

Calculate the correlation between z_1 and z_2.

3 The following are factor loadings on two-factors for six variables:

Variable	F_1	F_2
1	.9	.3
2	.8	.3
3	.7	.3
4	.6	.6
5	.5	.6
6	.4	.6

Calculate *(a)* the communalities, *(b)* the uniqueness for each variable, *(c)* the contribution of each factor to the total communality, *(d)* the percent contribution of each factor to the total communality.

4 For the data of Exercise 3 above, write down the table of reproduced correlation coefficients.

5 The reliability coefficient for variable 6 in Exercise 3 is .80. For this variable calculate that part of the variance that is error variance.

6 In a normalized configuration the vectors are of unit length. Normalized factor loadings are obtained by dividing the factor loadings for any variable by the square root of the communality. Calculate the normalized factor loadings for the data of Exercise 3.

7 The loadings for a particular variable on two factors, F_1 and F_2, are, respectively, .3 and .5. These loadings may be plotted as a point with reference to two orthogonal reference axes. Calculate the distance of the point from the origin.

8 The following are loadings for six variables on two factors:

Variable	F_1	F_2
1	.2	.2
2	.3	.3
3	.4	.4
4	.2	−.2
5	.3	−.3
6	.4	−.4

Rotate these factors to what appears to be the most parsimonious position. Calculate the rotated factor loadings.

ANSWERS TO EXERCISES

1

Variable	g	s
1	.8	.6000
2	.9	.4359
3	.7	.7141
4	.2	.9798
5	.2	.9798

2 .505

3 **a** .90, .73, .58, .72, .61, .52
b .10, .27, .42, .36, .39, .48
c 2.71, 1.35 **d** 66.75, 33.25

4

	1	**2**	**3**	**4**	**5**	**6**
1						
2	.81					
3	.72	.65				
4	.72	.66	.60			
5	.63	.58	.53	.66		
6	.54	.50	.46	.60	.56	

5 .20

6 F_1: .949, .936, .919, .707, .640, .555
F_2: .316, .351, .394, .707, .768, .832

7 .5831

8 Rotated loadings are:
F_1: .2828, .4243, .5657, . . . , . . . , . . .
F_2: . . . , . . . , . . . , .2828, .4243, .5657

REFERENCES

Allen, Mary J., and Wendy M. Yen, 1979: "Introduction to Measurement Theory," Brooks/ Cole Publishing Co., Monterey, Calif.

Bancroft, T. A., 1968: "Topics in Intermediate Statistical Method," The Iowa State University Press, Ames, Iowa.

Box, G. E. P., 1953: Non-normality and Tests on Variance, *Biometrika,* **40:**318–335.

———, 1954: Some Theorems on Quadratic Forms applied in the Study of Analysis of Variance Problems, *The Annals of Mathematical Statistics,* **25:**290–302, 484–498.

Bradley, James V., 1968: "Distribution Free Statistical Tests," Prentice-Hall, Inc., Englewood Cliffs, N.J.

Camilli, Gregory, and Kenneth D. Hopkins, 1978: Applicability of Chi-Square to 2 × 2 Contingency Tables with Small Expected Cell Frequencies, *Psychological Bulletin,* **88:**163–167.

———, and Kenneth D. Hopkins, 1979: Testing for Association in 2 × 2 Tables with Very Small Sample Sizes, *Psychological Bulletin,* **86:**1011–1014.

Carroll, John B., 1953: An Analytical Solution for Approximating Simple Structure in Factor Analysis, *Psychometrika,* **18:**23–38.

Cochran, William G., and Gertrude M. Cox, 1957: "Experimental Designs," 2d ed., John Wilcy & Sons, Inc., New York.

Cohen, Jacob, 1968: Multiple Regression as a General Data-Analytic System, *Psychological Bulletin,* **70:**426–443

de Finetti, Bruno, 1978: Probability: Interpretations, in William H. Kruskal and Judith M. Turner (eds.), "International Encyclopedia of Statistics," The Free Press, New York, 97–107.

Duncan, D. B., 1955: Multiple Range and Multiple *F*-Tests, *Biometrics,* **11:**1–42.

———, 1957: Multiple Range Tests for Correlated and Heteroscedastic Means, *Biometrics,* **13:**164–174.

Edwards, Allen L., 1973: "Statistical Methods," 3d ed., Holt, Rinehart and Winston, Inc., New York.

———, 1979: "Multiple Regression and the Analysis of Variance and Covariance," W. H. Freeman and Co., San Francisco.

Feldt, L. S., and M. W. Mahmoud, 1958: Power Function Charts for Specification of Sample Size in Analysis of Variance, *Psychometrika,* **23:**201–210.

Ferguson, George A., 1954: The Concept of Parsimony in Factor Analysis, *Psychometrika,* **19:**281–290.

———, 1965: "Nonparametric Trend Analysis," McGill University Press, Montreal.

Finney, D. J., 1948: The Fisher-Yates Test of Significance in 2 × 2 Contingency Tables, *Biometrika,* **35:**145–156.

Fisher, R. A., 1970: "Statistical Methods for Research Workers," 14th ed., Hafner Publishing Company, Inc., New York.

———, and F. Yates, 1963: "Statistical Tables for Biological, Agricultural, and Medical Research," 6th ed., Hafner Publishing Company, Inc., New York.

Friedman, M., 1937: The Use of Ranks to Avoid the Assumption of Normality Implicit in the Analysis of Variance, *Journal of the American Statistical Association,* **32:**675–701.

———, 1940: A Comparison of Alternative Tests of Significance for the Problem of *m* Rankings, *Annals of Mathematical Statistics,* **11:**86–92.

Geisser, S., and S. Greenhouse, 1958: Extension of Box's Results on the Use of the *F* Distribution in Multivariate Analysis, *Annals of Mathematical Statistics,* **29:**855–891.

Glass, Gene V., P. D. Peckham, and J. R. Sanders, 1972: Consequences of Failure to Meet Assumptions Underlying the Fixed Effects Analysis of Variance and Covariance, *Review of Educational Research,* **42:**237–288.

———, and Julian C. Stanley, 1970: "Statistical Methods in Education and Psychology," Prentice-Hall, Inc., Englewood Cliffs, N.J.

Gorsuch, Richard L., 1974: "Factor Analysis," W. B. Saunders Company, Philadelphia.

Gulliksen, H., 1950: "Theory of Mental Tests," John Wiley & Sons, Inc., New York.

Harman, Harry H., 1967: "Modern Factor Analysis," University of Chicago Press, Chicago.

Harris, Richard J., 1975: "A Primer of Multivariate Statistics," Academic Press, Inc., New York.

Herzberg, Paul A., 1969: The Parameters of Cross-Validation, *Psychometrika, Monograph Supplement,* no. 16.

Holzinger, Karl J., and Frances Swineford, 1937: The Bi-factor Method, *Psychometrika,* **2:**41–54.

Hotelling, H., 1935: The Most Predictable Criterion, *Journal of Educational Psychology,* **26:**139–142.

Huynh, H., 1978: Some Approximate Tests for Repeated Measurement Designs, *Psychometrika,* **43:**161–175.

———, and L. S. Feldt, 1970: Conditions under which Mean Square Ratios in Repeated Measurement Designs Have Exact F-distributions, *Journal of the American Statistical Association,* **65:**1582–1589.

———, and L. S. Feldt, 1976: Estimation of the Box Correction for Degrees of Freedom from Sample Data in Randomized Block and Split-Plot Designs, *Journal of Educational Statistics,* **1:**69–82.

———, and G. K. Mandeville, 1979: Validity Conditions in Repeated Measures Designs, *Psychological Bulletin,* **86:**964–973.

Jackson, R. W. B., and George A. Ferguson, 1941: "Studies in the Reliability of Tests," Bulletin 12, University of Toronto, Department of Educational Research, Toronto.

Johnson, Palmer O., 1949: "Statistical Methods in Research." Prentice-Hall, Inc., Englewood Cliffs, N.J.

Kaiser, Henry F., 1958: The Varimax Criterion for Analytic Rotation in Factor Analysis, *Psychometrika,* **23:**187–200.

______, 1960: Directional Statistical Decisions, *Psychological Review,* **67:**160–167.

Kendall, M. G., 1952: "The Advanced Theory of Statistics," vol. 1, 5th ed., Charles Griffin & Company, Ltd., London.

______, and Allan Stuart, 1968–1973: "The Advanced Theory of Statistics," 3 vols, Hafner Publishing Company, Inc., New York.

______, 1970: "Rank Correlation Methods," 4th ed., Charles Griffin & Company, Ltd., London.

Keppel, Geoffrey, 1973: "Design and Analysis: A Researcher's Handbook," Prentice-Hall, Inc., Englewood Cliffs, N.J.

Kerlinger, Fred N., and Elazar J. Pedhazur, 1973: "Multiple Regression in Behavioral Research," Holt, Rinehart and Winston, Inc., New York.

Keuls, M., 1952: The Use of the Studentized Range in Connection with the Analysis of Variance, *Euphytica,* **1:**112–122 (not seen).

Kruskal, William H., and Judith M. Tanur (eds.), 1978: "International Encyclopedia of Statistics," The Free Press, New York.

______, and W. A. Wallis, 1952: Use of Ranks in One-Criterion Variance Analysis, *Journal of the American Statistical Association,* **47:**583–621.

Kuder, G. F., and M. W. Richardson, 1937: The Theory and Estimation of Test Reliability, *Psychometrika,* **2:**151–160.

Lawley, D. N., 1940: The Estimation of Factor Loadings by the Method of Maximum Likelihood, *Proceedings of the Royal Society of Edinburgh,* **60:**44–82.

______, and A. E. Maxwell, 1963: "Factor Analysis as a Statistical Method," Butterworth & Co., Ltd., London.

Lord, F. M., 1955a: Estimating Test Reliability, *Educational and Psychological Measurement,* **15:**325–336.

______, 1955b: Sampling Fluctuations Resulting from the Sampling of Test Items, *Psychometrika,* **20:**1–22.

______, 1967: A Paradox in the Interpretation of Group Comparisons, *Psychological Bulletin,* **68:**304–305.

______, 1969: Statistical Adjustments when Comparing Preexisting Groups, *Psychological Bulletin,* **72:**336–337.

______, and Melvin R. Novick, 1968: "Statistical Theories of Mental Test Scores," Addison-Wesley Publishing Company, Inc., Reading, Mass.

Magnusson, David, 1967: "Test Theory," Addison-Wesley Publishing Company, Inc., Reading, Mass.

Mann, H. B., and D. R. Whitney, 1947: On a Test of Whether One of Two Random Variables is Stochastically Larger than the Other, *Annals of Mathematical Statistics,* **18:**50–60.

McNemar, Quinn, 1947: Note on the Sampling Error of the Differences between Correlated Proportions or Percentages, *Psychometrika,* **12:**153–157.

______, 1969: "Psychological Statistics," 4th ed., John Wiley & Sons, Inc., New York.

Mulaik, S. A., 1972: "The Foundations of Factor Analysis," McGraw-Hill Book Company, New York.

Myers, Jerome L., 1979: "Fundamentals of Experimental Design," 3rd ed. Allyn and Bacon, Inc., Boston.

Newman, D., 1939: The Distribution of the Range in Samples from a Normal Population in terms of an Independent Estimate of Standard Deviation, *Biometrika* **31:**20–30.

Novick, M. R., and P. H. Jackson, 1974: "Statistical Methods for Educational and Psychological Research," McGraw-Hill Book Company, New York.

Rankin, N. O. 1974: The Harmonic Mean Method for One-Way and Two-Way Analyses of Variance, *Biometrika,* **61:**117–129.

Scheffé, H., 1953: A Method for Judging All Contrasts in the Analysis of Variance, *Biometrika,* **40:**87–104.

———, 1959: "The Analysis of Variance," John Wiley & Sons, Inc., New York.

Siegel, Sidney, 1956: "Nonparametric Statistics," McGraw-Hill Book Company, New York.

Smith, H. Fairfield, 1957: Interpretation of Adjusted Treatment Means and Regression in Analysis of Covariance, *Biometrics,* 13:282–305.

Stavig, Gordon R., 1978: The Median Controversy, *Perceptual and Motor Skills,* **47:**1177–1178.

———, and Jean D. Gibbons, 1977: Comparing the Mean and the Median as Measures of Centrality, *International Statistical Review,* **45:**63–70.

Thomson, Godfrey, H., 1950: "The Factorial Analysis of Human Ability," 5th ed., University of London Press, Ltd., London.

Thorndike, Robert M., 1978: "Correlational Procedures for Research," Gardner Press, Inc., New York.

Thurstone, L. L., 1947: "Multiple Factor Analysis," The University of Chicago Press, Chicago.

Tukey, John W., 1949: Comparing Individual Means in the Analysis of Variance, *Biometrics,* **5:**99–114.

———, 1977: "Exploratory Data Analysis," Addison-Wesley Publishing Company, Reading, Mass.

van Dantzig, David, 1978: Exploratory Data Analysis, in William A. Kruskal and Judith M. Tanur (eds.), *International Encyclopedia of Statistics,* The Free Press, New York, 97–107.

Weisberg, Herbert I., 1979: Statistical Adjustments and Uncontrolled Studies, *Psychological Bulletin,* **86:**1149–1163.

Welch, B. L., 1938: The Significance of the Differences between Two Means when the Population Variances Are Unequal, *Biometrika,* **29:**350–362.

Wilk, M. B., and O. Kempthorne, 1955: Fixed, Mixed, and Random Models, *Journal of the American Statistical Association,* **50:**1144–1167.

Wilkenson, Leland, 1979: Tests of Significance in Stepwise Regression, *Psychological Bulletin,* **86:**168–174.

Winer, B. J., 1971: "Statistical Principles in Experimental Design," 2d ed., McGraw-Hill Book Company, New York.

Woo, T. L., 1928: Dextrality and Sinistrality of Hand and Eye, 2d memoir, *Biometrika,* **20A:**79–148.

APPENDIX TABLES

Table A

Ordinates and areas of the normal curve (In terms of σ units)

$\frac{x}{\sigma}$	*Area*	*Ordinate*	$\frac{x}{\sigma}$	*Area*	*Ordinate*	$\frac{x}{\sigma}$	*Area*	*Ordinate*
00	.0000	.3989	.50	.1915	.3521	1.00	.3413	.2420
.01	.0040	.3989	.51	.1950	.3503	1.01	.3438	.2396
.02	.0080	.3989	.52	.1985	.3485	1.02	.3461	.2371
.03	.0120	.3988	.53	.2019	.3467	1.03	.3485	.2347
.04	.0160	.3986	.54	.2054	.3448	1.04	.3508	.2323
.05	.0199	.3984	.55	.2088	.3429	1.05	.3531	.2299
.06	.0239	.3982	.56	.2123	.3410	1.06	.3554	.2275
.07	.0279	.3980	.57	.2157	.3391	1.07	.3577	.2251
.08	.0319	.3977	.58	.2190	.3372	1.08	.3599	.2227
.09	.0359	.3973	.59	.2224	.3352	1.09	.3621	.2203
.10	.0398	.3970	.60	.2257	.3332	1.10	.3643	.2179
.11	.0438	.3965	.61	.2291	.3312	1.11	.3665	.2155
.12	.0478	.3961	.62	.2324	.3292	1.12	.3686	.2131
.13	.0517	.3956	.63	.2357	.3271	1.13	.3708	.2107
.14	.0557	.3951	.64	.2389	.3251	1.14	.3729	.2083
.15	.0596	.3945	.65	.2422	.3230	1.15	.3749	.2059
.16	.0636	.3939	.66	.2454	.3209	1.16	.3770	.2036
.17	.0675	.3932	.67	.2486	.3187	1.17	.3790	.2012
.18	.0714	.3925	.68	.2517	.3166	1.18	.3810	.1989
.19	.0753	.3918	.69	.2549	.3144	1.19	.3830	.1965
.20	.0793	.3910	.70	.2580	.3123	1.20	.3849	.1942
.21	.0832	.3902	.71	.2611	.3101	1.21	.3869	.1919
.22	.0871	.3894	.72	.2642	.3079	1.22	.3888	.1895
.23	.0910	.3885	.73	.2673	.3056	1.23	.3907	.1872
.24	.0948	.3876	.74	.2703	.3034	1.24	.3925	.1849
.25	.0987	.3867	.75	.2734	.3011	1.25	.3944	.1826
.26	.1026	.3857	.76	.2764	.2989	1.26	.3962	.1804
.27	.1064	.3847	.77	.2794	.2966	1.27	.3980	.1781
.28	.1103	.3836	.78	.2823	.2943	1.28	.3997	.1758
.29	.1141	.3825	.79	.2852	.2920	1.29	.4015	.1736
.30	.1179	.3814	.80	.2881	.2897	1.30	.4032	.1714
.31	.1217	.3802	.81	.2910	.2874	1.31	.4049	.1691
.32	.1255	.3790	.82	.2939	.2850	1.32	.4066	.1669
.33	.1293	.3778	.83	.2967	.2827	1.33	.4082	.1647
.34	.1331	.3765	.84	.2995	.2803	1.34	.4099	.1626
.35	.1368	.3752	.85	.3023	.2780	1.35	.4115	.1604
.36	.1406	.3739	.86	.3051	.2756	1.36	.4131	.1582
.37	.1443	.3725	.87	.3078	.2732	1.37	.4147	.1561
.38	.1480	.3712	.88	.3106	.2709	1.38	.4162	.1539
.39	.1517	.3697	.89	.3133	.2685	1.39	.4177	.1518
.40	.1554	.3683	.90	.3159	.2661	1.40	.4192	.1497
.41	.1591	.3668	.91	.3186	.2637	1.41	.4207	.1476
.42	.1628	.3653	.92	.3212	.2613	1.42	.4222	.1456
.43	.1664	.3637	.93	.3238	.2589	1.43	.4236	.1435
.44	.1700	.3621	.94	.3264	.2565	1.44	.4251	.1415
.45	.1736	.3605	.95	.3289	.2541	1.45	.4265	.1394
.46	.1772	.3589	.96	.3315	.2516	1.46	.4279	.1374
.47	.1808	.3572	.97	.3340	.2492	1.47	.4292	.1354
.48	.1844	.3555	.98	.3365	.2468	1.48	.4306	.1334
.49	.1879	.3538	.99	.3389	.2444	1.49	.4319	.1315
.50	.1915	.3521	1.00	.3413	.2420	1.50	.4332	.1295

FROM: J. E. Wert, *Educational statistics,* by courtesy of McGraw-Hill Book Company, New York.

Table A (*Continued*)

$\frac{x}{\sigma}$	*Area*	*Ordinate*	$\frac{x}{\sigma}$	*Area*	*Ordinate*	$\frac{x}{\sigma}$	*Area*	*Ordinate*
1.50	.4332	.1295	2.00	.4772	.0540	2.50	.4938	.0175
1.51	.4345	.1276	2.01	.4778	.0529	2.51	.4940	.0171
1.52	.4357	.1257	2.02	.4783	.0519	2.52	.4941	.0167
1.53	.4370	.1238	2.03	.4788	.0508	2.53	.4943	.0163
1.54	.4382	.1219	2.04	.4793	.0498	2.54	.4945	.0158
1.55	.4394	.1200	2.05	.4798	.0488	2.55	.4946	.0154
1.56	.4406	.1182	2.06	.4803	.0478	2.56	.4948	.0151
1.57	.4418	.1163	2.07	.4808	.0468	2.57	.4949	.0147
1.58	.4429	.1145	2.08	.4812	.0459	2.58	.4951	.0143
1.59	.4441	.1127	2.09	.4817	.0449	2.59	.4952	.0139
1.60	.4452	.1109	2.10	.4821	.0440	2.60	.4953	.0136
1.61	.4463	.1092	2.11	.4826	.0431	2.61	.4955	.0132
1.62	.4474	.1074	2.12	.4830	.0422	2.62	.4956	.0129
1.63	.4484	.1057	2.13	.4834	.0413	2.63	.4957	.0126
1.64	.4495	.1040	2.14	.4838	.0404	2.64	.4959	.0122
1.65	.4505	.1023	2.15	.4842	.0395	2.65	.4960	.0119
1.66	.4515	.1006	2.16	.4846	.0387	2.66	.4961	.0116
1.67	.4525	.0989	2.17	.4850	.0379	2.67	.4962	.0113
1.68	.4535	.0973	2.18	.4854	.0371	2.68	.4963	.0110
1.69	4545	.0957	2.19	.4857	.0363	2.69	.4964	.0107
1.70	.4554	.0940	2.20	.4861	.0355	2.70	.4965	.0104
1.71	.4564	.0925	2.21	.4864	.0347	2.71	.4966	.0101
1.72	.4573	.0909	2.22	.4868	.0339	2.72	.4967	.0099
1.73	.4582	.0893	2.23	.4871	.0332	2.73	.4968	.0096
1.74	.4591	.0878	2.24	.4875	.0325	2.74	.4969	.0093
1.75	.4599	.0863	2.25	.4878	.0317	2.75	.4970	.0091
1.76	.4608	.0848	2.26	.4881	.0310	2.76	.4971	.0088
1.77	.4616	.0833	2.27	.4884	.0303	2.77	.4972	.0086
1.78	.4625	.0818	2.28	.4887	.0297	2.78	.4973	.0084
1.79	.4633	.0804	2.29	.4890	.0290	2.79	.4974	.0081
1.80	.4641	.0790	2.30	.4893	.0283	2.80	.4974	.0079
1.81	.4649	.0775	2.31	.4896	.0277	2.81	.4975	.0077
1.82	.4656	.0761	2.32	.4898	.0270	2.82	.4976	.0075
1.83	.4664	.0748	2.33	.4901	.0264	2.83	.4977	.0073
1.84	.4671	.0734	2.34	.4904	.0258	2.84	.4977	.0071
1.85	.4678	.0721	2.35	.4906	.0252	2.85	.4978	.0069
1.86	.4686	.0707	2.36	.4909	.0246	2.86	.4979	.0067
1.87	.4693	.0694	2.37	.4911	.0241	2.87	.4979	.0065
1.88	.4699	.0681	2.38	.4913	.0235	2.88	.4980	.0063
1.89	.4706	.0669	2.39	.4916	.0229	2.89	.4981	.0061
1.90	.4713	.0656	2.40	.4918	.0224	2.90	.4981	.0060
1.91	.4719	.0644	2.41	.4920	.0219	2.91	.4982	.0058
1.92	.4726	.0632	2.42	.4922	.0213	2.92	.4982	.0056
1.93	.4732	.0620	2.43	.4925	.0208	2.93	.4983	.0055
1.94	.4738	.0608	2.44	.4927	.0203	2.94	.4984	.0053
1.95	.4744	.0596	2.45	.4929	.0198	2.95	.4984	.0051
1.96	.4750	.0584	2.46	.4931	.0194	2.96	.4985	.0050
1.97	.4756	.0573	2.47	.4932	.0189	2.97	.4985	.0048
1.98	.4761	.0562	2.48	.4934	.0184	2.98	.4986	.0047
1.99	.4767	.0551	2.49	.4936	.0180	2.99	.4986	.0046
2.00	.4772	.0540	2.50	.4938	.0175	3.00	.4987	.0044

Table B

Critical values of *t*

	Level of significance for one-tailed test					
	.10	.05	.025	.01	.005	.0005
	Level of significance for two-tailed test					
df	.20	.10	.05	.02	.01	.001
1	3.078	6.314	12.706	31.821	63.657	636.619
2	1.886	2.920	4.303	6.965	9.925	31.598
3	1.638	2.353	3.182	4.541	5.841	12.941
4	1.533	2.132	2.776	3.747	4.604	8.610
5	1.476	2.015	2.571	3.365	4.032	6.859
6	1.440	1.943	2.447	3.143	3.707	5.959
7	1.415	1.895	2.365	2.998	3.499	5.405
8	1.397	1.860	2.306	2.896	3.355	5.041
9	1.383	1.833	2.262	2.821	3.250	4.781
10	1.372	1.812	2.228	2.764	3.169	4.587
11	1.363	1.796	2.201	2.718	3.106	4.437
12	1.356	1.782	2.179	2.681	3.055	4.318
13	1.350	1.771	2.160	2.650	3.012	4.221
14	1.345	1.761	2.145	2.624	2.977	4.140
15	1.341	1.753	2.131	2.602	2.947	4.073
16	1.337	1.746	2.120	2.583	2.921	4.015
17	1.333	1.740	2.110	2.567	2.898	3.965
18	1.330	1.734	2.101	2.552	2.878	3.922
19	1.328	1.729	2.093	2.539	2.861	3.883
20	1.325	1.725	2.086	2.528	2.845	3.850
21	1.323	1.721	2.080	2.518	2.831	3.819
22	1.321	1.717	2.074	2.508	2.819	3.792
23	1.319	1.714	2.069	2.500	2.807	3.767
24	1.318	1.711	2.064	2.492	2.797	3.745
25	1.316	1.708	2.060	2.485	2.787	3.725
26	1.315	1.706	2.056	2.479	2.779	3.707
27	1.314	1.703	2.052	2.473	2.771	3.690
28	1.313	1.701	2.048	2.467	2.763	3.674
29	1.311	1.699	2.045	2.462	2.756	3.659
30	1.310	1.697	2.042	2.457	2.750	3.646
40	1.303	1.684	2.021	2.423	2.704	3.551
60	1.296	1.671	2.000	2.390	2.660	3.460
120	1.289	1.658	1.980	2.358	2.617	3.373
∞	1.282	1.645	1.960	2.326	2.576	3.291

FROM: Table III of R. A. Fisher and F. Yates, *Statistical tables for biological, agricultural, and medical research,* published by Oliver & Boyd Ltd., Edinburgh, by permission of the authors and the publishers.

Table C

Critical values of chi square

Probability under H_0 that $\chi^2 \geq$ chi square

df	.99	.98	.95	.90	.80	.70	.50	.30	.20	.10	.05	.02	.01	.001
1	.00016	.00063	.0039	.016	.064	.15	.46	1.07	1.64	2.71	3.84	5.41	6.64	10.83
2	.02	.04	.10	.21	.45	.71	1.39	2.41	3.22	4.60	5.99	7.82	9.21	13.82
3	.12	.18	.35	.58	1.00	1.42	2.37	3.66	4.64	6.25	7.82	9.84	11.34	16.27
4	.30	.43	.71	1.06	1.65	2.20	3.36	4.88	5.99	7.78	9.49	11.67	13.28	18.46
5	.55	.75	1.14	1.61	2.34	3.00	4.35	6.06	7.29	9.24	11.07	13.39	15.09	20.52
6	.87	1.13	1.64	2.20	3.07	3.83	5.35	7.23	8.56	10.64	12.59	15.03	16.81	22.46
7	1.24	1.56	2.17	2.83	3.82	4.67	6.35	8.38	9.80	12.02	14.07	16.62	18.48	24.32
8	1.65	2.03	2.73	3.49	4.59	5.53	7.34	9.52	11.03	13.36	15.51	18.17	20.09	26.12
9	2.09	2.53	3.32	4.17	5.38	6.39	8.34	10.66	12.24	14.68	16.92	19.68	21.67	27.88
10	2.56	3.06	3.94	4.86	6.18	7.27	9.34	11.78	13.44	15.99	18.31	21.16	23.21	29.59
11	3.05	3.61	4.58	5.58	6.99	8.15	10.34	12.90	14.63	17.28	19.68	22.62	24.72	31.26
12	3.57	4.18	5.23	6.30	7.81	9.03	11.34	14.01	15.81	18.55	21.03	24.05	26.22	32.91
13	4.11	4.76	5.89	7.04	8.63	9.93	12.34	15.12	16.98	19.81	22.36	25.47	27.69	34.53
14	4.66	5.37	6.57	7.79	9.47	10.82	13.34	16.22	18.15	21.06	23.68	26.87	29.14	36.12
15	5.23	5.98	7.26	8.55	10.31	11.72	14.34	17.32	19.31	22.31	25.00	28.26	30.58	37.70
16	5.81	6.61	7.96	9.31	11.15	12.62	15.34	18.42	20.46	23.54	26.30	29.63	32.00	39.29
17	6.41	7.26	8.67	10.08	12.00	13.53	16.34	19.51	21.62	24.77	27.59	31.00	33.41	40.75
18	7.02	7.91	9.39	10.86	12.86	14.44	17.34	20.60	22.76	25.99	28.87	32.35	34.80	42.31
19	7.63	8.57	10.12	11.65	13.72	15.35	18.34	21.69	23.90	27.20	30.14	33.69	36.19	43.82
20	8.26	9.24	10.85	12.44	14.58	16.27	19.34	22.78	25.04	28.41	31.41	35.02	37.57	45.32
21	8.90	9.92	11.59	13.24	15.44	17.18	20.34	23.86	26.17	29.62	32.67	36.34	38.93	46.80
22	9.54	10.60	12.34	14.04	16.31	18.10	21.34	24.94	27.30	30.81	33.92	37.66	40.29	48.27
23	10.20	11.29	13.09	14.85	17.19	19.02	22.34	26.02	28.43	32.01	35.17	38.97	41.64	49.73
24	10.86	11.99	13.85	15.66	18.06	19.94	23.34	27.10	29.55	33.20	36.42	40.27	42.98	51.18
25	11.52	12.70	14.61	16.47	18.94	20.87	24.34	28.17	30.68	34.38	37.65	41.57	44.31	52.62
26	12.20	13.41	15.38	17.29	19.82	21.79	25.34	29.25	31.80	35.56	38.88	42.86	45.64	54.05
27	12.88	14.12	16.15	18.11	20.70	22.72	26.34	30.32	32.91	36.74	40.11	44.14	46.96	55.48
28	13.56	14.85	16.93	18.94	21.59	23.65	27.34	31.39	34.03	37.92	41.34	45.42	48.28	56.89
29	14.26	15.57	17.71	19.77	22.48	24.58	28.34	32.46	35.14	39.09	42.56	46.69	49.59	58.30
30	14.95	16.31	18.49	20.60	23.36	25.51	29.34	33.53	36.25	40.26	43.77	47.96	50.89	59.70

FROM: Table IV of R. A. Fisher and F. Yates: *Statistical tables for biological, agricultural, and medical research,* published by Oliver & Boyd, Ltd., Edinburgh, by permission of the authors and the publishers.

Table D
Critical values of the correlation coefficient

	Level of significance for one-tailed test			
	.05	**.025**	**.01**	**.005**
	Level of significance for two-tailed test			
df	**.10**	**.05**	**.02**	**.01**
1	.988	.997	.9995	.9999
2	.900	.950	.980	.990
3	.805	.878	.934	.959
4	.729	.811	.882	.917
5	.669	.754	.833	.874
6	.622	.707	.789	.834
7	.582	.666	.750	.798
8	.549	.632	.716	.765
9	.521	.602	.685	.735
10	.497	.576	.658	.708
11	.476	.553	.634	.684
12	.458	.532	.612	.661
13	.441	.514	.592	.641
14	.426	.497	.574	.623
15	.412	.482	.558	.606
16	.400	.468	.542	.590
17	.389	.456	.528	.575
18	.378	.444	.516	.561
19	.369	.433	.503	.549
20	.360	.423	.492	.537
21	.352	.413	.482	.526
22	.344	.404	.472	.515
23	.337	.396	.462	.505
24	.330	.388	.453	.496
25	.323	.381	.445	.487
26	.317	.374	.437	.479
27	.311	.367	.430	.471
28	.306	.361	.423	.463
29	.301	.355	.416	.456
30	.296	.349	.409	.449
35	.275	.325	.381	.418
40	.257	.304	.358	.393
45	.243	.288	.338	.372
50	.231	.273	.322	.354
60	.211	.250	.295	.325
70	.195	.232	.274	.303
80	.183	.217	.256	.283
90	.173	.205	.242	.267
100	.164	.195	.230	.254

FROM: R. A. Fisher and F. Yates, *Statistical tables for biological, agricultural, and medical research,* Oliver & Boyd, Ltd., Edinburgh, by permission of the authors and publishers.

Table E

Transformation of r to z_r

r	z_r	r	z_r	r	z_r	r	z_r	r	z_r
.000	.000	.200	.203	.400	.424	.600	.693	.800	1.099
.005	.005	.205	.208	.405	.430	.605	.701	.805	1.113
.010	.010	.210	.213	.410	.436	.610	.709	.810	1.127
.015	.015	.215	.218	.415	.442	.615	.717	.815	1.142
.020	.020	.220	.224	.420	.448	.620	.725	.820	1.157
.025	.025	.225	.229	.425	.454	.625	.733	.825	1.172
.030	.030	.230	.234	.430	.460	.630	.741	.830	1.188
.035	.035	.235	.239	.435	.466	.635	.750	.835	1.204
.040	.040	.240	.245	.440	.472	.640	.758	.840	1.221
.045	.045	.245	.250	.445	.478	.645	.767	.845	1.238
.050	.050	.250	.255	.450	.485	.650	.775	.850	1.256
.055	.055	.255	.261	.455	.491	.655	.784	.855	1.274
.060	.060	.260	.266	.460	.497	.660	.793	.860	1.293
.065	.065	.265	.271	.465	.504	.665	.802	.865	1.313
.070	.070	.270	.277	.470	.510	.670	.811	.870	1.333
.075	.075	.275	.282	.475	.517	.675	.820	.875	1.354
.080	.080	.280	.288	.480	.523	.680	.829	.880	1.376
.085	.085	.285	.293	.485	.530	.685	.838	.885	1.398
.090	.090	.290	.299	.490	.536	.690	.848	.890	1.422
.095	.095	.295	.304	.495	.543	.695	.858	.895	1.447
.100	.100	.300	.310	.500	.549	.700	.867	.900	1.472
.105	.105	.305	.315	.505	.556	.705	.877	.905	1.499
.110	.110	.310	.321	.510	.563	.710	.887	.910	1.528
.115	.116	.315	.326	.515	.570	.715	.897	.915	1.557
.120	.121	.320	.332	.520	.576	.720	.908	.920	1.589
.125	.126	.325	.337	.525	.583	.725	.918	.925	1.623
.130	.131	.330	.343	.530	.590	.730	.929	.930	1.658
.135	.136	.335	.348	.535	.597	.735	.940	.935	1.697
.140	.141	.340	.354	.540	.604	.740	.950	.940	1.738
.145	.146	.345	.360	.545	.611	.745	.962	.945	1.783
.150	.151	.350	.365	.550	.618	.750	.973	.950	1.832
.155	.156	.355	.371	.555	.626	.755	.984	.955	1.886
.160	.161	.360	.377	.560	.633	.760	.996	.960	1.946
.165	.167	.365	.383	.565	.640	.765	1.008	.965	2.014
.170	.172	.370	.388	.570	.648	.770	1.020	.970	2.092
.175	.177	.375	.394	.575	.655	.775	1.033	.975	2.185
.180	.182	.380	.400	.580	.662	.780	1.045	.980	2.298
.185	.187	.385	.406	.585	.670	.785	1.058	.985	2.443
.190	.192	.390	.412	.590	.678	.790	1.071	.990	2.647
.195	.198	.395	.418	.595	.685	.795	1.085	.995	2.994

FROM: Allen L. Edwards, *Statistical methods*, 2d ed., Holt, Rinehart, and Winston, Inc., New York, 1967.

Table F

Critical values of F^*

5 percent (roman type) and 1 percent (bold-face type) points for the distribution of F^*

Degrees of freedom for denominator	*Degrees of freedom for numerator* 1	2	3	4	5	6	7	8	9	10	11	12	14	16	20	24	30	40	50	75	100	200	500	∞
1	161	200	216	225	230	234	237	239	241	242	243	244	245	246	248	249	250	251	252	253	253	254	254	254
	4052	**4999**	**5403**	**5625**	**5764**	**5859**	**5928**	**5981**	**6022**	**6056**	**6082**	**6106**	**6142**	**6169**	**6208**	**6234**	**6258**	**6286**	**6302**	**6323**	**6334**	**6352**	**6361**	**6366**
2	18.51	19.00	19.16	19.25	19.30	19.33	19.36	19.37	19.38	19.39	19.40	19.41	19.42	19.43	19.44	19.45	19.46	19.47	19.47	19.48	19.49	19.49	19.50	19.50
	98.49	**99.01**	**99.17**	**99.25**	**99.30**	**99.33**	**99.34**	**99.36**	**99.38**	**99.40**	**99.41**	**99.42**	**99.43**	**99.44**	**99.45**	**99.46**	**99.47**	**99.48**	**99.48**	**99.49**	**99.49**	**99.49**	**99.50**	**99.50**
3	10.13	9.55	9.28	9.12	9.01	8.94	8.88	8.84	8.81	8.78	8.76	8.74	8.71	8.69	8.66	8.64	8.62	8.60	8.58	8.57	8.56	8.54	8.54	8.53
	34.12	**30.81**	**29.46**	**28.71**	**28.24**	**27.91**	**27.67**	**27.49**	**27.34**	**27.23**	**27.13**	**27.05**	**26.92**	**26.83**	**26.69**	**26.60**	**26.50**	**26.41**	**26.35**	**26.27**	**26.23**	**26.18**	**26.14**	**26.12**
4	7.71	6.94	6.59	6.39	6.26	6.16	6.09	6.04	6.00	5.96	5.93	5.91	5.87	5.84	5.80	5.77	5.74	5.71	5.70	5.68	5.66	5.65	5.64	5.63
	21.20	**18.00**	**16.69**	**15.98**	**15.52**	**15.21**	**14.98**	**14.80**	**14.66**	**14.54**	**14.45**	**14.37**	**14.24**	**14.15**	**14.02**	**13.93**	**13.83**	**13.74**	**13.69**	**13.61**	**13.57**	**13.52**	**13.48**	**13.46**
5	6.61	5.79	5.41	5.19	5.05	4.95	4.88	4.82	4.78	4.74	4.70	4.68	4.64	4.60	4.56	4.53	4.50	4.46	4.44	4.42	4.40	4.38	4.37	4.36
	16.26	**13.27**	**12.06**	**11.39**	**10.97**	**10.67**	**10.45**	**10.27**	**10.15**	**10.05**	**9.96**	**9.89**	**9.77**	**9.68**	**9.55**	**9.47**	**9.38**	**9.29**	**9.24**	**9.17**	**9.13**	**9.07**	**9.04**	**9.02**
6	5.99	5.14	4.76	4.53	4.39	4.28	4.21	4.15	4.10	4.06	4.03	4.00	3.96	3.92	3.87	3.84	3.81	3.77	3.75	3.72	3.71	3.69	3.68	3.67
	13.74	**10.92**	**9.78**	**9.15**	**8.75**	**8.47**	**8.26**	**8.10**	**7.98**	**7.87**	**7.79**	**7.72**	**7.60**	**7.52**	**7.39**	**7.31**	**7.23**	**7.14**	**7.09**	**7.02**	**6.99**	**6.94**	**6.90**	**6.88**
7	5.59	4.74	4.35	4.12	3.97	3.87	3.79	3.73	3.68	3.63	3.60	3.57	3.52	3.49	3.44	3.41	3.38	3.34	3.32	3.29	3.28	3.25	3.24	3.23
	12.25	**9.55**	**8.45**	**7.85**	**7.46**	**7.19**	**7.00**	**6.84**	**6.71**	**6.62**	**6.54**	**6.47**	**6.35**	**6.27**	**6.15**	**6.07**	**5.98**	**5.90**	**5.85**	**5.78**	**5.75**	**5.70**	**5.67**	**5.65**
8	5.32	4.46	4.07	3.84	3.69	3.58	3.50	3.44	3.39	3.34	3.31	3.28	3.23	3.20	3.15	3.12	3.08	3.05	3.03	3.00	2.98	2.96	2.94	2.93
	11.26	**8.65**	**7.59**	**7.01**	**6.63**	**6.37**	**6.19**	**6.03**	**5.91**	**5.82**	**5.74**	**5.67**	**5.56**	**5.48**	**5.36**	**5.28**	**5.20**	**5.11**	**5.06**	**5.00**	**4.96**	**4.91**	**4.88**	**4.86**
9	5.12	4.26	3.86	3.63	3.48	3.37	3.29	3.23	3.18	3.13	3.10	3.07	3.02	2.98	2.93	2.90	2.86	2.82	2.80	2.77	2.76	2.73	2.72	2.71
	10.56	**8.02**	**6.99**	**6.42**	**6.06**	**5.80**	**5.62**	**5.47**	**5.35**	**5.26**	**5.18**	**5.11**	**5.00**	**4.92**	**4.80**	**4.73**	**4.64**	**4.56**	**4.51**	**4.45**	**4.41**	**4.36**	**4.33**	**4.31**
10	4.96	4.10	3.71	3.48	3.33	3.22	3.14	3.07	3.02	2.97	2.94	2.91	2.86	2.82	2.77	2.74	2.70	2.67	2.64	2.61	2.59	2.56	2.55	2.54
	10.04	**7.56**	**6.55**	**5.99**	**5.64**	**5.39**	**5.21**	**5.06**	**4.95**	**4.85**	**4.78**	**4.71**	**4.60**	**4.52**	**4.41**	**4.33**	**4.25**	**4.17**	**4.12**	**4.05**	**4.01**	**3.96**	**3.93**	**3.91**
11	4.84	3.98	3.59	3.36	3.20	3.09	3.01	2.95	2.90	2.86	2.82	2.79	2.74	2.70	2.65	2.61	2.57	2.53	2.50	2.47	2.45	2.42	2.41	2.40
	9.65	**7.20**	**6.22**	**5.67**	**5.32**	**5.07**	**4.88**	**4.74**	**4.63**	**4.54**	**4.46**	**4.40**	**4.29**	**4.21**	**4.10**	**4.02**	**3.94**	**3.86**	**3.80**	**3.74**	**3.70**	**3.66**	**3.62**	**3.60**

Table F *(Continued)*

Degrees of freedom for denominator	*Degrees of freedom for numerator* 1	2	3	4	5	6	7	8	9	10	11	12	14	16	20	24	30	40	50	75	100	200	500	∞
12	4.75 **9.33**	3.88 **6.93**	3.49 **5.95**	3.26 **5.41**	3.11 **5.06**	3.00 **4.82**	2.92 **4.65**	2.85 **4.50**	2.80 **4.39**	2.76 **4.30**	2.72 **4.22**	2.69 **4.16**	2.64 **4.05**	2.60 **3.98**	2.54 **3.86**	2.50 **3.78**	2.46 **3.70**	2.42 **3.61**	2.40 **3.56**	2.36 **3.49**	2.35 **3.46**	2.32 **3.41**	2.31 **3.38**	2.30 **3.36**
13	4.67 **9.07**	3.80 **6.70**	3.41 **5.74**	3.18 **5.20**	3.02 **4.86**	2.92 **4.62**	2.84 **4.44**	2.77 **4.30**	2.72 **4.19**	2.67 **4.10**	2.63 **4.02**	2.60 **3.96**	2.55 **3.85**	2.51 **3.78**	2.46 **3.67**	2.42 **3.59**	2.38 **3.51**	2.34 **3.42**	2.32 **3.37**	2.28 **3.30**	2.26 **3.27**	2.24 **3.21**	2.22 **3.18**	2.21 **3.16**
14	4.60 **8.86**	3.74 **6.51**	3.34 **5.56**	3.11 **5.03**	2.96 **4.69**	2.85 **4.46**	2.77 **4.28**	2.70 **4.14**	2.65 **4.03**	2.60 **3.94**	2.56 **3.86**	2.53 **3.80**	2.48 **3.70**	2.44 **3.62**	2.39 **3.51**	2.35 **3.43**	2.31 **3.34**	2.27 **3.26**	2.24 **3.21**	2.21 **3.14**	2.19 **3.11**	2.16 **3.06**	2.14 **3.02**	2.13 **3.00**
15	4.54 **8.68**	3.68 **6.36**	3.29 **5.42**	3.06 **4.89**	2.90 **4.56**	2.79 **4.32**	2.70 **4.14**	2.64 **4.00**	2.59 **3.89**	2.55 **3.80**	2.51 **3.73**	2.48 **3.67**	2.43 **3.56**	2.39 **3.48**	2.33 **3.36**	2.29 **3.29**	2.25 **3.20**	2.21 **3.12**	2.18 **3.07**	2.15 **3.00**	2.12 **2.97**	2.10 **2.92**	2.08 **2.89**	2.07 **2.87**
16	4.49 **8.53**	3.63 **6.23**	3.24 **5.29**	3.01 **4.77**	2.85 **4.44**	2.74 **4.20**	2.66 **4.03**	2.59 **3.89**	2.54 **3.78**	2.49 **3.69**	2.45 **3.61**	2.42 **3.55**	2.37 **3.45**	2.33 **3.37**	2.28 **3.25**	2.24 **3.18**	2.20 **3.10**	2.16 **3.01**	2.13 **2.96**	2.09 **2.89**	2.07 **2.86**	2.04 **2.80**	2.02 **2.77**	2.01 **2.75**
17	4.45 **8.40**	3.59 **6.11**	3.20 **5.18**	2.96 **4.67**	2.81 **4.34**	2.70 **4.10**	2.62 **3.93**	2.55 **3.79**	2.50 **3.68**	2.45 **3.59**	2.41 **3.52**	2.38 **3.45**	2.33 **3.35**	2.29 **3.27**	2.23 **3.16**	2.19 **3.08**	2.15 **3.00**	2.11 **2.92**	2.08 **2.86**	2.04 **2.79**	2.02 **2.76**	1.99 **2.70**	1.97 **2.67**	1.96 **2.65**
18	4.41 **8.28**	3.55 **6.01**	3.16 **5.09**	2.93 **4.58**	2.77 **4.25**	2.66 **4.01**	2.58 **3.85**	2.51 **3.71**	2.46 **3.60**	2.41 **3.51**	2.37 **3.44**	2.34 **3.37**	2.29 **3.27**	2.25 **3.19**	2.19 **3.07**	2.15 **3.00**	2.11 **2.91**	2.07 **2.83**	2.04 **2.78**	2.00 **2.71**	1.98 **2.68**	1.95 **2.62**	1.93 **2.59**	1.92 **2.57**
19	4.38 **8.18**	3.52 **5.93**	3.13 **5.01**	2.90 **4.50**	2.74 **4.17**	2.63 **3.94**	2.55 **3.77**	2.48 **3.63**	2.43 **3.52**	2.38 **3.43**	2.34 **3.36**	2.31 **3.30**	2.26 **3.19**	2.21 **3.12**	2.15 **3.00**	2.11 **2.92**	2.07 **2.84**	2.02 **2.76**	2.00 **2.70**	1.96 **2.63**	1.94 **2.60**	1.91 **2.54**	1.90 **2.51**	1.88 **2.49**
20	4.35 **8.10**	3.49 **5.85**	3.10 **4.94**	2.87 **4.43**	2.71 **4.10**	2.60 **3.87**	2.52 **3.71**	2.45 **3.56**	2.40 **3.45**	2.35 **3.37**	2.31 **3.30**	2.28 **3.23**	2.23 **3.13**	2.18 **3.05**	2.12 **2.94**	2.08 **2.86**	2.04 **2.77**	1.99 **2.69**	1.96 **2.63**	1.92 **2.56**	1.90 **2.53**	1.87 **2.47**	1.85 **2.44**	1.84 **2.42**
21	4.32 **8.02**	3.47 **5.78**	3.07 **4.87**	2.84 **4.37**	2.68 **4.04**	2.57 **3.81**	2.49 **3.65**	2.42 **3.51**	2.37 **3.40**	2.32 **3.31**	2.28 **3.24**	2.25 **3.17**	2.20 **3.07**	2.15 **2.99**	2.09 **2.88**	2.05 **2.80**	2.00 **2.72**	1.96 **2.63**	1.93 **2.58**	1.89 **2.51**	1.87 **2.47**	1.84 **2.42**	1.82 **2.38**	1.81 **2.36**
22	4.30 **7.94**	3.44 **5.72**	3.05 **4.82**	2.82 **4.31**	2.66 **3.99**	2.55 **3.76**	2.47 **3.59**	2.40 **3.45**	2.35 **3.35**	2.30 **3.26**	2.26 **3.18**	2.23 **3.12**	2.18 **3.02**	2.13 **2.94**	2.07 **2.83**	2.03 **2.75**	1.98 **2.67**	1.93 **2.58**	1.91 **2.53**	1.87 **2.46**	1.84 **2.42**	1.81 **2.37**	1.80 **2.33**	1.78 **2.31**
23	4.28 **7.88**	3.42 **5.66**	3.03 **4.76**	2.80 **4.26**	2.64 **3.94**	2.53 **3.71**	2.45 **3.54**	2.38 **3.41**	2.32 **3.30**	2.28 **3.21**	2.24 **3.14**	2.20 **3.07**	2.14 **2.97**	2.10 **2.89**	2.04 **2.78**	2.00 **2.70**	1.96 **2.62**	1.91 **2.53**	1.88 **2.48**	1.84 **2.41**	1.82 **2.37**	1.79 **2.32**	1.77 **2.28**	1.76 **2.26**
24	4.26 **7.82**	3.40 **5.61**	3.01 **4.72**	2.78 **4.22**	2.62 **3.90**	2.51 **3.67**	2.43 **3.50**	2.36 **3.36**	2.30 **3.25**	2.26 **3.17**	2.22 **3.09**	2.18 **3.03**	2.13 **2.93**	2.09 **2.85**	2.02 **2.74**	1.98 **2.66**	1.94 **2.58**	1.89 **2.49**	1.86 **2.44**	1.82 **2.36**	1.80 **2.33**	1.76 **2.27**	1.74 **2.23**	1.73 **2.21**

FROM: G. W. Snedecor and William G. Cochran, *Statistical methods,* 6th ed., Iowa State University Press, Ames, Iowa, 1967.

Table F *(Continued)*

25	4.24	3.38	2.99	2.76	2.60	2.49	2.41	2.34	2.28	2.24	2.20	2.16	2.11	2.06	2.00	1.96	1.92	1.87	1.84	1.80	1.77	1.74	1.72	1.71
	7.77	**5.57**	**4.68**	**4.18**	**3.86**	**3.63**	**3.46**	**3.32**	**3.21**	**3.13**	**3.05**	**2.99**	**2.89**	**2.81**	**2.70**	**2.62**	**2.54**	**2.45**	**2.40**	**2.32**	**2.29**	**2.23**	**2.19**	**2.17**
26	4.22	3.37	2.98	2.74	2.59	2.47	2.39	2.32	2.27	2.22	2.18	2.15	2.10	2.05	1.99	1.95	1.90	1.85	1.82	1.78	1.76	1.72	1.70	1.69
	7.72	**5.53**	**4.64**	**4.14**	**3.82**	**3.59**	**3.42**	**3.29**	**3.17**	**3.09**	**3.02**	**2.96**	**2.86**	**2.77**	**2.66**	**2.58**	**2.50**	**2.41**	**2.36**	**2.28**	**2.25**	**2.19**	**2.15**	**2.13**
27	4.21	3.35	2.96	2.73	2.57	2.46	2.37	2.30	2.25	2.20	2.16	2.13	2.08	2.03	1.97	1.93	1.88	1.84	1.80	1.76	1.74	1.71	1.68	1.67
	7.68	**5.49**	**4.60**	**4.11**	**3.79**	**3.56**	**3.39**	**3.26**	**3.14**	**3.06**	**2.98**	**2.93**	**2.83**	**2.74**	**2.63**	**2.55**	**2.47**	**2.38**	**2.33**	**2.25**	**2.21**	**2.16**	**2.12**	**2.10**
28	4.20	3.34	2.95	2.71	2.56	2.44	2.36	2.29	2.24	2.19	2.15	2.12	2.06	2.02	1.96	1.91	1.87	1.81	1.78	1.75	1.72	1.69	1.67	1.65
	7.64	**5.45**	**4.57**	**4.07**	**3.76**	**3.53**	**3.36**	**3.23**	**3.11**	**3.03**	**2.95**	**2.90**	**2.80**	**2.71**	**2.60**	**2.52**	**2.44**	**2.35**	**2.30**	**2.22**	**2.18**	**2.13**	**2.09**	**2.06**
29	4.18	3.33	2.93	2.70	2.54	2.43	2.35	2.28	2.22	2.18	2.14	2.10	2.05	2.00	1.94	1.90	1.85	1.80	1.77	1.73	1.71	1.68	1.65	1.64
	7.60	**5.42**	**4.54**	**4.04**	**3.73**	**3.50**	**3.33**	**3.20**	**3.08**	**3.00**	**2.92**	**2.87**	**2.77**	**2.68**	**2.57**	**2.49**	**2.41**	**2.32**	**2.27**	**2.19**	**2.15**	**2.10**	**2.06**	**2.03**
30	4.17	3.32	2.92	2.69	2.53	2.42	2.34	2.27	2.21	2.16	2.12	2.09	2.04	1.99	1.93	1.89	1.84	1.79	1.76	1.72	1.69	1.66	1.64	1.62
	7.56	**5.39**	**4.51**	**4.02**	**3.70**	**3.47**	**3.30**	**3.17**	**3.06**	**2.98**	**2.90**	**2.84**	**2.74**	**2.66**	**2.55**	**2.47**	**2.38**	**2.29**	**2.24**	**2.16**	**2.13**	**2.07**	**2.03**	**2.01**
32	4.15	3.30	2.90	2.67	2.51	2.40	2.32	2.25	2.19	2.14	2.10	2.07	2.02	1.97	1.91	1.86	1.82	1.76	1.74	1.69	1.67	1.64	1.61	1.59
	7.50	**5.34**	**4.46**	**3.97**	**3.66**	**3.42**	**3.25**	**3.12**	**3.01**	**2.94**	**2.86**	**2.80**	**2.70**	**2.62**	**2.51**	**2.42**	**2.34**	**2.25**	**2.20**	**2.12**	**2.08**	**2.02**	**1.98**	**1.96**
34	4.13	3.28	2.88	2.65	2.49	2.38	2.30	2.23	2.17	2.12	2.08	2.05	2.00	1.95	1.89	1.84	1.80	1.74	1.71	1.67	1.64	1.61	1.59	1.57
	7.44	**5.29**	**4.42**	**3.93**	**3.61**	**3.38**	**3.21**	**3.08**	**2.97**	**2.89**	**2.82**	**2.76**	**2.66**	**2.58**	**2.47**	**2.38**	**2.30**	**2.21**	**2.15**	**2.03**	**2.04**	**1.98**	**1.94**	**1.91**
36	4.11	3.26	2.86	2.63	2.48	2.36	2.28	2.21	2.15	2.10	2.06	2.03	1.98	1.93	1.87	1.82	1.78	1.72	1.69	1.65	1.62	1.59	1.56	1.55
	7.39	**5.25**	**4.38**	**3.89**	**3.58**	**3.35**	**3.18**	**3.04**	**2.94**	**2.86**	**2.78**	**2.72**	**2.62**	**2.54**	**2.43**	**2.35**	**2.26**	**2.17**	**2.12**	**2.04**	**2.00**	**1.94**	**1.90**	**1.87**
38	4.10	3.25	2.85	2.62	2.46	2.35	2.26	2.19	2.14	2.09	2.05	2.02	1.96	1.92	1.85	1.80	1.76	1.71	1.67	1.63	1.60	1.57	1.54	1.53
	7.35	**5.21**	**4.34**	**3.86**	**3.54**	**3.32**	**3.15**	**3.02**	**2.91**	**2.82**	**2.75**	**2.69**	**2.59**	**2.51**	**2.40**	**2.32**	**2.22**	**2.14**	**2.08**	**2.00**	**1.97**	**1.90**	**1.86**	**1.84**
40	4.08	3.23	2.84	2.61	2.45	2.34	2.25	2.18	2.12	2.07	2.04	2.00	1.95	1.90	1.84	1.79	1.74	1.69	1.66	1.61	1.59	1.55	1.53	1.51
	7.31	**5.18**	**4.31**	**3.83**	**3.51**	**3.29**	**3.12**	**2.99**	**2.88**	**2.80**	**2.73**	**2.66**	**2.56**	**2.49**	**2.37**	**2.29**	**2.20**	**2.11**	**2.05**	**1.97**	**1.94**	**1.88**	**1.84**	**1.81**
42	4.07	3.22	2.83	2.59	2.44	2.32	2.24	2.17	2.11	2.06	2.02	1.99	1.94	1.89	1.82	1.78	1.73	1.68	1.64	1.60	1.57	1.54	1.51	1.49
	7.27	**5.15**	**4.29**	**3.80**	**3.49**	**3.26**	**3.10**	**2.96**	**2.86**	**2.77**	**2.70**	**2.64**	**2.54**	**2.46**	**2.35**	**2.26**	**2.17**	**2.08**	**2.02**	**1.94**	**1.91**	**1.85**	**1.80**	**1.78**
44	4.06	3.21	2.82	2.58	2.43	2.31	2.23	2.16	2.10	2.05	2.01	1.98	1.92	1.88	1.81	1.76	1.72	1.66	1.63	1.58	1.56	1.52	1.50	1.48
	7.24	**5.12**	**4.26**	**3.78**	**3.46**	**3.24**	**3.07**	**2.94**	**2.84**	**2.75**	**2.68**	**2.62**	**2.52**	**2.44**	**2.32**	**2.24**	**2.15**	**2.06**	**2.00**	**1.92**	**1.88**	**1.82**	**1.78**	**1.75**

Table F (*Continued*)

Degrees of freedom for denominator	*Degrees of freedom for numerator* 1	2	3	4	5	6	7	8	9	10	11	12	14	16	20	24	30	40	50	75	100	200	500	∞
46	4.05 **7.21**	3.20 **5.10**	2.81 **4.24**	2.57 **3.76**	2.42 **3.44**	2.30 **3.22**	2.22 **3.05**	2.14 **2.92**	2.09 **2.82**	2.04 **2.73**	2.00 **2.66**	1.97 **2.60**	1.91 **2.50**	1.87 **2.42**	1.80 **2.30**	1.75 **2.22**	1.71 **2.13**	1.65 **2.04**	1.62 **1.98**	1.57 **1.90**	1.54 **1.86**	1.51 **1.80**	1.48 **1.76**	1.46 **1.72**
48	4.04 **7.19**	3.19 **5.08**	2.80 **4.22**	2.56 **3.74**	2.41 **3.42**	2.30 **3.20**	2.21 **3.04**	2.14 **2.90**	2.08 **2.80**	2.03 **2.71**	1.99 **2.64**	1.96 **2.58**	1.90 **2.48**	1.86 **2.40**	1.79 **2.28**	1.74 **2.20**	1.70 **2.11**	1.64 **2.02**	1.61 **1.96**	1.56 **1.68**	1.53 **1.84**	1.50 **1.78**	1.47 **1.73**	1.45 **1.70**
50	4.03 **7.17**	3.18 **5.06**	2.79 **4.20**	2.56 **3.72**	2.40 **3.41**	2.29 **3.18**	2.20 **3.02**	2.13 **2.88**	2.07 **2.78**	2.02 **2.70**	1.98 **2.62**	1.95 **2.56**	1.90 **2.46**	1.85 **2.39**	1.78 **2.26**	1.74 **2.18**	1.69 **2.10**	1.63 **2.00**	1.60 **1.94**	1.55 **1.86**	1.52 **1.82**	1.48 **1.76**	1.46 **1.71**	1.44 **1.68**
55	4.02 **7.12**	3.17 **5.01**	2.78 **4.16**	2.54 **3.68**	2.38 **3.37**	2.27 **3.15**	2.18 **2.98**	2.11 **2.85**	2.05 **2.75**	2.00 **2.66**	1.97 **2.59**	1.93 **2.53**	1.88 **2.43**	1.83 **2.35**	1.76 **2.23**	1.72 **2.15**	1.67 **2.06**	1.61 **1.96**	1.58 **1.90**	1.52 **1.82**	1.50 **1.78**	1.46 **1.71**	1.43 **1.66**	1.41 **1.64**
60	4.00 **7.08**	3.15 **4.98**	2.76 **4.13**	2.52 **3.65**	2.37 **3.34**	2.25 **3.12**	2.17 **2.95**	2.10 **2.82**	2.04 **2.72**	1.99 **2.63**	1.95 **2.56**	1.92 **2.50**	1.86 **2.40**	1.81 **2.32**	1.75 **2.20**	1.70 **2.12**	1.65 **2.03**	1.59 **1.93**	1.56 **1.87**	1.50 **1.79**	1.48 **1.74**	1.44 **1.68**	1.41 **1.63**	1.39 **1.60**
65	3.99 **7.04**	3.14 **4.95**	2.75 **4.10**	2.51 **3.62**	2.36 **3.31**	2.24 **3.09**	2.15 **2.93**	2.08 **2.79**	2.02 **2.70**	1.98 **2.61**	1.94 **2.54**	1.90 **2.47**	1.85 **2.37**	1.30 **2.30**	1.73 **2.18**	1.68 **2.09**	1.63 **2.00**	1.57 **1.90**	1.54 **1.84**	1.49 **1.76**	1.46 **1.71**	1.42 **1.64**	1.39 **1.60**	1.37 **1.56**
70	3.98 **7.01**	3.13 **4.92**	2.74 **4.08**	2.50 **3.60**	2.35 **3.29**	2.23 **3.07**	2.14 **2.91**	2.07 **2.77**	2.01 **2.67**	1.97 **2.59**	1.93 **2.51**	1.89 **2.45**	1.84 **2.35**	1.79 **2.28**	1.72 **2.15**	1.67 **2.07**	1.62 **1.98**	1.56 **1.88**	1.53 **1.82**	1.47 **1.74**	1.45 **1.69**	1.40 **1.62**	1.37 **1.56**	1.35 **1.53**
80	3.96 **6.96**	3.11 **4.88**	2.72 **4.04**	2.48 **3.56**	2.33 **3.25**	2.21 **3.04**	2.12 **2.87**	2.05 **2.74**	1.99 **2.64**	1.95 **2.55**	1.91 **2.48**	1.88 **2.41**	1.82 **2.32**	1.77 **2.24**	1.70 **2.11**	1.65 **2.03**	1.60 **1.94**	1.54 **1.84**	1.51 **1.78**	1.45 **1.70**	1.42 **1.65**	1.38 **1.57**	1.35 **1.52**	1.32 **1.49**
100	3.94 **6.90**	3.09 **4.82**	2.70 **3.98**	2.46 **3.51**	2.30 **3.20**	2.19 **2.99**	2.10 **2.82**	2.03 **2.69**	1.97 **2.59**	1.92 **2.51**	1.88 **2.43**	1.85 **2.36**	1.79 **2.26**	1.75 **2.19**	1.68 **2.06**	1.63 **1.98**	1.57 **1.89**	1.51 **1.79**	1.48 **1.73**	1.42 **1.64**	1.39 **1.59**	1.34 **1.51**	1.30 **1.46**	1.28 **1.43**
125	3.92 **6.84**	3.07 **4.78**	2.68 **3.94**	2.44 **3.47**	2.29 **3.17**	2.17 **2.95**	2.08 **2.79**	2.01 **2.65**	1.95 **2.56**	1.90 **2.47**	1.86 **2.40**	1.83 **2.33**	1.77 **2.23**	1.72 **2.15**	1.65 **2.03**	1.60 **1.94**	1.55 **1.85**	1.49 **1.75**	1.45 **1.68**	1.39 **1.59**	1.36 **1.54**	1.31 **1.46**	1.27 **1.40**	1.25 **1.37**
150	3.91 **6.81**	3.06 **4.75**	2.67 **3.91**	2.43 **3.44**	2.27 **3.14**	2.16 **2.92**	2.07 **2.76**	2.00 **2.62**	1.94 **2.53**	1.89 **2.44**	1.85 **2.37**	1.82 **2.30**	1.76 **2.20**	1.71 **2.12**	1.64 **2.00**	1.59 **1.91**	1.54 **1.83**	1.47 **1.72**	1.44 **1.66**	1.37 **1.56**	1.34 **1.51**	1.29 **1.43**	1.25 **1.37**	1.22 **1.33**
200	3.89 **6.76**	3.04 **4.71**	2.65 **3.88**	2.41 **3.41**	2.26 **3.11**	2.14 **2.90**	2.05 **2.73**	1.98 **2.60**	1.92 **2.50**	1.87 **2.41**	1.83 **2.34**	1.80 **2.28**	1.74 **2.17**	1.69 **2.09**	1.62 **1.97**	1.57 **1.88**	1.52 **1.79**	1.45 **1.69**	1.42 **1.62**	1.35 **1.53**	1.32 **1.48**	1.26 **1.39**	1.22 **1.33**	1.19 **1.28**
400	3.86 **6.70**	3.02 **4.66**	2.62 **3.83**	2.39 **3.36**	2.23 **3.06**	2.12 **2.85**	2.03 **2.69**	1.96 **2.55**	1.90 **2.46**	1.85 **2.37**	1.81 **2.29**	1.78 **2.23**	1.72 **2.12**	1.67 **2.04**	1.60 **1.92**	1.54 **1.84**	1.49 **1.74**	1.42 **1.64**	1.38 **1.57**	1.32 **1.47**	1.28 **1.42**	1.22 **1.32**	1.16 **1.24**	1.13 **1.19**
1000	3.85 **6.66**	3.00 **4.62**	2.61 **3.80**	2.38 **3.34**	2.22 **3.04**	2.10 **2.82**	2.02 **2.66**	1.95 **2.53**	1.89 **2.43**	1.84 **2.34**	1.80 **2.26**	1.76 **2.20**	1.70 **2.09**	1.65 **2.01**	1.58 **1.89**	1.53 **1.81**	1.47 **1.71**	1.41 **1.61**	1.36 **1.54**	1.30 **1.44**	1.26 **1.38**	1.19 **1.28**	1.13 **1.19**	1.08 **1.11**
∞	3.84 **6.64**	2.99 **4.60**	2.60 **3.78**	2.37 **3.32**	2.21 **3.02**	2.09 **2.80**	2.01 **2.64**	1.94 **2.51**	1.88 **2.41**	1.83 **2.32**	1.79 **2.24**	1.75 **2.18**	1.69 **2.07**	1.64 **1.99**	1.57 **1.87**	1.52 **1.79**	1.46 **1.69**	1.40 **1.59**	1.35 **1.52**	1.28 **1.41**	1.24 **1.36**	1.17 **1.25**	1.11 **1.15**	1.00 **1.00**

Table G

Critical values of ρ, the Spearman rank correlation coefficient

	Significance level (one-tailed test)	
N	**.05**	**.01**
4	1.000	
5	.900	1.000
6	.829	.943
7	.714	.893
8	.643	.833
9	.600	.783
10	.564	.746
12	.506	.712
14	.456	.645
16	.425	.601
18	.399	.564
20	.377	.534
22	.359	.508
24	.343	.485
26	.329	.465
28	.317	.448
30	.306	.432

FROM: E. G. Olds, Distributions of sums of squares of rank differences for small numbers of individuals. *Annals of Mathematical Statistics,* **9:**133–148, 1938; The 5% significance levels for sums of squares of rank differences and a correction, *Annals of Mathematical Statistics,* **20:**117–118, 1949; with the kind permission of the author and the publisher.

Table H

Probabilities associated with values as large as observed values of S in the Kendall rank correlation coefficient

	Values of N					*Values of N*		
S	4	5	8	9	S	6	7	10
0	.625	.592	.548	.540	1	.500	.500	.500
2	.375	.408	.452	.460	3	.360	.386	.431
4	.167	.242	.360	.381	5	.235	.281	.364
6	.042	.117	.274	.306	7	.136	.191	.300
8		.042	.199	.238	9	.068	.119	.242
10		.0083	.138	.179	11	.028	.068	.190
12			.089	.130	13	.0083	.035	.146
14			.054	.090	15	.0014	.015	.108
16			.031	.060	17		.0054	.078
18			.016	.038	19		.0014	.054
20			.0071	.022	21		.00020	.036
22			.0028	.012	23			.023
24			.00087	.0063	25			.014
26			.00019	.0029	27			.0083
28			.000025	.0012	29			.0046
30				.00043	31			.0023
32				.00012	33			.0011
34				.000025	35			.00047
36				.0000028	37			.00018
					39			.000058
					41			.000015
					43			.0000028
					45			.00000028

FROM: M. G. Kendall, *Rank correlation methods,* 2d ed., Charles Griffin & Co. Ltd., London, 1955.

Table I

Critical and quasi-critical lower-tail values of W_+ (and their probability levels) for Wilcoxon's signed-rank test

N	$\alpha = .05$		$\alpha = .025$		$\alpha = .01$		$\alpha = .005$	
5	0	.0313						
	1	.0625						
6	2	.0469	0	.0156				
	3	.0781	1	.0313				
7	3	.0391	2	.0234	0	.0078		
	4	.0547	3	.0391	1	.0156		
8	5	.0391	3	.0195	1	.0078	0	.0039
	6	.0547	4	.0273	2	.0117	1	.0078
9	8	.0488	5	.0195	3	.0098	1	.0039
	9	.0645	6	.0273	4	.0137	2	.0059
10	10	.0420	8	.0244	5	.0098	3	.0049
	11	.0527	9	.0322	6	.0137	4	.0068
11	13	.0415	10	.0210	7	.0093	5	.0049
	14	.0508	11	.0269	8	.0122	6	.0068
12	17	.0461	13	.0212	9	.0081	7	.0046
	18	.0549	14	.0261	10	.0105	8	.0061
13	21	.0471	17	.0239	12	.0085	9	.0040
	22	.0549	18	.0287	13	.0107	10	.0052
14	25	.0453	21	.0247	15	.0083	12	.0043
	26	.0520	22	.0290	16	.0101	13	.0054
15	30	.0473	25	.0240	19	.0090	15	.0042
	31	.0535	26	.0277	20	.0108	16	.0051
16	35	.0467	29	.0222	23	.0091	19	.0046
	36	.0523	30	.0253	24	.0107	20	.0055
17	41	.0492	34	.0224	27	.0087	23	.0047
	42	.0544	35	.0253	28	.0101	24	.0055
18	47	.0494	40	.0241	32	.0091	27	.0045
	48	.0542	41	.0269	33	.0104	28	.0052
19	53	.0478	46	.0247	37	.0090	32	.0047
	54	.0521	47	.0273	38	.0102	33	.0054
20	60	.0487	52	.0242	43	.0096	37	.0047
	61	.0527	53	.0266	44	.0107	38	.0053

FROM: Table II in F. Wilcoxon, S. K. Katti, and Roberta A. Wilcox, *Critical values and probability levels for the Wilcoxon rank sum test and the Wilcoxon signed rank test,* American Cyanamid Company (Lederle Laboratories Division, Pearl River, N.Y.) and The Florida State University (Department of Statistics, Tallahassee, Fla.), August, 1963, with permission of the authors and publishers with changes only in notation.

Table I (*Continued*)

N	$\alpha = .05$		$\alpha = .025$		$\alpha = .01$		$\alpha = .005$	
21	67	.0479	58	.0230	49	.0097	42	.0045
	68	.0516	59	.0251	50	.0108	43	.0051
22	75	.0492	65	.0231	55	.0095	48	.0046
	76	.0527	66	.0250	56	.0104	49	.0052
23	83	.0490	73	.0242	62	.0098	54	.0046
	84	.0523	74	.0261	63	.0107	55	.0051
24	91	.0475	81	.0245	69	.0097	61	.0048
	92	.0505	82	.0263	70	.0106	62	.0053
25	100	.0479	89	.0241	76	.0094	68	.0048
	101	.0507	90	.0258	77	.0101	69	.0053
26	110	.0497	98	.0247	84	.0095	75	.0047
	111	.0524	99	.0263	85	.0102	76	.0051
27	119	.0477	107	.0246	92	.0093	83	.0048
	120	.0502	108	.0260	93	.0100	84	.0052
28	130	.0496	116	.0239	101	.0096	91	.0048
	131	.0521	117	.0252	102	.0102	92	.0051
29	140	.0482	126	.0240	110	.0095	100	.0049
	141	.0504	127	.0253	111	.0101	101	.0053
30	151	.0481	137	.0249	120	.0098	109	.0050
	152	.0502	138	.0261	121	.0104	110	.0053
31	163	.0491	147	.0239	130	.0099	118	.0049
	164	.0512	148	.0251	131	.0105	119	.0052
32	175	.0492	159	.0249	140	.0097	128	.0050
	176	.0512	160	.0260	141	.0103	129	.0053
33	187	.0485	170	.0242	151	.0099	138	.0049
	188	.0503	171	.0253	152	.0104	139	.0052
34	200	.0488	182	.0242	162	.0098	148	.0048
	201	.0506	183	.0252	163	.0103	149	.0051
35	213	.0484	195	.0247	173	.0096	159	.0048
	214	.0501	196	.0257	174	.0100	160	.0051
36	227	.0489	208	.0248	185	.0096	171	.0050
	228	.0505	209	.0258	186	.0100	172	.0052
37	241	.0487	221	.0245	198	.0099	182	.0048
	242	.0503	222	.0254	199	.0103	183	.0050
38	256	.0493	235	.0247	211	.0099	194	.0048
	257	.0509	236	.0256	212	.0104	195	.0050
39	271	.0493	249	.0246	224	.0099	207	.0049
	272	.0507	250	.0254	225	.0103	208	.0051
40	286	.0486	264	.0249	238	.0100	220	.0049
	287	.0500	265	.0257	239	.0104	221	.0051

Table I *(Continued)*

N	α = .05		α = .025		α = .01		α = .005	
41	302	.0488	279	.0248	252	.0100	233	.0048
	303	.0501	280	.0256	253	.0103	234	.0050
42	319	.0496	294	.0245	266	.0098	247	.0049
	320	.0509	295	.0252	267	.0102	248	.0051
43	336	.0498	310	.0245	281	.0098	261	.0048
	337	.0511	311	.0252	282	.0102	262	.0050
44	353	.0495	327	.0250	296	.0097	276	.0049
	354	.0507	328	.0257	297	.0101	277	.0051
45	371	.0498	343	.0244	312	.0098	291	.0049
	372	.0510	344	.0251	313	.0101	292	.0051
46	389	.0497	361	.0249	328	.0098	307	.0050
	390	.0508	362	.0256	329	.0101	308	.0052
47	407	.0490	378	.0245	345	.0099	322	.0048
	408	.0501	379	.0251	346	.0102	323	.0050
48	426	.0490	396	.0244	362	.0099	339	.0050
	427	.0500	397	.0251	363	.0102	340	.0051
49	446	.0495	415	.0247	379	.0098	355	.0049
	447	.0505	416	.0253	380	.0100	356	.0050
50	466	.0495	434	.0247	397	.0098	373	.0050
	467	.0506	435	.0253	398	.0101	374	.0051

Table J

Coefficients of orthogonal polynomials

k	*Polynomial*	*Coefficients*										Σc_{ij}^2
3	Linear	−1	0	1								2
	Quadratic	1	−2	1								6
	Linear	−3	−1	1	3							20
4	Quadratic	1	−1	−1	1							4
	Cubic	−1	3	−3	1							20
	Linear	−2	−1	0	1	2						10
5	Quadratic	2	−1	−2	−1	2						14
	Cubic	−1	2	0	−2	1						10
	Quartic	1	−4	6	−4	1						70
	Linear	−5	−3	−1	1	3	5					70
6	Quadratic	5	−1	−4	−4	−1	5					84
	Cubic	−5	7	4	−4	−7	5					180
	Quartic	1	−3	2	2	−3	1					28
	Linear	−3	−2	−1	0	1	2	3				28
7	Quadratic	5	0	−3	−4	−3	0	5				84
	Cubic	−1	1	1	0	−1	−1	1				6
	Quartic	3	−7	1	6	1	−7	3				154
	Linear	−7	−5	−3	−1	1	3	5	7			168
	Quadratic	7	1	−3	−5	−5	−3	1	7			168
8	Cubic	−7	5	7	3	−3	−7	−5	7			264
	Quartic	7	−13	−3	9	9	−3	−13	7			616
	Quintic	−7	23	−17	−15	15	17	−23	7			2184
	Linear	−4	−3	−2	−1	0	1	2	3	4		60
	Quadratic	28	7	−8	−17	−20	−17	−8	7	28		2772
9	Cubic	−14	7	13	9	0	−9	−13	−7	14		990
	Quartic	14	−21	−11	9	18	9	−11	−21	14		2002
	Quintic	−4	11	−4	−9	0	9	4	−11	4		468
	Linear	−9	−7	−5	−3	−1	1	3	5	7	9	330
	Quadratic	6	2	−1	−3	−4	−4	−3	−1	2	6	132
10	Cubic	−42	14	35	31	12	−12	−31	−35	−14	42	8580
	Quartic	18	−22	−17	3	18	18	3	−17	−22	18	2860
	Quintic	−6	14	−1	−11	−6	6	11	1	−14	6	780

Table K

Critical lower-tail values of R_1 for rank test for two independent samples ($N_1 < N_2$)

N_2	$N_1 = 1$: 0.001	0.005	0.010	0.025	0.05	0.10	$2\bar{R}_1$	$N_1 = 2$: 0.001	0.005	0.010	0.025	0.05	0.10	$2\bar{R}_1$	N_2
2							4						—	10	2
3							5						3	12	3
4							6					—	3	14	4
5							7					3	4	16	5
6							8					3	4	18	6
7							9				—	3	4	20	7
8						—	10				3	4	5	22	8
9						1	11				3	4	5	24	9
10						1	12				3	4	6	26	10
11						1	13				3	4	6	28	11
12						1	14			—	4	5	7	30	12
13						1	15			3	4	5	7	32	13
14						1	16			3	4	6	8	34	14
15						1	17			3	4	6	8	36	15
16						1	18			3	4	6	8	38	16
17						1	19			3	5	6	9	40	17
18					—	1	20		—	3	5	7	9	42	18
19					1	2	21		3	4	5	7	10	44	19
20					1	2	22		3	4	5	7	10	46	20
21					1	2	23		3	4	6	8	11	48	21
22					1	2	24		3	4	6	8	11	50	22
23					1	2	25		3	4	6	8	12	52	23
24					1	2	26		3	4	6	9	12	54	24
25	—	—	—	—	1	2	27	—	3	4	6	9	12	56	25

N_2	$N_1 = 3$: 0.001	0.005	0.010	0.025	0.05	0.10	$2\bar{R}_1$	$N_1 = 4$: 0.001	0.005	0.010	0.025	0.05	0.10	$2\bar{R}_1$	N_2
3					6	7	21								
4				—	6	7	24			—	10	11	13	36	4
5				6	7	8	27		—	10	11	12	14	40	5
6			—	7	8	9	30		10	11	12	13	15	44	6
7			6	7	8	10	33		10	11	13	14	16	48	7
8		—	6	8	9	11	36		11	12	14	15	17	52	8
9		6	7	8	10	11	39	—	11	13	14	16	19	56	9
10		6	7	9	10	12	42	10	12	13	15	17	20	60	10
11		6	7	9	11	13	45	10	12	14	16	18	21	64	11
12		7	8	10	11	14	48	10	13	15	17	19	22	68	12
13		7	8	10	12	15	51	11	13	15	18	20	23	72	13
14		7	8	11	13	16	54	11	14	16	19	21	25	76	14
15		8	9	11	13	16	57	11	15	17	20	22	26	80	15
16	—	8	9	12	14	17	60	12	15	17	21	24	27	84	16
17	6	8	10	12	15	18	63	12	16	18	21	25	28	88	17
18	6	8	10	13	15	19	66	13	16	19	22	26	30	92	18
19	6	9	10	13	16	20	69	13	17	19	23	27	31	96	19
20	6	9	11	14	17	21	72	13	18	20	24	28	32	100	20
21	7	9	11	14	17	21	75	14	18	21	25	29	33	104	21
22	7	10	12	15	18	22	78	14	19	21	26	30	35	108	22
23	7	10	12	15	19	23	81	14	19	22	27	31	36	112	23
24	7	10	12	16	19	24	84	15	20	23	27	32	38	116	24
25	7	11	13	16	20	25	87	15	20	23	28	33	38	120	25

FROM: Table I in L. R. Verdooren, Extended tables of critical values for Wilcoxon's test statistic, *Biometrika,* 50, 177–186, 1963, with permission of the author and editor, with changes in notation.

The table shows (i) values, R_1, which are just significant at the probability level quoted at the head of column, (ii) twice the mean $2\bar{R}_1$. A blank space indicates that no value is significant at that level. The upper critical value is $2\bar{R}_1 - R_1$.

Table K (*Continued*)

$N_1 = 5$

N_2	0.001	0.005	0.010	0.025	0.05	0.10	$2\bar{R}_1$
5		15	16	17	19	20	55
6		16	17	18	20	22	60
7	—	16	18	20	21	23	65
8	15	17	19	21	23	25	70
9	16	18	20	22	24	27	75
10	16	19	21	23	26	28	80
11	17	20	22	24	27	30	85
12	17	21	23	26	28	32	90
13	18	22	24	27	30	33	95
14	18	22	25	28	31	35	100
15	19	23	26	29	33	37	105
16	20	24	27	30	34	38	110
17	20	25	28	32	35	40	115
18	21	26	29	33	37	42	120
19	22	27	30	34	38	43	125
20	22	28	31	35	40	45	130
21	23	29	32	37	41	47	135
22	23	29	33	38	43	48	140
23	24	30	34	39	44	50	145
24	25	31	35	40	45	51	150
25	25	32	36	42	47	53	155

$N_1 = 6$

0.001	0.005	0.010	0.025	0.05	0.10	$2\bar{R}_1$	N_2
—	23	24	26	28	30	78	**6**
21	24	25	27	29	32	84	**7**
22	25	27	29	31	34	90	**8**
23	26	28	31	33	36	96	**9**
24	27	29	32	35	38	102	**10**
25	28	30	34	37	40	108	**11**
25	30	32	35	38	42	114	**12**
26	31	33	37	40	44	120	**13**
27	32	34	38	42	46	126	**14**
28	33	36	40	44	48	132	**15**
29	34	37	42	46	50	138	**16**
30	36	39	43	47	52	144	**17**
31	37	40	45	49	55	150	**18**
32	38	41	46	51	57	156	**19**
33	39	43	48	53	59	162	**20**
33	40	44	50	55	61	168	**21**
34	42	45	51	57	63	174	**22**
35	43	47	53	58	65	180	**23**
36	44	48	54	60	67	186	**24**
37	45	50	56	62	69	192	**25**

$N_1 = 7$

N_2	0.001	0.005	0.010	0.025	0.05	0.10	$2\bar{R}_1$
7	29	32	34	36	39	41	105
8	30	34	35	38	41	44	112
9	31	35	37	40	43	46	119
10	33	37	39	42	45	49	126
11	34	38	40	44	47	51	133
12	35	40	42	46	49	54	140
13	36	41	44	48	52	56	147
14	37	43	45	50	54	59	154
15	38	44	47	52	56	61	161
16	39	46	49	54	58	64	168
17	41	47	51	56	61	66	175
18	42	49	52	58	63	69	182
19	43	50	54	60	65	71	189
20	44	52	56	62	67	74	196
21	46	53	58	64	69	76	203
22	47	55	59	66	72	79	210
23	48	57	61	68	74	81	217
24	49	58	63	70	76	84	224
25	50	60	64	72	78	86	231

$N_1 = 8$

0.001	0.005	0.010	0.025	0.05	0.10	$2\bar{R}_1$	N_2
40	43	45	49	51	55	136	**8**
41	45	47	51	54	58	144	**9**
42	47	49	53	56	60	152	**10**
44	49	51	55	59	63	160	**11**
45	51	53	58	62	66	168	**12**
47	53	56	60	64	69	176	**13**
48	54	58	62	67	72	184	**14**
50	56	60	65	69	75	192	**15**
51	58	62	67	72	78	200	**16**
53	60	64	70	75	81	208	**17**
54	62	66	72	77	84	216	**18**
56	64	68	74	80	87	224	**19**
57	66	70	77	83	90	232	**20**
59	68	72	79	85	92	240	**21**
60	70	74	81	88	95	248	**22**
62	71	76	84	90	98	256	**23**
64	73	78	86	93	101	264	**24**
65	75	81	89	96	104	272	**25**

$N_1 = 9$

N_2	0.001	0.005	0.010	0.025	0.05	0.10	$2\bar{R}_1$
9	52	56	59	62	66	70	171
10	53	58	61	65	69	73	180
11	55	61	63	68	72	76	189
12	57	63	66	71	75	80	198
13	59	65	68	73	78	83	207
14	60	67	71	76	81	86	216
15	62	69	73	79	84	90	225
16	64	72	76	82	87	93	234
17	66	74	78	84	90	97	243
18	68	76	81	87	93	100	252
19	70	78	83	90	96	103	261
20	71	81	85	93	99	107	270
21	73	83	88	95	102	110	279
22	75	85	90	98	105	113	288
23	77	88	93	101	108	117	297
24	79	90	95	104	111	120	306
25	81	92	98	107	114	123	315

$N_1 = 10$

0.001	0.005	0.010	0.025	0.05	0.10	$2\bar{R}_1$	N_2
65	71	74	78	82	87	210	**10**
67	73	77	81	86	91	220	**11**
69	76	79	84	89	94	230	**12**
72	79	82	88	92	98	240	**13**
74	81	85	91	96	102	250	**14**
76	84	88	94	99	106	260	**15**
78	86	91	97	103	109	270	**16**
80	89	93	100	103	113	280	**17**
82	92	96	103	110	117	290	**18**
84	94	99	107	113	121	300	**19**
87	97	102	110	117	125	310	**20**
89	99	105	113	120	128	320	**21**
91	102	108	116	123	132	330	**22**
93	105	110	119	127	136	340	**23**
95	107	113	122	130	140	350	**24**
98	110	116	126	134	144	360	**25**

Table K *(Continued)*

N_2				$N_1 = 11$								$N_1 = 12$				N_2
	0.001	0.005	0.010	0.025	0.05	0.10	$2\bar{R}_1$		0.001	0.005	0.010	0.025	0.05	0.10	$2\bar{R}_1$	
11	81	87	91	96	100	106	253									
12	83	90	94	99	104	110	264		98	105	109	115	120	127	300	**12**
13	86	93	97	103	108	114	275		101	109	113	119	125	131	312	**13**
14	88	96	100	106	112	118	286		103	112	116	123	129	136	324	**14**
15	90	99	103	110	116	123	297		106	115	120	127	133	141	336	**15**
16	93	102	107	113	120	127	308		109	119	124	131	138	145	348	**16**
17	95	105	110	117	123	131	319		112	122	127	135	142	150	360	**17**
18	98	108	113	121	127	135	330		115	125	131	139	146	155	372	**18**
19	100	111	116	124	131	139	341		118	129	134	143	150	159	384	**19**
20	103	114	119	128	135	144	352		120	132	138	147	155	164	396	**20**
21	106	117	123	131	139	148	363		123	136	142	151	159	169	408	**21**
22	108	120	126	135	143	152	374		126	139	145	155	163	173	420	**22**
23	111	123	129	139	147	156	385		129	142	149	159	168	178	432	**23**
24	113	126	132	142	151	161	396		132	146	153	163	172	183	444	**24**
25	116	129	136	146	155	165	407		135	149	156	167	176	187	456	**25**

N_2				$N_1 = 13$								$N_1 = 14$				N_2
	0.001	0.005	0.010	0.025	0.05	0.10	$2\bar{R}_1$		0.001	0.005	0.010	0.025	0.05	0.10	$2\bar{R}_1$	
13	117	125	130	136	142	149	351									
14	120	129	134	141	147	154	364		137	147	152	160	166	174	406	**14**
15	123	133	138	145	152	159	377		141	151	156	164	171	179	420	**15**
16	126	136	142	150	156	165	390		144	155	161	169	176	185	434	**16**
17	129	140	146	154	161	170	403		148	159	165	174	182	190	448	**17**
18	133	144	150	158	166	175	416		151	163	170	179	187	196	462	**18**
19	136	148	154	163	171	180	429		155	168	174	183	192	202	476	**19**
20	139	151	158	167	175	185	442		159	172	178	188	197	207	490	**20**
21	142	155	162	171	180	190	455		162	176	183	193	202	213	504	**21**
22	145	159	166	176	185	195	468		166	180	187	198	207	218	518	**22**
23	149	163	170	180	189	200	481		169	184	192	203	212	224	532	**23**
24	152	166	174	185	194	205	494		173	188	196	207	218	229	546	**24**
25	155	170	178	189	199	211	507		177	192	200	212	223	235	560	**25**

N_2				$N_1 = 15$								$N_1 = 16$				N_2
	0.001	0.005	0.010	0.025	0.05	0.10	$2\bar{R}_1$		0.001	0.005	0.010	0.025	0.05	0.10	$2\bar{R}_1$	
15	160	171	176	184	192	200	465									
16	163	175	181	190	197	206	480		184	196	202	211	219	229	528	**16**
17	167	180	186	195	203	212	495		188	201	207	217	225	235	544	**17**
18	171	184	190	200	208	218	510		192	206	212	222	231	242	560	**18**
19	175	189	195	205	214	224	525		196	210	218	228	237	248	576	**19**
20	179	193	200	210	220	230	540		201	215	223	234	243	255	592	**20**
21	183	198	205	216	225	236	555		205	220	228	239	249	261	608	**21**
22	187	202	210	221	231	242	570		209	225	233	245	255	267	624	**22**
23	191	207	214	226	236	248	585		214	230	238	251	261	274	640	**23**
24	195	211	219	231	242	254	600		218	235	244	256	267	280	656	**24**
25	199	216	224	237	248	260	615		222	240	249	262	273	287	672	**25**

N_2				$N_1 = 17$								$N_1 = 18$				N_2
	0.001	0.005	0.010	0.025	0.05	0.10	$2\bar{R}_1$		0.001	0.005	0.010	0.025	0.05	0.10	$2\bar{R}_1$	
17	210	223	230	240	249	259	595									
18	214	228	235	246	255	266	612		237	252	259	270	280	291	666	**18**
19	219	234	241	252	262	273	629		242	258	265	277	287	299	684	**19**
20	223	239	246	258	268	280	646		247	263	271	283	294	306	702	**20**
21	228	244	252	264	274	287	663		252	269	277	290	301	313	720	**21**
22	233	249	258	270	281	294	680		257	275	283	296	307	321	738	**22**
23	238	255	263	276	287	300	697		262	280	289	303	314	328	756	**23**
24	242	260	269	282	294	307	714		267	286	295	309	321	335	774	**24**
25	247	265	275	288	300	314	731		273	292	301	316	328	343	792	**25**

Table K (*Continued*)

N_2	$N_1 = 19$ 0.001	0.005	0.010	0.025	0.05	0.10	$2\bar{R}_1$	$N_1 = 20$ 0.001	0.005	0.010	0.025	0.05	0.10	$2\bar{R}_1$	N_2
19	267	283	291	303	313	325	741								
20	272	289	297	309	320	333	760	298	315	324	337	348	361	820	**20**
21	277	295	303	316	328	341	779	304	322	331	344	356	370	840	**21**
22	283	301	310	323	335	349	798	309	328	337	351	364	378	860	**22**
23	288	307	316	330	342	357	817	315	335	344	359	371	386	880	**23**
24	294	313	323	337	350	364	836	321	341	351	366	379	394	900	**24**
25	299	319	329	344	357	372	855	327	348	358	373	387	403	920	**25**

N_2	$N_1 = 21$ 0.001	0.005	0.010	0.025	0.05	0.10	$2\bar{R}_1$	$N_1 = 22$ 0.001	0.005	0.010	0.025	0.05	0.10	$2\bar{R}_1$	N_2
21	331	349	359	373	385	399	903								
22	337	356	366	381	393	408	924	365	386	396	411	424	439	990	**22**
23	343	363	373	388	401	417	945	372	393	403	419	432	448	1012	**23**
24	349	370	381	396	410	425	966	379	400	411	427	441	457	1034	**24**
25	356	377	388	404	418	434	987	385	408	419	435	450	467	1056	**25**

N_2	$N_1 = 23$ 0.001	0.005	0.010	0.025	0.05	0.10	$2\bar{R}_1$	$N_1 = 24$ 0.001	0.005	0.010	0.025	0.05	0.10	$2\bar{R}_1$	N_2
23	402	424	434	451	465	481	1081								
24	409	431	443	459	474	491	1104	440	464	475	492	507	525	1176	**24**
25	416	439	451	468	483	500	1127	448	472	484	501	517	535	1200	**25**

N_2	$N_1 = 25$ 0.001	0.005	0.010	0.025	0.05	0.10	$2\bar{R}_1$
25	480	505	517	536	552	570	1275

Table L
Critical values of the Studentized range statistic

		k = number of means or steps between ordered means								
df for S_w^2	$1-\alpha$	2	3	4	5	6	7	8	9	10
1	.95	18.0	27.0	32.8	37.1	40.4	43.1	45.4	47.4	49.1
	.99	90.0	135	164	186	202	216	227	237	246
2	.95	6.09	8.3	9.8	10.9	11.7	12.4	13.0	13.5	14.0
	.99	14.0	19.0	22.3	24.7	26.6	28.2	29.5	30.7	31.7
3	.95	4.50	5.91	6.82	7.50	8.04	8.48	8.85	9.18	9.46
	.99	8.26	10.6	12.2	13.3	14.2	15.0	15.6	16.2	16.7
4	.95	3.93	5.04	5.76	6.29	6.71	7.05	7.35	7.60	7.83
	.99	6.51	8.12	9.17	9.96	10.6	11.1	11.5	11.9	12.3
5	.95	3.64	4.60	5.22	5.67	6.03	6.33	6.58	6.80	6.99
	.99	5.70	6.97	7.80	8.42	8.91	9.32	9.67	9.97	10.2
6	.95	3.46	4.34	4.90	5.31	5.63	5.89	6.12	6.32	6.49
	.99	5.24	6.33	7.03	7.56	7.97	8.32	8.61	8.87	9.10
7	.95	3.34	4.16	4.69	5.06	5.36	5.61	5.82	6.00	6.16
	.99	4.95	5.92	6.54	7.01	7.37	7.68	7.94	8.17	8.37
8	.95	3.26	4.04	4.53	4.89	5.17	5.40	5.60	5.77	5.92
	.99	4.74	5.63	6.20	6.63	6.96	7.24	7.47	7.68	7.87
9	.95	3.20	3.95	4.42	4.76	5.02	5.24	5.43	5.60	5.74
	.99	4.60	5.43	5.96	6.35	6.66	6.91	7.13	7.32	7.49
10	.95	3.15	3.88	4.33	4.65	4.91	5.12	5.30	5.46	5.60
	.99	4.48	5.27	5.77	6.14	6.43	6.67	6.87	7.05	7.21
11	.95	3.11	3.82	4.26	4.57	4.82	5.03	5.20	5.35	5.49
	.99	4.39	5.14	5.62	5.97	6.25	6.48	6.67	6.84	6.99
12	.95	3.08	3.77	4.20	4.51	4.75	4.95	5.12	5.27	5.40
	.99	4.32	5.04	5.50	5.84	6.10	6.32	6.51	6.67	6.81
13	.95	3.06	3.73	4.15	4.45	4.69	4.88	5.05	5.19	5.32
	.99	4.26	4.96	5.40	5.73	5.98	6.19	6.37	6.53	6.67
14	.95	3.03	3.70	4.11	4.41	4.64	4.83	4.99	5.13	5.25
	.99	4.21	4.89	5.32	5.63	5.88	6.08	6.26	6.41	6.54
16	.95	3.00	3.65	4.05	4.33	4.56	4.74	4.90	5.03	5.15
	.99	4.13	4.78	5.19	5.49	5.72	5.92	6.08	6.22	6.35
18	.95	2.97	3.61	4.00	4.28	4.49	4.67	4.82	4.96	5.07
	.99	4.07	4.70	5.09	5.38	5.60	5.79	5.94	6.08	6.20
20	.95	2.95	3.58	3.96	4.23	4.45	4.62	4.77	4.90	5.01
	.99	4.02	4.64	5.02	5.29	5.51	5.69	5.84	5.97	6.09
24	.95	2.92	3.53	3.90	4.17	4.37	4.54	4.68	4.81	4.92
	.99	3.96	4.54	4.91	5.17	5.37	5.54	5.69	5.81	5.92
30	.95	2.89	3.49	3.84	4.10	4.30	4.46	4.60	4.72	4.83
	.99	3.89	4.45	4.80	5.05	5.24	5.40	5.54	5.56	5.76
40	.95	2.86	3.44	3.79	4.04	4.23	4.39	4.52	4.63	4.74
	.99	3.82	4.37	4.70	4.93	5.11	5.27	5.39	5.50	5.60
60	.95	2.83	3.40	3.74	3.98	4.16	4.31	4.44	4.55	4.65
	.99	3.76	4.28	4.60	4.82	4.99	5.13	5.25	5.36	5.45
120	.95	2.80	3.36	3.69	3.92	4.10	4.24	4.36	4.48	4.56
	.99	3.70	4.20	4.50	4.71	4.87	5.01	5.12	5.21	5.30
∞	.95	2.77	3.31	3.63	3.86	4.03	4.17	4.29	4.39	4.47
	.99	3.64	4.12	4.40	4.60	4.76	4.88	4.99	5.08	5.16

FROM: Table II.2 in *The probability integrals of the range and of the Studentized range,* prepared by H. Leon Harter, Donald S. Clemm, and Eugene H. Guthrie. These tables are published in WADC Tech. Rep. 58–484, vol. 2, 1959, Wright Air Development Center, and are reproduced with the kind permission of the authors.

Epitaph for a Statistician

With no applause from saint or devil
He was significant at the .05 level.
He squeezed life's data, but when done
He failed to reach the .01.

INDEX